Culture and Art of Scientific Discoveries

Balazs Hargittai
Editor

Culture and Art of Scientific Discoveries

A Selection of István Hargittai's Writings

Editor
Balazs Hargittai
Chemistry Department
Saint Francis University
Loretto, PA
USA

ISBN 978-3-319-96688-5
Library of Congress Control Number: 2019933170

Cover illustration: "Thinker" by Imre Varga, 1968 (it stands at 29 Medve Street, 2nd District, Budapest, photo by István Hargittai)

This Springer imprint is published by the registered company Springer Nature Switzerland AG
The registered company address is: Gewerbestrasse 11, 6330 Cham, Switzerland

István and Magdolna Hargittai in Stockholm, December 10, 2011 (by unknown photographer).

To Mother

Preface

Disclosure
István Hargittai is my father.
Sometimes I refer to him as István
to imply objectivity—
an impossible goal.

This volume offers a selection from among István Hargittai's nontechnical papers. István first appeared in print in 1964 in a Hungarian literary magazine. He responded to an article by the Hungarian-born American physicist and recent Nobel laureate Eugene P. Wigner about the limits of science. This first article was a forerunner of the present collection as it was not a research paper but was still about science. For the present collection, I have selected some seventy papers about the interaction between science and art, the nature of scientific discoveries, quasicrystals, symmetry, and personalia. Some are editorial contributions; others are book reviews; yet others are invited contributions on a variety of topics. The time span is three and a half decades with the earliest from 1983 and the latest from 2017. Most of these contributions have appeared originally in the following periodicals: *Chemical Intelligencer*, *Hungarian Quarterly*, *Israel Journal of Chemistry*, *Journal of Chemical Education*, *Journal of Molecular Structure*, *Leonardo*, *Mathematical Intelligencer*, *Nature*, *Physics World*, *Scientific American*, *Structural Chemistry*, and *Symmetry*.

István's principal research interest has been the determination and modeling of molecular structure. His technique of choice was gas-phase electron diffraction and at one time he was a leading figure internationally in innovation and dissemination of this technique. His other interests have included the concept and utilization of symmetry and twentieth-century science history, and in both areas, he produced contributions of lasting value. His personal interactions with hundreds of the most prominent scientists of our time were manifested in his famous interviews. They appeared in the six-volume *Candid Science* book series and in various periodicals. I hasten to note that much of these activities have been carried out in close interaction with another Hargittai, Magdolna, István's partner, friend, often coauthor, and wife of 50 years—my mother.

I have been involved in their activities initially as their interested son and, lately, as an enthusiastic participant. The idea of compiling and editing a selection of my father's writings had the added attraction of continuing our working on something together after we had recently coedited two volumes, *Culture of Chemistry* and *Science of Crystal Structures*.

The entries I selected—with István's consent, to be sure—I divided into three sections, namely, Art & Science and Symmetry, Discoveries and Personalia, and Book Reviews. There should have been a fourth, an autobiographical narrative, but he was feeling increasingly reluctant doing it. In 2004, he published a book with considerable autobiographical content (*Our Lives: Encounters of a Scientist*). Here, I give a short biographical note with emphasis of the first period of István's life.

István was born in 1941 in Budapest into a happy, "assimilated" Jewish family. His father was a successful lawyer whom the anti-Jewish laws of Hungary made into a slave laborer and who was killed in 1942. István was not 3 years old yet when the train of cattle carriages started his journey to Auschwitz. By some fatal error, another train carrying a selected group of people had been sent to Auschwitz instead of Austria. István's train was to be a substitute; it moved some distance backward and went to Vienna. István, his mother and 10 year-old brother survived, though István's grandmother did not. After liberation they returned to István's mother's hometown in southeastern Hungary where he grew up having a loving stepfather (without having been officially adopted) who had lost his wife and son in another camp in Austria.

Hungary soon became a Soviet-type totalitarian state. Although István excelled in his studies, in 1955 the local authorities prevented him from entering the local high school on account of his late shopkeeper grandfather. Upon appeal and 6 weeks of delay, the higher authorities allowed István to continue his studies in Budapest. In 1959, this ordeal repeated when he faced hurdles in getting admitted to study chemistry at Eötvös Loránd University in Budapest. This time it was on account of his stepfather for he was a former shopkeeper. On both occasions István was declared "class-alien," the most disadvantaged social category. István obstinately fought for changing this label to "intellectual," which he should have had in the first place on account of his late father. He succeeded, but the social categorization was rapidly losing its significance. From 1961, István continued at Lomonosov Moscow State University and graduated in 1965 with a master's degree equivalent.

Back in Hungary he worked for the Hungarian Academy of Sciences and created an area of research that was new in his country. He became internationally recognized and earned his PhD-equivalent degree and then his DSc degree, and for quite a period he was a leading figure in his field. István and Magdolna (Magdi—my mother) married in 1967 and since then they have worked together, did much research independently, and developed side projects jointly that they both enjoyed and that added to their fame. Such projects were the application of the symmetry concept in chemistry and almost everything else, twentieth-century science history, interviewing famous scientists, and writing books. István gave invited lectures in over thirty countries, received visiting scientists from all over the world, gave courses as visiting professor in a number of countries, and founded and has edited an international journal, *Structural Chemistry*, since 1990. It was at the time of the political changes, in 1989–90, that he was invited to his professorship at the Budapest University of Technology and Economics and to be in charge of a large department. For years, he was the founding head of the George A. Olah PhD School. He has been recognized internationally with named lectureships, medals, honorary doctorates, and academy memberships. My parents are still active, still working mostly on joint projects, and still appreciating each other and everything what they are doing—they involve themselves only in projects they can enjoy.

It was fun growing up in this family, but it was also quite demanding. Our parents never spelled out their expectations to me and my sister, Eszter, who was born 3 years after me. Their example sufficed keeping us on our toes. They started taking us on their foreign trips very early on and we became involved in many of their projects, not so much by design, but just as it happened. They kept telling us that they wanted us to be happy and that did not necessarily mean worldly successes. We have built up our own lives in academia and our responsibilities in our personal lives. On a different level, the four of us still make a closely knit entity.

This compilation offers a cross section of István's multifaceted activities apart from his principal research work. It provides a glimpse into his and other scientists' interests and concerns around the turn of the twentieth and twenty-first centuries. In addition to the selection of the entries, I chose an additional image to introduce each entry in order to broaden the scope of the collection. Again, I consulted with István about these choices, especially since most of the images were from among his own photographs. I stress again that this compilation covers what might be termed side products of a noted scientist. However, the accelerating progress in science makes the turnover period for research results short; thus, it may happen that this side product will have a longer lifetime than his research achievements.

Saint Francis University, Loretto, PA, USA   Balazs Hargittai
Spring 2018

Also by the Editor

B. Hargittai, I. Hargittai, *Wisdom of the Martians of Science: In Their Own Words with Commentaries* (World Scientific, 2016)

B. Hargittai, I. Hargittai, Eds., *Culture of Chemistry: The Best Articles on the Human Side of 20th-Century Chemistry from the Archives of the Chemical Intelligencer* (Springer, 2015)

I. Hargittai, B. Hargittai, Eds., *Science of Crystal Structures: Highlights in Crystallography* (Springer, 2015)

B. Hargittai, M. Hargittai, I. Hargittai, *Great Minds: Reflections of 111 Top Scientists* (Oxford University Press, 2014)

B. Hargittai, I. Hargittai, *Candid Science V: Conversations with Famous Scientists* (Imperial College Press, 2005)

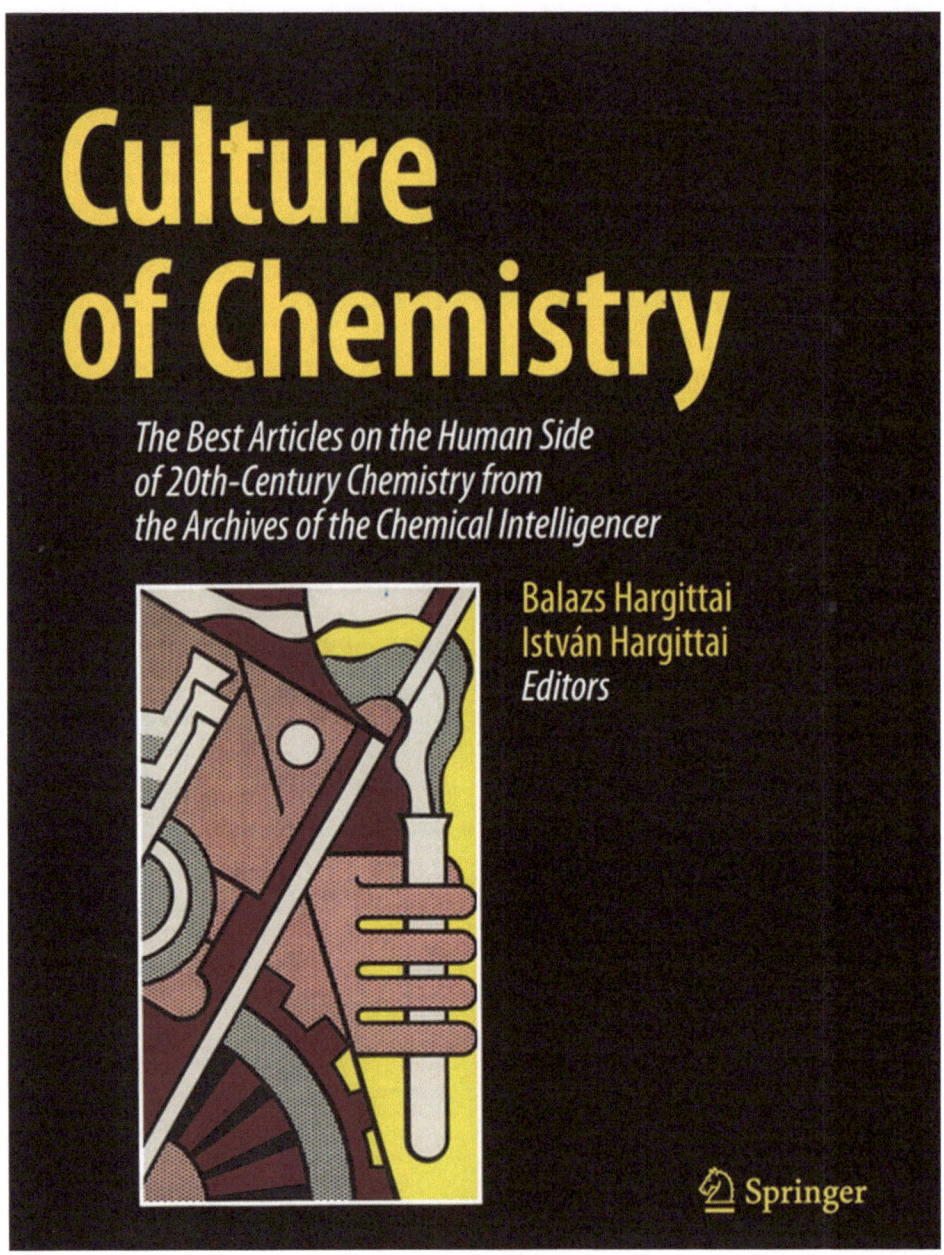

Cover of B. Hargittai and I. Hargittai, Eds., *Culture of Chemistry* (Springer 2015) with a detail of Roy Lichtenstein's Peace through Chemistry 1970 (courtesy of the late Roy Lichtenstein, © Estate of Roy Lichtenstein).

Selection of Authored Books by István Hargittai[1]

I. Hargittai, M. Hargittai, *New York Scientific: A Culture of Inquiry, Knowledge, and Learning* (Oxford University Press, 2017)

B. Hargittai, I. Hargittai, *Wisdom of the Martians of Science: In Their Own Words with Commentaries* (World Scientific, 2016)

I. Hargittai, M. Hargittai, *Budapest Scientific: A Guidebook* (Oxford University Press, 2015)

B. Hargittai, M. Hargittai, I. Hargittai, *Great Minds: Reflections of 111 Top Scientists* (Oxford University Press, 2014)

I. Hargittai, *Buried Glory: Portraits of Soviet Scientists* (Oxford University Press, 2013)

I. Hargittai, *Drive and Curiosity: What Fuels the Passion for Science* (Prometheus, 2011)

I. Hargittai, *Judging Edward Teller: A Closer Look at One of the Most Influential Scientists of the Twentieth Century* (Prometheus, 2010)

M. Hargittai, I. Hargittai, *Visual Symmetry* (World Scientific, 2009)

I. Hargittai, *The DNA Doctor: Candid Conversations with James D. Watson* (World Scientific, 2007)

I. Hargittai, *The Martians of Science: Five Physicists Who Changed the Twentieth Century* (Oxford University Press, 2006, 2008)

I. Hargittai, *Our Lives: Encounters of a Scientist* (Akadémiai Kiadó, 2004)

I. Hargittai, *The Road to Stockholm: Nobel Prizes, Science, and Scientists* (Oxford University Press, 2002, 2003)

B. Hargittai, I. Hargittai, M. Hargittai, *Candid Science I–VI: Conversations with Famous Scientists* (Imperial College Press, 2000–2006)

I. Hargittai, M. Hargittai, *In Our Own Image: Personal Symmetry in Discovery* (Kluwer/ Plenum, 2000; Springer, 2012)

I. Hargittai, M. Hargittai, *Symmetry: A Unifying Concept* (Shelter, 1994; Random House, 1996)

R.J. Gillespie, I. Hargittai, *The VSEPR Model of Molecular Geometry* (Allyn and Bacon, 1991; Dover, 2012)

I. Hargittai, M. Hargittai, *Symmetry through the Eyes of a Chemist* (VCH, 1986, 1987; Plenum, 1995; Springer 2009, 2010)

---

[1] These books are in English; some of them have also appeared in Hungarian, Russian, German, Swedish, Italian, Chinese, Japanese, and Farsi.

Selection of Edited Books by István Hargittai

B. Hargittai, I. Hargittai, Eds., *Culture of Chemistry: The Best Articles on the Human Side of 20th-Century Chemistry from the Archives of the Chemical Intelligencer* (Springer, 2015)

I. Hargittai, B. Hargittai, Eds., *Science of Crystal Structures: Highlights in Crystallography* (Springer, 2015)

I. Hargittai, T. C. Laurent, Eds., *Symmetry 2000*, Volumes I and II (Portland Press, 2002)

A. Domenicano, I. Hargittai., Eds., Strength from Weakness: Structural Consequences of Weak Interactions in Molecules, Supramolecules, and Crystals (Kluwer Academic, 2002)

M. Hargittai, I. Hargittai, Series Eds., *Advances in Molecular Structure Research*, Vols 1–6 (JAI Press, 1995–2000)

I. Hargittai, Ed., *Fivefold Symmetry* (World Scientific, 1992)

I. Hargittai, C. A. Pickover, Eds., *Spiral Symmetry* (World Scientific, 1992)

A. Domenicano, I. Hargittai, Eds., *Accurate Molecular Structures* (Oxford University Press, 1992)

I. Hargittai, Ed., *Quasicrystals, Networks, and Molecules of Fivefold Symmmetry* (VCH Publishers, 1990

I. Hargittai, Ed., *Symmetry 2, Unifying Human Understanding* (Pergamon Press, 1989)

I. Hargittai, M. Hargittai, Eds., *Stereochemical Applications of Gas-Phase Electron Diffraction.* Volume A: *The Electron Diffraction Technique.* Volume B: *Structural Information for Selected Classes of Compounds* (VCH Publishers, 1988)

I. Hargittai, B. K. Vainshtein, Eds., *Crystal Symmetries*: *Shubnikov Centennial Papers* (Pergamon Press, 1988)

I. Hargittai, Ed., *Symmetry Unifying Human Understanding* (Pergamon Press, 1986)

I. Hargittai, W. J. Orville-Thomas, Eds., *Diffraction Studies on Non-Crystalline Substances* (Elsevier, 1981)

Contents

About the Editor

Balazs Hargittai PhD (University of Minnesota), is Professor and Chair of Chemistry at Saint Francis University, Loretto, Pennsylvania. He is also the Founding Director of the Office of Student Research of Saint Francis University and the Founding Editor of *Spectrum*, the student research periodical of Saint Francis University. He has been active in publishing and editorial activities in addition to his research and teaching. He was coauthor of *Candid Science V* (Imperial College Press), *Great Minds—Reflections of 111 Top Scientists* (Oxford University Press), and *Wisdom of the Martians of Science: In Their Own Words with Commentaries* (World Scientific) and coeditor of *Culture of Chemistry* (Springer) and *Science of Crystal Structures* (Springer). His wife, Michele, is a PhD scientist, and they have a son, Matthew, and a daughter, Stephanie, both attending school.

Cover of B. Hargittai and I. Hargittai, *Wisdom of the Martians of Science: In Their Own Words with Commentaries* (World Scientific, 2016).

Béla Bartók's bust by András Beck on the Művész Walkway, Margaret Island, in Budapest (photograph by Istvan Hargittai). The great composer and musicologist Béla Bartók (1881–1945), Istvan's favorite, often used hidden symmetries in his compositions. In 1981, Istvan contributed to a discussion concerning Bartók in the *International Herald Tribune* ("First the Music"). In the August 6 issue he wrote concerning a suggestion published earlier in the newspaper about the possible effects of having concerts, which would not identify either the composer or the performers. Istvan wrote, among others: "Bartók stressed that we should direct our interest in the actual work of art rather than focus on the name of its creator. He cited the joy that is derived looking at a cathedral or a painting or listening to a poem without having any knowledge about the architect, artist or author. He was wondering whether it might be advantageous to perform musical works without any mention of the composer's names. This was but another expression of his legendary modesty and dedication, some of the many qualities of his human greatness. . . . knowing about Bartók may add to the pleasures of listening to his music. But, surely, it is his music that comes first."

Introduction

As far as I remember, our dinner conversations at home revolved mostly around science. That does not mean chemical reactions or other technical topics; rather, ideas, meetings, books, and people. Symmetry stands out especially, because our parents encouraged us, my sister, Eszter, and me, to participate in the discussion of symmetry. This was probably the easiest to get involved in. Our family outings were also peppered by observations of symmetries. They came in a great variety. There was a symmetrical flower and there was some construction equipment being reflected in the near-still water surface of the river Danube. This I remember in particular, because we stood there, on the bank of the Danube and a breeze arrived and disturbed the mirror image and it took a while before it reappeared after the breeze had gone. Eszter and I had gotten quite hooked on symmetry and this pleased our parents. They liked to explain things to us. Eventually, they developed a book about symmetry for children and used some of our drawings in the book.

As I was compiling this volume, I recognized names who visited us in our home in Budapest; there were always visitors, chemists, physicists, and medical researchers, and even more artists. Father liked to take me with him to art studios, and I quickly learned to notice patterns. Eszter was four years old when we as family started going on foreign trips. We always found more fascinating things to admire than what our guide books showed. This inclination to observe and discover is well illustrated in the papers I selected for Section 1 in this volume, which is about art & science and symmetry. Even the geographical spread is remarkable. Just in the order of the papers, the locations include Paris, Beijing, Rome, Liège, Uppsala, Istanbul, London, and Zurich.

The journey to Belgium especially stuck in my memory. We were on our way to the University of Connecticut for a whole year (it then became two years), and made a stop in Liège. My parents attended a large crystallography conference and Father was to give one of the plenary lectures. It was the eighth European Crystallography Meeting. He should have given such a talk a year before, in Jerusalem, at the seventh European Crystallography Meeting, but was denied permission to leave Hungary. Almost to the last minute he was let to believe that he might be allowed to go, but then he was not. There was some press about his no show in Israel. Father was disappointed; this would have been his first visit and it was only after the political changes of 1989/1990 when he could finally visit Israel. The organizers were understanding about this debacle and repeated the invitation for the next meeting in Liège. We went to listen to Father's talk with Mother; it was in a huge auditorium and there was a crowd. Until then, we did not realize that Father was a teacher—we equated public speaking with teaching. He spoke without notes for almost an hour and when he paused sometimes, Eszter and I felt terrified that he might have forgotten what he was supposed to say. From Liège we travelled straight to the United States; for Eszter and me it was our first visit. We would not have thought that so many more visits would follow and that years and years later we would ourselves become American citizens. We did, that is, Eszter and me, not our parents.

In Connecticut, Father was visiting professor of physics for one year and of chemistry for the second year. Mother sometimes had part-time positions as a research associate. Our parents always had side projects and in Connecticut, they produced a book, *Symmetry through the Eyes of a Chemist*. It appeared in 1986 for the first time while the third edition came out in 2009. The book reviewed the utility of the symmetry concepts in a whole range of branches of chemistry and was copiously illustrated. The first edition got excellent reviews and the book has been used as textbook in some universities. When many years later I arrived for my graduate studies at Northwestern University, it gave me a warm feeling that one of the recommended books in our course work was *Symmetry through the Eyes of a Chemist*. Father had yet another symmetry project in Connecticut. He organized his first major edited, interdisciplinary volume, *Symmetry: Unifying Human Understanding*. There was a great deal of interactions with scientists and artists.

Father's involvement with symmetry originated from an earlier visit to the United States, before I was born. In 1969, he spent a year as a visiting research associate at the Department of Physics of the University of Texas at Austin. There he spent quite some time with Eugene P. Wigner, the Hungarian-born American Nobel laureate during Wigner's brief visit there. Wigner personally introduced Father to some of the intricacies of the symmetry concept and even more important, he inoculated Father with a lasting and contagious sensitivity towards symmetry.

JOURNAL OF

Chemical Education

FEBRUARY 1983

Using Degas To Teach
Projection Drawings

Cover of the February 1, 1983, issue of *Journal of Chemical Education* from I. Hargittai, "Degas' Dancers: An Illustration for Rotational Isomers." Reprinted with permission, © 1983, American Chemical Society.

Degas' Dancers: An Illustration for Rotational Isomerism[a]

István Hargittai

The representation of rotational isomerism by projection drawings is difficult for some students to grasp because of its abstract nature. Two drawings by Degas, "End of the Arabesque"[1] and "Seated Dancer Adjusting Her Shoes,"[2] provide an opportunity to introduce the concepts of staggered and eclipsed conformations of A_2B-BC_2 molecules in a concrete, interesting (and aesthetic) way.

The two dancers are shown in Fig. 1 as drawn by Lantos[3] after Degas. Stylized contour drawings are presented in Fig. 2 in order to facilitate understanding the transition from the dancers to the chemical formulas. A staggered and an eclipsed forms of the A_2B-BC_2 molecule are shown by the usual projection representation in Fig. 3.

The projections represent the view along the B–B bond. The dancer's body then corresponds to this bond. The plane bisecting the B–B bond is shown by the circle, and it is represented by the dancer's skirt. The dancer's arms and legs refer to the B–A and B–C bonds, respectively. Even the bouquet in the right hand of the dancer showing the staggered conformation may have a useful function: it is viewed as a different substituent, and may help to understand more complicated rotational isomerism. My experience is that showing Degas' drawings not only enlivens a lecture on conformational problems but also facilitates the introduction of the subject.

[a]Originally published in *Journal of Chemical Education* 1983, 60:94. Reproduced with permission, © 1983 American Chemical Society.

[1] Louvre, Musee de l'Impressionisme, Paris.

[2] The Hermitage, Leningrad.

[3] The author appreciates the kindness of the artist Mr. Ferenc Lantos, Pécs, Hungary, who prepared the drawings of Fig. 1 after postcards of the pictures by Degas.

I. Hargittai (✉)
Department of Inorganic and Analytical Chemistry, Budapest
University of Technology and Economics, Budapest, Hungary
e-mail: istvan.hargittai@gmail.com

Fig. 1 (Left) Lantos' drawing after Degas' "End of the Arabesque." (Right) Lantos' drawing after Degas' "Seated Dancer Adjusting Her Shoes." Full color reproductions of the originals are available In editions of Degas' work.

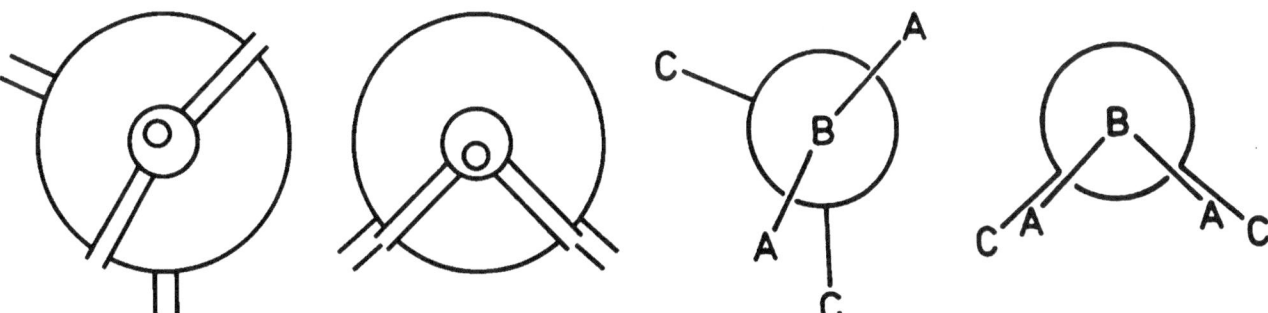

Fig. 2 Contour drawings of the dancers shown in Fig. 1 illustrating the preparation of the projection representation.

Fig. 3 Staggered and eclipsed rotational isomers of the A2BBC2 molecule by projection representing the view along the B–B bond.

ISSN 0021-9584

JOURNAL OF

Chemical Education

DECEMBER 1984

A Winter's Entertainment

Published by the DIVISION OF CHEMICAL EDUCATION OF THE AMERICAN CHEMICAL SOCIETY

Cover of the December 1, 1984, issue of *Journal of Chemical Education*. The art is an actual needlework by one of the editors of the journal, from I. Hargittai and G. Lengyel, "The Seven One-Dimensional Space-Group Symmetries Illustrated by Hungarian Folk Needlework." Reprinted with permission, © 1984, American Chemical Society.

The Seven One-Dimensional Space-Group Symmetries Illustrated by Hungarian Folk Needlework[a]

István Hargittai and Gyorgyi Lengyel

The idea of infinite translations is a crucial point in teaching crystallography and symmetry. However, for some students it is difficult to grasp because of its abstract nature. Experience shows that analogies from outside crystallography greatly facilitate the understanding of this concept.

The seven one-dimensional classes are the simplest space-group symmetries. They are illustrated here by patterns of genuine Hungarian needlework. This kind of needlework is a real "one-sided band" and is ideally suited for this purpose.

Figure 1 shows a consistent system of an asymmetric motif, a black triangle, corresponding to the seven symmetry classes of the one-dimensional space groups. In the table, the symmetry elements are enumerated together with a brief description of the corresponding needlework presented in Fig. 2.

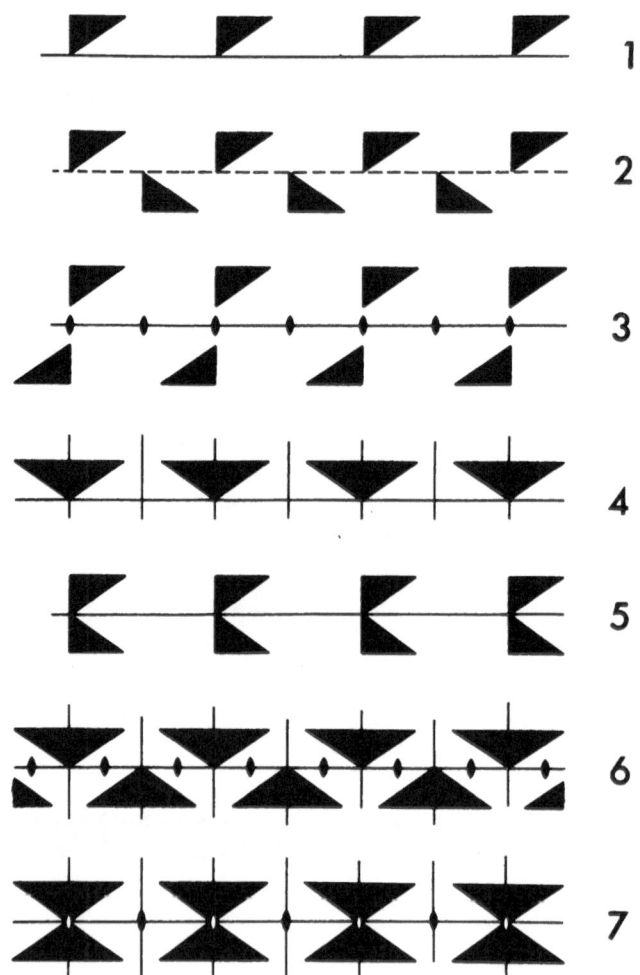

Fig. 1

[a]Originally published in *Journal of Chemical Education* 1984, 61:1033–1034. Reproduced with permission, © 1984 American Chemical Society.

I. Hargittai (✉)
Department of Inorganic and Analytical Chemistry, Budapest University of Technology and Economics, Budapest, Hungary
e-mail: istvan.hargittai@gmail.com

G. Lengyel
Council of Popular Arts and Crafts, Budapest, Hungary

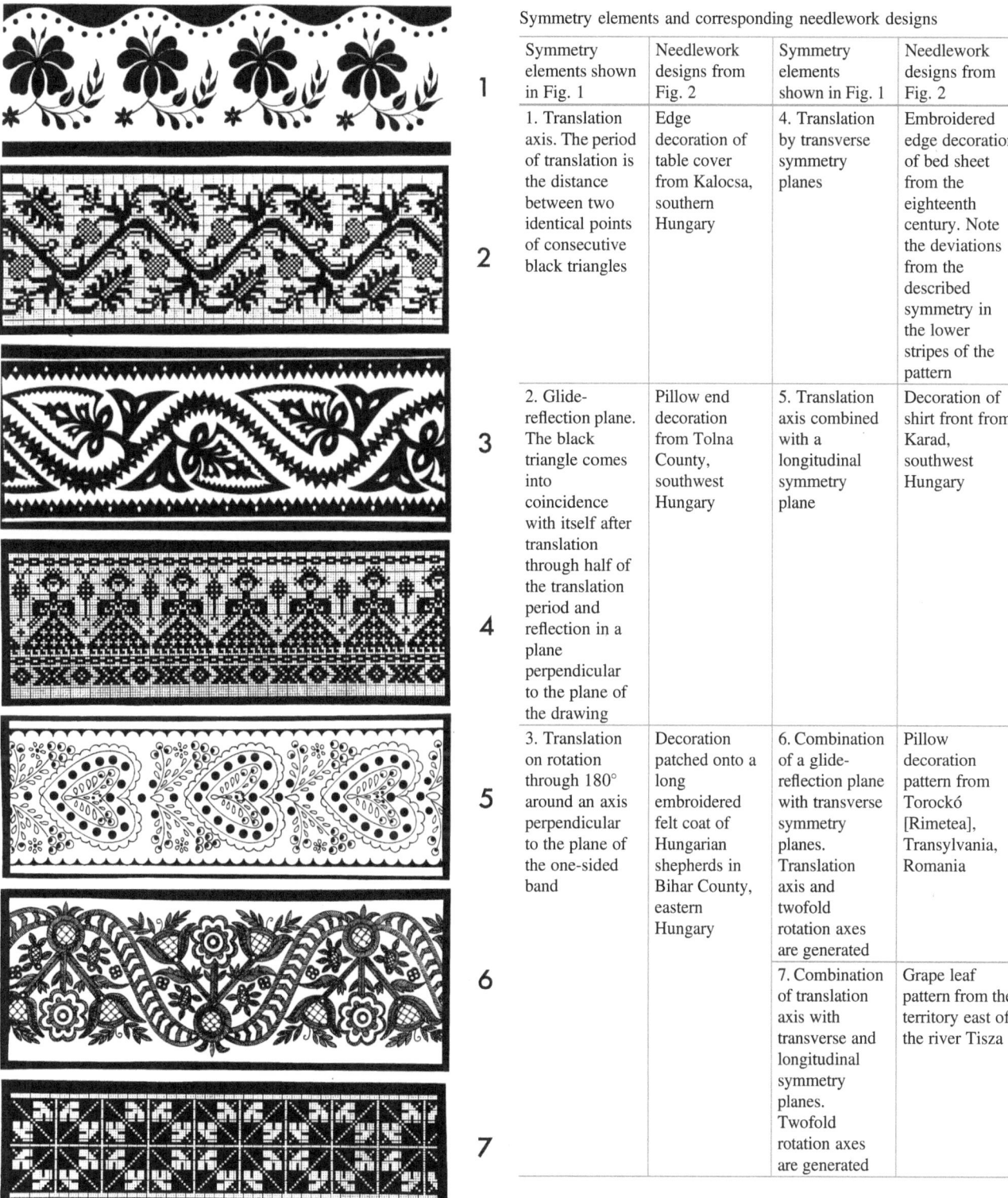

Fig. 2

Symmetry elements and corresponding needlework designs

Symmetry elements shown in Fig. 1	Needlework designs from Fig. 2	Symmetry elements shown in Fig. 1	Needlework designs from Fig. 2
1. Translation axis. The period of translation is the distance between two identical points of consecutive black triangles	Edge decoration of table cover from Kalocsa, southern Hungary	4. Translation by transverse symmetry planes	Embroidered edge decoration of bed sheet from the eighteenth century. Note the deviations from the described symmetry in the lower stripes of the pattern
2. Glide-reflection plane. The black triangle comes into coincidence with itself after translation through half of the translation period and reflection in a plane perpendicular to the plane of the drawing	Pillow end decoration from Tolna County, southwest Hungary	5. Translation axis combined with a longitudinal symmetry plane	Decoration of shirt front from Karad, southwest Hungary
3. Translation on rotation through 180° around an axis perpendicular to the plane of the one-sided band	Decoration patched onto a long embroidered felt coat of Hungarian shepherds in Bihar County, eastern Hungary	6. Combination of a glide-reflection plane with transverse symmetry planes. Translation axis and twofold rotation axes are generated	Pillow decoration pattern from Torockó [Rimetea], Transylvania, Romania
		7. Combination of translation axis with transverse and longitudinal symmetry planes. Twofold rotation axes are generated	Grape leaf pattern from the territory east of the river Tisza

István Hargittai/Magdolna Hargittai

Symmetry through the Eyes of a Chemist

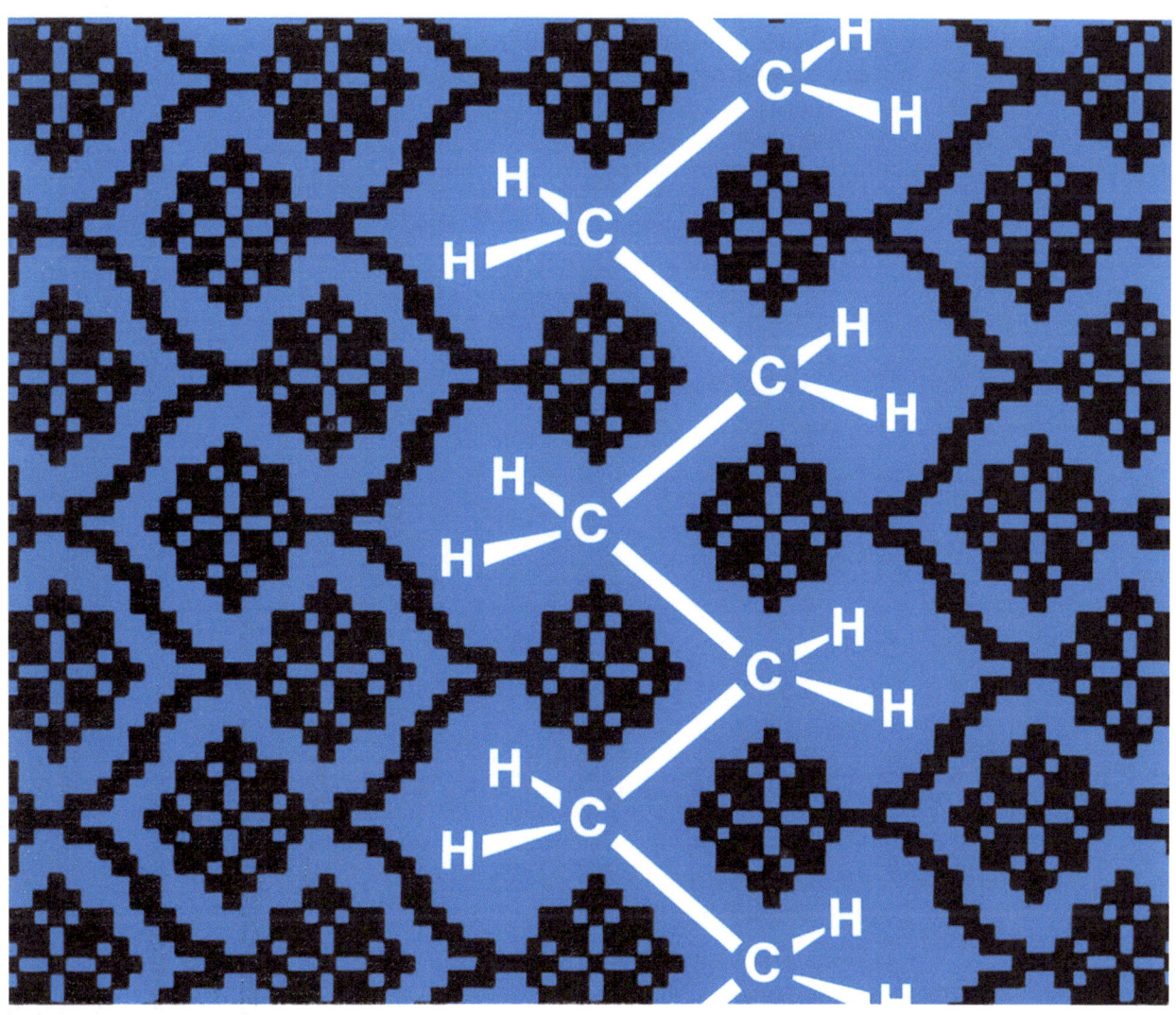

Cover of the first edition of Symmetry through the Eyes of a Chemist. The model of the polymeric molecule is embedded in the pattern of a pillow-slip decoration with scrolling stem motif, which was much used in Hungary around the (18/19) turn of the century.

The Seventeen Two-Dimensional Space-Group Symmetries in Hungarian Needlework[a]

István Hargittai and Gyorgyi Lengyel

We have recently demonstrated all seven one-dimensional space-group symmetries through examples found in Hungarian needlework [1]. The utility of these analogies is obvious in teaching crystallography and symmetry. Several colleagues have urged us to compile and communicate a similar system for the 17 two-dimensional space groups.

The two-dimensional space groups are more complex than the one-dimensional ones, but they are also considerably closer to the three-dimensional space-groups of the crystals. As is well known, there are 230 of the latter, but unfortunately no needlework analogies can be presented for them.

Seventeen Hungarian needlework are shown below together with corresponding systems of an asymmetric motif, the black triangle. Some of the most important symmetry elements are also indicated on them. More detailed descriptions can be found in books, including Buerger's classic, "Elementary Crystallography," [2] which has in some ways inspired the present work.

A brief description of the 17 pieces of needlework is given in the captions (Fig. 1).

[a]Originally published in *Journal of Chemical Education* 1985, 62:35–36. Reproduced with permission © 1985 American Chemical Society.

I. Hargittai (✉)
Department of Inorganic and Analytical Chemistry, Budapest University of Technology and Economics, Budapest, Hungary
e-mail: istvan.hargittai@gmail.com

G. Lengyel
Council of Popular Arts and Crafts, Budapest, Hungary

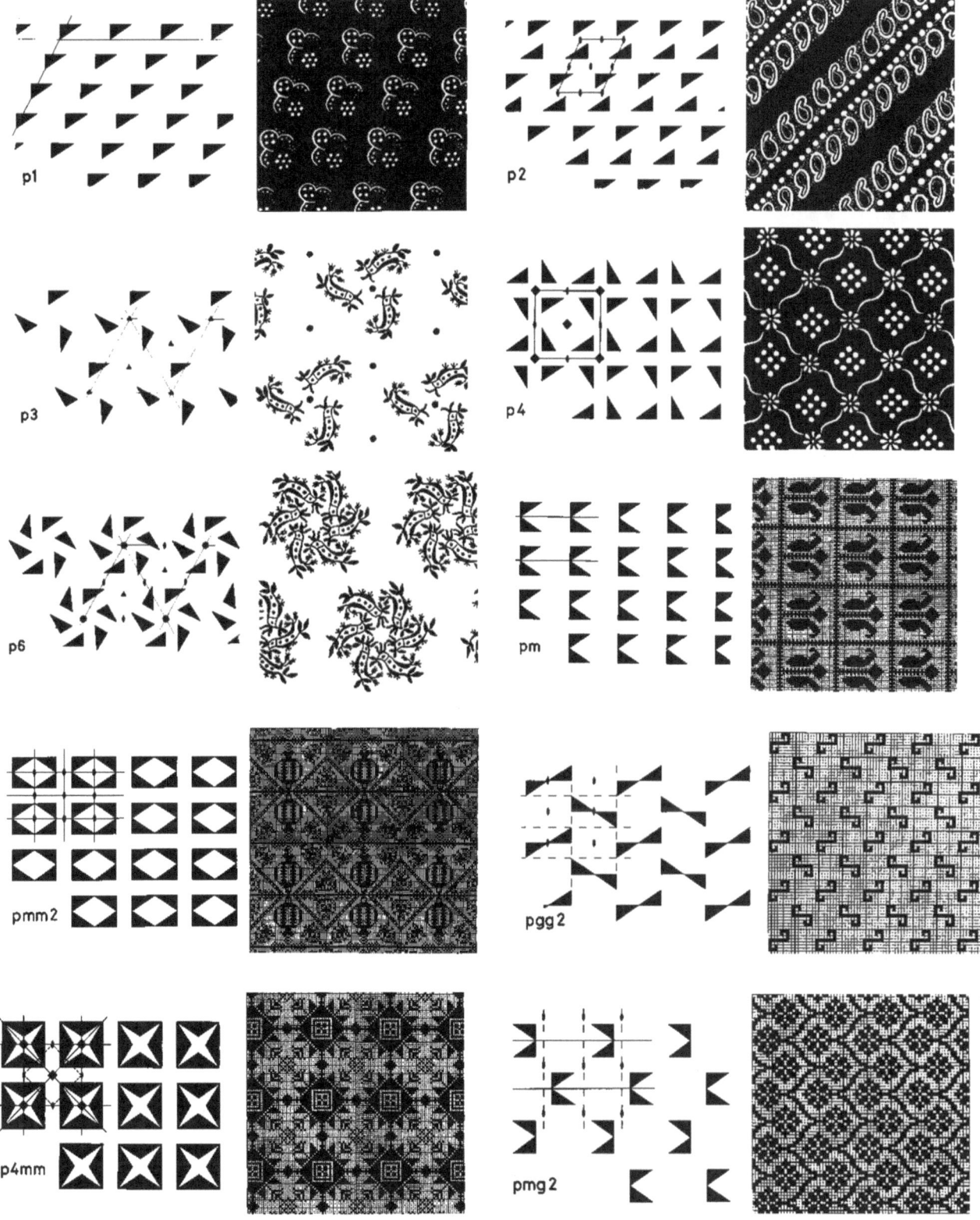

Fig. 1 17 parts: p1 and p4: Patterns of indigo-dyed decorations on textiles for clothing. Sellye, Baranya County, 1899. p2: Indigo-dyed decoration with palmette motif for curtains. Currently very popular pattern. p3 and p6: Decorations with characteristic bird motifs from peasant vests. Northern Hungary. pm: Decoration with tulip motif for table-cloth. Cross-stitched needlework from the turn of the century. pmm2: Bed-sheet border decoration with pomegranate motif. Northwest Hungary, nineteenth century. p4mm: Pillow-slip decoration with stars. Cross-stitched

Fig. 1 (continued) needlework, Transylvania, nineteenth century. p6mm, p3m1, and p31m: Decorations with characteristic bird motifs from peasant vests. Northern Hungary. cm: Pillow-slip decoration with peacock tail motif. Cross-stitched needlework. Much used throughout Hungary around the turn of the century. cmm2: Bed-sheet border decoration with cockscomb motif. Cross-stitched needlework. Somogy County, nineteenth century. pg: From a pattern book of indigo-dyed decorations. Pápa, Veszprém County, 1856. pgg2: Children's bag decoration. Transylvania, turn of the century. pmg2: Pillow-slip decoration with scrolling stem motif. Much used throughout Hungary around the turn of the century. p4gm: Blouse-arm embroidery. Bács-Kiskun County, nineteenth century.

References

1. Hargittai, I., and Lengyel, Gy., J. Chem. Educ., 61, 1033 (1984).
2. Buerger, M.J., "Elementary Crystallography," Wiley, New York, 1963.

VOLUME 17
NUMBER 3
SUMMER 1995
$8.95

The Mathematical
Intelligencer

Lion and
Buckyball

Springer
PRINTED ON ACID-FREE PAPER

Cover of the Summer 1995 issue of *The Mathematical Intelligencer* from I. Hargittai, "Fullerene Geometry under the Lion's Paw." © 1995, Springer.

Fullerene Geometry Under the Lion's Paw[a]

István Hargittai

... the spherical is the form of all forms most perfect, having need of no articulation; and the spherical is the form of greatest volumetric capacity, best able to contain and circumscribe all else; and all the separated parts of the world—I mean the sun, the moon, and the stars—are observed to have spherical form; and all things tend to limit themselves under this form—as appears in drops of water and other liquids—whenever of themselves they tend to limit themselves. So no one may doubt that the spherical is the form of the world, the divine body.

From Copernicus, *De Revolutionibus Orbium Caelestium*, 1543. Heaven, in fact, is often depicted as a sphere in sculptures. The sphere may also be a representation of the Globe, and it is said to symbolize power as well.

Two lions stand guard in front of the Spanish Chamber of Deputies (*Congreso de los Diputados*). One of them has a sphere under the right paw and the other under the left paw (Fig. 1). The surfaces of these spheres are smooth, without any decoration.

It has been common practice in China to have lion sculptures in front of important (and not so important) buildings. These lions also appear in pairs. The female has a baby lion under the left paw and the male has a sphere under the right paw. The female lion is apparently teasing the baby lion while the sphere under the male's paw is said to represent a ball made of strips of silk, which was a favorite toy in ancient China.

Figure 2 shows a pair of bronze lions in front of the *Gate of Supreme Harmony* (Taihemen) in the Forbidden City, Beijing. It was made during the reign of the Ming Dynasty (1368–1644). An elaborate regular hexagonal decoration of the surface of the sphere is under the male lion's paw. It is not possible, however, to cover the surface of the sphere by a regular hexagonal pattern. Indeed, considerable chunks of the sphere are hidden by the lion's paw and by the stand itself on which the lion and the sphere stand. Other lions with similar decorations of the sphere are found in many other places.

An interesting pair of lions whose male partner has a sphere under the paw with a different decoration is shown in Fig. 3, with sphere in a close-up in Fig. 4. This pair is in front of the *Gate of Heavenly Purity* (Qianqingmen) in the Forbidden City and dates back to the reign of Qian Long (1736–1796) of the Qing Dynasty. The surface of this sphere is decorated by a hexagonal pattern which, however, is interspersed by pentagonal shapes. Such a pattern can indeed cover the complete surface of a sphere.

Mathematicians have known, of course, that one can close an even-number of vertices with any number of hexagons (except one), provided 12 pentagons are included in the network (see, e.g., [1]). An important recent discovery in chemistry is related to such structures. When Kroto and co-workers observed [2] the great relative abundance of C_{60} molecules in their laser vaporization cluster-beam experiment, a search followed for the structure of this extraordinarily stable species. Kroto [3] describes eloquently how his previous encounters with Buckminster Fuller's work, and in particular the Geodesic Dome as the U.S. Exhibition Hall at the Montreal Expo, assisted him and his colleagues to arrive at the highly symmetrical truncated icosahedral structure (Fig. 5). A visit to the Forbidden City might have been similarly instructive and beneficial.

All photographs in this Note were taken by the author in 1993. I am grateful to Miss Jing Wei, student of Peking University, for her kind assistance in gathering information about the lion sculptures in the Forbidden City.

[a]Originally published in *The Mathematical Intelligencer* 1995, 17 (3):34–36

I. Hargittai (✉)
Department of Inorganic and Analytical Chemistry, Budapest
University of Technology and Economics, Budapest, Hungary
e-mail: istvan.hargittai@gmail.com

 Springer

Fig. 1 Two lions in front of the Chamber of Deputies (*Congreso de los Diputados*), Madrid, Spain.

Fig. 2 Two bronze lions in front of the *Gate of Supreme Harmony* (TAIHEMEN) in the Forbidden City, Beijing, China.

Fig. 3 Two gold-plated lions in front of the *Gate of Heavenly Purity* (QIANQINGMEN) in the Forbidden City, Beijing, China.

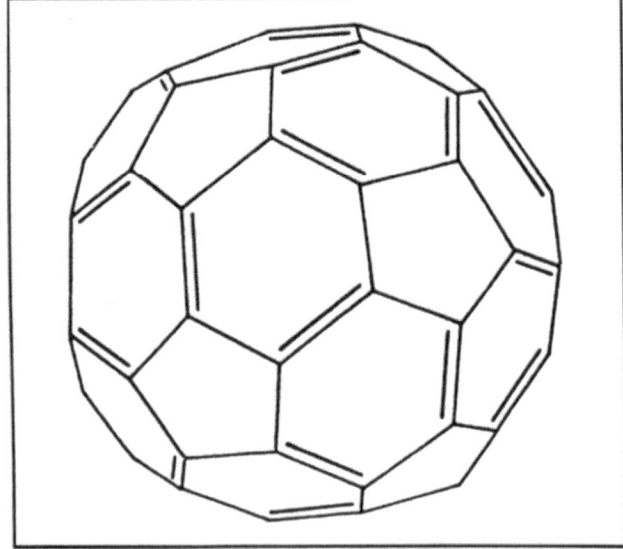

Fig. 4 Close-up of the sphere under the male lion's paw; several pentagonal shapes are seen interspersed in the hexagonal pattern decorating the surface of the sphere.

Fig. 5 Structure [4] proposed for the super-stable C_{60} all-carbon molecule. Each vertex is occupied by a carbon atom. The single and double lines represent two different carbon–carbon linkages. This structure has been proved by a variety of physical and computational techniques (see, e.g., [5]).

References

1. Gasson, P. C. *Geometry of Spatial Forms.* Ellis Horwood, Chichester, 1983, pp. ix–x.
2. Kroto, H. W., Heath, J. R., O'Brien, S. C., Curl, R. E, Smalley, *R. E. Nature* 1985, 318, 162.
3. Kroto, H. W. *Angew. Chem. Int. Ed. Engl.* 1992, 31,111.
4. Kroto, H. W. In *Symmetry 2: Unifying Human Understanding;* Hargittai, I., Ed.; Pergamon Press: Oxford, 1989; pp. 417-423.
5. Hargittai, I. *Per. Mineral.* 1992, 61, 9.

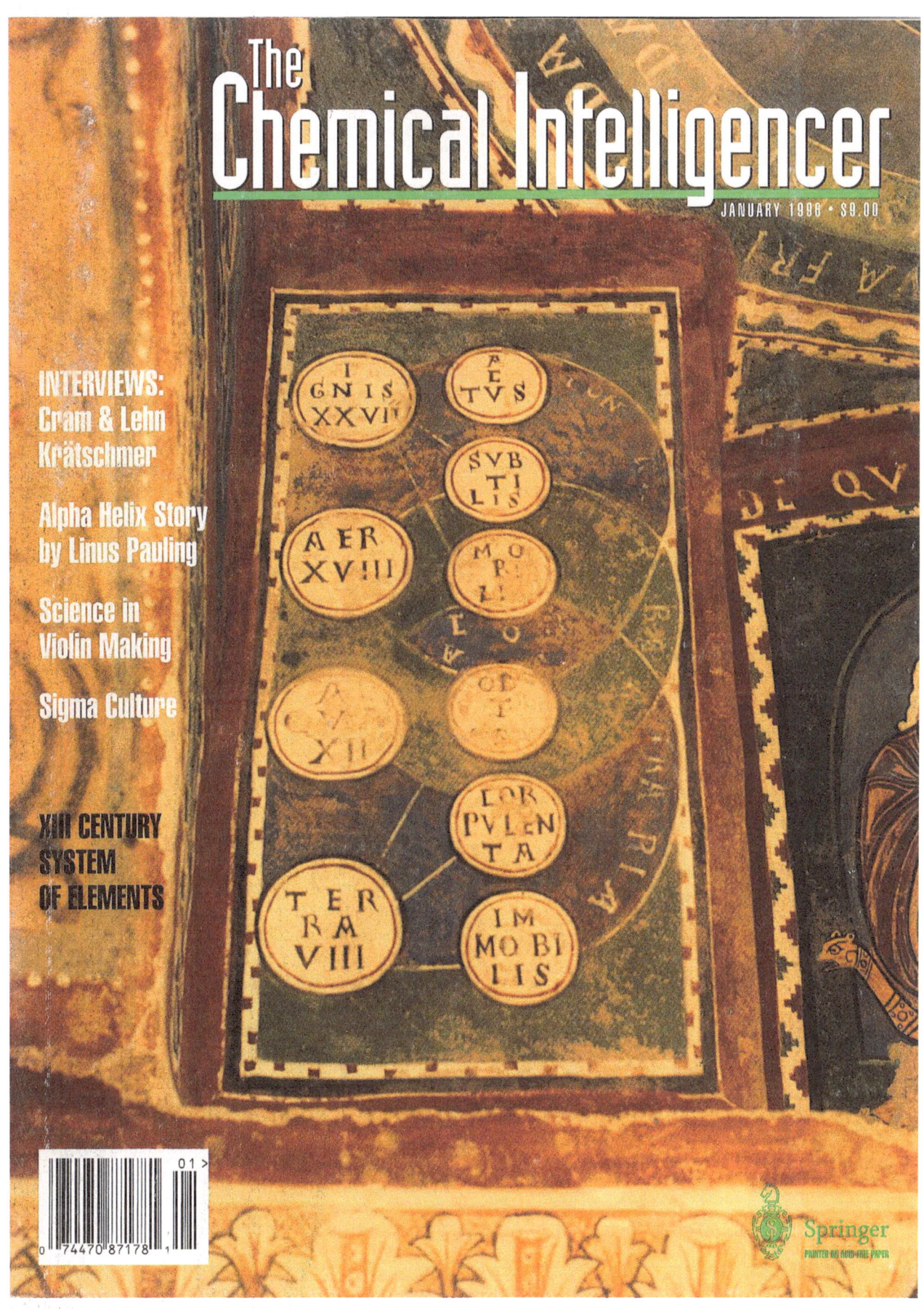

The **Chemical Intelligencer**

JANUARY 1996 • $9.00

INTERVIEWS:
Cram & Lehn
Krätschmer

Alpha Helix Story
by Linus Pauling

Science in
Violin Making

Sigma Culture

XIII CENTURY
SYSTEM
OF ELEMENTS

I
GN IS
XXVII

AE
TVS

SVB
TI
LIS

A ER
XVIII

MO
RI
LI

A
V
XII

OD
T
E

COR
PVLEN
TA

TER
RA
VIII

IM
MO BI
LIS

Springer

PRINTED ON ACID-FREE PAPER

Cover of the January 1996 issue of *The Chemical Intelligencer* from I. Hargittai and A. Domenicano, "System of Elements in Anagni." © 1996, Springer.

System of Elements in Anagni[a]

István Hargittai and Aldo Domenicano

Anagni is an ancient little town, beautifully situated on top of a hill about 60 km southeast of Rome, off the Rome–Naples motorway. Originally a Hernic settlement, it was conquered by the Romans in 306 B.C. Anagni became wealthy and important in the thirteenth century, during which it gave four popes to the Roman Catholic church.

Anagni Cathedral (Fig. 1) was built between 1072 and 1104, originally in the Romanesque style. Gothic elements were added later in the thirteenth century. A famous feature of the cathedral is its mosaic floors, created by the Cosma family in the first half of the thirteenth century.

To the chemical tourist though, the most interesting feature may be some of the frescoes covering the walls and ceiling of the crypt, built in the same period as the upper church. These twelfth- and thirteenth-century frescoes are due to Benedictine painters of the Roman-Byzantine school. They blend religious topics and representations of the physical world, namely, medicine, astrology, and alchemy [1, 2]. In one of the 21 vaults, a human figure symbolizes the allegory of life in relation to the astronomical cycles. The four ages of man are presented in relation to the four seasons and *the four elements*. The fresco is thought to have been inspired by Platonic cosmology (Plato's teachings were spread in southern Italy by the Salerno medical school). Another fresco displays two physicians, Hippocrates (fourth

Fig. 1 Anagni Cathedral (the transept and two apses). (Photograph taken by I. Hargittai, June 1995).

century B.C.) and Galenus (second century A.D.), sitting together as Teacher and Disciple.

Next to the two physicians, there is a diagram of the four elements (Fig. 2), Earth, Water, Air, and Fire, and six properties, *immobile, corpulent, obtuse, mobile, subtle,* and *acute*. The straight connecting lines indicate correspondence (e.g., fire is mobile, subtle, and acute) whereas the curved lines connect opposite qualities. There are Roman numerals beneath the names of the elements: for Earth, $8 = 2^3$; for Fire, $27 = 3^3$; for Water, $12 = 3 \times 2^2$; for Air, $18 = 2 \times 3^2$. The equality containing these numbers, i.e., $8/12 = 18/27$, unifies the whole universe in its perfection according to Platonic philosophy [3]. This relationship may be generalized as $x^3/[(x + 1)x^2] = x(x + 1)^2/(x + 1)^3$.

A detailed description of the six properties and their relationship to the four elements, corresponding closely to the Anagni diagram, was already given by Chalcidius (ca. fourth century A.D.), a Latin philosopher who translated and commentated Plato's *Timaeus* [3].

[a]B. Hargittai and I. Hargittai (eds.), *Culture of Chemistry: The Best Articles on the Human Side of 20th-Century Chemistry from the Archives of the Chemical Intelligencer*, DOI 10.1007/978-1-4899-7565-2_13, © Springer Science+Business Media New York 2015.

I. Hargittai (✉)
Department of Inorganic and Analytical Chemistry, Budapest University of Technology and Economics, Budapest, Hungary
e-mail: istvan.hargittai@gmail.com

A. Domenicano
University of L'Aquila, L'Aquila, Italy

Fig. 2 System of four elements in the crypt of Anagni Cathedral (twelfth- or thirteenth-century fresco). (a) Scheme after Ref. [1]; (b) photograph (taken by I. Hargittai, June, 1995).

Acknowledgments Alan L. Mackay (Birkbeck College, University of London) suggested that we visit Anagni Cathedral to see its mosaic floors. The Cathedral authorities graciously gave us permission to take one photograph in the crypt.

References

1. Ribaudo, C.; Scascitelli, S. *Anagni: Historical and Artistic Guide to the City:* ITER: Subiaco, 1990.
2. Toesca, P. *Gli Affreschi della Cattedrale di Anagni;* Roma, 1902. Reprinted, with color photographs, by ITER: Subiaco, 1990.
3. Plato, *Timaeus.* XXXI–XXXII.

The central section of the Petrovsky Palace, 40 Leningradsky Avenue, Moscow (photograph by I. Hargittai). A star polyhedron (in close-up, top image) decorates the top of each of the two towers at the entrance. The Petrovsky Palace was designed by the architect Matvei F. Kazakov and built in 1776–1782 by the order of Catherine II. Years before he photographed the Petrovsky Palace, Istvan discovered a star polyhedron at the top of the Sacristy of St. Peter's Basilica in Vatican City and he wrote a note about it for *The Mathematical Intelligencer*.

Sacred Star Polyhedron[a]

István Hargittai

There is a beautiful star polyhedron at the top of the Sacristy of St. Peter's Basilica in Vatican City (Fig. 1). It was built by the architect Carlo Marchionni in the years 1776–1784. It is a great stellated dodecahedron, called also Kepler's great stellated dodecahedron (Fig. 2 [1]), with 2 of its 20 triangular pyramids left out to accommodate the vertical rod serving as the stand of the cross above the polyhedron. There are many other examples of star polyhedron decorations from even earlier times, such as at the top of the obelisks in St. Peter's Square and in the Rotunda Square in Rome, and on the gate in the Square of September 20 in Bologna (Fig. 3). The star polyhedron often stands on a pile of dome-shaped stones.

Fig. 1 Left: The Sacristy of St. Peter's Basilica in Vatican City; right: the star polyhedron at its top.

[a]Originally published in *The Mathematical Intelligencer* 1996, 18 (3):52–54

I. Hargittai (✉)
Department of Inorganic and Analytical Chemistry, Budapest University of Technology and Economics, Budapest, Hungary
e-mail: istvan.hargittai@gmail.com

Fig. 2 Great steUated dodecahedron. Photograph courtesy of Magnus J. Wenninger [1].

An octagonal star standing on top of a pile of dome-shaped stones was a characteristic motif in the coat of arms of the Chigi family of Pope Alexander VII (1655–1667). This motif is prominently displayed on the colonnades of St. Peter's Square (Fig. 4).

Giovanni Lorenzo Bernini (1598–1680) and Francesco Borromini (1599–1667) were leading architects of the Baroque period and their activities overlapped with the reign of Pope Alexander VII. The octagonal star and the coat of arms of the Chigi family are conspicuously present in many of their works. Figure 5 shows Sant Ivo's Church and three of its details by Borromini. Two of them display star polyhedra on piles of dome-shaped stones and octahedral stars. However, the decoration beneath the cross at the top of the tower is not a polyhedron but a sphere.

All photographs in this article were taken by the author in 1993. I am grateful to Anna Rita Campanelli and Aldo Domenicano (Rome), Lodovico Riva di Sanseverino (Bologna), and Magnus J. Wenninger (Collegeville, Minnesota) for assistance and advice.

Fig. 3 Left: Top of the obelisk in St. Peter's Square, Vatican City; center: top of the obelisk in Rotonda Square, Rome; right: one of the two side decorations of the gate in the Square of September 20, Bologna.

Fig. 4 Decoration from the top of the colonnade in St. Peter's Square, Vatican City.

Fig. 5 Three details of Sant Ivo's Church.

Khudu S. Mamedov (Mammadov, 1927–1988) Azerbaijani crystallographer in Baku, 1982 (photograph by I. Hargittai). Mamedov often applied antisymmetry in his periodic drawings and combined crystallography education and the preservation of cultural heritage.

Symmetry of Opposites: Antisymmetry[a]

István Hargittai and Magdolna Hargittai

Symmetry has long been identified with the properties of geometrical figures. The Russian crystallographer E. S. Fedorov, for example, gave it the following definition (quoted in [1]): "Symmetry is the property of geometrical figures of repeating their parts, or more precisely, their property of coinciding with their original position when in different positions." According to the Canadian geometer H. S. M. Coxeter [2], "when we say that a figure is 'symmetrical' we mean that there is a congruent transformation which leaves it unchanged as a whole, merely permuting its component elements."

However, it has also been recognized for a long time that symmetry, as observed in real nature, cannot be reduced entirely to this geometrical symmetry. K. Mislow and P. Bickart [3] observed in their epistemological note on chirality that "when one deals with natural phenomena, one enters 'a stage in logic in which we recognize the utility of imprecision.'" Material symmetry, devoid of the rigor of geometrical symmetry, has been viewed as applicable to material objects as well as abstractions with limitless implications [4]. Symmetry also connotes harmony of proportions, a rather vague notion according to Weyl [5]. Human ability to geometrize nongeometrical phenomena helps us see symmetry even in its "vague" and "fuzzy" variations [6, 7]. Thus Weyl [5] said Dürer "considered his canon of the human figure more as a standard from which to deviate than as a standard toward which to strive."

The vagueness and fuzziness of the broader interpretation of symmetry allow us to talk about degrees of symmetry. There must be a range of criteria, which may change from problem to problem, and may very well change in time as well. Today, Science is turning to the examination of the less orderly systems, yet symmetry considerations are not losing importance. On the contrary, their applications are gaining depth as well as breadth.

Chemistry [8], for example, is a science where the symmetry concept has played an increasing role, and not only in such areas as spectroscopy and crystallography but more recently even in such a seemingly nonexact field as organic synthesis. The so-called antisymmetry has become a seminal consideration in modern chemistry, for example, in the description of atomic and molecular orbitals of electronic structure and its changes and interactions during chemical reactions, and in the description of molecular vibrations.

"Operations of antisymmetry transform objects possessing two possible values of a given property from one value to the other" [9]. The simplest antisymmetry operation is color change. Let us first consider an identity operation and an antiidentity operation in Fig. 1. Move on then to antireflection in Fig. 2, and a few further examples of this in Fig. 3. Of course, geometrical symmetries are not restricted to reflection, and Fig. 4 presents examples combining color change with both reflection and rotation, after Shubnikov [1].

Figure 5 presents some op-art patterns. The first Vasarely picture, at least in this black-and-white version, illustrates simple color change between the upper and lower parts of the figure. The second Vasarely picture and the decorated car involve a change in the shape of the motifs in addition to the color change. Here we have alternative properties, color and shape, either of which can be changed into its opposite. We are also moving away from rigorous geometrical symmetry, and moving toward a wider application of the

[a]Originally published in *The Mathematical Intelligencer* 1994, 16 (2):60–66

I. Hargittai (✉)
Department of Inorganic and Analytical Chemistry, Budapest University of Technology and Economics, Budapest, Hungary
e-mail: istvan.hargittai@gmail.com

M. Hargittai
Hungarian Academy of Sciences, Budapest, Hungary

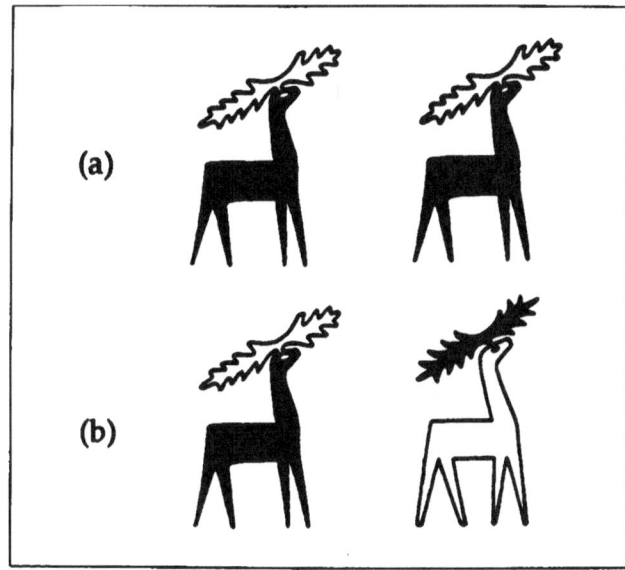

Fig. 1 (a) Identity operation and (b) antiidentity operation [8].

Fig. 2 Reflections (1/2 and 3/4) and antireflections (1/4 and 2/3) [8].

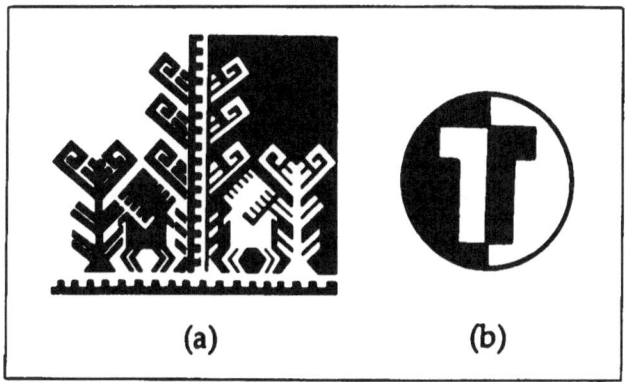

Fig. 3 Antireflections: (a) Hungarian batik design; (b) logo of Tungsram works.

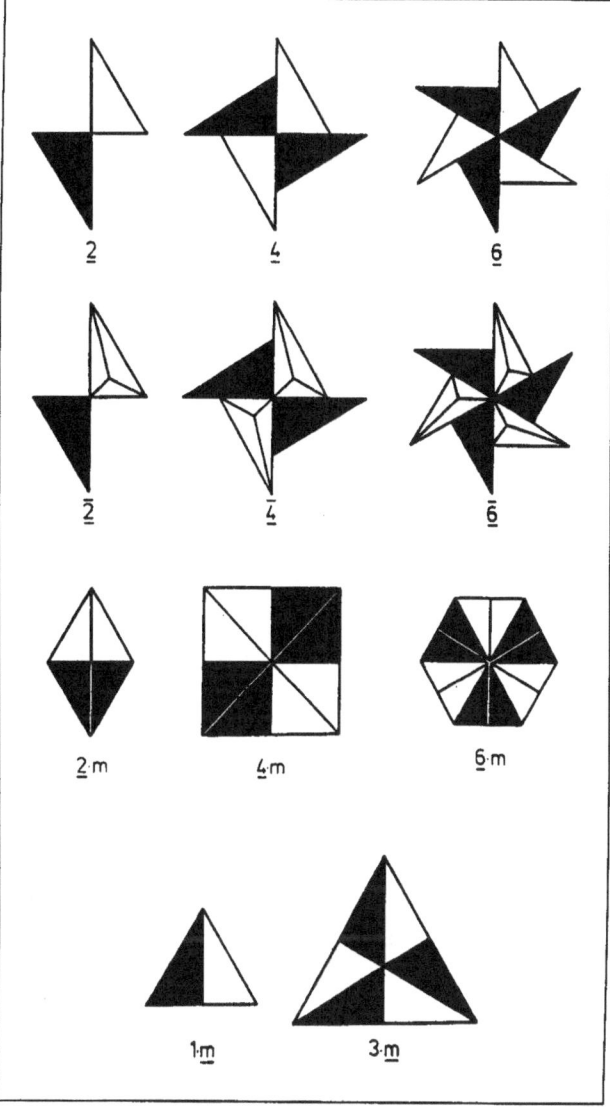

Fig. 4 Antisymmetry operations after Shubnikov [1]. $\underline{2}$, $\underline{4}$, $\underline{6}$, antirotation axes; $\underline{2}$, $\underline{4}$, $\underline{6}$, antireflection-rotation axes; $\underline{2}$ · m, $\underline{4}$ · m, $\underline{6}$ · m, antirotation axes combined with ordinary reflection planes; 1 · $\underline{m}$, 3 · $\underline{m}$, ordinary rotation axes combined with antireflection planes.

antisymmetry concept. The presence of a property turning into its opposite becomes the dominating effect; symmetry elements, such as reflection or rotation, may or may not accompany it.

Figure 6 shows the logo of a sporting goods store in Boston, Massachusetts. The antireflection plane relates winter and summer. Obviously, this store caters to both winter sports and summer sports fans. The color change in the self-serve/full-serve sign attracts attention in Fig. 7, but the concepts may also be considered to have an antisymmetrical relationship.

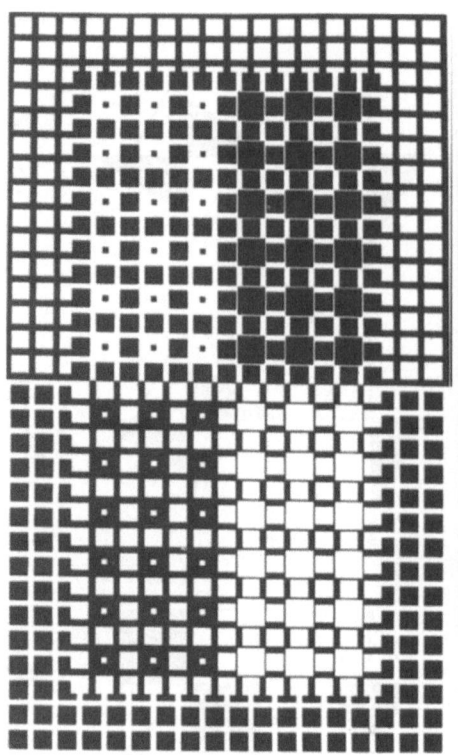 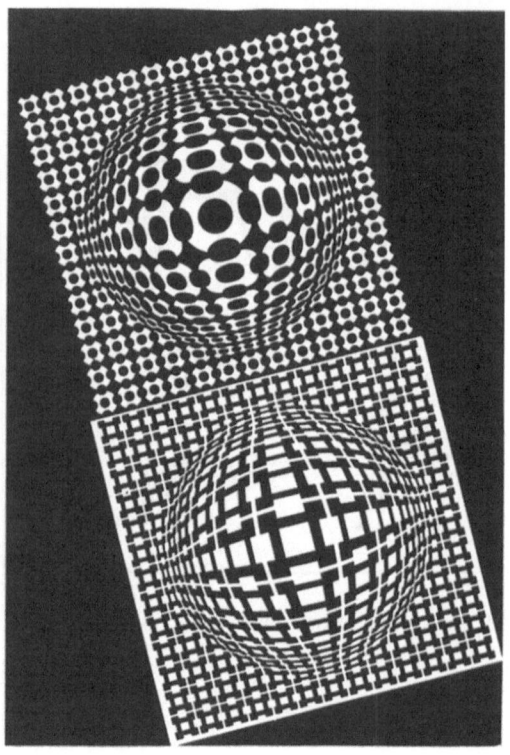

Fig. 5 Pictures by Victor Vasarely (courtesy of the artist).

Fig. 6 (top) Vasarely-like car decoration and (bottom) logo of sporting goods store in Boston, Massachusetts (photographs by the authors).

Fig. 7 Self-serve versus full-serve gas station in Oahu, Hawaii (photograph by the authors).

The Perestroika[1] poster of Fig. 8 displays color change only, and the implication is ironic: Forces against reform would like to reduce the significance of Perestroika to mere color changes.

Let us interrupt our visual examples for two literary examples. The first refers to some antisymmetrical

---

[1] The Russian word "Perestroika," re-structuring, was a buzz word by Soviet President Mikhail Gorbachev when he tried to save the Soviet system by introducing reform.

Fig. 8 "This is perestroika to some." An award-winning Soviet poster from 1987.

Fig. 9 Belgian holiday ad in Flemish and French from 1983 (photograph by the authors).

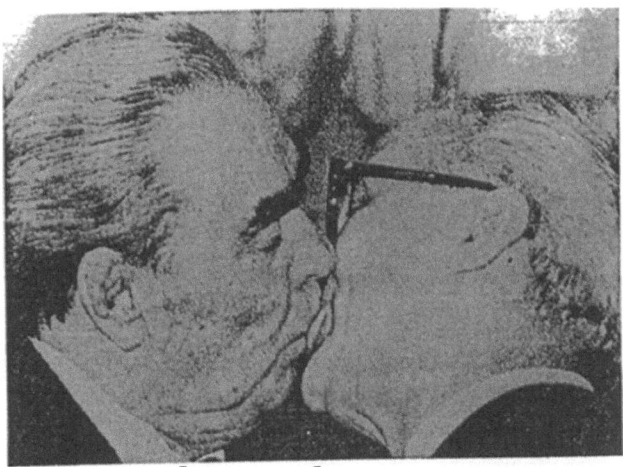

Fig. 10 Election poster by the (Hungarian) Alliance of Young Democrats (FIDESz), 1990. Upper half: Brezhnev and Honecker. Text in the middle: "Please, make your choice."

geographical relationships between, say, Western Europe and New Zealand. These locations can be connected by a straight diameter of the Earth going through its center. The noted American journalist James Reston [10] writes in his "Letter from Wellington. Search for the End of the Rainbow": ". . . Nothing is quite the same here. Summer is from December to March. It is warmer in the North Island and colder in the South Island. The people drive on the left rather than on the right. Even the sky is different—dark blue velvet with stars of the Southern Cross—and the fish love hooks." (He might have added, cyclones go clockwise there, as does water draining from a sink.)

The other example is taken from the Hungarian writer of the 1930s, Frigyes Karinthy, from a short story "Two diagnoses" [11]. The same person, Mr. Same, goes to see a physician at two different places on two different occasions. At the recruiting station he would obviously like to avoid getting drafted, whereas at the insurance company he would like to acquire the best possible terms for his policy. His answers to the identical questions of the physicians are related by antisymmetry.

At the recruiting station

Mr. Same:	*Broken-looking, sad, ruined human wreckage, feeble masculinity, haggard eyes, unsteady movement.*
Physician:	How old are you?
Mr. Same:	Old. . . very old, indeed.
Physician:	Your I.D. says you're thirty two.
Mr. Same:	*With pain.* To be old is not to be far from the cradle—but near the coffin.
Physician:	Are you ever dizzy?

Fig. 11 Viennese dancing school ad (photograph by the authors).

Fig. 12 Military jets and a sea gull, off Bodo, Norway, 1981 (photograph by the authors).

Mr. Same: Don't mention dizziness, please, Doctor, or else I'll collapse at once. I always have to walk in the middle of the street, because if I look down from the curb, I become dizzy at once.

Fig. 13 E Brisse: (a) "Northwest Territories"; (b) "Canada" (From Ref. [13], reproduced by permission).

Fig. 14 M. C. Escher: "Dogs" (From Ref. [16], reproduced by courtesy of the International Union of Crystallography).

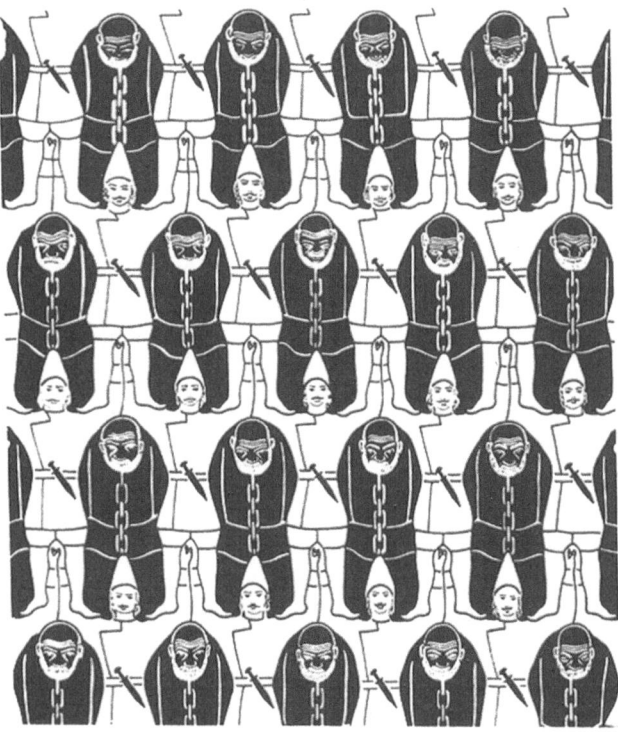

Fig. 16 Drawings by Kh. Mamedov: "Unity" (From Ref. [17], courtesy of Professor Mamedov).

Fig. 15 Drawings by Kh. Mamedov: "Girls" (From Ref. [17], courtesy of Professor Mamedov).

At the insurance company

Mr. Same:	*Young athlete with straightened back, flashing eyes.*
Physician:	How old are you?
Mr. Same:	*Coyly,* Oh, my gosh, I'm almost ashamed of it... I'm so silly...
Physician:	Your I.D. says you're thirty two.
Mr. Same:	To be young is not to be near the cradle, but far from the coffin.
Physician:	Are you ever dizzy?
Mr. Same:	Quite often, sorry to say. Every time I'm aboard an airplane and it's upside-down, and breaking to pieces. Otherwise, not...

Returning now to visual examples, Fig. 9 shows a Belgian travel ad, and the changing property is the language, Flemish/French. The horizontal antireflection is very approximate in

Fig. 10 on the election poster by the Alliance of Young Democrats at the time of the 1990 Hungarian elections.[2] The Viennese dancing school ad (Fig. 11) relates an elephant's legs and a girl's by antisymmetry, obviously for the ability to dance. Figure 12 shows two military jets and a sea gull off Bodø, Norway, a military base, and they may imply the polarity of war and peace.

A few examples of translational antisymmetry are shown above. Apparently, the first systematic discussion of the 46 - two-color two-dimensional patterns was communicated by H. J. Woods in 1936, in a work recently saved from oblivion by D. W. Crowe [12]. Woods pointedly called these two-color patterns "counterchange" patterns. The first two of our illustrations (Fig. 13) are by a Canadian crystallographer, F. Brisse [13]. In one, the polar bear is subjected to a twofold rotational antisymmetry and then translation in two directions. In the other, the two-dimensional space group of the pattern, disregarding color change, would be *p4gm*. This pattern has already been used by G. Polya [14] among his representations of the 17 two-dimensional space groups. It may also be found as a typical decoration in Islamic geometrical patterns [15]. However, in Brisse's pattern there is a two-color change during a complete revolution. There is then translation in two directions. Further simple color changes are involved in the

---

[2] On the poster, Leonid Brezhnev was the Soviet President and Erich Honecker the East German communist leader.

Fig. 17 Symmetric (a, when the road chosen is parallel to the mirror) and antisymmetric (b, when we chose a road perpendicular to the mirror) consequences of reflection (drawing courtesy of architect G. Doczi [8]).

next two figures. M. C. Escher's famous "Dogs" [16] is an excellent illustration of closest packing (Fig. 14). The color change is combined with glide lines. Reflection is also involved in generating Kh. Mamedov's "Girls" in Fig. 15 [17]. The Azerbaijani crystallographer's other drawing "Unity" (Fig. 16) once again combines geometrical symmetries with a conceptual opposition: young versus old.

The symmetric and antisymmetric consequences of reflection for two movements are illustrated in Fig. 17. Suppose we walk alongside a long wall of mirror (Fig. 17a). Our mirror image will be walking with us; the two velocities will be the same. Now walk from a distance toward the mirror, perpendicular to it (Fig. 17b). In this case, our mirror image will have a different velocity from ours. The speed will be the same again, but the direction will be the opposite. If we don't stop in time, we shall collide.

We conclude our discussion by mentioning A. Koestler's concept of bisociation. According to Koestler [18], the connection in thought association is made between thoughts on the same plane, whereas bisociation refers to connection of thoughts from different planes. Thus, bisociation may be considered to be the antisymmetric partner of thought association. Let us just quote one example from Koestler: "The Prince, travelling through his domain, noticed a man in the cheering crowd who bore a striking resemblance to himself. He beckoned him over and asked: 'Was your mother ever employed in my palace?' 'No Sire,'—the man replied.—'But my father was.'"

References

1. A. V. Shubnikov, *Simmetriya i antisimmetriya konechnikh figur*, Izv. Akad. Nauk SSSR, Moscow, 1951. English translation: A. V. Shubnikov, N. V. Belov et al., *Colored Symmetry: A Series of Publications from the Institute of Crystallography, Academy of Sciences of the U.S.S.R., Moscow, 1951–1958,* W. T. Holster, ed., Pergamon Press, New York, 1964.

2. H. S. M. Coxeter, *Regular Polytopes,* 3rd ed., Dover, New York, 1973.

3. K. Mislow and P. Bickart, *Israel J. Chem.* 15 (1976/77), 1.

4. I. Hargittai, *Limits of perfection, Computers Math. Appl. 12B* (1986), 1–17; also in *Symmetry: Unifying Human Understanding,* I. Hargittai, ed., Pergamon Press, New York, 1986, pp. 1–17.

5. H. Weyl, *Symmetry,* Princeton University Press, Princeton, NJ, 1952.

6. I. Hargittai, The joy of symmetry, *Computers Math. Appl.* 17 (1989), 1067–1072; also in *Symmetry 2: Unifying Human Understanding,* I. Hargittai, ed., Pergamon Press, Oxford, 1989, pp. 1067–72.

7. I. Hargittai, Real turned ideal through symmetry, *Symmetrie in Geistes- und Naturwissenschaft,* R. Wille, ed., Springer, Berlin, 1988, pp. 131–161.

8. I. Hargittai and M. Hargittai, *Symmetry through the Eyes of a Chemist,* VCH, Weinheim and New York, 1986.

9. A. L. Mackay, *Acta Crystallogr.* 10 (1957), 543.

10. J. Reston, Letter from Wellington. Search for the end of the rainbow, *International Herald Tribune,* 7 May, 1981, p. 4.

11. E Karinthy, Two diagnoses, *Selected Works,* Szepirodalmi, Budapest, 1962 (in Hungarian).

12. D. W. Crowe, The mosaic patterns of H. J. Woods, *Computers Math. Appl.* 12B (1986), 407–411; also in *Symmetry: Unifying Human Understanding,* I. Hargittai, ed., Pergamon Press, New York, 1986, pp. 407–411.

13. E Brisse, La symétrie bidimensionnelle et le Canada, *Can. Mineral.* 19 (1981), 217-224.

14. G. Polya, Uber die Analogie der Kristallsymmetrie in der Ebene. Z. *Krist.* 60 (1924), 278–282.

15. I. El-Said and A. Parman, *Geometric Concepts in Islamic Art,* World of Islam Festival Publ. Co., London, 1976.

16. C. H. Macgillavry, *Symmetry Aspects of M. C. Escher's Periodic Drawings,* Bohn, Scheltema & Holkema, Utrecht, 1976.

17. Kh. Mamedov, I. R. Amiraslanov, G. N. Nadzhafov, and A. A. Muzhaliev, *Decorations Remember,* Azerneshr, Baku, 1981 (in Azerbaijani).

18. A. Koestler, The *Act of Creation,* Macmillan, New York, 1964.

Next to me, a waterwheel—an example of rotational symmetry—many years ago in Budapest (photograph by Istvan Hargittai).

Symmetry and Perception: Logos of Rotational Point-Groups Induce the Feeling of Motion[a]

Magdolna Hargittai and István Hargittai

In addition to being aesthetically pleasing, the symmetric design of decorations can induce the feeling of motion or the feeling of stopping motion (see the references [1–3]).

Polar one-dimensional space-group border decorations (frieze patterns) can direct the movement of people in underpasses or airline terminals. Two-dimensional space-group patterns of rotational symmetry only have been suggested for decorating dance halls; those containing symmetry planes have been suggested for decorating the sites of serious meetings. Glide-reflection may induce the feeling of confusion.

In this article we suggest that point-groups also have the capability of inducing a feeling of motion, and that certain symmetries in company logos may be better suited to convey the essence of company activities than others.

First, let us consider a four-bladed propeller (Fig. 1). It has fourfold rotational symmetry and no symmetry plane. Having rotational symmetry only corresponds to its function, as do the rotational symmetries of other rotating parts in machinery, such as propellers, turbine wheels, windmills, or children's pinwheels.

Logos themselves do not rotate physically, but they may best convey the essence of the company's activities if their symmetries induce consistent feelings in observers. Thus a railway company, or travel companies in general, may be best represented by a logo with rotational symmetry only, and even more specifically, by twofold rotational symmetry.

[a]Originally published in *The Mathematical Intelligencer* 1997, 19:355–358.

M. Hargittai (✉)
Hungarian Academy of Sciences and Eötvös University, Budapest, Hungary
e-mail: hargittaim@udens.elte.hu

I. Hargittai
Department of Inorganic and Analytical Chemistry, Budapest University of Technology and Economics, Budapest, Hungary
e-mail: istvan.hargittai@gmail.com

Fig. 1 Four-bladed propeller displayed in front of the Budapest Technical Museum. (All photographs in this article are by the authors).

There is always motion, and the motion is back and forth: the train is taking you there and bringing you back, again and again. Our sampler of examples in Fig. 2 includes logos of railway companies and other transportation companies, such as subways, tourist bureaus, bus companies, and expediters.

Of course, we are not suggesting that a transportation company with a logo containing mirror planes would perform its function any worse. We are suggesting, though, that a logo of only rotational symmetry conveys the essence of transportation companies better than a logo with mirror planes.

Banks very frequently have logos of rotational symmetry only and no symmetry planes. A sampler of examples is shown in Fig. 3. Here the abstraction is of even higher degree, as banks and other financial institutions do not represent or

Fig. 2 Sampler of logos of transportation companies (all of two-fold rotational symmetry).

Osterreichische Verkehrakreditbank (Linz, Austria), Bank in Stockholm, Banco Mello (Portugal)
American Service Bank

Banca Popolare di Ancona (Rome), Sicilcassa (Palermo, Italy), Chase Manhattan Bank, and a bank in
Illinois

Korea Housing Bank (Seoul), Bank in Tokyo, and Frost Bank (Austin, Texas)

Fig. 3 Sampler of bank logos.

perform physical motion. Yet turning around money is characteristic of them, and this activity may be the reason, if only subconsciously, why logos with rotational symmetry come to them so naturally. By the same token, we would suggest mirror-symmetric logos for insurance companies, health care services, retirement systems, and any other organizations where mobility is less desirable. We are not suggesting any rigorous correspondence between the symmetries of logos and the activities of the companies they represent, but there seems to be some correlation.

Note also that the logos of transportation companies, displayed in Fig. 2, are invariably of twofold symmetry, yet

the bank logos have no such characteristic number and show diversity in their rotational symmetries. This again seems natural, as there is a definite two-way directionality in the activities of transportation companies but a multiplicity of possibilities in directionality of bank activities (Fig. 4).

Our third and final category is recycling logos. They are, again, of only rotational symmetry, in keeping with the process of recycling—that is, turning around the wastes and producing new materials. Although threefold rotational symmetry is the most common, there is a variety in rotational symmetries. The variety of design is less than for banks, in keeping with the international and less competitive character of recycling.

Reynolds Aluminum Recycling and Bottles recycling (Italy)

Recycling (Washington, DC), New Hampshire recycling, University of Toronto recycling, Recycle Hawaii

Fig. 4 Sampler of recycling logos.

References

1. A.V. Shubnikov and V.A. Koptsik, *Symmetry in Science and Art*, Plenum Press, New York (1974). [Russian original: *Simmetriya v nauke i iskusstve,* Nauka, Moscow (1972)].

2. I. Hargittai and M. Hargittai, *Symmetry: A Unifying Concept*, Shelter Publications, Bolinas, California (1994).

3. I. Hargittai and M. Hargittai, *Symmetry through the Eyes of a Chemist*, Second Edition, Plenum Press, New York (1995). [Latest edition: M. Hargittai and I. Hargittai, *Symmetry through the Eyes of a Chemist,* 3rd Edition, Springer (2009 and 2010)]

George Pólya (1887–1985; courtesy of Gerald L. Alexanderson). Pólya was born in Budapest and became a world renowned mathematician. He did his high school studies in the Berzsenyi Gimnázium in downtown Budapest. It was one of the elite schools from which a number of internationally recognized personalities graduated, such as the Nobel laureate Dennis Gabor, the computational specialist John Kemeny, the tumor biologist George Klein, the financier George Soros, and many others. Pólya attended the universities of Budapest and Vienna and did his doctoral work under Leopold Fejér. He was a mathematics professor at the Zurich Federal Institute of Technology until 1940. Then, fearing the spread of Nazism, he moved to the United States and was professor at Stanford University. In 1924, Pólya published a set of the 17 two-dimensional plane groups Istvan reproduced in the following article.

Symmetry in Crystallography[a]

István Hargittai

The science of crystals involves symmetry. Symmetry is also an excellent link to other fields of human endeavor. The first scientific crystallographer, Johannes Kepler, came to the idea of close packing when he was considering the symmetry of snow crystals. When Louis Pasteur observed crystal and molecular chirality, he opened a Pandora's Box of the notion of the dissymmetry of the universe. Since the start of X-ray crystallography in 1912, emphasis has been on single-crystal symmetry, and the field has moved from triumph to triumph. In the late 1920s, however, interest in less than perfect structures developed, leading to the establishment of molecular biology. Helical symmetries were found to characterize life's most important molecules. Symmetry considerations were decisive in these discoveries, which stimulated the expansion of the symmetry concept. In the mid-1980s, the belief that fivefold symmetry was a noncrystallographic symmetry crumbled, and the concept of the crystal had to be revised. Crystallography has now become the science of structures. Symmetry has helped crystallography to influence the arts. This tends to unify our culture—a side effect of the enormous work of uncovering the secrets of matter for the betterment of human life.

[a]Originally published in *Acta Crystallographica* 1998, A54:697–706 and in: H. Schenk (ed.), *Crystallography across the Sciences: A Celebration of 50 Years of Acta Crystallographica and the IUCr.* (International Union of Crystallography 1998), pp. 697–706

I. Hargittai (✉)
Institute of General and Analytical Chemistry, Budapest Technical University, Budapest, Hungary

Department of Inorganic and Analytical Chemistry, Budapest University of Technology and Economics, Budapest, Hungary
e-mail: istvan.hargittai@gmail.com

Introduction

With the appearance of combinatorial chemistry, we have lost count of the number of new substances produced in the laboratory. Does this mean that we are losing sight of the structure of matter because its variations are too numerous? We should not fear this because there are patterns in the structures, appearing as symmetry, and the search for pattern is the most characteristic scientific approach in uncovering the secrets of nature. The patterns of elementary particles and those of the chemical elements are well established yet patterns are becoming discernible only in outline for the structures of substances. With about a quarter of a million crystal structures determined so far, the prediction of the crystal structure of a new substance is still elusive.

Eugene Wigner [1] made a brief speech at the Stockholm City Hall in December 1963 on the occasion of the presentation of his Nobel Prize in Physics. This is what he said, when he talked about the inspiration received from his teacher, Michael Polanyi: 'He taught me, among other things, that science begins when a body of phenomena is available which shows some coherence and regularities, that science consists in assimilating these regularities and in creating concepts which permit expressing these regularities in a natural way. He also taught me that it is this method of science rather than the concepts themselves (such as energy) which should be applied to other fields of learning'. What Polanyi taught Wigner was to recognize patterns, and the main tool was the symmetry concept.

The determination of structure by X-ray diffraction is based on symmetry, which exists in the internal arrangement of the building elements of the structure. Thus, there are two aspects of symmetry underlying much of recent structural research. One is the symmetry of the building element of the structure and the other is the limited number of rules needed to generate all structures.

Crystallography had initially evolved as a science of crystals. Then the application of X-ray diffraction gave a tremendous emphasis to the structure of individual molecules. These molecules are embedded in a matrix of other molecules in the closest proximity. Yet the fascination with their structures and the emerging regularities among them had, for a while, pushed back the interest in the interactions between the molecules themselves. The appearance of supramolecular chemistry in general and the recognition that the molecular crystal is a supermolecule *par excellence* [2], in particular, has brought back the interest in crystal chemistry and, more generally, in materials crystallography.

Focusing on molecular structures and their variations has also provided enormous benefits. Murray-Rust [3] estimated that Linus Pauling [4], at the time of the first edition of *The Nature of the Chemical Bond*, possessed one hundredth of one per cent of the structural chemistry information that was available 50 years later, yet his observations and generalizations have been found to apply to almost all the rest.

The present article illustrates the role of the symmetry concept in the science of structures and the contribution of crystallography to the enhancement of the symmetry concept as a research tool. This concept has been a bridging tool between the most diverse fields of human endeavor [5–7]. As a set of examples, packing, biological structures, the recent discoveries related to fivefold symmetry, and chirality are chosen. We comment on the role of the symmetry concept in countering the effects of narrow specialization and in bringing science into human proximity for a broader audience. Materials of a forthcoming book have aided the preparation of the present article [8].

Packing

The importance of symmetry in structure does not mean that the highest symmetry is the most advantageous. Lucretius [2] proclaimed about two millenia ago in his *De Rerum Natura* that 'Things whose textures have a mutual correspondence, that cavities fit solids, the cavities of the first the solids of the second, the cavities of the second the solids of the first, form the closest union'. In modern science, Kepler [9] recognized that the origin of the shape and symmetry of snowflakes is the internal arrangement of the building elements of water. This observation may be considered as the start of scientific crystallography. Lord Kelvin's (William Thomson's) mostly forgotten geometry [10] was a return to Lucretius's fundamental observation.

As Lord Kelvin was building up the arrangement of molecular shapes, he examined two fundamental variations (Fig. 1). In one, the molecules are all oriented in the same

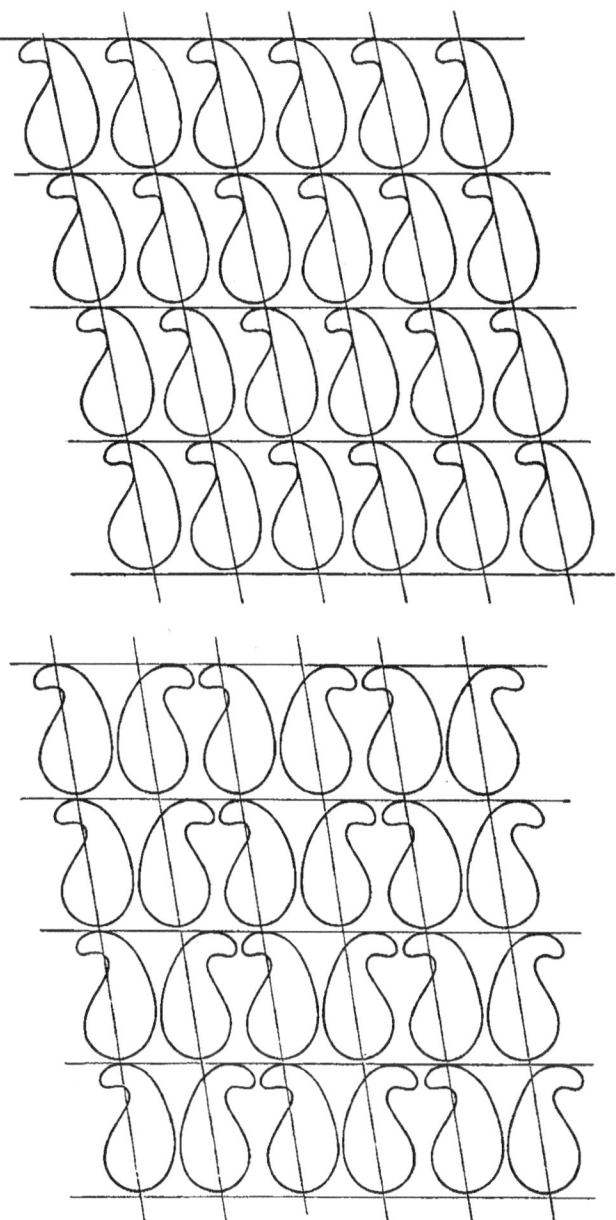

Fig. 1 Arrangements of molecular shapes by Lord Kelvin [10].

way, while, in the other, the rows of molecules are alternately oriented in two different ways. Kelvin considered the puzzle of the boundary of each molecule as a purely geometrical problem. This is the point where his successors introduced considerations for inter-molecular interactions and, ultimately, Aleksandr I. Kitaigorodskii 'dressed the molecules in the fur-coat of van der Waals domains'.

Lord Kelvin was using nearly rectilinear shapes for partitioning the plane but he did not let his molecules quite touch one another. Otherwise, he created a modern representation of molecular packing in the plane, including the recognition of complementariness in packing.

Fig. 2 Truncated octahedron by Lord Kelvin [10].

Then he came to extending the division of continuous two-dimensional space into the third dimension. He restricted his examinations to polyhedra and found one of the five space-filling parallelohedra, which were discovered by E. S. Fedorov as capable of filling the space in parallel orientation without gaps or overlaps (Fig. 2). The Fedorov polyhedra are the cube, the hexagonal prism, the rhombic dodecahedron, an elongated rhombic dodecahedron with eight rhombic and four hexagonal faces, and the truncated octahedron. Figure 3 shows the truncated octahedron filling space (after [11]).

Fedorov was one of the three scientists who determined the number (230) of three-dimensional space groups. The other two were Arthur Schoenflies and the amateur William Barlow. Barlow considered oriented motifs, and 'his method was hanging pairs of gloves on a rack to make space-group models'. It was a truly empirical approach. "He bought gloves by the gross, so the story goes, mystifying the sales lady by answering 'I don't care' to her question, 'What size, sir?'" [12].

Fedorov also derived the 17 two-dimensional plane groups but their best known presentation is by George Pólya [13] who illustrated them with patterns that completely fill the surface without gaps or overlaps (Fig. 4). Today we would call them Escher-like patterns [14].

An important contribution appeared in 1940 from the structural chemist Linus Pauling and the physicist-turned-biologist Max Delbruck [15], dealing with the nature of intermolecular forces in biological processes. They suggested precedence for interaction between complementary parts, rather than the importance of interaction between identical parts. They argued that the intermolecular interactions of van der Waals attraction and repulsion, electrostatic interactions, hydrogen-bond formation etc. give stability to a system of two molecules with complementary rather than identical structures in juxtaposition. Accordingly, complementariness should be given primary consideration in discussing intermolecular interactions.

Considerations of complementarity in molecular packing culminated in the works of Kitaigorodskii [16]. His most important contribution was the prediction that three-dimensional space groups of lower symmetry should be much more frequent than those of higher symmetry among crystal structures. This was a prediction at a time when few crystal structures had been determined experimentally.

Kitaigorodskii's realisation of the complementary packing of molecules was not intuition; he arrived at this principle by empirical investigation. Today his findings appear simple, almost self-evident, a sure sign of a truly fundamental contribution.

When Kitaigorodskii finally came to the idea of using identical but arbitrary shapes, he started by probing into the

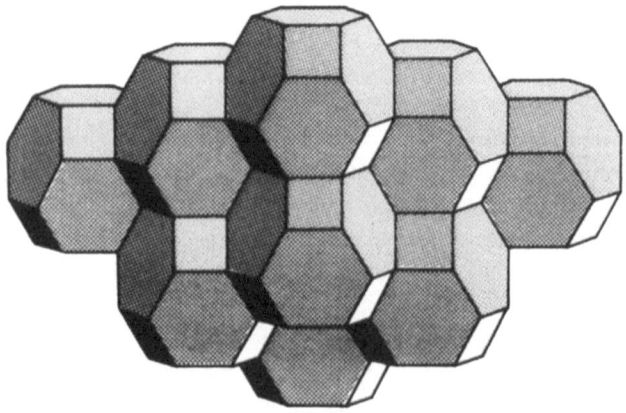

Fig. 3 Space filling by truncated octahedra by Weyl [11]. Reprinted with permission. Copyright (1946) Princeton University Press.

Fig. 4 The 17 two-dimensional plane groups by Pólya [13].

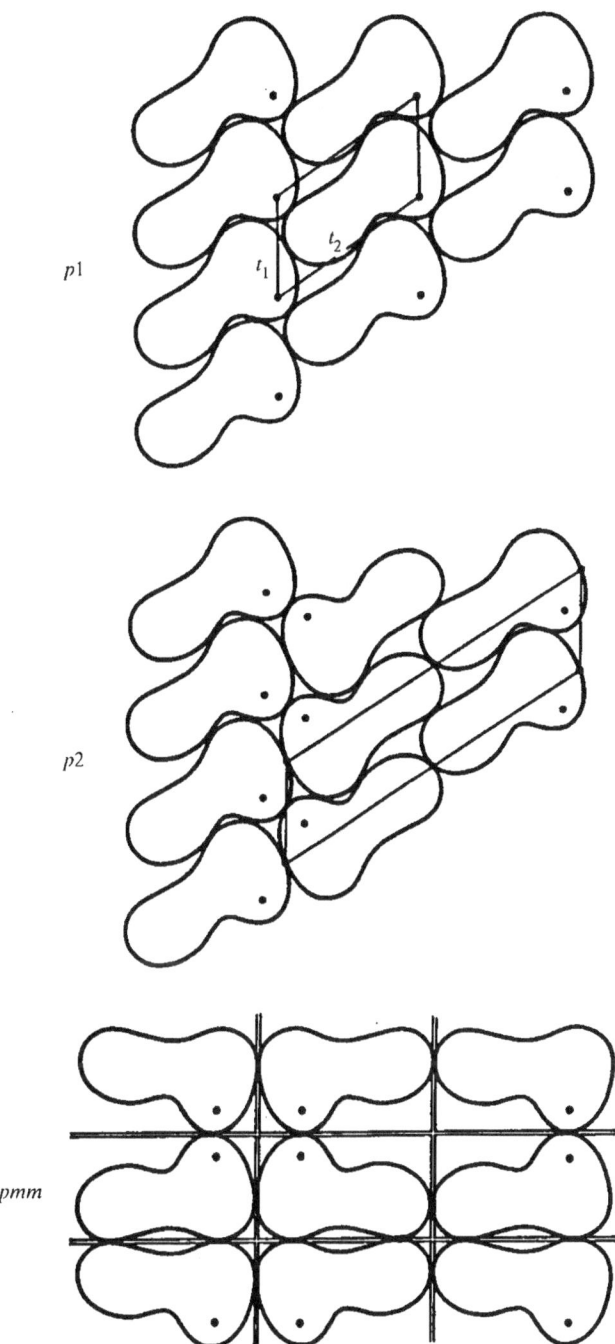

Fig. 5 Sampler of molecular packing arrangements in the plane by Kitaigorodskii [16].

best possible arrangements in the plane. Figure 5 presents a sampler of the arrangements considered by Kitaigorodskii [16]. He established the symmetry of two-dimensional layers that allow a coordination number of six at an arbitrary tilt angle of the molecules with respect to the tilt axes of the layer unit cell. He found that such an arrangement will always be among those that have the densest packing. In the general case for molecules of *arbitrary* shape, there are only two kinds of such layers. One has inversion centers and is associated with a nonorthogonal lattice. The other has a rectangular net, from which the associated lattice is formed by translations, plus a second-order screw axis parallel to the translation. The next task was to select the space groups for which such layers are possible. This was of great interest since it answered the question as to why there is a high occurrence of a few space groups among the crystals while many of the 230 groups hardly ever occur.

Biological Structures

While single-crystal studies were still on the climb and most modern techniques of structure elucidation were still in the making, research on biologically important macromolecules had also begun. In the early 1920s, Polanyi found (cf. [19]) that the X-ray diffraction from cellulose fibers indicated the presence of crystallites oriented in the direction of the fiber axis. The first proteins subjected to X-ray diffraction were protein fibers. In the early 1930s, W. T. Astbury and his co-workers published a series of papers on the X-ray studies of hair, wool and related fibers [20–22]. They observed that stretched moist hair showed a drastic change in its X-ray diffraction pattern, compared with dry un-stretched hair. This was interpreted as two forms of the polypeptide chain, ß-keratin and α-keratin, today known as ß-pleated sheet and the α-helix. One of Astbury's co-workers, H. J. Woods, studied extensively the symmetry properties of textile decorations [23].

Linus Pauling [24] decided to determine the atomic arrangement of α -keratin, using his knowledge of structural chemistry in addition to Astbury's X-ray diffraction patterns. The effort cost about 15 years and led to the discovery of the α-helix. It was a spectacular example of pattern recognition and modeling. In the course of this work, Pauling utilized the structural information on small molecules determined by gas-phase electron diffraction and the resonance theory, and deduced the planarity of peptide bonding. He also disregarded nonessential features, such as the differences in the side chains of the various amino acids and the discrepancy between the 5.1 Å repeat distance along the axis measured from Astbury's patterns and the 5.4 Å repeat distance that came out of his own modeling. Finally, he remembered a mathematical theorem that the most general operation relating an asymmetric object to another copy is a rotation-translation equivalent to a helix when repeated. Thus, helical symmetry made its entry into the description of biological systems [25, 26] although it was not for the first time that it was used to describe assemblies of identical units. Eventually, Cochran et al. [27] worked out the theory of diffraction of the polypeptide helix. Astbury's observation of the 5.1 Å repeat distance was correct and, eventually, Pauling and Francis Crick explained [28], independently, this discrepancy by a slight additional coiling of the helices. Because of the non-integer screw, a shift by slight coiling facilitates their best packing, providing a nice example of symmetry breaking by a weak interaction. Shortly before Pauling's discovery, Bragg et al. [29] proposed about 20 polypeptide structures, none of them correct, and not only because they rigorously adhered to the 5.1 Å repeat distance but also because they did not observe the planarity of peptide bonding [30].

Although crystallographic work on biological macromolecules had begun in the 1920s, the great debate about colloids *versus* polymers in biological systems raged on for some time. It was only in 1953 that H. Staudinger was awarded the Nobel Prize in Chemistry for his fundamental studies of macromolecules. The Nobel Prize for 1954 went to Linus Pauling, stressing his contribution to the understanding the nature of the chemical bond. By then, he had published a triple helix for DNA, which proved to be a wrong structure. The correct double-helix structure of DNA was communicated by groups of Cambridge and London scientists [17, 31, 32].

The double-helix structure had important novel features. One was that it had two helical chains, each coiling around the same axis but having opposite direction. The two helices going in opposite directions, and thus complementing each other, is a simple consequence of the twofold symmetry with the twofold axis being perpendicular to the axis of the double helix. The other novel feature was the manner in which the two chains are held together by the purine and pyrimidine bases. 'They are joined in pairs, as a single base from one chain being hydrogen-bonded to a single base from the other chain, so that the two lie side by side with identical z-coordinates. One of the pair must be a purine and the other a pyrimidine for bonding to occur'. A little later, Watson and Crick [17] add that 'if the sequence of bases on one chain is given, then the sequence on the other chain is automatically determined'. Thus, symmetry and complementarity appear most beautifully in this model but the paper culminates in a final remark which sounds like a symmetry description of a simple rule to generate a pattern, 'It has not escaped our notice that the specific pairing we have postulated immediately suggests a possible copying mechanism for the genetic material'. This is a far from casual remark; on the contrary, a lot of consideration had been distilled into this sentence [33]. Watson and Crick [17] illustrated their brief note with a purely diagrammatic figure (Fig. 6) of elegant simplicity, showing the two chains related by a twofold axis of rotation perpendicular to the axis of the helices. The structure since has been immortalized in sculptures (see the one, for example, in Fig. 7), on medals and stamps (examples are shown in Fig. 8), and by other means.

There are four different nucleotides in the DNA double helix but even four building elements can permute in virtually infinite possibilities if the chain is long enough, and the DNA molecules are very long. Thus it seemed likely to Watson and Crick [18] that the precise sequence of the bases is the code carrying the genetic information. The double-helix structure offers a simple visually appealing way of self-duplication. Once the hydrogen bonds are broken, each of the chains may reassemble a new partner chain from among the nucleotides available in their surroundings. Basically, this is the mechanism that has been accepted ever since and utilized with outstanding results in various applications such as, for example, the polymerase chain reaction invented by Kary Mullis [34].

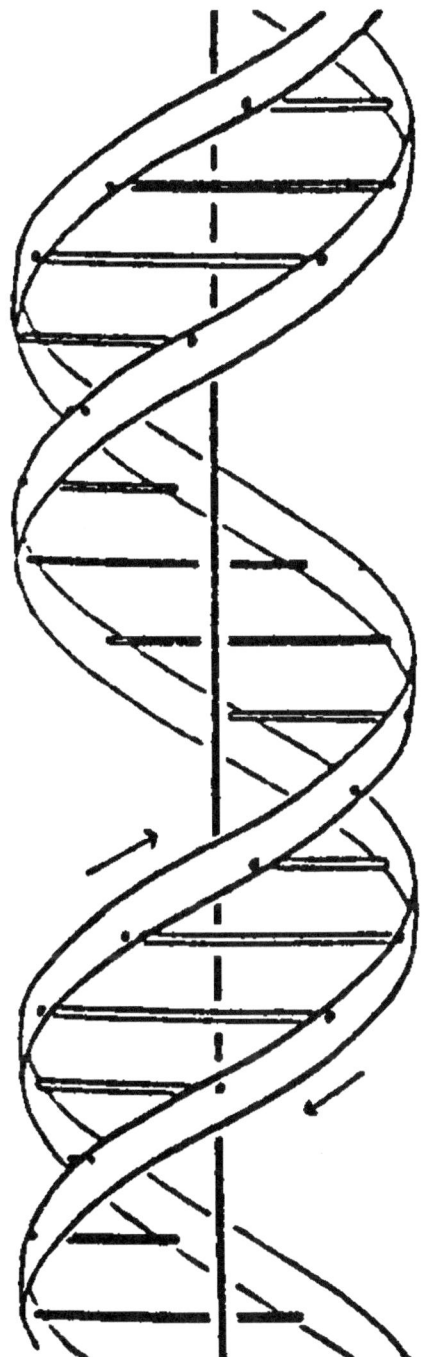

Fig. 6 Diagrammatic representation of the double helix by Watson and Crick [17, 18]. Reprinted with permission. Copyright (1953) Macmillan Magazines Ltd.

Pentagonal Synergy

To some extent, the success of X-ray diffraction in single-crystal structure determination has hindered research in areas of less-ordered materials. However, some of the best scientists have paid a lot of attention to these both in materials science and in biological structures. J. D. Bernal was one of the pioneers in both areas. His interest in liquid structures was expressed by Nikolai Belov [35] as: 'His last enthusiasm was for the laws of lawlessness'.

There was a curious absence of integer number residues in the a-helix structure, in the unit cell along the fiber direction, which was a sign of formal crystallography breaking down. Bernal [36] commented upon this in the following way: "We clung to the rules of crystallography, constancy of angles and so forth, the limitation of symmetry rotations of two-, three-, four-, and sixfold, which gave us the 230 space groups, as long as we could. Bragg hung onto them, and I'm not sure whether Perutz didn't too, up to a point, and it needed Pauling to break with them with his irrational helix".

In view of Bernal's interest in generalized crystallography, it is curious that at one point in his career he actually refrained from studying less-ordered rather than more-ordered systems. He and W. T. Astbury apparently divided crystallographic areas between themselves. In the words of Bernal [37]: "A strategic mistake may be as bad as a factual error. So it turned out to be with me. Faithful to my gentleman's agreement with Astbury, I turned from the study of the amorphous nucleic acids to their crystalline components, the nucleosides".

Nonetheless, Bernal had great influence in extending traditional crystallography into the science of structures. A sure sign of the expansion has been the gaining importance of fivefold symmetry in it. It is remarkable that two outstanding discoveries of the mid-1980s in materials, the fullerenes and the icosahedral quasicrystals, are both related to fivefold symmetry [38].

The stability of the truncated icosahedral C_{60} molecule was initially predicted by Eiji Osawa [39] on the basis of purely symmetry considerations. When the conspicuous relative stability of C_{60} was observed, Kroto et al. [40], not knowing of Osawa's prediction, were looking for a highly symmetrical structure. Although they eventually 'rediscovered' the truncated icosahedron (an artistic representation is shown in Fig. 9), they also reached out to R. B. Fuller's geodesic dome and thereby established a most valuable linkage between structural chemistry and design science. This was not the first time Fuller's ideas had facilitated structural research. Caspar and Klug [41] also acknowledged the inspiration received from Fuller's physical geometry in their discovery of the icosahedral virus structures.

The quasicrystal discovery *could* be described in the following *fictional* way: "For centuries excellent minds, including Johannes Kepler and Albrecht Du rer, have tried to employ regular pentagons for covering the extended surface with a pattern of repetitive fivefold symmetry without gaps or overlaps. In the early 1970s, Roger Penrose [42] came up with such a pattern. Alan Mackay [43] extended this pattern into the third dimension, and has urged experimentalists to be on the lookout for such solids in their experiments. Taking up

Fig. 7 Sculpture of the double helix by the sculptor Bror Marklund. Photograph by the author.

others with various theoretical models. As a result of these *concerted* activities, the science of structures has fast expanded".

Alas, this is not the way it happened. In reality, the story of the quasicrystal discovery [46] illustrates a development when many different threads of far-away origins come together for a unique moment of great importance, only to diverge again in many different directions. The moment may be an experiment or a sudden realisation of the significance of data or it may be a longer period in time. In the quasicrystal discovery, it was the period from Dan Shechtman's original observation in April 1982 to the end of 1984 when the wider world of science learned about the discovery and took over. It is noted though that the observation of incommensurately modulated structures [47, 48] had already challenged the periodicity paradigm. It was, however, salvaged by bringing these disturbing experiments into line, as if following a prescription by Kuhn [49] in *The Structure of Scientific Revolutions* (see [50]).

The discovery of quasicrystals has led to a paradigm change in crystallography, expressed even in a proposal for a new definition of what is a crystal by one of the IUCr's commissions: 'any substance is a crystal if it has a diffraction pattern with Bragg spots'.

Mackay (see [46]) has called attention to the rather careless original definition of crystallinity which needlessly excluded substances such as what we call today quasicrystals. In this sense, the discovery was a kind of legalistic discovery. This happens when the human classification system is more restrictive than the laws of nature and discoveries appear to break the laws that had been artificially constructed in the first place.

Pejorative words, such as deviation, imperfect, distortion, deformation, disordered etc., may be a consequence of such human imperfection, rather than nature's. This also applies to the various degrading and upgrading adjectives of symmetry in pseudosymmetry, subsymmetry, supersymmetry and suchlike. Molecules and atoms do not follow human-made rules of symmetry in their arrangements; rather, our symmetry rules reflect our observations.

Dissymmetry

Mackay's challenge, Dan Shechtman et al. [44] made such an observation. He used metal alloys of various compositions in rapid solidification and anticipated that this rapid solidification of the alloys would produce the predicted structures. Shechtman's experimental observations were published promptly and were embraced instantly by the leading scientists of structure. Figure 10 shows a quasicrystal. Shechtman's experimental observations were also interpreted right away by Dov Levine and Paul Steinhardt [45] and many

Louis Pasteur's 1848 discovery of molecular and crystal chirality (Fig. 11) was a rich starting point for many branches which grew from a common root. The specific chirality of biological molecules has puzzled scientists and philosophers alike ever since. This is the question that Vladimir Prelog [51] called 'molecular theology'. It was a great achievement of crystallography when Bijvoet et al. [52] determined the sense of chirality of molecules. Originally, Emil Fischer [53] had arbitrarily assigned an absolute configuration to sugars,

Fig. 10 Flower-like icosahedral quasicrystal in a quenched Al/Mn sample. Photograph courtesy of Dr Ágnes Csanády (Budapest).

Fig. 8 The double helix on a medal of the Pontifical Academy and on a Swedish stamp.

with a 50% chance of being correct and, luckily, indeed it proved correct. By now the absolute configuration has been established for relatively simple as well as for large biological molecules.

Pasteur [54] was aware of the possible implications of chirality; in his words, 'Is it not necessary and sufficient to admit that at the moment of the elaboration of the primary principles in the vegetable organism, [a dissymmetric) force is present? ... Do these [dissymmetric) actions, possibly placed under cosmic influences, reside in light, in electricity, in magnetism, or in heat? Can they be related to the motion of the earth, or to the electric currents by which physicists explain the terrestrial magnetic poles?' The most general symmetry statement, by Pierre Curie [55], must have relied a great deal on Pasteur's observations: *"c'est la dissymétrie*

qui crée le phénomène", 'dissymmetry creates the phenomenon'.

When Lee and Yang [56] predicted the nonconservation of parity in certain interactions of fundamental particles, and it was immediately confirmed by a series of experiments, the notion of the 'asymmetric universe' received general acceptance. In the wake of the violation of parity discovery, J. B. S. Haldane [57] graciously returned to Pasteur's conclusion, *"L'universest dissymétrique"*. Almost as a follow-up, such diverse areas of science as particle physics and astrophysics are being joined today in the search for fundamental forces in nature.

There are practical consequences of understanding the mechanism of chiral discrimination in organisms.

Fig. 9 Truncated icosahedron sticking out of the wall above the entrance into the 'Hall with the Fountain' at the Topkapi Saray in Istanbul. Photograph by the author.

Fig. 11 Louis Pasteur's chiral models of enantiomeric crystals in the Pasteur Institute, Paris. Photograph by the author.

Accumulated knowledge has included some tragic experiences. By now, research, characterization, manufacturing and marketing of enantiomers as potential drugs are rigorously legislated [58].

Appeal

The examples selected above have demonstrated various applications of the symmetry concept in crystallography. The fruitful interplay between them has also contributed to the development of the concept. The examples have also demonstrated the connecting ability of symmetry. Packing considerations are of importance not only to crystallography but to mathematics as well. Helical symmetry is a link between crystallography and molecular biology, fivefold symmetry between crystallography and materials science, chirality between crystallography and both medicine and physics, and examples relating to chemistry also abound [59].

There is yet another important area of human endeavor, the arts, where the symmetry concept provides a link for crystallography. Escher's periodic drawings [60] and sculptures resembling quasicrystals, helices and double helices in various artifacts all help crystallographers to reach outside their specialization and help non-crystallographers grasp the discoveries of the science of structures. Perhaps, however, nowhere so much as in education [61] does the symmetry concept help understand and appreciate our material world from the smallest molecule to the largest biological system and draw the most thrilling intellectual experience from it.

I appreciate the comments by Professor Alan L. Mackay, FRS (Birkbeck College, University of London), and by an anonymous reviewer, on the manuscript.

References

1. Wigner, E. P. (1967). *Symmetries and Reflections: Scientific Essays.* Indiana University Press.
2. Dunitz, J. D. (1996). *The Crystal as a Supramolecular Entity. Perspectives in Supramolecular Chemistry*, Vol. 2, edited by G. R. Desiraju, pp. 1-30. Chichester: John Wiley & Sons.
3. Murray-Rust, P. (1992). *Computer Modelling of Biomolecular Processes*, edited by J. Goodfellow & D. S. Moss, p. 19. New York: Ellis Horwood.
4. Pauling, L. (1939). *The Nature of the Chemical Bond and the Structure of Molecules and Crystals: an Introduction to Modern Structural Chemistry*, 1st ed. Ithaca, New York: Cornell University Press.
5. Hargittai, I. (1986). Editor. *Symmetry 2: Unifying Human Understanding.* Oxford: Pergamon Press.
6. Hargittai, I. (1989). Editor. *Symmetry: Unifying Human Understanding.* New York: Pergamon Press.
7. Hargittai, I. & Hargittai, M. (1994). *Symmetry: a Unifying Concept.* Bolinas, CA: Shelter Publications.
8. Hargittai, I. & Hargittai, M. (1999). *In Our Own Image: Personal Symmetry in Discovery.* New York: Plenum Press.
9. Kepler, J. (1611). *Strena Seu de Nive Sexangula.* Frankfurt am Main: Godefridum Tampach.
10. Kelvin, Lord (1904). Baltimore Lectures on Molecular Dynamics and the Wave Theory of Light, Appendix H, pp. 618-619. London: C. J. Clay & Sons.
11. Weyl, H. (1952). *Symmetry.* Princeton University Press.
12. Senechal, M. (1990). *Historical Atlas of Crystallography*, edited by J. Lima-de-Faria, pp. 43-59. Dordrecht: Kluwer Academic Publishers.
13. Pólya, G. (1924). *Z. Kristallogr.* 60, 278-282.
14. Schattschneider, D. (1990). *Visions of Symmetry: Notebooks, Periodic Drawings, and Related Work of M. C. Escher.* New York: W. H. Freeman.
15. Pauling, L. & Delbruck, M. (1940). *Science*, 92, 77-79.
16. Kitaigorodskii, A. I. (1971). *Molekulyarnie Kristalli.* Moscow: Nauka. (In Russian.)
17. Watson, J. D. & Crick, F. H. C. (1953a). *Nature (London)*, 171, 737-738.
18. Watson, J. D. & Crick, F. H. C. (1953b) *Nature (London)*, 171, 964-967.
19. Morawetz, H. (1994). *Herman Francis Mark: May 3, 1895-April 6, 1992. Biographical Memoires.* Washington, DC: National Academy of Sciences.
20. Astbury, W. T. & Sisson, W. A. (1935). *Proc. R. Soc. London Ser. A*, 150, 333-351.
21. Astbury, W.T. & Street, A. (1932). *Philos. Trans. R. Soc. London Ser. A*, 230, 75-101.
22. Astbury, W. T. & Woods, H. J. (1934). *Philos. Trans. R. Soc. London Ser. A*, 232, 333-394.
23. Crowe, D. W. (1986). *Symmetry: Unifying Human Understanding*, edited by I. Hargittai, pp. 407-411. New York: Pergamon Press.
24. Pauling, L. (1996). *Chem. Intell.* 2(1), 32-38.
25. Pauling, L. & Corey, R. B. (1950). *J. Am. Chem. Soc.* 72, 5349.
26. Pauling, L., Corey, R. B. & Branson, H. R. (1951). *Proc. Natl Acad. Sci. USA*, 37, 205-211.
27. Cochran, W., Crick, F. H. C. & Vand, V. (1952). *Acta Cryst.* 5, 581-586.
28. Crick, F. (1988). *What Mad Pursuit: a Personal View of Scientific Discovery.* New York: Basic Books.
29. Bragg, L., Kendrew, J. C. & Perutz, M. F. (1950). *Proc. R. Soc. London Ser. A*, 303, 321-357.
30. Perutz, M. (1997). Conversation with I. Hargittai, scheduled to be published in *The Chemical Intelligencer.*
31. Franklin, R. E. & Gosling, R. G. (1953). *Nature (London)*, 171, 740-741.
32. Wilkins, M. H. F., Stokes, A. R. & Wilson, H. R. (1953). *Nature (London)*, 171, 738-740.
33. Watson, J. D. (1994). *The Polymerase Chain Reaction*, edited by K. B. Mullis, F. Ferre' & R. A. Gibbs, pp. v-viii. Boston: Birkha user.
34. Mullis, K. B. & Faloona, F. A. (1987). *Methods Enzymol.* 155, 335-350.
35. Belov, N. (1991). In *A Dictionary of Scientific Quotations*, edited by A. L. Mackay. Bristol: Adam Hilger.
36. Olby, R. (1994). *The Path to the Double Helix: the Discovery of DNA.* New York: Dover Publications.
37. Bernal, J. D. (1968). *Labour Mon.* pp. 323-326.
38. Hargittai, I. (1990). Editor. *Quasicrystals, Networks, and Molecules of Fivefold Symmetry.* New York: VCH Publications.
39. Osawa, E. (1970). *Kagaku*, 25, 854-863.
40. Kroto, H. W., Heath, J. R., O'Brien, S. C., Curl, R. F. & Smalley, R. E. (1985). *Nature (London)*, 318, 162-163.

41. Caspar, D. L. D. & Klug, A. (1962). *Cold Spring Harbor Symp. Quant. Biol.* 27, 1-24.
42. Gardner, M. (1977). *Sci. Am.* 236, 110.
43. Mackay, A. L. (1982). *Physica (Utrecht)*, 114A, 609-613.
44. Shechtman, D., Blech, I., Gratias, D. & Cahn, J. W. (1984). *Phys. Rev. Lett.* 53, 1951-1953.
45. Levine, D. & Steinhardt, P. J. (1984). *Phys. Rev. Lett.* 53, 2477-2480.
46. Hargittai, I. (1997). *Chem. Intell.* 3(4), 25-49.
47. Janner, A. & Janssen, T. (1979). *Physica (Utrecht)*, A99, 47-76.
48. de Wolff, P. M. & van Aalst, W. (1972). *Acta Cryst.* A28, S111.
49. Kuhn, T. S. (1970). *The Structure of Scientific Revolutions*, 2nd ed., enlarged. The University of Chicago Press.
50. Cahn, J. (1995. *Epilogue. Proceedings of the 5th International Conference on Quasicrystals*, edited by C. Janot & R. Mosseri, pp. 807-810. Singapore: World Scientific.
51. Prelog, V. (1976). *Science*, 193, 17-24.
52. Bijvoet, J. M., Peerdeman, A. F. & van Bommel, A. J. (1951) *Nature (London)*, 168, 271-272.
53. Fischer, E. (1894). *Ber. Dtsch. Chem. Ges.* 27, 2985.
54. Pasteur, L. (1897). *Researches on the Molecular Asymmetry of Natural Organic Products.* Alembic Club Reprints No. 14. Edinburgh: W. F. Clay.
55. Curie, P. (1894). *J. Phys. (Paris)*, 3, 393-415.
56. Lee, T. D. & Yang, C. N. (1956). *Phys. Rev.* 104, 254-258.
57. Haldane, J. B. S. (1960). *Nature (London)*, 185, 87.
58. Richards, A. & McCague, R. (1997). *Chem. Ind.* pp. 422-425.
59. Hargittai, I. & Hargittai, M. (1995). *Symmetry through the Eyes of a Chemist*, 2nd ed. New York: Plenum Press.
60. MacGillavry, C. H. (1976). *Symmetry Aspects of M. C. Escher's Periodic Drawings.* Utrecht: Bohn, Scheltema & Holkema. (Present distributor Kluwer Academic Publishers, Dordrecht.)
61. Hargittai, M. & Hargittai, I. (1998). *Uppta ck Symmetri*! (*Discover Symmetry!*) Stockholm: Natur och Kultur. (In Swedish.)

Istvan and Magdolna Hargittai (*left*) and Zipora and Dan Shechtman (*right*) in December 2011 at the Royal Swedish Academy of Sciences in Stockholm during the 2011 Nobel award celebrations (by unknown photographer).

"There Is No Such Animal (אין חיה כזו)": Lessons of a Discovery[a,b]

István Hargittai

Do not consider it proof just because it is written in books...

Maimonides (attributed)

Abstract

The discovery of quasicrystals by Dan Shechtman in the early 1980s was a conspicuous event in materials science not only because it led to the production of a plethora of new materials but also because it signified the demise of a dogma in the science of condensed phase materials concerning symmetry restrictions. Having the discovery recognized was not easy and it required stamina on Shechtman's part. The story of the quasicrystal discovery offers a set of lessons that might be useful to remember in similar situations.

Quasicrystals are not only important for science; they are also visually attractive! (Fig. 1). The story of Dan Shechtman's discovery of quasicrystals gives us an opportunity to ponder about the nature of scientific discoveries. This was the subject of my after-dinner talk on January 12, 2011, at the Technion in Haifa, on the occasion of Shechtman's seventieth birthday celebrations. The title of the presentation was "Lessons of the Quasicrystals Discovery."

As the readers of Structural Chemistry know, quasicrystal structures have order but no periodicity; thus they are different from crystals having both order and periodicity and from amorphous materials having neither. Thus, quasicrystals are in between the two other kinds of materials. Crystals and amorphous bodies (glass, for example) have been known for a very long time, but what we call today quasicrystals used to be considered impossible until Shechtman's experiment. Shechtman discovered the quasicrystals; he did not invent them, because they have always been around without our knowing about them.

The official date of the quasicrystals discovery was marked by the appearance of Shechtman's and his co-authors' report in Physical Review Letters at the end of 1984 [2]. However, the actual date of discovery was April 8, 1982, when Shechtman noticed that one of his electron diffraction patterns of a manganese–aluminum alloy displayed tenfold symmetry. He knew that it was "impossible," if from nothing else, because many years before he had passed a university examination proving that there was no such thing. His own disbelief was marked by his words (in Hebrew), "There is no such animal," hence the title of this contribution. According to the rules of classical crystallography, fivefold, sevenfold, eightfold, etc., symmetries are impossible in the condensed phase.

What happened was an example *par excellence* of what Louis Pasteur expressed as "Chance favors the prepared mind." Danny could have dismissed his serendipitous experiment as some artifact, but he did not. Also, a weaker character might have sooner or later succumbed to the pressure of friendly advice and not so friendly ridicule that he had been exposed to when he insisted on the validity of his discovery. Shechtman's going against the dogma of symmetry restrictions in the condensed phase reminded me of the statement attributed to Maimonides: "Do not consider it proof just because it is written in books..."

For a long time illustrious scientists and artists such as Johannes Kepler and Albrecht Dürer, tried to create patterns in which regular pentagons covered a surface without gaps or overlaps, but they did not succeed. Then, the British

[a]Originally published in *Structural Chemistry* 2011, 22(4):745–748.
[b]This Editorial is based on an invited contribution to IGGERET, the Hebrew-language publication of the Israel Academy of Sciences and Humanities

I. Hargittai (✉)
Department of Inorganic and Analytical Chemistry, Budapest University of Technology and Economics, Budapest, Hungary
e-mail: istvan.hargittai@gmail.com

Fig. 1 Flowerlike icosahedral quasicrystals in quenched aluminum–manganese sample (By Ágnes Csanády, Budapest and Hans-Ude Nissen, Zurich; Courtesy of Ágnes Csanády) [1].

mathematician Roger Penrose created a pattern in which a pentagon was surrounded by five other pentagons within a larger pentagon, and the pattern continued by including larger and larger pentagons. He used not only the whole pentagons, but parts of them, too, and produced an appealing pattern, which covered the surface with regular pentagons, but of gradually changing sizes (Fig. 2). Another British scientist,

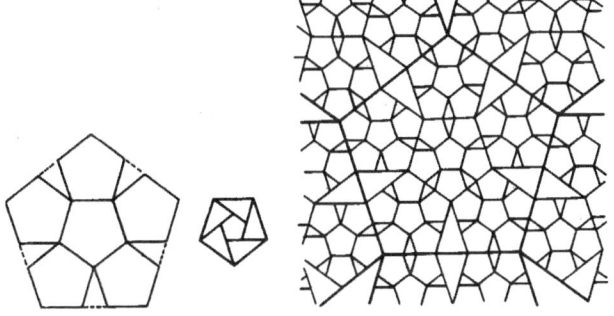

Fig. 2 Tiling with regular pentagons of changing sizes after Alan Mackay [3].

the crystallographer Alan Mackay was intrigued by Penrose's pattern, and simulated on this pattern a diffraction experiment. He figured that if he could produce a diffraction pattern from this planar pattern, the same should be possible even from three-dimensional structures.

When in the fall of 1982 Mackay told me about his simulation, Shechtman had already performed his crucial experiment, but we did not know about it. Neither did Shechtman know about Mackay's attempts. Shechtman, who earned his B.S., M.Sc., and Ph.D. degrees at the Technion, was a visiting scientist at the then U.S. National Bureau of Standards (NBS) where he was experimenting with aluminum–manganese alloys. That is when he ran his experiment that shocked him and that later would shock the world of science of condensed materials. But it took another 2 years before this happened.

Once Shechtman had completed his experiment, he became a very lonely person as every scientific discoverer does: the discoverer knows something nobody else does. At the time of his crucial experiment Shechtman was alone in the laboratory, but felt the urge to share his excitement with somebody else, and he ran out into the corridor, but nobody was there, so he returned to his electron microscope and performed a series of additional experiments. Other scientists, including Louis Pasteur have described similar experience of having the urge to share the new knowledge at the moment of discovery.

Right after his experiment Shechtman started inquiring what others knew about the possibility of whether or not what he had observed might be real. Sadly, however, scientists, who were authorities in his field—where he had not made his name yet—dismissed his claims. In one aspect this was also fortunate as nobody tried to expropriate his discovery. Finally, he found an expert in X-ray diffraction, Ilan Blech, who was willing to listen to him, and the two started making models that might produce the diffraction pattern that Shechtman had observed.

There was yet another direction of related investigation, at the physics department at the University of Pennsylvania in Dov Levine's Ph.D. work under Paul Steinhardt's supervision. Levine's dissertational work produced a theoretical model that in time would prove very suitable to interpret Shechtman's experimental observation. Levine was anxious to publish his findings, but his advisor was reluctant, probably fearing the reaction that a suggestion showing an established dogma invalid might confront them. Both Dan Shechtman and Dov Levine are currently professors at the Technion (Fig. 3), Shechtman at the Department of Materials Engineering and Levine at the Department of Physics.

For Shechtman it was not easy to bring out his discovery in a scientific paper. A more established scientist might have rushed to publish such a new finding to avoid losing the claim of priority. On the other hand, a more established scientist might have been very cautious, too, in coming out with a

Fig. 3 Dan Shechtman (on the *right*) and Dov Levine in Shechtman's office at the Technion in 1996 (Photograph by I. Hargittai).

discovery that others thought impossible. Shechtman was convinced that he had a novel observation, but it bothered him that he could not offer a credible explanation for it, and this restrained him from trying to publish it.

When Shechtman finally decided to submit a manuscript in which he and Blech described his experiment, it was done in a tentative manner. The information was there but an unsuspecting reader might have not noticed it, since it was buried under a mountain of information about alloys. It read like a report in metallurgy yet they submitted it to a physics publication, the Journal of Applied Physics. However, the editor of the journal wrote back by return of mail that the material of the manuscript was not suitable for the journal and—incredibly—that it would not interest physicists. It is not that rare that it is difficult to publish a seminal discovery. Of course, bad papers are rejected. But often this is the fate of manuscripts that report radically new results. As the Shechtman–Blech manuscript was more a report in metallurgy than in physics, they re-submitted it to a metallurgical periodical, and it appeared there, eventually [4].

Shechtman knew that he should considerably improve the presentation of his discovery. He turned for help to John Cahn, a veteran scientist in the section where he worked at NBS. Cahn for two long years did not believe in the discovery, but now he changed his mind and the grateful Shechtman invited Cahn to be a co-author of the report. Then, Cahn added yet another name, Denis Gratias, a young French mathematical crystallographer who helped in shaping the mathematical format of the article, and Shechtman added Blech's name out of loyalty, hence the four authors on the paper [2].

They sent the new manuscript to the prestigious periodical, *Physical Review Letters*, where it was received on October 9, 1984. The manuscript was accepted for publication without delay, and appeared in the issue dated November

12, 1984. It had a somewhat innocuous title, "Metallic Phase with Long-Range Orientational Order and No Translational Symmetry" [2]. The impact was tremendous, as if a floodgate had been lifted. Scientists, especially theoreticians, had been working on related problems and those who were versed in the literature could easily connect their work to the Penrose pattern, Mackay's simulated diffraction diagram, and Shechtman's discovery.

Dov Levine's and Paul Steinhardt's purely theoretical paper quickly followed in the same journal. Their manuscript arrived on November 2, 1984, that is less than 2 weeks before the Shechtman paper appeared, and this was no coincidence. They learned about the Shechtman manuscript back in October from a Harvard professor to whom Cahn had sent a copy. The Levine–Steinhardt report had an elegant title, "Quasicrystals: A New Class of Ordered Structures," thus it gave a name to the new structures [5]. The name was easy to remember, it well expressed the essence of the new substance, and it stuck. There are some who argue that giving a name to a discovery may be as important as making the discovery itself. A scientific discovery is made sooner or later and if not by this scientist, then by another scientist. In contrast, giving it a name is a unique act that belongs to an individual. Still, it is the discovery rather than its name that impacts science and the application of its fruits.

It is a question of contention to what extent should Shechtman's finding be considered a real discovery. To some, the identification of quasicrystals merely pointed to the inadequacy of how crystals had been defined. To this we might ask, if it was merely an easy correction, why nobody had bothered to make it before Shechtman. In reality, this discovery expanded our concept of materials by adding a new class to them. It brought about a paradigm change. There had been earlier attempts to challenge the dogma, but any prior "disturbing" experiment could be brought into line and the dogma could be left in place.

The main critique of the quasicrystal concept came from the great chemist Linus Pauling. He had distinguished himself not only with exceptional knowledge of structures of materials, but also with fundamental discoveries that broke with previous dogmas! His refusal of Shechtman's discovery made Shechtman's case even harder because this came from a scientist of Pauling's reputation. But Shechtman persevered. Pauling died in 1994 without having accepted Shechtman's discovery. Pauling's many followers found it easier to accept the new concept afterwards. The way it happened strengthened what Max Planck wrote in his scientific biography: "A new scientific truth does not triumph by convincing its opponents and making them see the light, but rather because its opponents eventually die, and a new generation grows up that is familiar with it" [6].

There are a few general lessons from the quasicrystal discovery that include the following:

- Start with a practical problem. In this case it was the quest for alloys with special properties.
- Play with it. In this case Shechtman did not limit his testing of alloy compositions that might have been expected to become useful.
- Do not discard the unexpected. When we get what we expect it is not a discovery; it happens when we get something we did not expect. Shechtman certainly did not discard the unexpected.
- Expose what you found to wide scrutiny. This Shechtman did impeccably.
- Coin a name for it. This happened here, too, though by others than the discoverer.
- Publish it conspicuously. This did not happen right away, but eventually it did.
- Talk about it to all that will listen, and try in every way easing your loneliness. Public Relations is also important in science.
- Stay with your discovery while it brings new knowledge and leave it only when it becomes routine. In this, Shechtman might have stayed with it a little longer although after a while he returned to it.

The quasicrystal story is yet another example of the unique role symmetry considerations play in many scientific discoveries [7]. This is part of the visual effects quasicrystals exert that led to their impact not only in science but in the visual arts and design as well [8]. The discovery opened the door to a world of "structures beyond crystals" [9]. Today Shechtman is the undisputed discoverer of quasicrystals, which is the recognition not only of his ingenuity, but also of his stamina and perseverance [10]. The discovery caused a paradigm change in physics, chemistry, crystallography, and materials science. It deserves the highest recognition. Dan Shechtman is a most decorated scientist, but whether or not such recognition will further expand, it is difficult to predict. This author certainly feels that it should.

References

1. Csanády A, Papp K, Dobosy M, Bauer M (1990) Symmetry 1:75–79
2. Shechtman D, Blech I, Gratias D, Cahn JW (1984) Phys Rev Lett 53:1951–1953
3. Mackay AL (1981) Kristallografiya 26:910–919
4. Shechtman D, Blech I (1985) Metall Trans 16A:1005–1012
5. Levine D, Steinhardt PJ (1984) Phys Rev Lett 53:2477–2480
6. From Max Planck (1948) A Scientific Autobiography, quoted in W. F. Bynum, Roy Porter, eds, Oxford Dictionary of Scientific Quotations. OUP, Oxford, UK, 2005, p 494
7. Hargittai M, Hargittai I (2009; 2010) Symmetry through the eyes of a chemist, 3rd edn. Springer
8. Hargittai I, Hargittai M (2000) In Our Own Image: Personal Symmetry in Discovery. Kluwer, NY
9. Hargittai I (2010) J Mol Struct 976:81–86
10. Hargittai I (2011) Drive and curiosity: What fuels the passion for science. Prometheus, Amherst, NY

The Chemical Intelligencer

VOL. 3, NO. 4 / OCTOBER 1997

US $9.00 / CAN $12.50

INTERVIEWS:
H. Taube
R.A. Marcus

Quasicrystal
Discovery

Notes on
C.K. Ingold

American
"Ionists"

R.M. Roberts'
Serendipity

Springer

A flowerlike quasicrystal decorated the cover of the October 1997 issue of *The Chemical Intelligencer*, Ágnes Csanády (Budapest) prepared the quasicrystals in a quenched Al/Mn sample and Hans-Ude Nissen (Zurich) took its scanning electron micrograph.

Dan Shechtman's Quasicrystal Discovery in Perspective[a]

István Hargittai

Abstract

Dan Shechtman's discovery of quasicrystals brought about a paradigm change in chemistry, physics, materials science, and other areas of science and engineering. Although superficially it could be looked at as a serendipitous event, Shechtman's curiosity and drive played equal parts with serendipity in this discovery. Shechtman was a lonely discoverer, again, seemingly detached from the main stream of generalized crystallography for which his contribution was a milestone. Generalized crystallography is the science of structures without restrictions— "structures beyond crystals (Hargittai, J Mol Struct 976, 81–86, 2010)." The discovery of quasicrystals can be seen as written into the history of ideas that have much extended our views about the tools of our scientific inquiry and the materials we aim at producing and utilizing. This review augments a recent Editorial in the August 2011 issue of Structural Chemistry about the lessons of the quasicrystal discovery (Hargittai, Struct Chem, 22, 745–748, 2011) and a book chapter about Dan Shechtman's traits as a discoverer and about his road to the discovery (Hargittai, Drive and Curiosity: What Fuels the Passion for Science. Prometheus Books, Amherst, 2011).

Introduction

Dan Shechtman (Fig. 1), winner of the 2011 Nobel Prize in Chemistry for the discovery of quasicrystals, exemplifies how curiosity and drive can lead to major scientific breakthroughs. Shechtman was preparing and investigating rapidly solidified aluminum-manganese alloys possessing properties that would make them useful for applications. He examined them under the electron microscope, and he varied their compositions within reasonable limits, looking for the most useful ones. At one point he reached the limit of the manganese content above which he knew the alloys would become too brittle for application and where he was supposed to limit his inquiry toward the larger manganese contents. This is what he should have done in a purely applied laboratory. In 1981, Shechtman had arrived for his first sabbatical at the National Bureau of Standards [NBS; today, National Institute of Standards and Technology (NIST)]. His stay was sponsored by the US Defense Advanced Project Agency (DARPA, later, ARPA). When he started his studies, the person with whom he was supposed to check his plans for research told him to feel free to go in any direction he found worthwhile. This instruction gave Shechtman freedom when he reached the upper reasonable limit of manganese content. He did not feel he had to stop, and indeed, he started probing alloys with ever-increasing manganese content to satisfy his curiosity. Both his conditions of work and his personal traits carried Shechtman in this direction.

Shechtman's drive manifested itself when he did not let benevolent colleagues, as well as those who ridiculed him, talk him out of pursuing the idea that what he had observed was what classical crystallography had deemed impossible symmetry (Fig. 2). This drive kept him functioning in an intellectually belligerent world. Linus Pauling, the most authoritative chemist of his time, with great renown as far as structural chemistry was concerned, also found Shechtman's claims impossible. Despite Pauling's own reputation as innovative and a maverick, he could not come to terms with Shechtman's interpretation of the diffraction photographs. For example, Pauling in his quest for the protein structures was not bothered by the non-integer repetition of amino acid units along the molecular axis, because the presence of intra-chain hydrogen bonding precluded integer repetition [1]. It would be hard to imagine a more powerful

[a]Originally published in *Israel Journal of Chemistry* 2011, 51:1144–1152. Reproduced by permission

I. Hargittai (✉)
Department of Inorganic and Analytical Chemistry, Budapest University of Technology and Economics, Budapest, Hungary
e-mail: istvan.hargittai@gmail.com

Fig. 1 Dan Shechtman in 2007 in Budapest; photo by and © I. Hargittai.

opponent to recognizing Shechtman's discovery than Linus Pauling, but even this could not stop Shechtman's drive.

Shechtman was honored with many awards for his discovery, among which the Aminoff Prize occupied a special place because it was awarded by one of the most authoritative bodies of science, the Royal Swedish Academy of Sciences, expressly for recognition in the field of crystallography. This was in the year 2000, and many thought that while this was a very special distinction, it was also a subtle way to position Shechtman's discovery among important events in crystallography, without elevating it to the category of discoveries of more general significance. It has happened, but very rarely, that an Aminoff Prize laureate would later be awarded a Nobel Prize. I doubt that Shechtman did this consciously, but he dressed too formally for the prize-awarding ceremony, as if it were an event of higher importance. The unwritten

Fig. 2 Flowerlike icosahedral quasicrystal in a quenched Al/Mn sample, courtesy of Agnes Csanady, Budapest. Used with permission.

Fig. 3 *Candid Science V* book cover highlighting John Conway, Roger Penrose, Alan Mackay, and Dan Shechtman (© I. Hargittai).

dress code for the event prescribed a much less formal appearance. Secondly, he started his presentation by listing three discoveries related to new materials, of which two had already been awarded a Nobel Prize (high-temperature superconductivity and buckminsterfullerene). The third was the discovery of quasicrystals, and the implication was obvious.

Shechtman's Nobel Prize finally arrived in 2011. It is significant that he received it unshared, and—although some might have thought that the circle of awardees could have been expanded—no displeasure was expressed among the scientific community following this judgment. At this point, however, it is equally appropriate to view Shechtman's discovery in the context of the intellectual process that led to the development of what is called "generalized crystallography." The most august scientific body has now put its "stamp of approval" on this development [2].

The Story

The quasicrystal story begins with John Desmond Bernal, who was the first to recognize the confining nature of classical crystallography, and he initiated generalized crystallography (Fig. 3). He noticed that there are arrangements, especially among the low-coordination cases, both among organic and inorganic structures, where the classical restrictions of symmetry to two-, three-, four-, and sixfoldedness no longer hold [3]. He stressed that icosahedral geometry is not capable of

Fig. 4 J. Desmond Bernal and his model of an ideal monoatomic liquid; courtesy of John Finney, University College London. Used with permission.

forming regular extended arrangements, although it could provide close-packed structures. The absence of long-range order would account for the much greater variation of properties of such structures than the corresponding classical crystals. Bernal's conclusion was, "We clung to the rules of crystallography..., which gave us the 230 space groups, as long as we could. Bragg hung on to them, and I'm not sure whether Perutz didn't too, up to a point, and it needed Pauling to break with them with his irrational helix [4]."

Looking back to Bernal's teachings (Fig. 4) and the developments since, up to the quasicrystal discovery, a *fictional* story could be compiled of how the discovery *might* have happened—although it did not go this way [5]:

> For centuries, excellent minds, including Johannes Kepler and Albrecht Dürer, have tried to employ regular pentagons for covering the extended surface with a pattern of repetitive fivefold symmetry without gaps or overlaps. In the early 1970 s, finally, Roger Penrose came up with such a pattern. Alan Mackay extended this pattern into the third dimension, and, by showing it was possible theoretically, he urged experimentalists to be on the lookout for such solids in their experiments. Taking up

Mackay's challenge, Dan Shechtman then made such an observation. Shechtman used metal alloys of various compositions in rapid solidification. He anticipated that this rapid solidification of the alloys would produce the predicted structures. His experimental observations were published promptly and were embraced instantly by the leading scientists dealing with structures. His experimental observations were also interpreted right away by Paul Steinhardt and many others with various theoretical models. As a result of these *concerted* activities, the science of structures has fast expanded considerably.

In reality, everything was different: there were no concerted efforts, Shechtman was not aware of the previous attempts, and he made his observations serendipitously. Also, there was a long gestation period, two and a half years between April, 1982, and the fall of 1984, before Shechtman could publish his findings. That is when the broader scientific community learned about his discovery and responded with an avalanche of papers. The peculiarity of fivefold symmetry in this story is explained in Mackay's statement [6]:

> The main significance of fivefold symmetry for science is that it furnishes us with an explicit example of frustration, which has proved a most fertile concept in the physics of condensed matter... Neither we or nature can have everything simultaneously—not all things are possible,... We have only the freedom of necessity. "Nature must obey necessity", as Shakespeare (*Julius Caesar* IV: iii), Democritos, Monod, Bernal, and many others have also recognized. Science probes the limits of necessity and, in the case of fivefold symmetry, has found a corridor that leads us to a new territory.

My personal interest in fivefold symmetry remained at the hobby level, because in my research of molecular structures there was no restriction on this or other symmetries. But I found the issue intriguing and invited Alan Mackay to talk to us in Budapest about fivefold symmetry. In September, 1982, he gave us two lectures on this topic (Fig. 5) and issued a warning that we should be aware of the possibility of extended structures of fivefold symmetry, because if we thought them impossible,

Fig. 5 Alan L. Mackay lecturing on fivefold symmetry in September 1982, in Budapest; photo by and © I. Hargittai.

they might go unnoticed and unrecognized. Mackay did not know, and, obviously, neither did we, that by then Dan Shechtman had already observed such structures. I will always remember our amazement at what Mackay told us, especially looking back; it felt as if we were present at creation.

Mackay was always interested in non-commensurate structures, and he considered simple things, like printing wallpaper. "...[S]uppose you are printing two motifs from two rollers of different diameter. Then you get a non-repeating pattern. I wasn't able to think of producing an aperiodic two-dimensional pattern in this way.... I was really interested in hierarchic patterns... It came directly from Bernal... I produced a hierarchic pattern of pentagons [7]." Mackay heard about the Penrose pattern, and contacted Penrose to discuss it. Mackay's interest in hierarchic structures and Penrose's interest in forcing aperiodicity turned out to be very similar.

Roger Penrose started playing around with tile shapes and tiling problems. He was interested, for example, in the shape Maurits C. Escher used in his picture titled *Ghosts*. Penrose showed Escher his tiling, but for the time being these were periodic patterns. Then he became interested in hierarchical tiling and noticed a logo in a letterhead. The logo had a pentagon in the middle, surrounded by five others within a larger pentagon. He decided to iterate this, and sought a way of filling the gaps in a systematic way. In Penrose's words [8]:

> The only interesting thing is how to fill the gaps up. Thus I produced this pattern, which I designed partly to show somebody, who'd been in hospital, just as an amusement. A little later I realized that you could actually force that pattern by making it a jigsaw. There are pentagons, little rhombuses, five-sided, what I call jester's caps, which are half of them. The problem was to find a way forcing that pattern by some local matching rules. Having three versions of the pentagons and one of each of the others you could force it, so it was a six-piece tiling, which was non-periodic and which happened to have this fivefold quasisymmetry. But I wasn't thinking particularly to refute crystallography. It was just like an amusement.

Once Penrose had produced this tiling pattern, he published an article about it in 1974 in the Bulletin of the Institute of Mathematics and Its Applications [9]. The paper grew out of a lecture he gave on aesthetics. His lecturing about his patterns prompted him to think about possible applications in crystallography. Penrose thought that a generalization might be possible, and fivefold symmetry and icosahedral symmetry might occur in crystals. He thought an obstacle would be the impossibility of spotting mistakes and such events would prevent continuation. There were no local assembly rules and this is why he thought that it would be impossible to spot natural occurrences of what later became known as quasicrystals. This was at the time of our conversation in March, 2000, in Oxford (Fig. 6). Within a decade, though, quasicrystals were found in nature [10].

Penrose's paper in the obscure mathematical journal did not generate much interest. However, when Martin Gardner

Fig. 6 Roger Penrose in his office in Oxford in 2000; photo by and © I. Hargittai.

wrote about the Penrose patterns in *Scientific American*, interest was aroused [11]. Gardner had started corresponding with Penrose and he decided that these patterns deserved more exposure. The cover of the *Scientific American* issue in which the Gardner article appeared was designed by the mathematician John Conway [12]. At the time of the preparation of the magazine cover, Conway and Gardner conjectured about the possibility of crystallization, but they never published anything about their discussion, which Conway later regretted. In his words, "I remember that I wondered to myself how many different substances have been studied with respect to crystallization, and my guess was less than ten to the seventh power. Then I thought of the probability that something will crystallize in this manner and one in ten to the seventh power seemed to be a reasonable guess; therefore such crystallization should happen [13]."

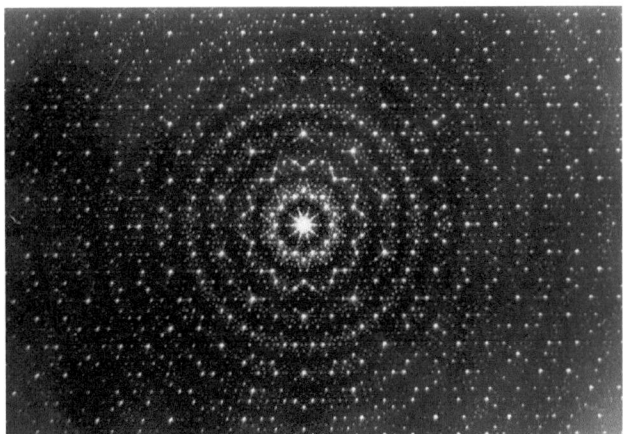

Fig. 7 Simulated "electron diffraction pattern of three-dimensional Penrose tiling" in 1982, prepared for Professor Alan L. Mackay by Dr. G. Harburn at Cardiff University; courtesy of Alan Mackay, London. Used with permission.

Fig. 8 Dan Shechtman's electron diffraction pattern of an aluminum-manganese alloy with tenfold symmetry; courtesy of Dan Shechtman, Haifa. Used with permission.

Fig. 9 Dan Shechtman and Alan Mackay in 1995 in the Hargittais' home in Budapest; photo by and © I. Hargittai.

Alan Mackay continued to be intrigued by the possibility of the natural occurrence of three-dimensional Penrose patterns and, with assistance by others, he produced a simulated diffraction pattern from them (Fig. 7). These simulated patterns would be found to be similar to the diffraction patterns in Shechtman's experiments in which he discovered "forbidden" symmetry (Fig. 8). Shechtman and his colleagues were producing a series of aluminum–manganese alloys with increasing amounts of manganese in them. In Shechtman's own words from a conversation we recorded on May 14, 1995, in Balatonfüred, Hungary, during an international school on quasicrystals [14]:

> Eventually I ran wild, from a practical point of view, since beyond a few percents of manganese the rapidly solidified alloy becomes brittle and therefore useless. Among the alloy ribbons which I have prepared with Frank Biancaniello by melt spinning, there were alloys which contained over 25 weight percent manganese. On April 8, 1982, as I was doing electron microscopy on a rapidly solidified aluminum alloy which contained 25 % manganese, something very strange and unexpected happened. It is worthwhile to look at my TEM [transmission electron microscope] logbook records of that day. For plate number 1725 (Al-25 %Mn) I wrote, "10 Fold???" There were bright spots in the selected area diffraction pattern, equally spaced from the center and from one another. I counted them and repeated them and repeated the count in the other direction and said to myself: "There is no such animal," in Hebrew, Ein chaya kazo. I then walked out to the corridor to share it with somebody, but there was nobody there, so I returned to the microscope and in the next couple of hours performed a series of experiments. Most of the needed experiments were performed at that time. A few days later all my work was complete, and everything was ready for the announcement. Then it took two years to publish it.

Shechtman first consulted his NBS colleagues, but they told him that he had either observed something else, or suggested to him to refresh his knowledge of diffraction

theory. He knew what he was talking about and how powerful the dogma was about symmetry restrictions in the condensed state. He once had to prove it during an examination at the Technion. Shechtman knew that his observation was a lucky break, but was astonished over the years that a large number of knowledgeable scientists could not come up with an explanation. It might have helped him if he had known about Mackay's simulated diffraction experiment, but he did not (Fig. 9).

There was a long, lonely period for Shechtman, and only his stamina and perseverance saved him from giving up. Eventually, his loneliness was eased by Ilan Blech, another Technion scientist, and the two produced a manuscript which they sent to the *Journal of Applied Physics*; it was returned by the editor with a note saying that their report would not be of interest to physicists. Later Shechtman also judged this manuscript as poorly written; a variation of it appeared later in *Metallurgical Transactions* [15]. The principal report about Shechtman's observation appeared under his name with three co-authors in late fall 1984 [16].

The announcement of the discovery was followed by frantic activities and an extraordinary number of publications in the years that followed. It appeared as if the scientific world had been ready for the discovery; thus, for example, theoreticians published models right away following the publication by Shechtman and his colleagues. The report by Dov Levine and Paul Steinhardt stood out not only because of their speed and their attractive model, but also because they coined the name "quasicrystals," which then stuck [17].

The most conspicuous doubter of Shechtman's discovery was Linus Pauling (Fig. 10); it was not the experiments he doubted, but the interpretation. Shechtman had several encounters with Pauling, but Pauling would not budge. He suggested that the observation originated from twin crystals. It is worthwhile to quote a sample of Pauling's statements in order to appreciate the formidable barriers Shechtman was facing in getting his discovery accepted. Following the

Fig. 10 Linus Pauling in the early 1980s at Moscow State University; photo by and courtesy of Larissa Zasourskaya, Moscow. Used with permission.

success of a multidisciplinary symmetry volume in 1986, I edited a second volume in 1989, and Pauling wrote a paper for it with a long title: "Interpretation of So-called Icosahedral and Decagonal Quasicrystals of Alloys Showing Apparent Icosahedral Symmetry Elements as Twins of an 820-Atom Cubic Crystal." His stand was obvious already from the title, and he concluded his discussion with the following paragraph [18]:

> As a crystallographer, with 65 years of experience in X-ray crystallography, I am pleased that the problem of the so-called icosahedral quasicrystals has been resolved in this way. Crystallographers have believed for many years that crystals cannot have five-fold axes of symmetry. In my model the grains with apparent icosahedral symmetry consist of cubic crystals that have a conventional structure, but that have, by repeated twinning determined by the approximate icosahedral structure of the 104-atom clusters, arranged themselves into an aggregate of microcrystals that shows icosahedral symmetry.

In the fall of 1993, I asked the 92-year old Pauling again about his opinion of the quasicrystal discovery, and as it happened this may have been his last statement about this

Fig. 11 John Cahn at NIST in 1995; photo by and © I. Hargittai.

issue. Soon afterwards, he died. My questions referred to both the C60 and the quasicrystals discoveries, but of his responses I am quoting here only the one concerning the quasicrystals [19].

> *Question*: Recent discoveries such as the quasicrystals and the fullerenes seem to have caught the solid state and chemical communities by surprise. Were these exceptional events or should we be getting prepared to seeing more of these kinds of unexpected findings in the future?

> *Linus Pauling*: As to the quasicrystals, you know that I contend that icosahedral quasicrystals are icosahedral twins of cubic crystals containing very large icosahedral complexes of atoms. It is not surprising that these crystals exist. The first one to be discovered was the MgZnAl compound reported by my associates and me in 1952. We did not observe quasicrystals of this compound, but they have been observed since then.

John Cahn (Fig. 11) was a senior scientist at the time of Shechtman's stint at NBS, and for a while he also resisted accepting the quasicrystal discovery. He described how he heard about the discovery from Shechtman for the first time [20]:

> One day he came into my office, and said, "John, what do you think of a 10-fold axis?" I said, "Don't bother me, Danny, this is clearly twinning," and he said, "I don't think so." Then we discussed a number of experiments to decide this question. I didn't know much about twinning but I did know that through twinning you could get unexpected symmetries.

Two years after their first encounter about Shechtman's experiments, the two talked about it again when Shechtman returned for another visit at the NBS. By then the paper in *Metallurgical Transactions* [15] was already in production, and Shechtman showed the manuscript to Cahn. Cahn told Shechtman that the paper did not articulate Shechtman's discovery in any adequate way. Then, the following occurred, according to Cahn [21]:

> … in our conversation Danny at one point said, "If you feel so strongly about it, can you write this paper?" I said, "Danny, this is your work, you're making me an enormous gift." He said, "I don't mind." I began writing this paper for *Physical Review Letters*. I just wanted the data to speak for themselves, to show that they were not consistent with the paradigm of periodicity. The published paper is two and a half pages, and there are few things too many in it. One of the things I'm sorry about is that we said we couldn't fit the diffraction pattern to that of a periodic crystal; it couldn't be indexed. We should have said we cannot fit it to a periodic crystal up to a lattice parameter of a few nanometers. We should have been more specific because Linus Pauling noticed this and said that you can always fit something if you pick a large enough lattice parameter…

In the process of writing the paper, Cahn invited a young French theoretician, Denis Gratias, to join the team. This is how the four authors, Shechtman, Blech, Gratias, and Cahn, came together [16]. When the manuscript was ready, it had to be reviewed by the NIST Editorial Review Board. Since NBS had been burned in the polywater story, they were very careful. One of Cahn's friends warned him: "John, you have a wonderful reputation. Why ruin it by putting your name on something like such a paper [22]." Finally, however, the Board approved the manuscript and it could be sent off to the journal. As Cahn circulated preprints of the paper, it reached, among others, the theoretical physicist Paul Steinhardt (then) of the University of Pennsylvania who happened to be visiting at IBM at the time, and he showed it to his graduate student Dov Levine. Again, in Cahn's narrative: "... this was the first inkling that there was actually an explanation for the patterns we were seeing. Things moved very fast, and Steinhardt ... was rushing his paper with Levine to *Physical Review Letters* and it appeared about a month later. I remember when I saw Steinhardt's copy of our manuscript it was almost illegible because it was a copy of a copy of a copy [22]."

Levine summarized the essence of their paper as follows: "We sought to elucidate the symmetries of quasicrystals by generalizing the Penrose pattern. We showed that orientational symmetries forbidden to periodic crystals are allowed for structures with quasi-periodic translational symmetry [23]."

Conclusions

In October 1994, I was having a conversation in London with Alan Mackay about the significance of the quasicrystal discovery. Mackay considered it as part of the bigger picture, on the background of Bernal's teachings about generalized crystallography, and said that the discovery might be considered to be "a bogus discovery because it arose simply because our definitions of crystallinity were drawn up rather carelessly. Therefore, it's a kind of legalistic discovery. It's a discovery of a material which breaks the laws that were artificially constructed. They were not laws of nature; they were laws of the human classificatory system [24]." Of course, with such an approach many other important discoveries might be considered merely legalistic if they uncovered phenomena that had not been covered by previous human description of nature, like superconductivity. It was obvious that Mackay's intention was not to belittle Shechtman's discovery. When, on the same occasion, I asked him about the Nobel Prize, he considered the various kinds of Nobel Prize and related Shechtman's discovery to other discoveries that had already been awarded this distinction. He described the discoverer of quasicrystals as [25]

> someone who turns over a stone and finds something really important, and recognizes that he has got something really important, maybe like superconductivity or the scanning tunneling microscope or the Mossbauer effect. There isn't any enormous amount of work but someone was in the right place at the right time, and recognized what he's done. I think Shechtman would come in [this] category. There is actually some new evidence that Shechtman's discovery may be more important than it had been believed. It has been mostly followed by a tremendous amount of mathematics, an Ivory Tower of mathematics and little more. Now it appears, however, that the very

low thermal conductivity of quasicrystals may be useful for something more than the non-stick frying pan but also important as turbine blades, internal combustion engines, and so on. People are producing effectively quasicrystal surfaces by glazing metal with a laser. So Shechtman's discovery may be eventually related to a process of great economic importance.

Some have expressed surprise that Shechtman was awarded the Nobel Prize in chemistry rather than in physics. Apart from thinking in terms of Nobel Prize categories, or school subjects for that matter, his discovery could be assigned in modern terms to materials science, which is at the borderline between chemistry and physics with considerable overlap. The 2010 Nobel Prize in Physics for the discovery of graphene, for example, could have just as well been awarded in chemistry as in physics [26]. We are very much conditioned according to our school education, which with its "division into subjects creates the image of a compartmentalized world [27]," whereas "Nature is not organized in the way universities are [28]."

What truly matters is that Shechtman's discovery was par excellence the kind of achievement that, in Eugene P. Wigner's formulation, was the task of scientific inquiry (when he mentioned physics, it was not a compartmentalized branch of science, but Science itself). The chemical-engineer-turned-theoretical-physicist Wigner (Fig. 12) stated [29]:

> Physics does not endeavor to explain nature. In fact, the great success of physics is due to a restriction of its objectives: it only endeavors to explain the regularities in the behavior of objects. This renunciation of the broader aim, and the specification of the domain for which an explanation can be sought, now appears to us an obvious necessity. ...
>
> The regularities in the phenomena which physical science endeavors to uncover are called the laws of nature. The name is actually very appropriate. Just as legal laws regulate actions and behavior under certain conditions but not try to regulate all actions and behavior, the laws of physics also determine the behavior of its objects of interest only under certain well-defined conditions, but leave much freedom otherwise.

Fig. 12 Eugene P. Wigner and I. Hargittai in 1969, in Austin, Texas; by unknown photographer; © I. Hargittai.

In the main body of the present treatise much attention was paid to symmetry considerations. Indeed, "Symmetry is a stunning example of how rationally derived mathematical argument can be applied to descriptions of nature and lead to insights of the greatest generality [30]."

On a personal note, I was infinitely lucky that in 1969, while I was a research associate at the Department of Physics of the University of Texas at Austin, Wigner gave me one-on-one tutorials on symmetry during his stay in Austin. This experience has impacted me and those close to me during my entire research career, in which the determination and modeling of molecular structures have always been combined with symmetry considerations [31]. Part of this was the fascination with fivefold symmetry [32] and the sensitivity toward all its appearances around us [33]. It was also in 1969 in Austin that I met Michael Polanyi, the medical-doctor-turned-physical-chemist-turned-philosopher, who had also influenced his doctoral student, Eugene Wigner. Recognizing regularities in properties, be they structural or other, has always been a principal tool in chemistry. Suffice it to recall the discovery and development of the Periodic Table of the Elements. Wigner learned about the importance of observing regularities from Polanyi, and he stressed this in his brief statement at the Nobel Prize award banquet in 1963 in Stockholm [34]:

> I do wish to mention the inspiration received from Polanyi. He taught me, among other things, that science begins when a body of phenomena is available which shows some coherence and regularities, that science consists in assimilating these regularities and in creating concepts which permit expressing these regularities in a natural way. He also taught me that it is this method of science rather than the concepts themselves (such as energy) which should be applied to other fields of learning.

Returning to the "bigger picture," scientists and artists since Johannes Kepler and Albrecht Dürer have wondered about fivefold symmetry and both about its conspicuous presence and absence in nature. Classical crystallography and X-ray crystallography have had tremendous successes in uncovering the secrets of nature through the 1980s and beyond. J. Desmond Bernal and his disciples as well as others attempted to expand the science of structures to embrace more of fivefold symmetry and other "forbidden" symmetries in the extended world of solid state materials. Dan Shechtman's discovery arrived as an integral part of a unique succession of research and ingenuity [35–37].

Acknowledgments I appreciate the kind advice from Bob Weintraub of the Sami Shamoon College of Engineering and Irwin Weintraub of Beersheva in finalizing the text of this paper.

References

1. I. Hargittai, *Struct. Chem.* 2010, *21*, 1 – 7.

2. Our discussion will follow the train of thoughts of Chapter 7 of our book: I. Hargittai, M. Hargittai, *In Our Own Image: Personal Symmetry in Discovery*, Kluwer/Plenum, NY, 2000, pp. 144 – 179 and the interviews with some of the major players of the story, quoted in B. Hargittai, I. Hargittai, *Candid Science V: Conversations with Famous Scientists*, Imperial College Press, London, 2005, pp. 16 – 93.

3. J. D. Bernal, *Acta Physica Acad. Scient. Hungaricae* 1958, *8*, 269–276.

4. R. Olby, *The Path to the Double Helix: The Discovery of DNA*, Dover, NY, 1994, p. 289.

5. Hargittai and Hargittai, *In Our Own Image*, pp. 149–150.

6. A. L. Mackay, in *Quasicrystals, Networks, and Molecules of Fivefold Symmetry* (Ed.: I. Hargittai), VCH, New York, 1990, pp. 1 – 18.

7. Hargittai and Hargittai, *Candid Science V*, p. 70.

8. Ibid. p. 39.

9. R. Penrose, *Bulletin of the Institute of Mathematics and Its Applications* 1974, *10*, 266 – 271.

10. D. V. Talapin, E. V. Shevchenko, M. I. Bodnarchuk, X. Ye, J. Chen, C. B. Murray, *Nature* 2009, *461*, 964 – 967.

11. M. Gardner, *Sci. Am.* 1977, *236*, 110.

12. I. Hargittai, *Mathematical Intelligencer* 1997, *19(4)*, 36 – 40.

13. Hargittai and Hargittai, *Candid Science V*, p. 26.

14. Ibid., p. 85.

15. D. Shechtman, I. A. Blech, *Metallurgical Transactions* 1985, *16 A*, 1005 – 1012.

16. D. Shechtman, I. Blech, D. Gratias, J. W. Cahn, *Phys. Rev. Lett.* 1984, *53*, 1951 – 1953.

17. D. Levine, P. Steinhardt, *Phys. Rev. Lett.* 1984, *53*, 2477–2480.

18. L. Pauling in *Symmetry 2: Unifying Human Understanding* (Ed: I. Hargittai), Pergamon Press, Oxford, 1989 pp. 337–339.

19. I. Hargittai in *Candid Science: Conversations with Famous Chemists* (Ed: M. Hargittai), Imperial College Press, London, 2000, pp. 5– 6.

20. Hargittai and Hargittai, *In Our Own Image*, p. 166.

21. Ibid., p. 167.

22. Ibid., p. 168.

23. Ibid., p. 170.

24. I. Hargittai, *Chem. Intell.* 1997, *3*, 25 – 49.

25. Hargittai and Hargittai, *Candid Science V*, p. 72.

26. I. Hargittai, *Struct. Chem.* 2010, *21*, 1151 – 1154.

27. Hargittai and Hargittai, *In Our Own Image*, p. 2.

28. R. L. Ackoff in *A Dictionary of Scientific Quotations* (Ed: A. L. Mackay), Adam Hilger, Bristol, UK, 1991.

29. E. P. Wigner in *Nobel Lectures Including Presentation Speeches and Laureates' Biographies: Physics 1963–1970*, World Scientific, Singapore, 1998, pp. 6 – 17.

30. G. M. Edelman, *Bright Air, Brilliant Fire: On the Matter of Mind*, Basic Books, New York, 1992, p. 199.

31. M. Hargittai, I. Hargittai, *Symmetry through the Eyes of a Chemist*, Third Edition, Springer, 2009.

32. a I. Hargittai, *Fivefold Symmetry*, World Scientific, Singapore, 1992; b I. Hargittai, *Quasicrystals, Networks, and Molecules of Fivefold Symmetry*, VCH, New York, 1990.

33. M. Hargittai, I. Hargittai, *Visual Symmetry*, World Scientific, Singapore, 2009.

34. E. P. Wigner, *Symmetries and Reflections: Scientific Essays*, Indiana University Press, Bloomington, Indiana, 1963, pp. 262 – 263.

35. I. Hargittai, *J. Mol. Struct.* 2010, *976*, 81 – 86.

36. I. Hargittai, *Struct. Chem.* 2011, *22*, 745 – 748.

37. I. Hargittai, *Drive and Curiosity: What Fuels the Passion for Science*, Prometheus Books, Amherst, NY, 2011, Chapter 8, pp. 155 – 172.

Dan Shechtman is presenting his Nobel lecture in December 2011 in Stockholm (photograph by Magdolna Hargittai).

From an Electron Micrograph to a Postage Stamp[a]

István Hargittai

Abstract

Soon following Dan Shechtman's discovery of quasicrystals, Ágnes Csanády and her associates started producing beautiful quasicrystals of flowerlike morphology. The image of one of their specimen appeared on the Israeli postage stamp honoring Shechtman's discovery, his Nobel Prize, and the International Year of Crystallography.

Dan Shechtman was conducting experiments with alloys in the spring of 1982 at the National Bureau of Standards (NBS, as it was then) in Washington, DC. Shechtman had developed a technique for studying metallic powders by transmission electron microscopy at the Technion—Israel Institute of Technology. At NBS, he collaborated with associates of the metallurgy group in producing and analyzing rapidly solidified aluminum-iron and other aluminum alloys, including aluminum-manganese alloys. Frank Biancaniello was Shechtman's enthusiastic colleague in preparing alloys of a great variety of composition. Of the aluminum-manganese alloys, the practically useful ones contained only a few percent of manganese. However, it seemed that alloys much beyond the practically useful manganese content might be also of interest to study. As they put an alloy with 25% manganese content into the electron microscope, Shechtman made a most unexpected observation. The electron diffraction diagram showed tenfold symmetry. This happened on April 8, 1982, and it was for the first time that someone observed and recorded symmetry in the condensed state that classical crystallography deemed impossible in crystals.

Shechtman had a hard time getting his interpretation of his observation accepted by the crystallographic community and the broader scientific community. However, when he published his experiment in 1984 [1], an avalanche of studies and papers appeared, and many laboratories worldwide produced the new substance for which the name quasicrystals had been coined—short for quasiperiodic crystals. The story has been well documented (see, for example, [2–6]). From early on following Shechtman's discovery, I found that Ágnes Csanády and her colleagues produced the most beautiful quasicrystals at the development center of the Hungarian Aluminum Industry (Fig. 1).

Csanády and her group conducted extensive studies of the morphology of quasicrystals and the phase transformation of quasicrystals to crystals [7]. Some of the specimen selected for such investigation were quasicrystals of flowerlike morphology that Csanády and her colleagues started describing

In memoriam Oleg Shishkin in acknowledgment for his contributions to the science of structural chemistry and for his dedicated work for our journal Structural Chemistry.

[a]Originally published in *Structural Chemistry* 2016, 27:5–7.

I. Hargittai (✉)
Department of Inorganic and Analytical Chemistry, Budapest University of Technology and Economics, Budapest, Hungary
e-mail: istvan.hargittai@gmail.com

Fig. 1 Electron micrograph of flowerlike quasicrystals of a quenched aluminum-manganese alloy. The length of the full horizontal bar corresponds to 1μm (courtesy of Ágnes Csanády).

in 1987. They followed the phase transformation directly and observed that the nucleation of crystallization started on the surface of the icosahedral phase. As the new phase grew, the icosahedral phase kept shrinking. Here one has to be careful with semantics. At the time indeed they had to speak about phase transformation from the quasicrystal phase to the crystalline phase. Today, such a usage of terminology appears obsolete as the current definition of what a crystal is includes quasicrystals; then, this was not yet the case.

Above, I made reference to Csanády et al.'s paper in the periodical Symmetry [7]. This was a curious publication that did not survive its charter issue, but it was quite an issue. As I am mentioning this attempt for a uniquely interdisciplinary journal, Symmetry, it was exactly 25 years ago that its only issue appeared. It happened so that some of the communications in this issue were quite relevant to those interested in quasicrystals. I mention here a few only. Alan L. Mackay, major player in the quasicrystal story, wrote a thought-provoking essay, "Lucretius: Atoms and Opinion" [8]. Arthur Loeb and his co-authors discussed the icosahedron, pentagonal dodecahedron, and the rhombic triacontahedron [9]. Magnus J. Wenninger wrote about polyhedra and the golden number [10]. I single out a few additional contributions that were of a broader scope, from Erwin Chargaff [11], Herbert A. Hauptman [12], Jerome Karle [13], and Ernő Lendvai [14].

In May 1995, we organized an international school/conference on quasicrystals in the resort place at Lake Balaton, Balatonfüred, Hungary. There, in an unhurried atmosphere, we could learn, exchange ideas, and enjoy being part—at least as onlookers—of a fast emerging field. It was on this occasion when Ágnes Csanády could demonstrate personally her flowerlike quasicrystals to Dan Shechtman and everybody else (Fig. 2).

Shechtman discovered the quasicrystals on April 8, 1982. His report with co-authors appeared in November 1984 [1]. The Royal Swedish Academy of Sciences announced Shechtman's Nobel Prize in October 2011. The discovery was straightforward, but the scientific community, especially those deeply rooted in the teachings of classical crystallography, were slow in accepting it. By the time the International Year of Crystallography came about in 2014, the discovery of quasicrystals were among the stellar achievements of recent science.

When toward the end of 2013, the Israel Postal Company decided to issue a postage stamp to commemorate Dan Shechtman's Nobel Prize for the quasicrystal discovery and to honor the International Year of Crystallography (2014) they associated Csanády's appealing flowerlike quasicrystals with Shechtman's electron diffraction pattern. The Israel Postal Company made an excellent decision to put quasicrystals on a postage stamp and they chose one of the most beautiful representatives of such substance for display. The result was a scientifically sound and aesthetically pleasing image (Fig. 3).

Fig. 3 Israeli postage stamp (2013) honoring Dan Shechtman's discovery of quasicrystals and his Nobel Prize of 2011 as well as the International Year of Crystallography of 2014. The original image of the quasicrystals displayed on this postage stamp is that in Fig. 1.

Fig. 2 Ágnes Csanády and Dan Shechtman in 1995 in Balatonfüred (photo by I. Hargittai).

Csanády's quasicrystals and Shechtman's electron diffraction pattern were immortalized. Science has gained a tool for popularization in an unobtrusive and straightforward manner.

References

1. Shechtman D, Blech I, Gratias D, Cahn JW (1984) Metallic Phase with Long Range Orientational Order and No Translational Symmetry. Phys Rev Lett 53:1951–1953
2. Hargittai I (2010) Structures beyond crystals. J Mol Struct 976:81–86
3. Hargittai I (2011) "There is no such animal (כוז חיה אין)"—Lessons of a discovery. Struct Chem 22:745–748
4. Hargittai I (2011) Dan Shechtman's quasicrystal discovery in perspective. Israel J Chem 51:1144–1152
5. Hargittai B, Hargittai I (2012) Quasicrystal discovery—from NBS/NIST to Stockholm. Struct Chem 23:301–306
6. Hargittai B, Hargittai I (2005) Candid Science V: Conversations with Famous Scientists. Imperial College Press, London; see pp 36–93
7. Csanády Á, Papp K, Dobosy M, Bauer M (1990) Direct Observation of the Phase Transformation of Quasicrystals to Al_6Mn Crystals. Symmetry 1:75–79
8. Mackay AL (1990) Lucretius: Atoms and Opinions. Symmetry 1:3–17
9. Loeb AL, Gray JC, Mallinson PR (1990) On the Icosahedron, the Pentagonal Dodecahedron, and the Rhombic Triacontahedron. Symmetry 1:29–36
10. Wenninger MJ (1990) Polyhedra and the Golden Number. Symmetry 1:37–40
11. Chargaff E (1990) A Few Unscientific Thoughts About Symmetry. Symmetry 1:23–26
12. Hauptman HA (1990) The Role of Symmetry in Mathematical Discovery. Symmetry 1:27–28
13. Karle J (1990) An Application of Space Group Symmetry to Expected-Value Formulas in Crystallography. Symmetry 1:69074
14. Lendvai E (1990) Symmetries of Music, Part I. Symmetry 1:109–125

Alan L. Mackay preparing to attend a presentation at the Royal Institution in London (photograph by Istvan Hargittai).

Generalizing Crystallography: A Tribute to Alan L. Mackay at 90[a]

István Hargittai

In our preoccupation with finding out how atoms are arranged in space, we are in danger of losing sight of the whole picture.

Alan L. Mackay [1]

Crystallography is not just a scientific specialty, but is a way of life.

Alan L. Mackay [2]

Abstract

Alan L. Mackay, one of the rare generalists of our time, was a disciple and follower of J. Desmond Bernal. Mackay has contributed decisively to the development of the science of structures and taught generations to look at the broader picture when determining crystal and molecular structures. He was constantly seeking coherence and regularities in observations and in thought experiments and was aiming at creating concepts on the basis of those regularities. His inquiries prompted him to predict the existence of regular but not periodic crystal structures that are known today as quasicrystals.

Introduction

My first encounters with Alan L. Mackay (Fig. 1) were in the scientific literature. We met in person for the first time in 1981 in Ottawa during the Congress of the International Union of Crystallography. It was not a glorious occasion: I went up to him, introduced myself, we exchanged a few words; he then turned and left. I was surprised when a few weeks later I received a gracious letter from him that he was happy having made my acquaintance and urged me to visit him whenever I had an opportunity. A great interaction developed, including weeks of stays in each other's homes in Budapest and in London. We organized his first visit to Budapest in September 1982 and he gave three lectures on that occasion at the University of Budapest, including two on various aspects of fivefold symmetry. He said, among other things, that we should be aware of the possibility of extended structures of fivefold symmetry, although these were forbidden by the rules of classical crystallography. If we thought them impossible, they might go by us unnoticed and unrecognized.

By the time Mackay was delivering his talks on fivefold symmetry and issuing his warning about extended structures of fivefold symmetry, and without Mackay knowing about it, Dan Shechtman had already observed the first such extended structures—soon they became known as quasicrystals. Mackay did not merely think and speak about such structures, but he had published papers discussing them, complete with a simulated electron diffraction pattern. When I was listening to Mackay speaking about fivefold symmetry in September 1982, and, increasingly, in hindsight, I felt as if I were present at creation.

The universal importance of fivefold symmetry should not be exaggerated at the expense of other symmetries. However, because classical crystallography exiled it from its considerations as non-crystallographic symmetry, its comeback was all the more spectacular. It was remarkable that two outstanding discoveries in the mid-1980s, both in the science of materials, were related to fivefold symmetry. These were the fullerenes and the quasicrystals. Quoting Mackay [3]:

The main significance of fivefold symmetry for science is that it furnishes us with an explicit example of frustration, which has proved a most fertile concept in the physics of condensed matter.

[a]Originally published in *Structural Chemistry* 2017, 28:1–16.

I. Hargittai (✉)
Department of Inorganic and Analytical Chemistry, Budapest University of Technology and Economics, Budapest, Hungary
e-mail: istvan.hargittai@gmail.com

Fig. 1 Alan L. Mackay in 1982 in Budapest (photograph by I. Hargittai).

... Neither we nor nature can have everything simultaneously—not all things are possible ... We have only the freedom of necessity. 'Nature must obey necessity' as Shakespeare (*Julius Caesar IV:iii*), Democritus, Monod, Bernal, and many others have also recognized. Science probes the limits of necessity and, in the case of fivefold symmetry, has found a corridor that leads us to a new territory.

The Beginning

Alan Lindsay Mackay was born on September 6, 1926 in Wolverhampton, England. Both his parents were born in Glasgow. They were physicians, and lived in Wolverhampton in the English Midlands. Alan's father served as an infantry officer in World War I before he became a doctor and as a second in charge of a military field hospital in the Middle East in World War II. Alan's parents ran their own practice in the late 1920s and 1930s, which they sold in 1938. They then became consultants and, especially Alan's mother, served the community in various other capacities dictated by her social conscience. There was always professional talk at their table during their meals, which Alan found exciting. It was also understood that what he heard there could not be retailed outside their home. There were brothers and sisters who eventually dispersed to Australia and America.

Alan started his formal education in a small private school at the age of five, continuing at the Wolverhampton Grammar School from 1935 to 1940. He had to pass an entrance examination at the age of eight to get into this school. His school years overlapped with the Second World War. At the age of thirteen, he was a messenger in the Auxiliary Fire Service. From 1940, he was sent to a boarding school—Oundle School—after he passed another entrance examination. There was talk of a possible German invasion. Alan

stayed at Oundle until 1944 and received there an excellent science education. There were difficulties in life during the war, but not in education. His teachers had first class degrees in science and mathematics—teaching was a sought-after profession during and after the Depression. Just to characterize the level of instruction, his chemistry teacher one day demonstrated periodic chemical reactions, today called oscillating or Belousov-Zhabotinsky oscillating reactions. The concentrations of reactants and products undergo periodic changes in such a reaction and they offer a spectacular view if the participants have colors. No such reactions could occur under equilibrium conditions, but they can occur far from the equilibrium. Even 20 years later, Belousov found it difficult to get his manuscript describing such reactions accepted for publication.

School instruction included many demonstrations of experiments and an emphasis on practical applications of knowledge. Alan was a good student and was awarded various scholarships, which eased the financial burden on his parents. But there was never any doubt that he should study regardless of whether or not there were scholarships available. Alan developed an independent mind from an early age and he refers to this as that he was becoming an "internal immigrant."

Early on Alan had acquired a skeptical attitude and later he himself thought about the influences that must have moved him in this direction. He did remember one incident, when he was about five or six, and he told a girl of the same age that her parents had been lying to her over the nature of Father Christmas. In Alan's words, "I was very surprised to find how annoyed people were. It was like Gandhi's or H.G. Wells' experiments with truth. I discovered that you should not believe everything that grown-ups tell you nor say what you actually think. ... The tradition of my ancestors was to listen to what authority said and keep their doubts to themselves" [4].

This intellectual disposition of being an internal immigrant was probably strengthened by a predicament of gradual increased difficulty of hearing, which started becoming noticeable from 1955. On the other hand, Alan developed exceptional reading skills in at least half a dozen languages—he has been a voracious reader. He travelled a great deal, especially in Eastern Europe and from 1961, in Asia, including Japan, China, and Korea as well as India.

Start of Profession

Alan L. Mackay (Fig. 2) had earned excellent credentials and in October 1944, he went to Cambridge with a scholarship for the famed Trinity College. He focused on physics and chemistry and studied also electronics, mineralogy, and mathematics. Sir Lawrence Bragg was one of his professors along with

Fig. 2 The young Alan L. Mackay (courtesy of Robert H. Mackay).

Fig. 3 Alan and Sheila Mackay around 2000 in front of their home in London (photograph by I. Hargittai).

other famous scientists, such as the physical chemist and later Nobel laureate R.G. W. Norrish, the physical chemist Frederick Dainton, the inorganic chemist H.J. Emeleus, and others. He won the Percy Pemberton Prize and graduated in 1947.

In the summer of 1947, Alan went with a group of students to Yugoslavia to help build a railway and he has been actively interested in politics ever since. After graduation, in the years 1947–1949, Alan worked in the crystallography laboratory of Philips Electrical Ltd., and, while working for Philips, he earned his BSc degree in physics in 1948 as an external student. He decided to study for his PhD and he joined Birkbeck College of London University, and he has stayed at Birkbeck for the rest of his professional life. First he was there part-time, from 1949, in the crystallography laboratory of J. Desmond Bernal (1901–1971), later moving to full time. He defended his PhD thesis and was awarded the degree in 1951.

Mackay learned Russian in summer school, and there, he met Sheila, his future wife (Fig. 3). They married in 1951 and by 1961 they had three children, two boys and a girl, and moved to their home in North London where they stayed ever since.

Already by then, Alan's interests were broad and he published more broadly than would someone with a narrow specialization. This did not help his promotion in the university ranks. In this he followed his mentor's example although he learned also from Bernal that for his career broad interests counted as a disadvantage. Alan would be awarded his DSc degree in crystallography and studies of science in 1986. He was appointed Professor of Crystallography in the same year and became Professor Emeritus in 1991. In 1988, he was elected Fellow of the Royal Society (FRS).

Bernal's example was an inspiration for Alan ever since he had chosen Bernal's book, *The Social Function of Science*, as his prize for winning a competition in Cambridge. It would be difficult to imagine an environment more conducive to developing a generalist approach to science, and, in fact, to life, than Bernal's circle. Bernal was nicknamed "Sage" for he was supposed to know everything worth knowing. In the 1930s, Bernal was a member of the Club for Theoretical Biology, along with Joseph Needham, C. H. Waddington, and others. They dealt with such questions as the application of X-ray crystallography and other physical techniques to solving problems in biology. Already in the mid-1930s, Bernal had shone X-rays onto protein molecules and the fact that he could record interference patterns led him to believe that the structures of such large biological systems could be solved on the atomic level. Bernal was good in delegating tasks and he delegated the structure determination of large biological molecules to such disciples as the future Nobel laureates Dorothy Hodgkin, Max Perutz, and Aaron Klug. Bernal served as science advisor at the highest level during World War II. After the war, his communist politics and friendship with the Soviet Union were a serious impediment to his obtaining support for building up a research center that would have been adequate for implementing his far-reaching research ideas.

J. Desmond Bernal (Fig. 4) collected around him an excellent group of scientists in mathematics and computing, in the theory and experiment of X-ray crystallography, physical chemistry, both inorganic and organic structures, and his laboratory ran a skilled workshop. A stream of international visitors complemented his staff. Scientists like Norbert Wiener, Linus Pauling, André Lwoff, and H.S.M. Coxeter came and so did representatives of world culture, like Picasso and Paul Robeson. Bernal's associates felt they were "living in the center of the universe" [2]. Mackay realized from the start how privileged it was to be part of Bernal's circle of his closest associates. The combination of scientific, social, and political activities appealed to Mackay's own inclinations.

Fig. 4 J. Desmond Bernal about 1960 in London (photograph by and courtesy of Alan L. Mackay).

In 1956, Bernal invited Mackay to accompany him to Moscow. Bernal gave lectures on the origin of life at Aleksander I. Oparin's institute. Mackay had the opportunity to meet such giants of Soviet science as Petr L. Kapitza, Lev D. Landau, Igor E. Tamm and Vladimir A. Fock (of Hartree-Fock fame). Bernal and Mackay visited the Institute of Crystallography of the Soviet Academy of Sciences and met its director, Alexey V. Shubnikov and Shubnikov's co-workers, among them Boris K. Vainshtein and Zinovii G. Pinsker (Fig. 5). Mackay had already begun building up an international network of friends, especially with crystallographers at international meetings, and his interactions with the Moscow

Fig. 5 Alan L. Mackay (in the middle) in the company of Boris K. Vainshtein (left) and Zinovii G. Pinsker (right) in 1962 in Moscow (courtesy of Alan L. Mackay).

crystallographers were especially active. In 1962, he spent five months at the Institute of Crystallography in Moscow. Scientifically it was not a very fruitful stay, but for getting to know many colleagues and Soviet life, in a more realistic way than from propaganda materials, it was.

Research

Mackay's first research project was the structure analysis of a particular modification of solid calcium phosphate used in fluorescent tubes, which was of interest to Philips. The company had an array of various projects involving X-ray crystallography related to practical applications. When Mackay moved to Birkbeck College, he continued doing research on inorganic materials. He joined the section whose major concern was the properties of cement. When Bernal was at a committee of the Ministry of Works, he volunteered that he could find out why cement sets, and a whole research project developed from this assertion.

Icosahedral structures became the focus of Mackay's interest rather early. He had already met with the structure of beta-tungsten at Philips. Then, he found some interesting old papers at Birkbeck, evidence that there had been interest in these structures at the College before Mackay. Bernal also considered the icosahedral arrangement rather early, because it would prevent crystallization, and he thought that icosahedral coordination might give some clues to understanding the structure of liquids. Mackay was also aware of Pauling's interest in icosahedral structures. When Bernal was to go to Budapest to give a talk at the meeting honoring Zoltan Gyulai's 70th birthday, he asked Mackay to draw the figures. Bernal's talk was about the symmetry in solids and liquids. It was a most comprehensive presentation [5].

The icosahedral arrangement of atoms is interesting because it could also be a step in the progression from the isolated molecule to an extended structure. When a second icosahedral shell surrounds an icosahedron of 12 spheres about a sphere in the center, the size of this second shell is exactly twice the size of the first shell [6]. This second shell contains 42 spheres and lies over the first so that spheres are in contact along the fivefold axis. Further layers can be added in the same fashion.

The third layer is shown in the Figure and this is known as the Mackay polyhedron (Fig. 6) or Mackay icosahedron—an example of icosahedral packing of equal spheres. The layers of spheres succeed each other in cubic close packing sequence on each triangular face. Each sphere which is not on an edge or vertex touches only six neighbors, three above and three below. Each such sphere is separated by a distance of 5% of its radius from its neighbors in the plane of the face of the icosahedron. The whole assembly can be distorted to cubic close packing in the form of a cuboctahedron. The

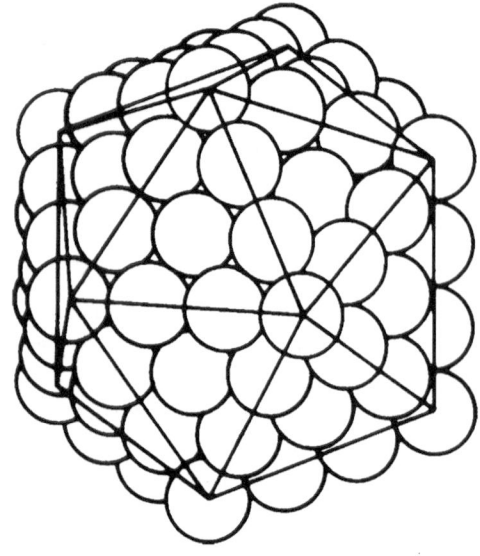

Fig. 6 The "Mackay polyhedron" emerging from the icosahedral packing of equal spheres. Only the third shell is visible (courtesy of Alan L. Mackay [6]).

Mackay icosahedron has "made tremendous impact on particle, cluster, intermetallics, and quasicrystal researchers...," [7] according to the late K.H. Kuo, the doyen of Chinese crystallographers. Kuo identified two basic concepts in Mackay's paper. One was the icosahedral shell structure consisting of concentric icosahedra displaying fivefold rotational symmetry. This structure occurs frequently and not only in various clusters, but also in intermetallic compounds and quasicrystals. The other concept, according to Kuo, was the hierarchic icosahedral structures due to the presence of a stacking fault in the face-centered-cubic packing of the successive triangular faces in the icosahedral shell structure.

Mackay questioned dogmas wherever and whenever he met them. This was especially so in the case of crystallography where the classical rules had worked so well but eventually proved increasingly to be limiting the scope of structures the subject embraced. Those rules limited the inclusion of novel kinds of structures that kept emerging as well as structures that had been abandoned by crystallographers; but the need arose to include them in a broader system. There was an obvious deficiency when the theoretical constraints of crystal symmetry were confronted with real crystals in that crystals are not infinite. The approach to discussing crystal symmetry used to be to think of the formation of a crystal through insertion of individual atoms or groups of atoms into the three-dimensional framework of symmetry elements, whereas in reality—as Mackay liked to point out—the symmetry elements emerge as a consequence of the structure being formed through the local interactions between individual atoms or other building elements. The concept of crystal symmetry itself became a target of Mackay's inquiry and he creatively deepened and expanded its meaning. When I asked him if he would like to select one of his papers for inclusion in the current special collection of articles, he chose the one titled "Crystal Symmetry" [8] reproduced in the Appendix.

Mackay compiled a list of concepts in two versions, showing the transition from the classical to the modern (Table 1). He has refined his list over the years, but the 1981 one demonstrates from a 35-year perspective how forward-looking his ideas were.

This list appeared in a paper, which Mackay titled De *nive quinquangula* (on the pentagonal snowflake), which was a direct reference to Johannes Kepler's treatise on the six-cornered snowflake [10].

Table 1 Mackay's compilation of classical versus modern concepts in 1981 (courtesy of Alan L. Mackay [9])

Classical concepts	Modern concepts
Absolute identity of components	Substitution and nonstoichiometry
Absolute identity of the environment of each unit	Quasi-identity and quasiequivalence
Operations of infinite range	Local elements of symmetry of finite range
"Euclidean" space elements (Plane sheets, straight lines)	Curved space elements. Membranes, micelles, helices. Higher structures by curvature of lower structures
Unique dominant minimum in free energy configuration space	One of many quasi-equivalent states; metastability recording arbitrary information (pathway); progressive segregation and specialization of information structure
Infinite number of units. Crystals	Finite numbers of units. Clusters; "crystalloids"
Assembly by incremental growth (one unit at a time)	Assembly by intervention of other components ("crystallise" enzyme). Information-controlled assembly. Hierarchic assembly
Single level of organization (with large span of level)	Hierarchy of levels of organization. Small span of each level
Repetition according to symmetry operations	Repetition according to program. Cellular automata
Crystallographic symmetry operations	General symmetry operations (equal "program statements")
Assembly by a single pathway in configuration space	Assembly by branched lines in configuration space. Bifurcations guided by "information", i.e., low-energy events of the hierarchy below

There were several threads in Alan's career that were rapidly coming together. In his words [11]:

I used to do science abstracts—for ten years I abstracted all the Russian papers on crystallography—and I remember abstracting a paper on the incommensurate arrangements of spins in iron oxides, in hematite. The period of the helical magnetic spin is not the same as the crystallographic period. So incommensurate structures were current before that time. Even much longer before that I thought of a simple thing about printing wall paper. Suppose your wall paper is simply printed from a roller. But suppose you are printing two motifs from two rollers of different diameter. Then you get a non-repeating pattern. I wasn't able to think of producing an aperiodic two-dimensional pattern in this way. I was only aware of the possibility of one-dimensional incommensurate patterns. I was really interested in hierarchic patterns and not in aperiodicity as such. It came directly from Bernal's suggestions and the polio virus project. I produced a hierarchic pattern, a hierarchic packing of pentagons. Then in 1974 I was getting some help in computing from Judith Daniels at the University College Computing Centre and, incidentally, showed her these patterns. She said that Roger Penrose had something like them. So I made an appointment with Roger Penrose (Fig. 7) and Robert, my son, and I went to see Penrose in Oxford, and he showed us the jigsaw puzzle, with the kits and darts and so on. Basically his concern was with forcing aperiodicity, and my concern was with hierarchic structures. It turned out to be very similar.

In the paper about the pentagonal snowflake, Mackay, à la Penrose, built up a regular, but non-periodic (he called it then "noncrystalline") structure from regular pentagons in a plane (Fig. 8).

It starts with a regular pentagon of given size, which we may call the zeroth-order pentagon. Six of these pentagons are combined to form a larger regular pentagon, the first-order pentagon. There are triangular gaps in this pentagon and Mackay filled these gaps with pieces from cutting up a seventh zeroth-order pentagon. This cutting up yielded five triangles and a smaller regular pentagon as left-over, which is

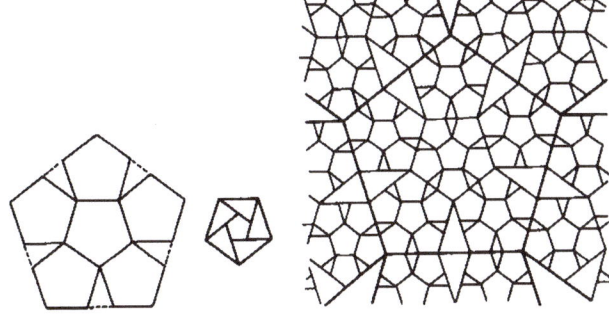

Fig. 8 Tiling with regular pentagons (courtesy of Alan L. Mackay [9]).

the pentagon of the order of −1. This design is repeated on an ever increasing scale.

After the meeting with Penrose, Alan's son Robert went back to his university at York where he was studying computer science and plotted a tiling on his pen-plotter (Fig. 9). We could call what he plotted a Mackay tiling as it was

Fig. 9 Robert H. Mackay's computer drawing of the formation of a "pentagonal snowflake" in 1975 [9] autographed by Roger Penrose in 2005 (courtesy of Robert H. Mackay).

Fig. 7 Roger Penrose and Alan L. Mackay (courtesy of Alan L. Mackay).

Fig. 10 Alan L. Mackay and Robert H. Mackay in April 2016 in London (photograph by and courtesy of Magdolna Hargittai).

different from the standard Penrose kites and darts. Robert (Fig. 10) started from pentagons of a certain size and as he kept going to larger and larger pentagons, he built up a pentagonal snowflake. Mackay included Robert's design in his paper on pentagonal snowflakes to give his considerations added emphasis.

Mackay was getting ready to make significant predictions concerning the possibilities of real three-dimensional structures with fivefold symmetry. At one point, he got the idea of producing a simulated diffraction pattern of the Penrose tiling [11]:

> First I just drew the Penrose type pattern and sent it to George Harburn in Cardiff who was a colleague of Charles Taylor who had a good optical diffractometer. I had stuck it into a laser beam here but you need a precise adjustment. You can do many

Fig. 11 Mackay's simulated "electron diffraction" pattern of a three-dimensional Penrose tiling (courtesy of Alan L. Mackay) [12].

beautiful things with the optical diffractometer that you can't see in the computer, with very fine detail; it is amazing. Then George Harburn made a second version which instead of consisting of lines, had dots; thus the diffraction pattern was not dominated by the streaks from the lines [11, p. 154]

Mackay wrote up and published his paper in which he communicated a simulated diffraction pattern (Fig. 11) [12].

It is remarkable, how, once again in a broader context, he was considering the characteristics of the pattern and the diffraction it generated [11]:

> I had also a theory about collagen, and had some patterns bearing on that. The theory was that collagen fibers are connected with the Fibonacci spiral. If you draw a Fibonacci spiral of circles along the spiral, then locally the pattern keeps changing between square packing and hexagonal close packing. This corresponds closely to the diffraction you infer from collagen fibers. Richard Welberry in Canberra, Australia, had a still better optical diffractometer and took some very good diffraction pictures from the Fibonacci spiral. Then [the botanist] Eriksson in Philadelphia showed that the diffraction pattern of the Fibonacci spiral was self-similar to the Fibonacci spiral itself. . . . This may point to a connection between phyllotaxis—the scattered leaf arrangement about stems—and internal structure on the atomic level" [11, p. 155].

Alan's story is a brilliant example of the importance of pursuing a lot of lines in research and look for their possible convergence. In this, Mackay followed Bernal's philosophy of asking a thousand questions rather than just one, because this way the probability of finding answers is greatly enhanced. Along the way, Mackay documented his findings. This was useful, because after the publication of Shechtman's experimental observation of quasicrystals in November 1984 [13], theoretical/modeling papers followed in rapid succession [14]. It could have been easy to distort the real succession of events related to the circumstances of the discovery. Indeed, one-sided reports did appear. For example, an account in one of the January issues of *The New York Times* stressed the priority of theoretical work, but failed to mention Mackay's modeling and simulation studies and even downplayed the experimental discovery itself. This prompted me to send in a "Letter to the Editor" in which I described Mackay's contributions, explicitly citing his two publications (*Physica* 1982, 114A:609–613 and *Soviet Physics Crystallography* 1981, 26:517–522). As far as I know the letter was not printed but it is well documented [11, pp. 171–172].

Mackay recognized the potential practical applications of quasicrystals early on. He thought that Shechtman's discovery may very well be more important than it had been believed. He recognized that the low thermal conductivity of quasicrystals may be utilized for nonstick frying pans, turbine blades, in internal combustion engines, and so on. A suitable technology might be to create quasicrystal surfaces by glazing metal with a laser. He foresaw great economic potential in the discovery.

Alan told me about this when I asked him about Shechtman's possible Nobel Prize, back in 1994. He had an interesting line of thought about the different kinds of Nobel Prize as he saw them. He characterized Shechtman's discovery as when someone turns over a stone and finds something truly important, maybe like superconductivity or the scanning tunneling microscope or the Mössbauer effect. There isn't an enormous amount of work but someone was in the right place at the right time, and recognized what he's found. In 1994, Mackay thought that Shechtman's Nobel Prize would come in this category.

The only reservation Mackay had in evaluating the importance of the discovery of quasicrystals was that it may have appeared more significant than it really was. He thought that the too restrictive definitions of classical crystallography lent a pivotal character to the discovery. Had the definitions of classical crystallography been broader and more inclusive, there would have been no need to bring about a paradigm change. However, as it happened, the discovery of quasicrystals did prove to be pivotal and it did bring a paradigm change about.

Mackay had truly predicted the existence of regular but non-periodic structures that Dan Shechtman (Fig. 12) then observed in his experiments. It would have been a wonderful

Fig. 12 Alan Mackay and Dan Shechtman in 1995 in the author's home in Budapest (photograph by I. Hargittai).

sequence of events had Shechtman and others known about Mackay's prediction and have embarked on looking for such structures and found them. The search for extended structures with fivefold symmetry had been going on for centuries and involved excellent minds, such as Johannes Kepler and Albrecht Dürer. Roger Penrose came up with such a pattern in two dimensions and Mackay crucially extended it to the third dimension, and urged experimentalists to be on the lookout for such structures. Nobody took up his challenge and when Shechtman made his observations, he was not aware of Mackay's predictions. Eventually though all these lines came together. In 2010, the American Physical Society awarded the Oliver Buckley Prize to Alan Mackay, jointly with Dov Levine and Paul Steinhardt for their contributions to the quasicrystal discovery. The next year Shechtman received the Nobel Prize in Chemistry.

Summing Up

Alan does not mind the adjective once applied to him by a colleague, "the well-known eclectic," and chose this word for the title of a selection of his writings, *Eclectica*, self-published for personal use in a handsome volume in 2009 (Fig. 13) [15]. In it, he reproduces many of his published papers and communicates a number of unpublished works as well. The volume is a rich source of information and ideas and here we will merely dip into it for a few selected entries to illustrate its scope and depth.

Appropriately the volume begins with a discussion of copyright—one of Mackay's pet projects. He has been an advocate of protecting the rights of scientist authors to their own intellectual productions versus the publishing companies. One of the solutions he found promising was for professional societies to start their own electronic journals with open access that would be supported by authors' fees. Currently the open access approach is gaining ground rapidly, but there may be a great divide between authors who can and those who cannot afford the often hefty fees for having their manuscript published in open access venues.

As we have seen above, the discovery of the Mackay polyhedron and his prediction of the structures today called quasicrystals, did not happen in isolation. Mackay had long been interested in structures that fell beyond the rigorous and confined system of classical crystallography. He has published at least three reviews under the title "Generalized Crystallography," the latest in 2002 [16]. He defined the aim of generalized crystallography as "to understand the properties of matter, inert and living, at our human scale, in terms of the arrangement and operation of atoms." He recognized the pioneering role of X-ray crystal structure analysis in this quest, but noted that as the array of techniques has become vast, it might be advisable to replace the term

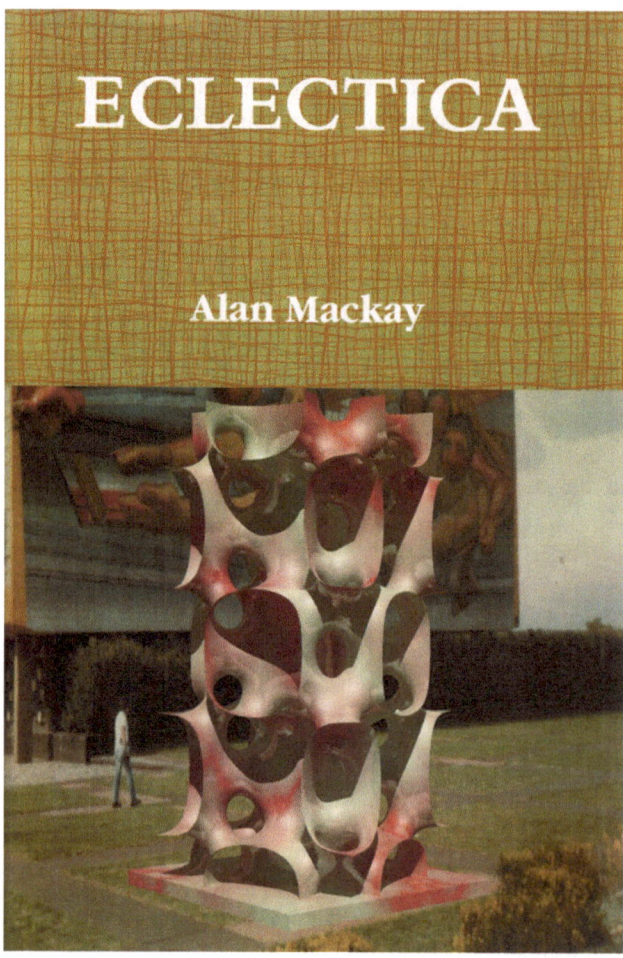

Fig. 13 The cover of Mackay's *Eclectica*. The art is a computer-creation by Alan L. Mackay, one in a long series of images inspired by his studies of minimal surfaces (courtesy of Alan L. Mackay [15]).

crystallography by *structural chemistry*. He also realized though that terms that had been embedded long in scientific literature would be hard to displace. This may be so unless the new term is glued to a fad as, for example, in the case of nanoscience and nanotechnology.

Concerning the pioneering role of X-ray crystallography, Mackay has written about the phenomenon of when a pioneering field becomes a brake on further progress. This happened with classical crystallography whose rigid system hindered the recognition of those structures that fall beyond this classical system. In short, its success became a barrier to progress. Of course, for this, blame should not be assigned to those who originally worked out the system, but it is our task to overcome the barriers that have been erected by the developments since. This kind of success turning into a brake is not unique to classical crystallography. When insulin was discovered for treating diabetes it was a great triumph of the biomedical sciences. It has then been gradually

recognized that the availability of this successful treatment, which is not a cure, might have diverted efforts and resources from continuing a quest for the cure of diabetes. Another example from the science of structures was the resistance to recognizing other techniques against the background of the enormously successful X-ray diffraction making it harder for electron crystallography and for neutron crystallography to become accepted and spread [17]. However, Mackay's teachings on generalized crystallography fell onto fertile ground; suffice it to mention a couple of additional contributions to the volume of *Structural Chemistry* dedicated to his 75th anniversary [18, 19].

Mackay's impact on the structural science community is hard to measure, but the impression is that it will be long lasting. He has impacted us through his writing and through personal interactions. In this connection it is notable that he adapted himself easily to local conditions on the occasion of his many visits. When he spent a longer period at the Institute of Crystallography in Moscow, he developed the habit of carrying a shopping bag with him. This was not only because the shops did not give out such bags to carry away their goods; but even more because one never knew what purchase might suddenly become available. After his return to London, he did not find it easy to give up the habit of having his shopping bag at readiness. Although his stay at the Institute of Crystallography in Moscow did not produce scientific results, his interactions with the Azerbaijani crystallographer Khudu Mamedov (1927–1988) greatly helped Mamedov to become well known in the West. Mamedov prepared periodic drawings that were reminiscent of Escher's patterns, but he used historical/cultural motifs from his region. Thus he created a unique interrelationship between art and science. Mamedov, perhaps in Mackay's style, used the term "crystallographic" in a broad sense. Mackay dedicated a talk to Mamedov's memory in 1991, "Form and pattern in Azerbaijani civilization," and its text is reproduced in *Eclectica*.

Mackay (Fig. 14) and Bernal co-authored a presentation entitled "Towards a science of science" for the 11th International Congress for the History of Science in Warsaw in 1965. They outlined what Science of Science was, why it was needed and the methods of their inquiry. Their program included practical recommendations, such as the establishment of departments of the history of science and the need for looking at science as a whole rather than always taking up merely its specificities. Further, they called for establishing the profession of science critic similarly to that of literary critic, and called for international cooperation as recognition of science as a world-wide activity. They also suggested experimental work in order to find the best means of science training and the like. They emphasized the importance of learning about non-European cultures where emphases were different from European cultures as illustrated, for example,

Fig. 14 Alan L. Mackay in 2011 in his study among many of his computer-generated drawings (photograph by I. Hargittai).

by a lower priority for written records, but a higher one for master-pupil relationships. This joint Mackay-Bernal presentation has been reproduced in a number of publications and in a number of languages, yet it is not easily accessible. Hence, it is very useful to have it in *Eclectica*. Mackay co-edited a volume on this topic and the idea of science of science permeated his activities throughout his entire career [20].

In the early 1980s Mackay ran a column called "Anecdotal evidence" in the journal *The Sciences* and the entries are reproduced in *Eclectica*. It suited him eminently, bringing together seemingly disparate ideas and facts. Even the titles reveal some aspects of his approach, such as "Science and Travel," "Rhyme and Reason," "How to write a best seller," "Mackay's *Michelin*," "Molecules and Moores" (referring to Henry Moore), "Message in a Bottle," and suchlike. The column served the readers of this unusual periodical well, but its editors liked to smooth over his often unorthodox style of writing; apparently the flavor of Mackay's writing was a little too much for them.

The *Eclectica* volume is concluded by a list of Mackay's work, including scientific publications (176 entries), miscellaneous publications (130), and book reviews (46). There is then a list of 30 unpublished papers, and 10 entries which he calls "indirect material," and those publications by others in which he figures, including the special issue in *Structural Chemistry* in 2002 dedicated to him [21].

Legacy

The full story of the quasicrystal discovery has yet to be written. At this point, I am offering my thoughts concerning only a tiny aspect of this story, viz. the demeanor of its

principal protagonists with respect to the loneliness of the scientific discoverer. With justifiable simplification, there were three protagonists in this story. Alan L. Mackay predicted the existence of quasicrystals. Dan Shechtman discovered them in his experiments. Dov Levine (a graduate student, then) and Paul J. Steinhardt (Levine's professor) coined the name quasicrystals and offered a theoretical interpretation of the structure of this new kind of matter. I have had opportunities of discussing the circumstances of their discoveries in person with Mackay, Shechtman, and Levine. I interacted with Steinhardt only via e-mail exchanges.

Alan L. Mackay's (Fig. 15) demeanor has been such that he was looking consciously for dismantling dogmas and scientific taboos. In doing so, he realized the indefensibility of the dogma of classical crystallography with respect to the prohibition of fivefold symmetry in extended structures. Once he recognized this, he voiced it in his publications and in his presentations. He did not have second thoughts about making a stand and risking his reputation. He did this at a time when there was a reduction of personnel at British universities and he could have been retired prematurely. In 1982, he was 56 years old, had been a Reader in Crystallography at Birkbeck College for quite some time. It would only be in 1986 that he was awarded a personal chair as Professor of Crystallography and was elected FRS in 1988. It seems that the loneliness of the scientific discoverer was his natural mode of existence.

In contrast, Dan Shechtman was not looking to do anything revolutionary. His interest in alloys was in finding compositions for improved practical applications. However, he possessed a good deal of curiosity and this was his driving force at his pre-discovery stage. This curiosity made him embark on testing metal compositions that could have not

Fig. 15 Alan L. Mackay and Istvan Hargittai in April 2016 in London (photograph by and courtesy of Magdolna Hargittai).

been expected to offer improved, or any, applications. Once he made the discovery and realized that it was revolutionary, he grew to the challenge and his stubborn nature helped him to see it through to general acceptance. In doing so, he invited the disapproval, even wrath, of the powers that be in science, for example that of the greatest chemist of his time. Shechtman conducted himself with dignified determination in his loneliness, but he was not enjoying it and welcomed any easing of this loneliness. He felt relief and gratitude when Ilan Blech joined him in co-authoring the first paper in which Shechtman—half-heartedly and half-buried among other materials—mentioned his discovery. When he was finally preparing the manuscript that reported unambiguously his discovery, he was happy and grateful that he found three co-authors who helped him formulate what he wanted to say and who eased his loneliness of the scientific discoverer.

Levine and Steinhardt were ready to publish their interpretation of the quasicrystal structure as soon as they had learned about the paper reporting its experimental observation. At this point, they did not have to face the loneliness of the scientific discoverer, because that burden had already fallen onto Shechtman, let alone Mackay. Had Levine and Steinhardt come out with their theoretical model before the experimental discovery, they might have felt the most acute loneliness, possibly even ridicule. From the immediacy of their publication following Shechtman's, we may suppose that they might have made their theoretical discovery some time before. Levine might have written a good thesis even on the basis of a failed model, but for Steinhardt, the risk would have been considerable and possibly sufficient to damage his reputation. He was a Professor of Physics at the University of Philadelphia, then; later on, at Princeton University. Steinhardt (not Levine) had the choice of taking the risk and face the loneliness of the discoverer or wait and see whether there might be a safer opportunity to strike out.

All this is my supposition only, but I see consistency with it in how things played out during those years in the second half of the 1980s.

The adjectives "consistent" and "rational" are among the many characterizing Alan L. Mackay, and they shine through the poem he composed recently that sounds like a parting gift:

Atoms and our Vision of the World
There are no gods.
We are alone.
I am thus two-fold alone
but I have the second sight of science.
As my eyes grow dim,
my mind sees the future.
I see a hand writing on the wall -
the wall surrounds a giant alembic -
built to win gas from coal.
The Chinese hand wrote large
the character which stands for entropy.

It questions the solid state of Earth.
Asking my computer, I find the words
"disordered hyperuniformity"
- today's myopic Vision of the World
glimpsed in the microcosm of atoms.
Death came to my wife of more than sixty years
Her flame went out. Her body was cremated -
Atoms to atoms – Lucretius saw truth.
But where is past history now?
Information increases locally from time to time -
but Entropy will win.
A.L.M. 30 August 2015 [22]

This was not the first time Mackay had expressed his views and sentiments through poetry. His published poems often express topics in crystallography and the science of structures [23]. He titled his collection of poems published in 1980 the Floating World after the works of Japanese artists who lived in the latter half of the eighteenth century and the first half of the nineteenth.

According to Mackay, "Scientists inhabit a kind of Floating World of their own, a kind of Global Village, in which they have friends, or friends of friends, everywhere. Rather, like members of a religious order, they can go to any laboratory dealing with their field of study and be hospitably received" [24]. Alan and Sheila Mackay certainly practiced this very hospitable attitude toward many members of the international scientific community.

Mackay has been much concerned with the ways to expand the science of structures to embrace systems with varying degrees of regularity. Here intentions and desires that cannot be formulated yet with exactitude can be expressed as a poem [25]:

We cruise through the hydrosphere
Our world is of water, like the sea,
But the molecules more sparsely spread,
Not independent, not touching
But somewhere in between,
Clustering, crystallizing, dispersing
In the delicate balance of radiation
And the adiabatic lapse rate.

Even when he is composing prose, it sometimes sounds like poetry. Consider this example: "Amorphous materials may be shapeless, but they are not without order. Order, like beauty, is in the eyes of the beholder. If you look only with X-ray diffraction eyes, then all you see is translational order, to wit crystals. . . . [T]here is a wide range of structures, between those of crystals and those of gases, . . . Other structures need not be failed crystals but are *sui generis*" [26]. (Italics in the original)

Contemplating Alan Mackay's legacy, it is often said that scientific discoveries, however important, are sooner or later overshadowed by new developments in science. So it is happening with Mackay's contributions to crystallography and the science of structures. However, his demeanor as a researcher and scientific discoverer will serve as inspiration for a long time.

Acknowledgments I thank John L. Finney, Magdolna Hargittai, Alan L. Mackay, and Robert H. Mackay for their kind assistance in the preparation of this manuscript.

References

1. Mackay AL (1975) Generalized Crystallography. Izv Jugosl centr krist (Zagreb) 10: 15–35
2. Mackay AL (1998) Pre-history: For the 50th Anniversary of the Crystallographic Laboratory of Birkbeck College. Unpublished notes of a presentation on November 26, 1998, in The Clore Lecture Theatre in London
3. Mackay AL (1990) Crystals and Five-fold Symmetry in I. Hargittai, ed, Quasicrystals, Networks, and Molecules of Five-fold Symmetry. VCH, New York, pp 1–18
4. Hargittai I (1997) Quasicrystal Discovery: A Personal Account. Chemical Intelligencer 3(4):25–49; actual quote, p 26
5. Bernal JD (1957) The Importance of Symmetry in Solids and Liquids. Acta Physica Acad Sci Hungaricae 8:269–276
6. Mackay AL (1962) A dense non-crystallographic packing of equal spheres. Acta Crystallographica 15:916–918
7. Kuo KH (2002) Mackay, Anti-Mackay, Double-Mackay, Pseudo-Mackay, and Related Icosahedral Shell Clusters. *Struct Chem* 13:221–230
8. Mackay AL (1976) Crystal Symmetry. Physics Bulletin November, 495–497
9. Mackay AL (1981) De Niva Quinquangula: On the pentagonal snowflake Kritallografiya (Soviet Physics Crystallography) 26:910–919 (517–522)
10. Kepler J (1611) Strena seu de nive sexangula. Godefridum Tampach, Francofurti ad Moenum. English translation, The Six-Cornered Snowflake, Clarendon Press, Oxford, 1966
11. Hargittai I, Hargittai M (2000) Excerpts from a conversation with Alan L. Mackay in 1994 in London, communicated in In Our Own Image: Personal Symmetry in Discovery. Kluwer/Plenum, New York
12. Mackay AL (1982) Crystallography and the Penrose Pattern. Physica 114A:609–613
13. Shechtman D, Blech I, Gratias D, Cahn JW (1984) Metallic Phase with Long Range Orientational Order and No Translational Symmetry. Phys Rev lett 53:1951–1953
14. Levine D, Steinhardt PJ (1984) Quasicrystals: A New Class of Ordered Structures. Phys Rev Lett 53:2477–2480
15. Mackay AL (2009) "Eclectica" from the writings of Alan Mackay, un-published (self-published), London
16. Mackay AL (2002) Generalized Crystallography. Struct Chem 13:215–220
17. Finney JL (2002) Crystallography without a lattice. Struct Chem 13:231–246
18. Ogawa T, Ogawa T (2002) Proportional representation system as generalized crystallography and science on form. Struct Chem 13:297–304
19. Terrones M, Terrones G, Terrones H (2002) Structure, chirality, and formation of giant icosahedral fullerenes and spherical graphitic onions. Struct Chem 13:373–384
20. Goldsmith M, Mackay AL (1964) The Science of Science. Souvenir Press, London.
21. Special issue honoring Alan L. Mackay in Struct Chem (2002), 13:213–412
22. Private communication from Alan L. Mackay by e-mail, August 30, 2015
23. Mackay AL (1980) The Floating World of Science: Poems. The RAM Press, London
24. Ibid., Introduction
25. Ibid., p. 28
26. Mackay AL (1987) Quasi-Crystals and Amorphous Materials. J Non-Cryst Solids 97&98:55–62

Aaron Klug with a model of the tobacco mosaic virus (TMV) structure in 2000, at the Laboratory of Molecular Biology in Cambridge, UK (photograph by Istvan Hargittai).

Ambiguity of Symmetry[a]

István Hargittai

Abstract

Beyond the universality of the symmetry concept there are different emphases on its application in different branches of science. Chemistry being between particle physics and astrophysics represents a bridge in and a utilitarian approach to the application of the symmetry concept, which has proved immensely fruitful in twentieth-century science. Some pivotal discoveries, especially in structural chemistry, molecular biology, and materials science, emerged by relaxing some of the stipulations of the classical teachings about symmetry. This highly personal presentation relies on ideas expressed by a number of notable individuals in recent science, among them J. Desmond Bernal, Francis Crick, Ronald J. Gillespie, Aleksandr I. Kitaigorodskii, Alan L. Mackay, George A. Olah, Linus Pauling, Roger Penrose, Dan Shechtman, James D. Watson, Steven Weinberg, and Eugene P. Wigner.

Introduction

Five years ago I published another essay in the *Israel Journal of Chemistry*, "Dan Shechtman's Quasicrystal Discovery in Perspective," and I concluded my presentation by mentioning my encounter with Eugene P. Wigner (Fig. 1) [1]. That I am starting the present essay with mentioning Wigner signifies a pleasant continuity.

Wigner's contributions were fundamental to the understanding of the importance of the symmetry concept for science. In his terminology, symmetries and invariances were interchangeable. According to Wigner, the invariances make it possible to formulate the laws of nature. He underlines that "the first and perhaps the most important theorem of invariance in physics" is that "absolute time and position are never essential initial conditions." [2] In everyday language, this means that the validity of physical laws of nature are independent of the location where and the point in time when we are considering them.

Wigner formulated the consecutiveness of invariances, laws of nature, and the events we observe and experience, that is, the physical phenomena:

Invariances → Laws of Nature → Physical Phenomena
(Symmetries)   (Events)

The concept of initial conditions has utmost importance even for dividing the sciences into disciplines in a meaningful way. Wigner says: "Other sciences which deal with what we physicists consider to be initial conditions, are, among others, geography and descriptive astronomy." [3] These examples explain what initial conditions mean. Events described by geography and descriptive astronomy depend on the place and time—initial conditions—hence they are not governed by laws of nature.

True laws of nature do not depend on the initial conditions. Sciences such as geography and descriptive astronomy "tell us only facts." Physics and mathematics are concerned with regularities. These are the two extremes. There are then sciences that are in between the two extremes, such as, for example, botany, zoology, and the medical sciences—according to Wigner. By geography, he means descriptive geography. As we move from descriptive geography toward physical geography, likewise, from descriptive geology to physical geology, and so on, we move from sciences solely concerned with initial conditions toward sciences concerned with regularities, that is, laws of nature. "Physics does not endeavor to explain nature. In fact, the great success of physics is due to a restriction of its objectives: it only endeavors to explain the regularities in the behavior of objects. . . . The regularities in the phenomena

[a]Originally published in *Israel Journal of Chemistry* 2016, 56:907–924. Reproduced by permission.

I. Hargittai (✉)
Department of Inorganic and Analytical Chemistry, Budapest University of Technology and Economics, Budapest, Hungary
e-mail: istvan.hargittai@gmail.com

Fig. 1 Eugene P. Wigner with the author in 1969 at the University of Texas at Austin.

Fig. 2 Eugene P. Wigner in the late 1960s (courtesy of the late Martha Wigner Upton).

which physical science endeavors to uncover are called the laws of nature." [4] This formulation reflects Wigner's modesty and his modesty may have helped him to recognize this profound limitation of natural sciences.

Before Wigner (Fig. 2), physicists used to use symmetry considerations for solving particular issues, whereas Wigner applied them in a most general way. One of the most profound messages Wigner conveyed to me was about the universality of the symmetry concept that its validity cuts across the disciplines. We need to keep this in mind in the discussion following here, because there will be peculiarities assigned here to various disciplines for an easier understanding and more efficient utilization of the symmetry principle.

Broken symmetry may be as important as symmetry itself. Such broken symmetries appear, for example, under the conditions above certain temperatures, and this has special significance for chemistry. As Steven Weinberg (Fig. 3) explains [5]:

> The laws that govern atoms are completely symmetrical with respect to direction. There's nothing in nature that says that one direction in the laboratory, whether it's east and west or up and down, is any different from any other direction. On the other hand, when atoms join to form a molecule, for example, when three oxygen atoms join to form an ozone molecule, that's a triangle that points in a definite direction. It breaks the rotational invariance of the laws of chemical attraction by forming a particular object that has not the full rotational symmetry but a smaller symmetry, just rotations by multiples of sixty degrees. If you had a more complicated molecule, there'd be no symmetry left, yet the underlying laws are perfectly symmetrical. Those molecules only exist below a certain temperature. You can always restore the symmetry by heating them sufficiently so the molecules break up into a gas. If you have a gas of monoatomic oxygen, without worrying about the walls, it is symmetrical; all directions are the same.

When the monoatomic gas has a distribution equivalent in all directions, perfect disorder emerges; thus we equate this perfect disorder with symmetry. Further, according to Weinberg [6]:

> Perfect disorder is symmetry. To have order, for example in a crystal, you break the symmetry. You only have symmetry by finite rotations. A crystal of salt is invariant when you change your point of view by rotation of 90 degrees around various axes. It's a cubic crystal. But if you have molten sodium chloride, then

Fig. 3 Steven Weinberg in 1998 at the physics department, University of Texas at Austin (photograph by I. Hargittai).

there's no preferred direction at all. You've created complete disorder, as far as the directions are concerned. People in condensed matter physics very often use the terms order and disorder rather than broken symmetry and restored symmetry although they are very closely related.

The symmetries that we talk about in elementary particle physics are not broken because of any particular object has formed. The physical state that breaks the symmetry is not a molecule or a crystal. It is empty space. The vacuum, although it's perfectly symmetrical with regard to rotations in space, or translations in space, is not invariant with respect to changes in your point of view about which particles are viewed. It's the vacuum that distinguishes the neutrino from the electron, or the weak interactions from the electromagnetic interactions. The reason that the photon is massless whereas the other particles on the same symmetry multiplet, the W and the Z particles, are very heavy, is because of the way they propagate through the vacuum.

At this point I have to refer to my long conversation with Yuval Ne'eman in 2000 in Stockholm. In the course of that conversation he told me that on the basis of symmetry considerations he had predicted the mass of the Higgs particle, which at that time had not yet been observed. Ne'eman said, "If and when the Higgs [particle] will be found and its mass measured, I would now like to advertise my theory and people to know that I had predicted it." [7] As Ne'eman died before the actual observation of the Higgs particle, I find it important to mention his prediction here.

Weinberg stressed the importance of symmetry of the laws of nature, but the symmetry of objects is also important and in chemistry and in molecular biology they play a distinct role [8]:

> It is important that the sugars in living things are right-handed and the amino acids are left-handed, but it's not the most fundamental about them. On the other hand, the symmetries of nature are the deepest things we know about nature. It's much easier to learn about the symmetries of a set of laws than about the laws themselves. For example, long before there was any clear understanding of the nuclear forces, it was clear that there was a symmetry that the nuclear forces obeyed that related neutrons and protons and it said that they behaved the same way with regard to the strong forces.

We shall mention the issue of chirality later in our discussion. At this point we once again bring up the relationship of crystals and symmetry to illustrate how different interpretations may there be depending on the kind of question we are asking about them. In chemistry, when crystallization occurs, there is translational symmetry on the inside in addition to other symmetries let alone the symmetry of the external shape. On a deeper level, as we speak about the external shape of a crystal, it already points to the breaking of translational symmetry, because translational symmetry does not include an ending of such symmetry; it should extend to infinity. In reality, although the crystal is finite, it is usually large enough to consider it infinite (from the point of view of a diffraction experiment, for example).

If we take the point of view in physics, there is a different approach, and this is what Weinberg emphasizes: "When you have a crystal, condensed from a liquid, the crystal breaks translational invariance. The crystal is in one location and if you translate the crystal by an infinitesimal amount, you have a different crystal, the atoms are clearly moved, the crystal has a definite location, it's here, not there. That means that translational invariance is a broken symmetry." [9]

In the examples that follow, mostly chemical structures will figure, and a utilitarian approach to symmetry. Emphasis will be on the application of the symmetry concept in a variety of discoveries in chemistry. The limits of the applicability or the utility of this concept will be indicated at places. We will mention examples in the complementarity in crystal structures; structure elucidation of large, biologically important molecules; determination and prediction of the molecular geometries and structural variations of simple molecules, among them the intriguing carbocations; the discovery of fullerenes and quasicrystals; and chirality.

Complementarity in Crystal Structures

For our discussion, the main interest of complementarity in crystal structures is in that the most symmetrical arrangements are by far not the most frequent among crystal structures. By now, with hundreds of thousands of crystal structures available in databases, the preeminence of complementarity is a well established fact. The Soviet-Russian crystallographer, Aleksandr I. Kitaigorodskii (Fig. 4) [10], predicted the relative frequencies of symmetry occurrences of all the 230 kinds of symmetry in crystal structures. He made his predictions long before hundreds of thousand

Fig. 4 Aleksandr I. Kitaigorodskii lecturing (courtesy of Irena Akhrem).

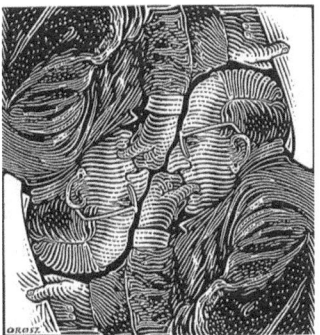

Fig. 5 "Complementary Kitaigorodskii" (drawing by and courtesy of István Orosz).

structures had become available. His initial goal was even more ambitious: it was finding regularities in how molecules build up crystals and on such a basis predicting crystal structures as soon as the composition of a new substance would become available. This goal has so far proved elusive.

In the late 1930s, the German physicist Pascual Jordan suggested that interactions between identical or nearly identical parts of molecules represent the advantageous mode of building up stable systems ("Zur Frage einer spezifischen Anziehung zwischen Genmolekülen," "To the question of a specific attraction between gene-molecules") [11]. In contrast, Linus Pauling and the physicist-turned-biologist Max Delbrück argued for precedence for interactions between complementary parts rather than identical ones [12]. The title of their short communication was "The Nature of the Intermolecular Forces Operative in Biological Processes." It is unlikely,

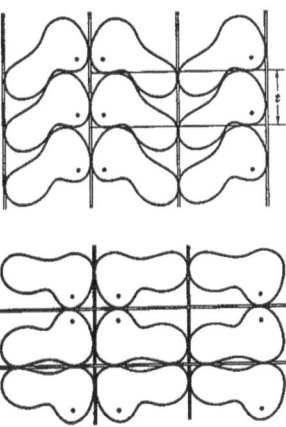

Fig. 7 The presence of symmetry planes in the space groups (top) pm and (bottom) pmm hinder densest packing. Both Figures 6 and 7 after Kitaigorodskii's *Molecular Crystals and Molecules* (Academic Press, New York, 1973).

especially under the war-time conditions, that Kitaigorodskii (Fig. 5) could have been familiar with the Pauling-Delbrück paper. Yet Kitaigorodskii, independently, declared a research program, "The close packing of molecules in crystals of organic compounds" in the then still existing English-language Soviet physics journal [13]. The program was based on his views on the preeminence of attractive interactions between complementary molecular shapes. He predicted that "the mutual location of molecules is determined by the requirements of the most close-packing."

Kitaigorodskii used his own so-called structure-finder, a simple stand to which he fastened wooden models of molecules of the same arbitrary shape and examined the densest packing of virtually all 230 symmetry variations. He found that the highest frequency occurrence of molecular packing should be that characterized by twofold rotational symmetry (Fig. 6). This is what corresponds to the complementary arrangement of molecules in which protrusions of one molecule meet with cavities of the other molecule. This is spatial complementarity. Molecular packing characterized with symmetry planes is not impossible, but it is rather disadvantageous for densest packing (Fig. 7). In other words, "lower symmetry packs better." [14] Today's wealth of data on hundreds of thousands of crystal structures has proved Kitaigorodskii's predictions demonstrating the correctness of his far-sighted vision of the interactions directing crystal architecture [15].

Alpha-Helix and Double Helix Structures

During the first half of the twentieth century, the existence of biological macromolecules was not yet generally accepted, but the efforts to establish the nature of biological substances

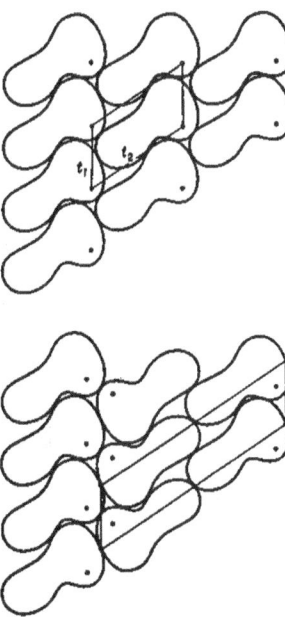

Fig. 6 Densest packing with space groups (top) p1 and (bottom) p2.

Fig. 8 Linus Pauling's autographed photograph (courtesy of the late Linus Pauling).

and to uncover their structures were going on. Michael Polanyi and Herman F. Mark subjected fibrous materials to X-ray diffraction in the 1920s, at the Kaiser Wilhelm Institute in Berlin. The experiments on cellulose indicated the presence of crystallites in cellulose that were oriented in the direction of the fiber axis. Furthermore, Polanyi and Mark observed characteristic changes when they stretched the cellulose fibers. X-ray crystallography was becoming the preeminent tool for the structure determination of biopolymers.

Linus Pauling (Fig. 8) joined early the quest for uncovering the structure of biopolymers. He had been the foremost structural chemist accumulating information about the structure and bonding of small molecules, including amino acids, the building blocks of proteins. Hemoglobin was the first protein that attracted Pauling's interest. As is well known, hemoglobin carries the oxygen in our organism. There was also a British group in Cambridge working on the structure of hemoglobin that had chosen this protein for their inquiry independent of Pauling's interest.

There were two types of proteins known; for example, hair, horn, porcupine quill, and fingernail belonged to one, and silk to the other. Hair in its normal state and in its wet state showed differences in their X-ray diffraction patterns according to the experiments by William T. Astbury. He called one alpha keratin, and the other beta keratin. The beta keratin was the stretched form and the X-ray pattern of the beta keratin state of hair was similar to the X-ray pattern of silk. Pauling decided to start his studies with alpha keratin. He launched a complex investigation in which he used all his accumulated knowledge of the structure of small molecules and all his knowledge about the correlation of geometrical features and bonding peculiarities of molecular structure. Understanding chemical bonding was as important as collecting X-ray diffraction data. Thus, knowing about the double-bond character of the peptide bond meant a drastic reduction of possible protein structures, because the bond configuration about such a bond had to be planar or nearly planar. This piece of information helped Pauling to reduce drastically the number of possible models. Nonetheless, at

the time—this was during the second half of the 1930s—Pauling still did not have enough information about the details of the protein structures to be able to propose a model that would be in agreement with all the X-ray diffraction evidence he possessed. Among the unknown factors, it was not known how the diversity among the building blocks—that is, among the amino acid units of the proteins—would influence the overall structure of the protein molecule.

About a decade later, Pauling continued his quest for the protein structures. At this time—this was in 1948—he decided to ignore the fact that the building block amino acids were different from each other, and assumed them to be equivalent. This was a huge simplification; in a way it was the introduction of translational symmetry in the protein chain where, rigorously considering it, it did not exist. His subsequent results justified this simplification. Once he could consider the protein chain more uniform than it was, he could apply to it a mathematical theorem according to which an asymmetric object can be converted into an equivalent asymmetric object by the application of rotation-translation. Subsequent and repeated application of this operation—and this is also prescribed by the mathematical theorem—produces a helix. This was a breakthrough and one of the factors that permitted Pauling to reach it was his bravely overlooking the absence of rigorous symmetries among the building blocks of the protein chain.

Pauling realized the fruitfulness of building models. He prepared a rudimentary drawing of the protein chain (Fig. 9)—using uniform amino acids—and determined the possible models emerging from folding the drawing in such a way that satisfied the possibility of hydrogen-bond formation that he had also found to be present in the desired structure [17]. The result showed one problem that did not seem to be surmountable, viz., the turn about the chain did not

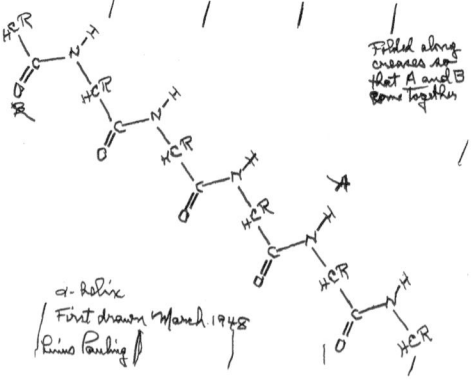

Fig. 9 Linus Pauling's sketch of the polypeptide chain in 1948. Pauling folded the paper along the creases and arrived at the alpha helix (courtesy of the late Dorothy Munro, Linus Pauling's long-time personal assistant) [16].

correspond to an integer number of amino acids. This, again, lowered the degree of symmetry of the emerging structure. In a bold move, Pauling decided to ignore this deficiency of his model as there really was no stipulation that such a symmetry requirement should be fulfilled. Once he went ahead, two models appeared satisfactory of which one could be discarded readily and the other was the model he called alpha helix. The Cambridge (UK) group that was working on the protein structures simultaneously with Pauling came up with numerous models none of which turned out to be acceptable. Pauling's competitors did not apply the simplifications that Pauling did with such a success.

There was at least one general lesson from Pauling's achievement in producing the alpha helix structure of proteins: "Pauling's approach to solving this complex problem was exemplary in focusing on what was essential and ignoring what had little consequence. When it turned out that the turn about the chain did not correspond to an integer number of amino acids, hinting at less than perfect symmetry, he did not let himself bothered by this. He thus expanded the realm of crystallography toward structures that were not part of classical crystallography yet included literally vital substances." [18]

Francis Crick (Fig. 10) and James D. Watson (Fig. 11) published their suggestion for the structure of deoxyribose nucleic acid (DNA) in April 1953 (Fig. 12). They wrote that their "structure has two helical chains each coiled round the same axis. . . . The two chains (but not their bases) are related by a *dyad* [twofold axis] *perpendicular to the fibre axis*. Both chains follow right-handed helices, but owing to the dyad the sequences of the atoms in the two chains run in opposite directions" [19] (emphasis by me). It may be argued that the mention of dyad here is equivalent to a twofold axis of rotation (C_2 symmetry), but one wonders why Watson and Crick were not more explicit about this feature of the

Fig. 11 James D. Watson in 2000 in the author's home in Budapest (photograph by I. Hargittai).

structure. There was some ambiguity about how Watson and Crick, each of them separately, handled the presence of symmetry in the DNA structure [20]. The impression has

Fig. 12 Artist's rendition of the double helix of DNA; Bror Marklund's sculpture in front of the Biomedical Center of Uppsala University (photograph by I. Hargittai in 1997; the area surrounding the sculpture has been built in since the time the snapshot was taken).

Fig. 10 Francis Crick in 2004 in La Jolla (photograph by I. Hargittai).

formed "that for Watson, the C_2 symmetry of the structure was not as revealing as it was for Crick. Back in 1951, he [Watson] wrote to Delbrück, 'Our method is to completely ignore the X-ray evidence [21].' In February 2004, Crick noted that Watson did not understand the significance of C_2 symmetry of the DNA structure." [22]

In his book, *What Mad Pursuit*, Crick was more self-critical with respect of the difficulties in recognizing the importance of C_2 symmetry and its implication for the DNA structure. He notes that discovering base-pairing was more the result of serendipity than logical thinking. It would have been more elegant to come to the right conclusion by logical thinking: "first to assume Chargaff's rules were correct and thus consider only the pairs suggested by these rules, and second, to look for the dyadic symmetry suggested by the C2 space group shown by the fiber patterns." [23]

At this point it is of interest to mention that Watson's apparent indifference to the C_2 symmetry of the DNA structure was not characteristic of his general demeanor towards symmetry. When the work on the DNA structure was temporarily halted in Cambridge, and he joined the investigation of tobacco mosaic virus, he did consider its symmetry. Donald Caspar described the story [24]:

> It was Jim Watson who recognized the helical symmetry in tobacco mosaic virus (TMV). This had grown out of the work on the DNA structure. [W. L.] Bragg was the director of the Cambridge laboratory where Watson and Crick were working on DNA, and he [Bragg] found out in 1952 that DNA research was also going on in Randall's lab at King's College in London. Bragg called Crick and Watson into his office and ordered a moratorium on the DNA work in his lab. At that time it was considered ungentlemanly in Great Britain to work on the same problem as your colleagues and to compete with them. That's

when Watson switched temporarily to the TMV problem. He recognized that TMV was a helix and applied to it the theory of diffraction by helical structures that Crick had worked out. The helical symmetry that Watson had inferred had turned out to be incorrect and Rosalind Franklin had got it right a few years later. But in 1952, DNA was more exciting than TMV, and when Bragg learned about the ongoing work on DNA by Linus Pauling in Pasadena, he lifted the moratorium and gave Watson and Crick full support to resume their DNA work.

The structure of TMV did have helical symmetry though different from the initial suggestion. It has a rod shape and the proteins envelop with a helical array a single-stranded RNA molecule (Fig. 13). At first sight biological macromolecules and other polymeric structures appear very similar. There is though an important difference. To build a model for a biological macromolecule, the starting point should correspond to a nucleation event. According to Aaron Klug (Fig. 14), "The key to biological specificity is a set of weak interactions. A polymer chemist could start building the model in the middle or at any other point." In contrast, for building the models of biological macromolecules, it is

Fig. 13 The model of TMV (courtesy of Aaron Klug).

Fig. 14 Aaron Klug with the TMV model in 2000 at the laboratory of Molecular Biology in Cambridge (photograph by I. Hargittai).

"important to find the special sequence for initiating nucleation." [25] Incidentally, preparing a model of TMV was a task for the world exhibition in Brussels in 1958. Rosalind Franklin was a major contributor to elucidating the TMV structure. By the time her group was preparing for the Brussels exhibition, she was already gravely ill and died in the same year as the world exhibition took place.

The terms helix and spiral are rigorously distinguishable, but every-day language often and even scientists sometimes use them interchangeably. Helical symmetry is when a constant amount of translation is accompanied by a constant amount of rotation. For spiral symmetry both the amounts of translation and accompanying rotation change gradually and regularly (Fig. 15). The biological molecules have helical rather than spiral symmetry whereas oscillating reactions accompanied by color changes may form beautiful spiral patterns (Fig. 16) [26].

The twofold rotational symmetry of the DNA double-helix structure is in beautiful correlation with the function of this biological macromolecule. Symmetry does not appear in such a directly visible way for many other biological systems. Sometimes it does, but the function is still not correlated in any perceivable way with it. An example is the attractive twofold symmetry of the photosynthesis reaction center, yet what, if anything, it means for the process of photosynthesis remains a puzzle (Fig. 17). Johann Deisenhofer described the moment of the discovery as follows [27]:

> It was extremely exciting to localize these features and build models for them. When I stepped back to see the arrangement,

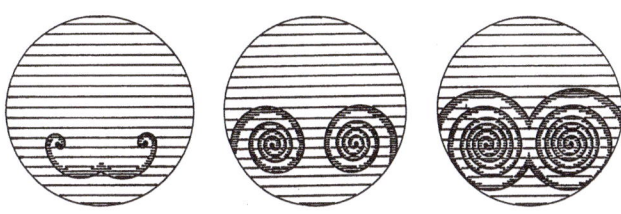

Fig. 16 Enhancing spiral pattern in a reacting Belousov-Zhabotinsky system (drawing by and courtesy of the late Endre Kőrös).

the unexpected observation about it was symmetric. There was symmetry in the arrangement of the chlorophyll that nobody had anticipated. Nobody, to this day, completely understands the purpose of this symmetry. I think it can be understood only on the basis of evolution. I think that the photosynthetic reaction started out as a totally symmetric molecule. Then it turned out to be preferable to disturb its symmetry, sticking to an approximate symmetry but changing subtly the two halves of the molecule. Because of the difference in properties of the two halves, the

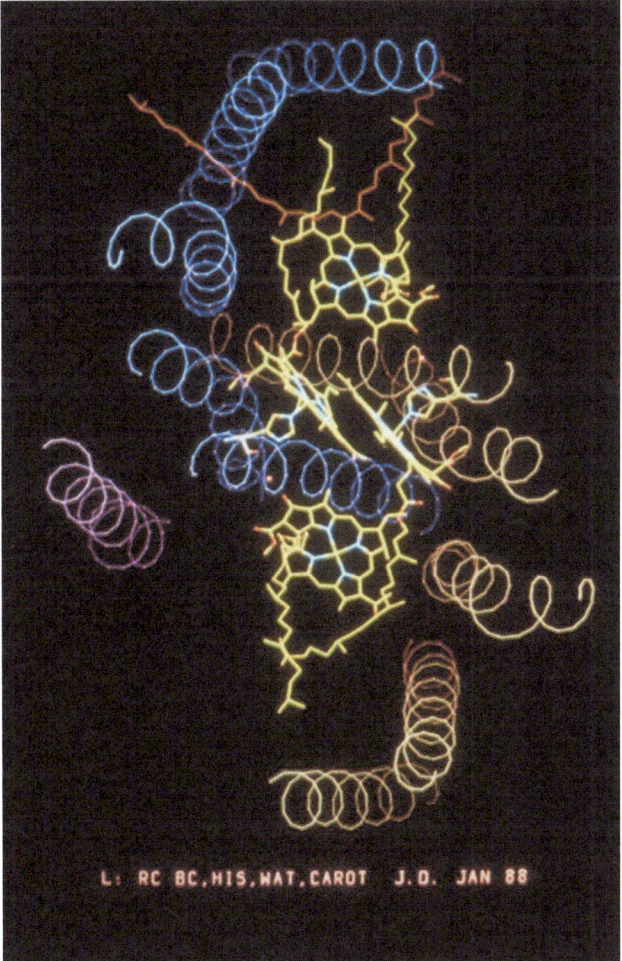

Fig. 15 Artist's rendition of a true double spiral, detail of "The Inner Light" by Gidon Graetz, in the garden of the Weizmann Institute in Rehovot (photograph by I. Hargittai).

Fig. 17 The structure of the photosynthetic reaction center with approximate C_2 symmetry (courtesy of Johann Deisenhofer).

Fig. 18 Ronald J. Gillespie in 1998 in Austin, Texas (photograph by I. Hargittai).

conclusion had been, before the structure came out, that there cannot be symmetry; that it has to be an asymmetric molecule. Now, when people looked at the structure, it looked totally symmetric to the naked eye. That realization was the high point I will never forget.

Structural Complications for Simple Molecules

Ronald J. Gillespie's (Fig. 18) *v*alence *s*hell *e*lectron *p*air *r*epulsion (VSEPR) model or theory predicts the geometry of the molecule on the basis of the number of electron domains (bonding pairs, lone pairs, multiple bonds) in the valence shell of its central atom [28]. The predicted shapes and symmetries depend not only on the general number of electron domains but to various extents also on the nature of those domains, whether they are single bonds, lone pairs or multiple bonds.

In all VSEPR considerations, spherical symmetry of the valence shell of the central atom is assumed and that all electron domains are at equal distances from the nucleus of the central atom (Fig. 19). The geometries may be determined as the ones assigned to the minimum of the potential energy. Its terms can be expressed as $V_{ij} = k/r_{ij}{}^n$ where k is a constant, r_{ij} is the distance between the points i and j in the spherical valence shell and the exponent n is large for strong and small for weak repulsions. The value of n is generally much larger than what it would be for merely electrostatic interactions,

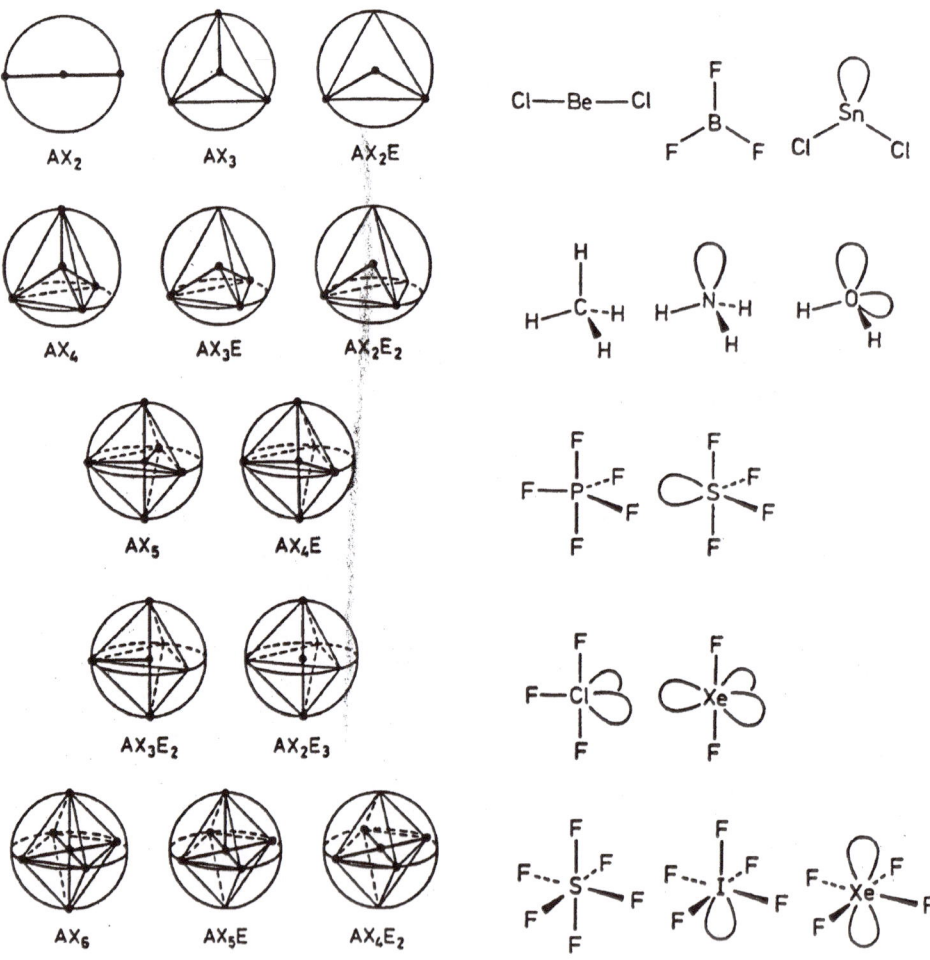

Fig. 19 A variety of VSEPR geometries.

but it is not known. The great advantage of the model is that it need not be known, because when it is larger than three, it no longer has any influence on the outcome of the geometry obtained by minimizing the potential energy. Here a few examples will illustrate the utility and limitations of the model in establishing molecular symmetries.

Early on the determination of the structure of xenon hexafluoride was a conspicuous success for the model. As soon as the substance was produced, some rudimentary molecular calculations predicted that the molecule has the highly symmetrical (O_h) regular octahedral shape. However, according to the VSEPR model, XeF_6 should be described as an AB_6E system where A is the central atom, the ligands B are linked to it by an electron pair each, and there is also a lone pair of electrons, E, in the valence shell of xenon. Hence, the molecule cannot have a regular octahedral shape. When Gillespie predicted a distorted geometry for XeF_6, the subsequent experimental studies suggested C_{3v} symmetry, or even C_{2v} symmetry, for the molecule that could be derived from the O_h structure by small distortions. Further work, however, yielded less unambiguous results. The suggestion was that it all depended on the stereochemical activity of the lone pair of electrons. In case of stereochemically active lone pair, the molecule is expected to display fluxional behavior, i.e., rapidly inter-converting series of configuration. In case of stereochemical non-activity, a rather rigid structure should be present. The case of XeF_6 that was such a convincing case for the utility of the VSEPR model has become an example of its limitations [29].

Applying rigorously the VSEPR arguments in cases where they should work impeccably, it is interesting to note that it is not always the highest symmetries that emerge in molecular structures. Thus, the seemingly analogous molecules OPF_3 and $OClF_3$ should be described in VSEPR formulation as CAB_3 and $C(E)AB_3$, respectively, where A is the central atom, C and B are ligands, and E is a lone pair of electrons in the valence shell of the chlorine atom. The bond configuration of OPF_3 has a distorted tetrahedral shape of C_{3v} symmetry whereas $OClF_3$ has a trigonal bipyramidal arrangement of the five electron domains and a bond configuration of C_s symmetry (Fig. 20). The electron domains whether they represent a single bonding pair, a multiple bond or a lone pair have different spatial requirements and accordingly,

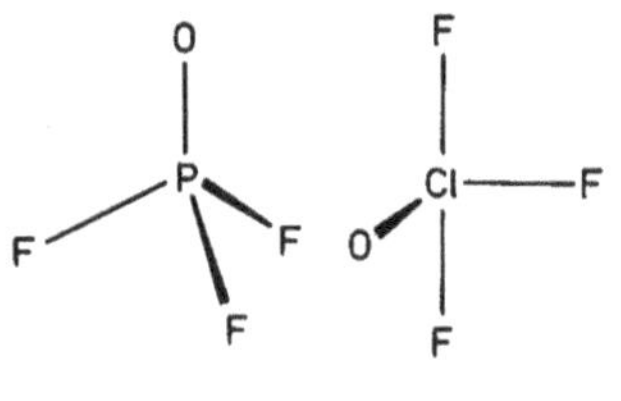

Fig. 20 OPF_3 is of C_{3v} symmetry and $OClF_3$ is of C_s symmetry.

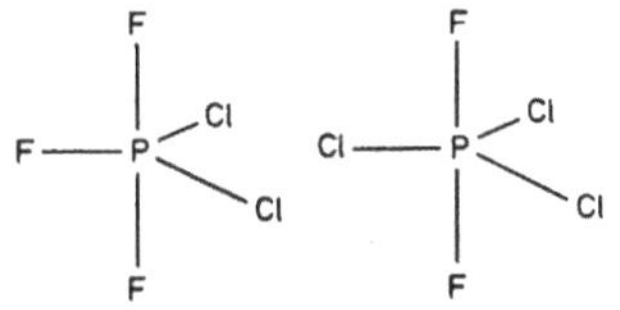

Fig. 21 In both trigonal bipyramidal structures of PF_3Cl_2 and PF_2Cl_3, the fluorines occupy axial and the chlorines equatorial positions. PF_3Cl_2 is of C_{2v} and PF_2Cl_3 is of D_{3h} symmetry.

repulsion strengths, in the valence shell. They are in decreasing order: lone pair, multiple bond, and single bond.

Further refinement of the model is necessary when, for example, single bonds connect ligands of different electronegativities. An example is the comparison of the molecular symmetries of F_3PCl_2 and F_2PCl_3 (Fig. 21). Both structures have C_3AB_2/C_2AB_3 description and, accordingly, trigonal bipyramidal bond configuration. However, the bonds leading to more electronegative ligands (fluorine) have smaller space requirements in the valence shell of the central atom than the bonds leading to less electronegative ligands (chlorine). Further, in the trigonal bipyramidal configuration, not all positions are equivalent for space requirements. The surroundings of the axial positions are more crowded than those of the equatorial positions. Hence, the bonds to fluorine are expected to be in axial positions and the bonds to chlorine in equatorial positions. The corresponding symmetries are, indeed, C_{2v} for F_3PCl_2 and D_{3h} for F_2PCl_3.

In the examples mentioned above we focused on the variations of molecular symmetry. Changing the nature of ligands may also cause characteristic changes in the bond lengths and bond angles in series of substituted derivatives without change in molecular symmetry. Sometimes it is fruitful to look at the variations in the distances between atoms not connected by chemical bonding, especially when the ligand atoms are large with respect to the central atom. Generally speaking, the VSEPR rules may work best for small ligands relative to the central atom. The relative weight of the non-bonded interactions in shaping the geometry of the molecule increases with increasing ligand size with respect to the central atom.

Thus, for example, retaining the constancy of such non-bonded distances may be looked at as the primary factor in the realization of certain structures. Here we single out the remarkable constancy of the O...O non-bonded distances of the sulfone groups in a series of substituted sulfone molecules, O_2SXY (Fig. 22). The O...O distance hardly change from being 2.48 Å in a series of free sulfone molecules while the lengths of the S=O bonds vary up to 0.05 Å and the O=S=O bond angles vary up to 5°, depending on the nature of the X and Y ligands [30]. The variations in the sulfone series could be visualized by a tetrahedron of the two oxygen atoms and X and Y ligands about the central sulfur atom with the two

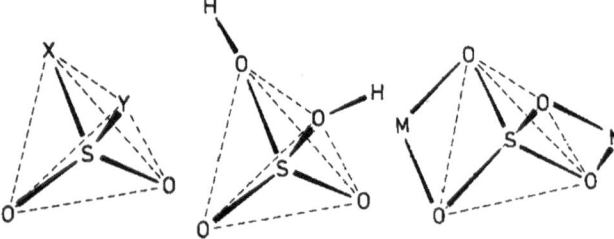

Fig. 22 Tetrahedral sulfur configurations, *from the left*, sulfones; sulfuric acid; alkali sulfates.

oxygen atoms firmly taking the positions of two of the vertices of the tetrahedron. Depending on the nature of the X and Y ligands, the central sulfur atom would be sliding along the bisector of the OSO angle, changing the SO bond lengths and OSO bond angles, but keeping the O...O distances unchanged. The recognition of the constancy of the O...O distances in extended series of sulfone molecules enhanced the possibilities of the combined application of electron diffraction and microwave spectroscopy in the accurate determination of related molecular structures.

Even systems as simple as metal dihalides may have complications in their molecular symmetry. The free

molecules of dihalides of alkaline earth metals used to be considered linear of $D_{\infty h}$ symmetry. We have learned, however, that this holds only for the dihalides of lighter metals, viz., beryllium and magnesium. Toward the dihalides of heavier metals, especially when combined with smaller halogens, the molecules are bent of C_{2v} symmetry. There are a few structures in between that are called quasi-linear. They are characterized with a small energy barrier on the bending potential energy distribution at the position of the linear configuration, but this energy barrier may even be below the ground-state energy level. Broad and flat minimum of the bending potential energy curve is typical for the quasi-linear molecules. They are floppy and very little energy input suffices to bend or straighten such molecules (Fig. 23).

The symmetry descriptions $D_{\infty h}$ and C_{2v} as well as quasi-linearity refer to the *minimum positions* on the bending potential energy curve. There is a related but rigorously distinguishable case when a molecule that is strictly linear in the minimum energy position appears nonetheless as bent from certain experimental structure elucidations. This happens most conspicuously when the bending vibrations of a metal dihalide appear as low-frequency, large-amplitude motion. Any experimental technique that determines the time-averaged structure and for which the interaction time is longer than the bending motion of

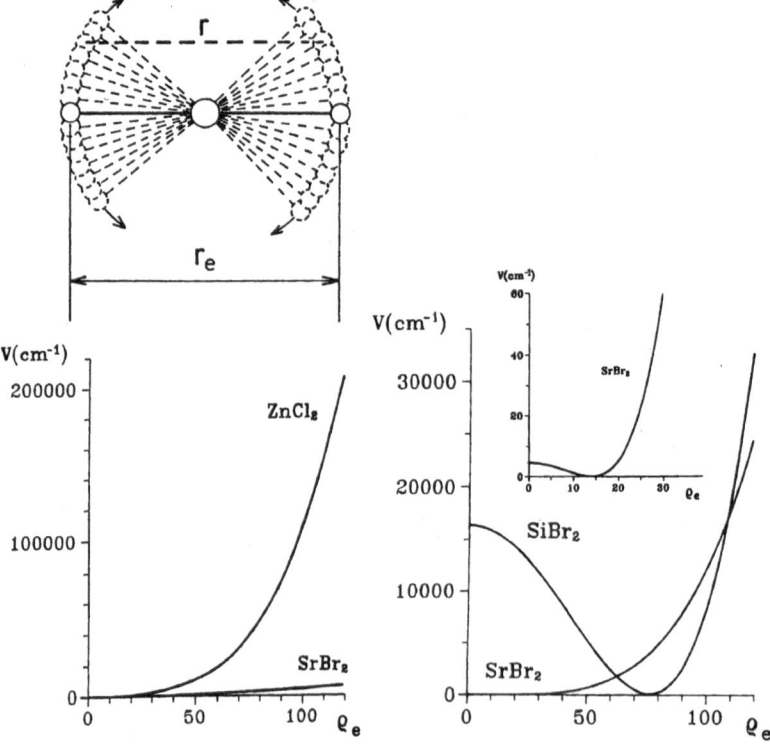

Fig. 23 Bending motions and a sampler of potential energy functions. Top: bending vibrations of a linear triatomic molecule, where r is the instantaneous distance between the end atoms and r_e is the equilibrium distance of the linear

configuration ($r < r^e$). Bottom: Comparison of bending potential functions for linear and bent models of symmetric triatomic molecules.

the molecule, will yield a bent geometry even when the molecule is linear in the minimum energy position. In terms of the halogen-halogen non-bonded distance, in the linear configuration, this will be exactly twice the bond length; in any bent position of the molecule, the halogen-halogen distance will be shorter than twice the bond length. However, this is only an apparent deviation from linearity. Such a difference between molecular shapes and symmetries of the minimum position structure (called also equilibrium structure) and the average structure is characteristic not only of the triatomic AB_2 molecules, but any polyatomic molecule. Again, the higher the probability of low-frequency, large-amplitude deformation motion the larger the difference expected between the equilibrium and average structures.

Generally speaking, fluxional molecular behavior decreases the probability of unambiguous determination and description of molecular shape and symmetry. Permutational isomerism is an example of fluxional behavior. R. Stephen Berry discovered it for trigonal bipyramidal structures, and it is called Berry pseudorotation (Fig. 24) [31]. Permutational isomerism is when different structures with different symmetries of the same molecule interconvert. Identical atoms permute among nonequivalent sites in these processes. The VSEPR model suggests only a slight energy preference

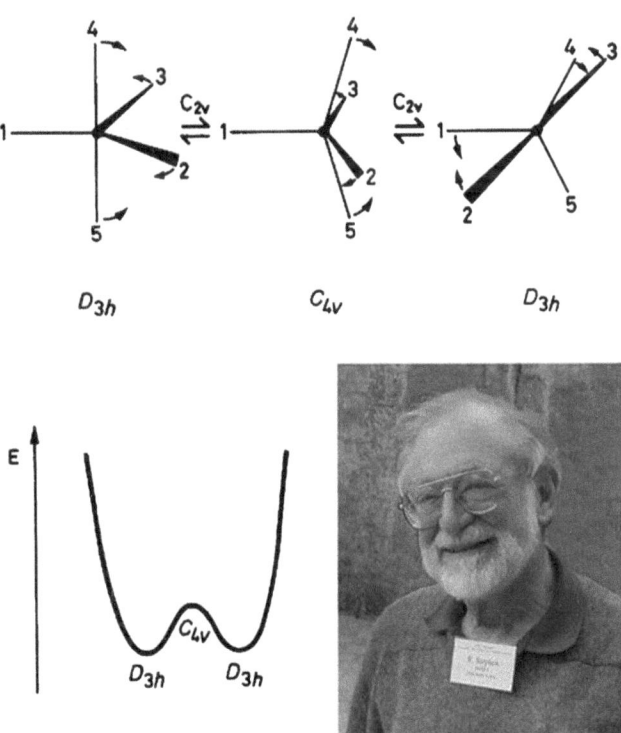

Fig. 24 Berry-pseudorotation of PF5-type molecules with a potential energy function and R. Stephen Berry in 2001 in Erice, Italy (photograph by I. Hargittai).

for the trigonal bipyramidal model (D_{3h}) over the square pyramidal one (C_{4v}) and they easily interconvert.

Above, we have already considered in passing the important relationship of the interaction time required by the physical phenomena on which the experimental technique is based and the life time of the structure being measured. The interaction times are relatively long for the techniques of NMR spectroscopy and very short for the diffraction techniques. Vibrational spectroscopy is somewhere in between. Thus, for example, the rapidly inter-converting AB_5 trigonal bipyramidal geometries, NMR spectroscopy may yield information only about the average A-B bond whereas electron diffraction may distinguish the axial and the equatorial A-B bonds.

Berry-pseudorotation introduces a great deal of ambiguity into the description of structures analogous to PF_5. It plays an important role in the chemistry of large, biologically important molecules as well as in the physical-philosophical considerations of the concept of molecular structure. Frank Westheimer found far-reaching implications of Berry-pseudorotation when he and his students investigated the hydrolysis of phosphate ester. In this process, the four-coordinate phosphorus becomes five-coordinate as it goes into its transition state, then becomes five-coordinate through pseudorotation, and then returns to four-coordinate. The driving force of this process is in the difference in bond strengths. In the five-coordinate situation, the bonds are weaker in the axial positions than in the equatorial positions. The formation of a weaker axial bond precedes its rearrangement into a stronger equatorial bond, while another equatorial bond becomes a weaker axial bond and breaks off. This is a pivotal event for important biochemical processes [32].

The philosophical implication of Berry-pseudorotation may be formulated in this way: It appears to be a paradox that in the process of pseudorotation, identical nuclei occupy observably nonequivalent sites. Quantum mechanics prescribes that identical particles, electrons, for example, but not only electrons, have to be indistinguishable and the wave function for identical particles reflects this indistinguishability. In contrast, in chemistry, we operate with the distinguishability of different sites in molecules. The considerations of the time scales, interaction time versus the life-time of a structure, resolve this apparent paradox.

Above we have considered the appearance of a bent average geometry for a linear molecule as a result of averaging over all configurations during bending vibrations. We have also mentioned that the interaction time in NMR spectroscopy may be too long to distinguish between axial and equatorial bonds in triginal bipyramidal systems. It is possible, however, to conduct a series of NMR experiments with changing relationship of the two time scales, demonstrating, for example, the coalescence of NMR lines that in a fast experiment correspond to different chemical shifts and in a slow experiment correspond to the average [33].

Symmetry-lowering may happen due to the Jahn-Teller effect. A non-linear symmetrical nuclear configuration in a degenerate electronic state is unstable and gets distorted—this is how it removes the electronic degeneracy until it achieves a non-degenerate state, according to the formulation of the Jahn-Teller effect [34]. A typical case of Jahn-Teller distortions is the structure of crystalline manganese trifluoride. The six fluoride ions surround the manganese ion in six-coordination in its structure. In a regular octahedral arrangement the six manganese-fluoride distances would be uniform; but this is not the case. Rather than having O_h symmetry, there is the lower D_{4h} symmetry with two different manganese-fluorine distances.

Recently, the Jahn-Teller effect was demonstrated with reliable geometrical parameters of a free molecule in gaseous manganese trifluoride [35]. For this molecule, the highest possible symmetry would be D_{3h}, but the Jahn-Teller effect lowers it to C_{2v} symmetry. Rather than having three 120 degrees bond angles, there are two of 106 and one of 148 degrees; also, one of the Mn-F bonds is shorter than the other two. The distortion stabilizes the molecule. These are not apparent, but real geometrical changes that are present in the minimum position of the potential energy distribution and characterize the corresponding equilibrium structure.

The complexity of the CH_5^+ structure rivals that of the XeF_6 molecule. This carbocation has had special significance in organic chemistry as its discovery was related to the beginning of a whole new direction in the discipline [36]. This new direction has transformed hydrocarbon chemistry from a rather inert kind of domain into the source of exceptional wealth of new substances. It became possible by the application of superacids that stabilized the otherwise short-lived carbocations and enabled the otherwise un-reactive covalent carbon-carbon and carbon-hydrogen bonds to become reactive. The discoverer George A. Olah (Fig. 25) could rightly conclude: "The realization of the

Fig. 25 George A. Olah in 1995 at the Budapest University of Technology and Economics (photograph by I. Hargittai).

electron donor ability of shared electron pairs could one day rank equal in importance with G. N. Lewis' realization of the electron donor unshared pairs." [37]

The carbocation CH_5^+ contains a five-coordinated carbon—note, however, that it is not a hypervalent carbon, only hypercoordinated. Thus it could be viewed as containing five electron domains that each is somewhat poorer in electrons than a two-electron covalent bond. In that case, the five-coordination and five electron domain carbocation could have a trigonal bipyramidal arrangement of D_{3h} symmetry, according to the predictions of the VSEPR model. This turned out to be not the case.

Early quantum chemical computations predicted a C_s symmetry structure for CH_5^+, which would correspond in Olah's description to the presence of three two-electron two-center bonds and one two-electron three-center bond [38]. This may be a structure of either having a high-degree of localization, or having a fluxional character by exchanging the positions of the two-electron two-center bonds and the two-electron three-center bond. Provided that the C_s symmetry structure is in a sufficiently deep energy minimum, it could be observable in experiments if the life-time of this structure is long enough as compared with the interaction time of the experimental technique employed. There had been attempts to apply the concept of pseudorotation to the highly fluxional CH_5^+, however, if the CH_3^+ plus H_2 description holds, pseudorotation would not be the right approach to its description.

Recent high-resolution spectroscopic experiments on CH_5^+ have suggested the presence of structures corresponding to Olah's description (Fig. 26) [39]. We have to keep in mind, however, that all the spectroscopic evidence are also consistent with a highly fluxional character of CH_5^+: "the five proton swarm around the central carbon" [40]. Olah's model of three two-electron covalent bonds and one two-electron three-center bond may thus be only one of the models that can be singled out from among a multitude of models. They, in their totality as an average may be considered as a more realistic representation of this carbocation that Oka called the "enfant terrible" of structural chemistry [41].

Although the VSEPR model could not predict unambiguously the geometry of CH_5^+ some analogous structures appear consistent with VSEPR predictions. The geometry of monopositively charged carbocation $\{[(C_6H_5)_3PAu]_5C\}^+$ containing five-coordinate carbon is trigonal bipyramidal (Fig. 27). According to Olah, this gold complex is an isolobal analog of CH_5^+, hence the isolobal analogy would suggest a trigonal bipyramidal geometry for CH_5^+, which is not the case (see above). A similar discussion could apply to the CH_6^{2+} carbocation and the $\{[(C_6H_5)_3PAu]_6C\}^{2+}$ carbocation. Six equivalent electron domains would favor a regular octahedral geometry of O_h symmetry and the six bonding directions point to the vertices of a regular octahedron in the gold complex. In contrast, the computations have suggested the presence of two

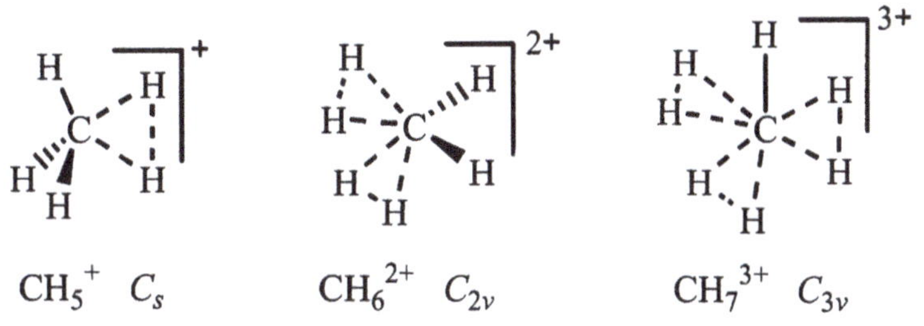

Fig. 26 Two-electron two-center bonds and two-electron three-center bonds in protonated alkanes [42].

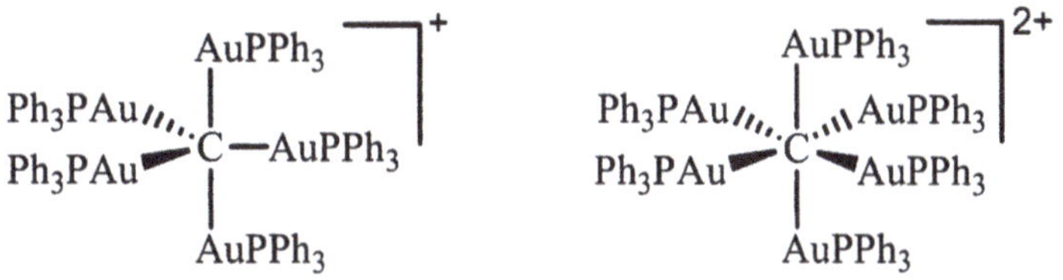

CH_5^+ C_s CH_6^{2+} C_{2v} CH_7^{3+} C_{3v}

Fig. 27 The trigonal bipyramidal monopositively charged carbocation $\{[(C_6H_5)_3PAu]_5C\}^+$ and the octahedral dipositively charged carbocation $\{[(C_6H_5)_3PAu]_6C\}^{2+}$ [43].

two-electron covalent bonds and two two-electron three-center bonds for the CH_6^{2+} carbocation. There is then the CH_7^{3+} carbocation, for which one two-electron covalent bond and three two-electron three-center bonds would apply by analogy.

Fivefold Symmetry in Extended Structures

Fivefold symmetry is just as common in the world of molecules as any other symmetry. This is not the case, however, in extended structures. Fivefold symmetry was an excluded symmetry in classical crystallography. Two important discoveries in the 1980s, both in materials science—one in the world of molecules and the other in extended structures—were related to fivefold symmetry. Each of the two discoveries was eventually awarded by a chemistry Nobel Prize.

The discovery of buckminsterfullerene (first of its existence [44], then, its production [45]) made waves due to the beauty of its structure and the fact that it was a heretofore unknown modification of carbon. The C_{60} molecule is of truncated icosahedral shape. There is a presence of fullerene-type structures and their fragments in nanotubes. Considering today's importance of nanoscience and nanotechnology, even a symbolic impact by the buckminsterfullerene discovery in this development is noteworthy.

The icosahedral arrangement of atoms has interested researchers because they considered it as containing some of the clues of the puzzle of the progression from isolated molecules to extended systems. The icosahedral arrangement caught J. Desmond Bernal's (Fig. 28) eye early on. He was interested in the structure of liquid water and the icosahedral arrangement was viewed as the one preventing the

Fig. 28 J. Desmond Bernal giving a speech (photograph by and courtesy of Alan L. Mackay).

crystallization of water. Linus Pauling also showed distinct interest in icosahedral structures.

Alan L. Mackay (Fig. 29) enveloped a sphere with an icosahedral shell consisting of 12 spheres and enveloped this structure by another shell and the second shell was arranged over the first so that the spheres were in contact along fivefold axes. When he added a third shell, the structure already contained 147 spheres, and this is what has been known as the Mackay icosahedron or Mackay polyhedron [47]. When icosahedra are packed together, like in the Mackay polyhedron, they gradually curve up to form a closed system. In addition to the Mackay polyhedron, another example is the icosahedral polyoma virus (Fig. 30) [48].

In a parallel development, Roger Penrose (Fig. 31) has invented a two-dimensional pattern of hierarchic tessellation of the plane, which was regular, that is, it was constructed by

Fig. 29 Alan L. Mackay in 1982 in Budapest (photograph by I. Hargittai).

Fig. 30 Icosahedral polyoma virus drawn after Adolph et al. [46].

Fig. 31 Roger Penrose in 2000 in Oxford, UK (photograph by I. Hargittai).

Penrose, when he was still a graduate student, had had some interactions with Bernal in the mid-1950s; the story borders the mysterious as Penrose remembered decades later: "He [Bernal] came to see me completely out of the blue, just because he was looking for people who might have ideas, to do with these pentagons, and so on." [50]

Mackay took Penrose's effort one step further and produced a simulated electron diffraction pattern of a three-dimensional Penrose pattern [51]. Mackay warned that if we exclude the possibility of extended structures with five-fold symmetry, we may experience it, yet ignore it [52]. Fortunately, this is not how it played out although it could have.

In 1982, at the then National Bureau of Standards, Dan Shechtman (Fig. 32), a visiting scientist from the Technion, was experimenting with electron diffraction of a great variety of manganese-aluminum alloys. He obtained a diffraction pattern that could be interpreted as an extended structure of tenfold symmetry—clearly "in violation" of the rules of classical crystallography [53]. When the experimental observation was properly documented, it turned out that there was instant theoretical interpretation and even a catchy name for this new state of matter [54]. Shechtman's perseverance brought this discovery to triumph but not before he had to face the disbelief and even ridicule of establishment

Fig. 32 In 1984, at NBS, from left to right, Dan Shechtman, Frank Biancaniello, Denis Gratias, John Cahn, Leonid Bendersky, and Robert Schaefer. Photograph by H. Mark Helfer/NIST; courtesy of NIST. Biancaniello prepared the alloy samples and created a broad range of allow compositions. John Cahn and Dennis Gratias were two of Shechtman's three co-authors on his seminal paper reporting the discovery.

well-defined rules, but it was not periodic [49]. Penrose had seen a logo of a pentagon surrounded by same-size pentagons, and Penrose started iterating it and augmented the gaps of the pattern by parts obtained by cutting up additional pentagons.

scientists. The story has been well documented by several authors, including the present one [55].

In this case, it was not the ambiguity of a concept but the ambiguity of the scope and definition of classical crystallography that was what had to be sorted out. It was not trivial though. We have seen above how innovative Linus Pauling could be in breaking down previous dogmas in his quest for the alpha helix. Decades later, he could not accept Shechtman's breaking down some other dogmas. Planck's words come to mind about the new ideas that keep appearing in science and that even great old scientists find unable to accept: "An important scientific innovation rarely makes its way by gradually winning over and converting its opponents: it rarely happens that Saul becomes Paul. What does happen is that its opponents gradually die out, and that the growing generation is familiarized with the ideas from the beginning." [56]

Chirality

This contribution is for a special issue in the *Israel Journal of Chemistry* honoring the fiftieth anniversary of the first successful chiral column separation of racemic amino acids by the late Emanuel Gil-Av of the Weizmann Institute. Chirality plays a fundamental role in many chemical events while, according to Weinberg, "chiral symmetries … are not fundamental symmetries underlying the laws of nature." [57] For the importance of differences in the properties of chiral pairs of molecular substances, suffice it to mention the thalidomide story. It had many more tragic consequences in Western Europe in the 1950s than in the United States. The difference was primarily due to an officer at the Food and Drug Administration, Frances O. Kelsey (Fig. 33), who was not satisfied with the knowledge about the substance to give it the green light to the US market. As it turned out, in the enantiomeric mixture of thalidomide,

Fig. 34 Vladimir Prelog's *ex libris* plate by Hans Erni with Prelog's dedication to the author (courtesy of the late Vladimir Prelog).

one enantiomer was teratogenic, the other was not (but even that transformed into the teratogenic isomer in the organism). For some time now, legislation has mandated that only enantiomerically pure pharmaceuticals can be marketed.

Yet another scientist to mention here is Vladimir Prelog, one of the founders of modern stereochemistry. His office at the Eidgenössische Technische Hochschule Zürich (Swiss Federal Institute of Technology at Zurich) was full of memorabilia of stereochemistry and chirality in particular. His *ex libris* was a drawing by Hans Erni, which has become well-known all over the world (Fig. 34). Prelog chose this drawing, because it represented all paraphernalia that describe chirality: human intelligence, a left and a right hand, and two enantiomorphous tetrahedra. Erni prepared more than one version of this drawing, but the one Prelog chose for his books was peculiar. The two hands appear as if the were turned around, inverted that can be imagined as a consequence of the two arms being crossed [58]. Other versions of Erni's drawing were displayed in Prelog's office with the two hands being non-inverted, parallel. Further examples of pairs of hands appear in Figs. 35 and 36.

Fig. 33 Frances O. Kelsey in 2000 in her office at the Food and Drug Administration in Rockville, MD (photograph by and courtesy of Magdolna Hargittai).

Fig. 35 Heterochiral pair of hands in the old Jewish cemetery in Prague (photograph by I. Hargittai).

Fig. 36 Homochiral pair of hands, "The Cathedral," by Auguste Rodin in the Rodin Museum in Paris (photograph by and courtesy of Magdolna Hargittai).

Chirality may not be a fundamental property underlying the laws of nature, yet it has vital consequences for our lives and for life in general. It intrigued Lewis Carroll's *Alice* when she asked a question deep-rooted in chirality (Fig. 37). In the book, *Through the Looking Glass*, comparison of an image and its mirror reflection makes Alice wonder, "Perhaps Looking-glass milk is not good to drink ..." [59] Some time ago, in a brief paper titled "Eternal dissymmetry," we summarized some examples that illustrate how "the teachings of Louis Pasteur about chirality continue to instruct and inspire." [60]

Fig. 37 Sculptural group "'Curiouser and curiouser!' cried Alice" (by Jose de Creeft, 1959) in Central Park, near the Conservatory Lake (76th Street and Fifth Avenue), New York City, 2015 (photograph by I. Hargittai).

Conclusion

Ambiguity of symmetry impacts the validity of the symmetry concept and its applicability. Take, for example, molecular structure. Information about its symmetry is always interesting, often useful, and some times crucial. The reliability of the determination of molecular symmetry increases with increasing molecular rigidity and fluxional behavior enhances its ambiguity. Ambiguity may also exist in our various definitions, as it turned out, for example, for the one used to define what a crystal is. The old definition had to be replaced by a more inclusive one following the discovery of quasicrystals. The more comprehensive definition has rendered the label quasicrystal a misnomer or at least superfluous; nonetheless its usage has continued without causing any misunderstanding.

Another aspect of ambiguity is when perfect symmetry is damaged—is symmetry still there? The expression perfect symmetry does not make sense in a rigorously geometrical sense. There is symmetry or there is not. However, in real systems this is not the way we handle symmetry and in real systems the ambiguity may develop into arbitrariness. In other words, it depends on our tolerance—some times on our good will—whether we continue to consider something symmetrical whereas it no longer is according to stronger criteria.

Take, for example the sphere whose simple figure possesses infinite number of symmetries and about which Copernicus wrote: "... the spherical is the form of all forms most perfect, having need of no articulation; and the spherical is the form of greatest volumetric capacity, best able to contain and circumscribe all else; and all the separated parts of the world—I mean the sun, the moon, and the stars—are observed to have spherical form; and all things tend to limit themselves under this form—as appears in drops of water and other liquids—whenever of themselves they tend to limit themselves. So no one may doubt that the spherical is the form of the world, the divine body." [61]

For 30 years, Fritz Koenig's 7.6-m metallic sculpture "The Sphere" graced the plaza at the World Trade Center (Fig. 38). It was not a perfect sphere in the geometrical sense and it did not have all the symmetry elements of the sphere in the rigorous geometrical sense, yet nobody doubted its being a sphere. It symbolized world peace through world trade. The terror attack on September 11, 2001, badly damaged this sculpture. Defiance and resilience reconstructed the sculpture from its salvaged remains, and now it honors the victims of the terror attack. In this disfigured version, "The Sphere" has lost none of its grace and nobody has any problem in identifying it as being a sphere. Its sculptor noted its transformation from being a sculpture to becoming a monument (Fig. 39) [62].

Fig. 38 "The Sphere" by Fritz Koenig's at the World Trade Center in the mid-1980s (photograph by I. Hargittai).

Fig. 39 The sculpture-turned-monument in Battery Park, Manhattan, Fall 2014 (photograph by I. Hargittai).

There is a steady extension of our knowledge of materials, including systems under extreme conditions (created on our Planet or existing in Space), many of them disordered or partially disordered. Many others may yet emerge. Structural variations of the building blocks of living organisms have also expanded the scope of science about structures. We have learned to live with ambiguities and we are learning to value ambiguities in symmetry and elsewhere.

Acknowledgments I have drawn on the generous assistance of many in creating my views on and knowledge about symmetry over the years. I cannot list all the names of the individuals who have enriched my thinking about this topic (some, though by far not all, appear in the text above). I thank Steven Weinberg for having looked at my manuscript and for giving me permission to quote extensively from our 1998 conversation. I appreciate Robert Glaser's kindness of inviting me to participate in this special issue and Bob Weintraub's comments on the manuscript. I remember with gratitude the inspiring interactions with the late Yuval Ne'eman. Magdolna Hargittai's partnership, artistic talent, and scientific rigor have shaped the directions I have taken in this as in all my endeavors.

References

1. I. Hargittai, *Isr. J. Chem.* **2011**, *51*, 1144–1152.
2. E.P. Wigner, *Symmetries and Reflections: Scientific Essays*, Indiana University Press, Bloomington, Indiana, and London, **1967**, pp. 3–5.
3. *The Collected Works of Eugene P. Wigner, Volume VII: Historical and Biographical Reflections and Syntheses*, annotated and edited by Jagdish Mehra, Springer-Verlag, Berlin, Heidelberg, New York, **2001**, p. 429.
4. E.P. Wigner, *Symmetries and Reflections: Scientific Essays*, Indiana University Press, Bloomington, Indiana, and London, **1967**, p. 39 (from Wigner's Nobel lecture).
5. M. Hargittai, I. Hargittai, *Candid Science IV: Conversations with Famous Physicists*, Imperial College Press, London, **2004**, Chapter 2: "Steven Weinberg," pp. 20–31; actual quote, pp. 22–23.
6. Ibid., p. 23.
7. M. Hargittai, I. Hargittai, *Candid Science IV: Conversations with Famous Physicists*, Imperial College Press, London, **2004**, Chapter 3: "Yuval Ne'eman," pp. 32–63; actual quote, p. 51.
8. M. Hargittai, I. Hargittai, *Candid Science IV*, p. 23.
9. Ibid., p. 25.
10. There is a nice memorial volume, compiled by E. V. Leonova, *A. I. Kitaigorodskii: Uchonii, uchitel', drug* (in Russian, A. I. Kitaigorodskii: Scientist, Teacher, Friend), Moskvovedenie, 2011.
11. P. Jordan, *Physikalische Zeitschrift* **1938**, *39*, 711–714.
12. L. Pauling, M. Delbrück, *Science* **1940**, *92*, 77–79.
13. I. Kitaigorodskii, *J. Phys. (USSR)* **1945**, *9*, 351–352.
14. I. Hargittai, M. Hargittai, *In Our Own Image: Personal symmetry in discovery*, Kluwer/Plenum, New York, **2000**, p. 117.
15. For more on Kitaigorodskii's science and colorful personality, see, I. Hargittai, *Buried Glory: Portraits of Soviet Scientists*, Oxford University Press, New York, **2013**, pp. 250–266.
16. L. Pauling, *Chem. Intell.* **1996**, 2(1), 32–38 (posthumous publication, communicated by Dorothy Munro); reprinted in B. Hargittai, I. Hargittai, eds., *Culture of Chemistry*, Springer, New York, **2015**, pp. 161–167.
17. Ibid.
18. I. Hargittai, *Struct. Chem.* **2010**, *21*, 1–7.
19. J. D. Watson, F. H. C. Crick, *Nature* **1953**, *171*, 737–738.
20. I. Hargittai, *The DNA Doctor: Candid Conversations with James D. Watson*, World Scientific, Singapore, **2007**, p. 173.
21. Quoted from a letter by Watson to Delbrück, December 9, **1951**, in V. K. McElheny, *Watson and DNA: Making a Scientific Revolution*, Wiley, Chichester, UK, **2003**, p.42.
22. I. Hargittai, M. Hargittai, *Candid Science VI: More Conversations with Famous Scientists*, Imperial College Press, London, **2006**, Chapter 1: "Francis Crick," pp. 2–19; actual reference, p. 5.
23. F. Crick, *What Mad Pursuit: A Personal View of Scientific Discovery*, Basic Books, **1988**, pp. 65–66.
24. Conversation with Donald Caspar in Tallahassee, Florida, in 1996; unpublished records; the quoted paragraph appeared Hargittai, Hargittai, *In Our Own Image*, p. 96.
25. I. Hargittai, M. Hargittai, *Candid Science II: Conversations with Famous Biomedical Scientists*, Imperial College Press, London, **2002**, Chapter 20: "Aaron Klug," pp. 306–329; actual quote, p. 312.
26. E. Kőrös, "Oscillations, Waves, and Spirals in Chemical Systems." In I. Hargittai, C. Pickover, eds., *Spiral Symmetry*, World Scientific, Singapore, **1992**, pp. 221–249.
27. I. Hargittai, M. Hargittai, *Candid Science III: More Conversations with Famous Chemists*, Imperial College Press, London, **2003**, Chapter 24: "Johann Deisenhofer," pp. 342–353; actual quote, pp. 349–350.
28. See, e.g., R. J. Gillespie, I. Hargittai, *The VSEPR Model of Molecular Geometry*, Dover, Mineola, NY, **2012** (reprint edition; original edition, 1991).

29. For a more detailed discussion, see, I. Hargittai, D. K. Menyhárd, *J. Mol. Struct.* **2010**, *978*, 136–140.

30. I. Hargittai, *The Structure of Volatile Sulphur Compounds*, Reidel, Dordrecht, **1985**.

31. Initially, R. Stephen Berry did not fully recognize the importance of this phenomenon and described its discovery in a section of a large paper titled "Correlation of rates of intramolecular tunneling processes, with application to some group V compounds": R. S. Berry, *J. Chem. Phys.* **1960**, *32*, 933–938.

32. I. Hargittai, M. Hargittai, *Candid Science: Conversations with Famous Chemists*, Imperial College Press, London, **2000**, Chapter 4: Frank H. Westheimer," pp. 38–53; actual quote, pp. 49–50.

33. See more on this in a conversation with R. Stephen Berry in I. Hargittai, M. Hargittai, *Candid Science: Conversations with Famous Chemists*, Imperial College Press, London, **2000**, Chapter 33: "R. Stephen Berry," pp. 422–435; in particular, p. 426.

34. H.A. Jahn, E. Teller, *Proc. Ry. Soc.* **1937**, *A161*, 220–235.

35. M. Hargittai, B. Réffy, M. Kolonits, C.J. Marsden, J.-L. Heully, *J. Am. Chem. Soc.* **1997**, *119*, 9042–

36. G.A. Olah, *J. Am. Chem. Soc.* **1972**, *94*, 808–820.

37. G.A. Olah, "My search for carbocations and their role in chemistry." *Nobel Lectures in Chemistry 1991–1995*, World Scientific, Singapore, **1994**, pp. 149–176; actual quote and references, p. 173.

38. See, e.g., G. A. Olah, *A Life of Magic Chemistry: Autobiographical Reflections Including Post-Nobel Prize Years and the Methanol Economy*. Second updated edition (with Mathew T). John Wiley & Sons, Hoboken, NJ, **2015**, p. 158.

39. O. Asvany, K.M.T. Yamada, S. Brünken, A. Potapov, S. Schlemmer, *Science*, **2015**, *347*, 1346–1349.

40. T. Oka, *Science*, **2015**, *347*, 1313–1314.

41. Ibid.

42. Oláh Gy (2008) Fél évszázadot felölelő, hagyományos határokon átlépő kutatások. *A kémia újabb eredményei*, Volume 100, pp. 17–59; actual quote, p. 38.

43. Ibid.

44. H.W. Kroto, J.R. Heath, S.C. O'Brien, R.F. Curl, R.E. Smalley, *Nature*, **1985**, *318*, 162–163.

45. W. Krätschmer, L.D. Lamb, K. Fostiropoulos, D.R. Huffman, *Nature*, **1990**, *347*, 354–358.

46. K. W. Adolph, D. L. D. Caspar, C. J. Hollingshed, E. E. Lattman, W. C. Phillips, W. T. Murakami, *Science*, **1979**, *203*, 1117–1120.

47. A.L. Mackay, *Acta Crystallographica* **1962**, *15*, 916–918.

48. K. W. Adolph, D. L. D. Caspar, C. J. Hollingshed, E. E. Lattman, W. C. Phillips, W. T. Murakami, *Science*, **1979**, *203*, 1117–1120.

49. R. Penrose, *Bulletin of the Institute of Mathematics and Its Applications*, **1974**, *10*, 166–271.

50. B. Hargittai, I. Hargittai, *Candid Science V: Conversations with Famous Scientists*, Imperial College Press, London, **2005**, Chapter 3: "Roger Penrose," pp. 36–55; actual quote, p. 41.

51. A.L. Mackay, *Physica* **1982**, *114A*, 609–613.

52. A.L. Mackay, *Soviet Physics Crystallography* **1981**, *26*, 517–522.

53. D. Shechtman, I. Blech, D. Gratias, J.W. Cahn, *Phys. Rev. Lett.* 1984, 53, 1951–1953.

54. D. Levine, P.J. Steinhardt *Phys. Rev. Lett.* **1984**, *53*, 2477–2480.

55. (a) B. Hargittai, I. Hargittai, *Candid Science V: Conversations with Famous Scientists*, Imperial College Press, London, **2005**, see, pp. 36–93; (b) I. Hargittai, M. Hargittai, *In Our Own Image: Personal symmetry in discovery*, Kluwer/Plenum, New York, **2000**, see, pp. 144–179; (c) I. Hargittai, *J. Mol. Struct.* **2010**, *976*, 81–86; (d) I. Hargittai, *Struct. Chem.* **2011**, *22*, 745–748; (e) I. Hargittai,

Isr. J. Chem. **2011**, *51*, 1144–1152; (f) I. Hargittai, *Drive and Curiosity: What fuels the passion for science*, Prometheus, Amherst, NY, **2011**, see, pp. 155–172; (g) B. Hargittai, I. Hargittai, *Struct. Chem.* **2012**, *23*, 301–306.

56. A.L. Mackay, *A Dictionary of Scientific Quotations*, Adam Hilger, Bristol, **1991**, p. 59.

57. M. Hargittai, I. Hargittai, *Candid Science IV*, p. 24.

58. I. Hargittai, B. Hargittai, *Struct. Chem.* **2006**, *17*, 1–2.

59. L. Carroll, *Through the Looking Glass and what Alice found there*; See, e.g., in *The Complete Illustrated Works of Lewis Carroll*, Chancellor Press, London, **1982**, p. 127.

60. M. Hargittai, I. Hargittai, *Mendeleev Communications* **2003**, 91–92.

61. N. Copernicus, *De Revolutionibus Orbium Caelestium*, **1543**, as cited in G. Kepes, *The New Landscape in Art and Science*, Theobald & Co., Chicago, **1956**.

62. I. Hargittai, M. Hargittai, *New York Scientific: A Culture of Inquiry, Knowledge, and Learning*, Oxford University Press, Oxford, UK, **2017**.

István Hargittai is Professor of Chemistry Emeritus (Active) at the Budapest University of Technology and Economics. He is a member of the Hungarian Academy of Sciences and the Academia Europaea (London) and foreign member of the Norwegian Academy of Science and Letters. He is a PhD and DSc and has honorary doctorates from the Lomonosov Moscow State University, the University of North Carolina, and the Russian Academy of Sciences. His main research interest has been in the determination and modeling of molecular structures. He is the Editor-in-Chief of the international Springer journal *Structural Chemistry*. His recent books include *Wisdom of the Martians* (with B. Hargittai), *Budapest Scientific* (with M. Hargittai), *Great Minds* (with B. Hargittai and M. Hargittai), *Buried Glory*, *Drive and Curiosity*, *Judging Edward Teller*, *Martians of Science*, *The Road to Stockholm*, and *Symmetry through the Eyes of a Chemist* (with M. Hargittai). His most recent edited books are, both with Balazs Hargittai, *Culture of Chemistry* and *Science of Crystal Structures*.

Vladimir Prelog in 1995 in his office at the Zurich Federal Institute of Technology (photograph by Istvan Hargittai). Prelog is holding a gold-plated model of the backbone of tRNA (transfer ribonucleic acid) molecule According to private communications from Alex Rich and Jack Dunitz to Istvan Hargittai in 1998, when Prelog was given this model, he exclaimed, "God's signature!"

Prelog Centennial: Vladimir Prelog (1906–1998)[a]

István Hargittai and Balazs Hargittai

Vladimir Prelog in his office at ETH Zurich in 1995 (photo by I. Hargittai).

Prelog's familiar *ex libris* by Hans Erni with an inscription to one of the authors.

Vladimir Prelog was "the founder of modern stereochemistry, it was he who initiated and intellectually invigorated the current renaissance of this field ..." So wrote Kurt Mislow [1] in 1998, the first recipient of the Vladimir Prelog Medal and the first Prelog lecturer at the Swiss Federal Institute of Technology (ETH Zürich) in 1986. Among his many achievements we mention here only a few following Mislow

[1]. Prelog was the first who employed conformational analysis to ratwionalize the physical and chemical properties of medium-size-ring compounds. He made generalized observations concerning the steric effects in Grignard reactions. Together with Robert Cahn and Christopher Ingold, he proposed a terminology to specify the configuration of stereoisomers (the CIP system), and, together with William Klyne, a terminology to describe steric relations across single bonds. He did fundamental work on novel types of stereoisomers.

Prelog's innovations in nomenclature became popular and he often worked them out in unison with other luminaries of organic chemistry. At the end of 1953, beginning of 1954, an identical Note appeared in *Nature* and in *Science* concerning the nomenclature of bonds in cyclohexane [2]. Of the four authors of the Note—Barton, Hassel, Pitzer, and Prelog—three would eventually win the Nobel Prize in Chemistry. Prelog's Nobel Prize came in 1975 and it was

[a]Originally published in *Structural Chemistry* 2006, 17, 1–2.

I. Hargittai (✉)
Department of Inorganic and Analytical Chemistry, Budapest University of Technology and Economics, Budapest, Hungary
e-mail: istvan.hargittai@gmail.com

B. Hargittai
Department of Chemistry, Saint Francis University, Loretto, PA, USA

shared with John Cornforth. Prelog's citation said, "for his research into the stereochemistry of organic molecules and reactions."

According to Cornforth [3], Prelog "stood out from among his contemporaries . . . because he asked better questions and analyzed problems with greater clarity." Cornforth also noted that "Stereochemistry was the love of his life: he pursued it in many guises, devised rules for it, illuminated every aspect of it that he touched."

As Mislow noted [1], natural products chemistry was Prelog's first love and he remained true to it to the end of his days. However, his interest did not include proteins and nucleic acids. This is why one of his most brilliant disciples, Albert Eschenmoser [4] gently provoked him by saying, "Vlado, every year during which we did not work on DNA was a wasted year." Prelog did not rush to give an answer, but when he did, he put it in writing because he realized its importance for science history:

Zurich, October 3, 1995

Dear Albert

For some time you have prodded me to tell you, why the great Leopold [Ruzicka] and I did not recognize, in a timely fashion, that the nucleic acids are the most important natural products, and why did we waste our time on such worthless substances as the polyterpenes, steroids, alkaloids, etc.

My light-headed answer was that we considered the nucleic acids as dirty mixtures that we could not and should not investigate with our techniques. Further developments were, at least in part, to justify us.

As a matter of fact, for personal and pragmatic reasons, we never considered working on nucleic acids.

Yours

Vlado

In 1995, one of us recorded a conversation with Vladimir Prelog [5], in which he talked about his family background, youth, studies, about his interests, and sprinkled his narrative with anecdotes about which he was famous. His room was full of memorabilia of stereochemistry and chirality in particular. The most conspicuous was a series of Hans Erni's drawings, one of which became Prelog's *ex libris*. A peculiar feature of this drawing is that the two hands of the youth appear as if they were turned around, inverted. In a version of Erni's drawings, however, the two hands appeared to be non-inverted, being parallel. In the familiar version, the two hands can be imagined as a result of the two arms being crossed.

References

1. Mislow, K. Chem. Intelligencer 1998, 4(3), 51–54.
2. Barton, D. H. R.; Hassel, O.; Pitzer, K.S.; Prelog, V. Nature 1953, 172, 1096–1097; Science 1954, 119, 49.
3. Cornforth, J. Chem. Intelligencer 1998, 4(3), 50.
4. Hargittai, I. Albert Eschenmoser. In I. Hargittai, Candid Science III: More conversations with famous chemists. Imperial College Press: London, 2003, pp. 96–107.
5. Hargittai, I. Vladimir Prelog. In I. Hargittai, Candid Science: Conversations with famous chemists. Imperial College Press: London, 2000, pp. 138–147.

Pierre Curie and his hands by the graphic artist Istvan Orosz (Budapest). Orosz stressed handedness, that is, chirality to symbolize dissymmetry. Orosz prepared this drawing, along with some others, at my Parents' request for their book, Istvan Hargittai and Magdolna Hargittai, *In Our Own Image: Personal Symmetry in Discovery* (New York: Kluwer/Plenum, 2000).

Eternal Dissymmetry[a]

Magdolna Hargittai and István Hargittai

The teachings of Louis Pasteur about chirality continue to instruct and inspire [1].

Louis Pasteur was the first to suggest that molecules can be chiral. In his famous experiment 155 years ago, in 1848, he recrystallised a salt of tartaric acid and obtained two kinds of small crystals that were mirror images of each other. Pasteur prepared beautiful cardboard models, which have been preserved and are exhibited in the Pasteur Museum at the Pasteur Institute in Paris. Pasteur may have been motivated to make these models of large scale because Jean Baptiste Biot, the discoverer of optical activity, had very poor vision by the time of Pasteur's discovery [2] (Figs. 1 and 2).

As a moving moment in the history of science, it was left to Biot, rather old by then, to present Pasteur's findings to the French Academy of Sciences. The ever careful Biot first had Pasteur demonstrate his experiment to him in person. In Pasteur's description: [3]

> When [the solution] had furnished about 30 to 40 grams of crystals, he asked me to call at the Collége de France in order to collect them and isolate before him, by recognition of their crystallographic character, the right and the left crystals, requesting me to state once more whether I really affirmed that the crystals, which I should place at his right, would deviate [polarized light] to the right, and the others to the left. This done, he told me that he would undertake the rest. He prepared the solutions with carefully measured quantities, and when ready to examine them in the polarizing apparatus, he once more invited me to come into his room. He first placed in the apparatus

Fig. 1 Louis Pasteur's bust in front of the Pasteur Institute, Paris (photograph by the authors, © Hargittai Photo).

the more interesting solution, that which ought to deviate to the left. Without even making a measurement, he saw by the appearance of the tints of the two images, ordinary and extraordinary, in the analyzer, that there was a strong deviation to the left. Then, very visibly affected, the illustrious old man took me by the arm and said, 'My dear child, I have loved science so much throughout my life that this makes my heart throb.

According to the Nobel laureate George Wald [4], 'No other chemical characteristic is as distinctive of living organisms as is optical activity.' The roots of Pasteur's discovery reach farther back than Biot's work. According to

[a]Originally published in *Mendeleev Communications* 2003, 13:91–92. Reproduced with permission from Mendeleev Communications and Elsevier.

M. Hargittai (✉)
Structural Chemistry Research Group, Hungarian Academy of Sciences, Eötvös University, Budapest, Hungary
e-mail: hargittaim@ludens.elte.hu

I. Hargittai
Department of Inorganic and Analytical Chemistry, Budapest University of Technology and Economics, Budapest, Hungary

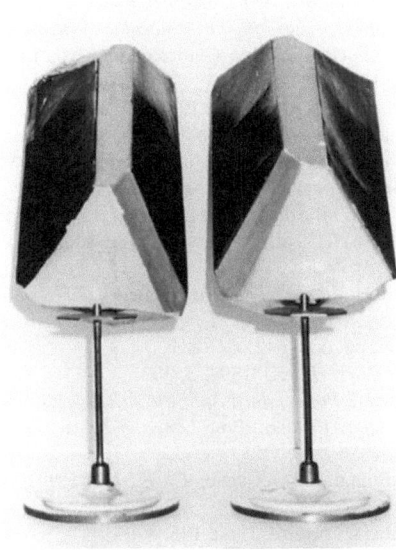

Fig. 2 Pasteur's models of enantiomeric crystals in the Museum of the Pasteur Institute, Paris (photograph by the authors, © Hargittai Photo).

J. D. Bernal [5], Pasteur's discovery arose at a meeting place of hitherto distinct disciplines. They were crystallography, physics and chemistry. He also showed that the fruits of the discovery benefited new branches in these sciences.

Of course, not only material objects can have chirality or handedness. Bach's *The Art of the Fugue* is a beautiful example. Handedness is also an area of symmetry that is charged with philosophical implications. Immanuel Kant [6] wrote about the puzzle of the isometric left and right hands that cannot be made to coincide in space and called the nonsuperposable mirror images 'incongruente Gegenstücke' (incongruent counterparts). Then, of course, Lord Kelvin [7] gave a definition for chirality that has stood the test of time, 'I call any geometrical figure, or group of points, chiral, and say that it has chirality if its image in a plane of mirror, ideally realised, cannot be brought to coincide with itself.'

The early success of the chirality concept culminated in Pierre Curie's statement in an 1894 paper [8], 'c'est la dissymétrie qui crée le phénomène' (dissymmetry creates the phenomenon). This most fundamental symmetry principle means that a phenomenon is expected to exist and can be observed only if certain elements are absent from the system. The forerunner of this principle was Franz Neumann's statement [9] in 1833 that 'the physical properties of crystals always conform to the symmetry of the crystal.'

Pierre Curie did not write much about symmetry and he did not live very long, but Marie Curie and the Russian crystallographer Aleksei Shubnikov did much to convey Pierre Curie's teachings on symmetry to a broader circle of scientists and thereby to help preserving the life of Louis Pasteur's teachings in this area of science.

Returning to Pasteur's story, it is important to stress that his discovery of molecular chirality did not happen out of nowhere. Pasteur himself stated that 'Dans les champs de l'observation, l'hasard ne favorise que les esprits préparés' (In the field of observation, chance only favors those minds that have been prepared). In fact, Pasteur's preparation for his discovery of molecular chirality was so perfect that to the famous biologist, Dubos [10], 'it appeared as if fate had brought together many influences to prepare Pasteur for his scientific adventure.' He was a well-trained chemist with a definite idea about the importance of molecular structure, and he was also a crystallographer, who viewed the crystals as carriers of chemical information. The chemical and physical methods appeared in unison in Pasteur's mind.

It was not only that various areas of science came together in Pasteur's discovery, the discovery gave rise to new branches in science as well. The emergence of stereochemistry was one of the consequences of Pasteur's discovery although it came a quarter of a century later. The discovery also brought about the realization that, in living organisms, biologically important substances occur in one of the two possible versions. This also led to the great question, 'How did it all start? What was the way one of the two was chosen?' This question deeply bothered Pasteur and a century later Vladimir Prelog called this a question of 'molecular theology' in his Nobel lecture. Pasteur is buried in the chapel of the Pasteur Institute and the key phrases of his scientific activities inscribed on the chapel walls include *dissymétrie moléculaire*.

Pasteur considered the asymmetric nature of living matter as a fundamental characteristic. Experimentation with molecular asymmetry was always on his mind. Even as late as 1886, he discussed the two asparagines [10], one of which is sweet while the other insipid. He suggested that the difference might be due to the difference in their actions on the two antipodes of the asymmetric constituents of the gustatory nerve. There are many conspicuous examples of different actions by enantiomeric isomers of various drugs. Suffice it to mention thalidomide, which was known as Contergan in Europe with which many tragedies were connected before it was withdrawn from the market. Since 1992, the U.S. FDA and the European Committee for Proprietary Medicinal Products have required manufacturers to research and characterise each enantiomorf of a potential drug [11].

In 1960, there was a short note in *Nature* [12], in which the British physiologist and geneticist John Haldane returned to Pasteur in the wake of the discovery of the violation of parity. The title was 'Pasteur and Cosmic Asymmetry,' and Haldane showed that the roots of Lee and Yang's [13] discovery were in Pasteur's notion 'The universe is dissymmetric.' [14] Parity was, of course, only the first example found to violate symmetry principles in the weak interaction. Next was the so-called CP symmetry violation (C stands for charge conjugation, meaning the change of a particle into an antiparticle, and P stands for parity. The combined CP symmetry means the change from a left-handed particle to a right-handed antiparticle, for example). The discovery of CP violation had conceptually profound implications concerning our ideas about the origin of the universe [15]. As T. D. Lee [16] has noted recently, 'The origin of these symmetry violations is still a mystery.' However, if we consider the combined CPT symmetry (here T stands for time reversal, which is a mirror symmetry with respect to time just as parity is a mirror symmetry with respect to space coordinates) that is not broken; CPT is a very solid symmetry.

The legacy of Louis Pasteur is rich in scientific achievements that have greatly contributed to the improvement of the quality of life. It is also rich in questions that are the best stimulants for scientific inquiry and they continue to help us in charting our labors in uncovering nature's secrets.

References

1. In preparing this communication we much relied on our book, I. Hargittai and M. Hargittai, *In Our Own Image: Personal Symmetry in Discovery*, Kluwer/Plenum, New York, 2000.
2. J. Applequist, *Am. Sci.*, 1987, Jan.-Feb., 59.
3. L. Pasteur, *Researches on the Molecular Asymmetry of Natural Organic Products*, Alembic Club Reprints No. 14. W. F. Clay, Edinburgh, 1897, p. 21.
4. G. Wald, *Ann. N. Y. Acad. Sci.*, 1957, **69**, 352.
5. J. D. Bernal, in *Science and Industry in the Nineteenth Century*, Routledge & Kegan Paul, London, 1953, pp. 181–219.
6. I. Kant, *Von dem ersten Grunde des Unterschiedes der Gegenden im Raume* (1768), in *Kant's gesammelte Schriften*, Königl. Preuss. Akad. Wissensch., vol. 2, Verlag Georg Reimer, Berlin, 1905, pp. 375–383.
7. Lord Kelvin, *Baltimore Lectures on Molecular Dynamics and the Wave Theory of Light*, C. J. Clay and Sons, London, 1904, Appendix H, pp. 602–642.
8. P. Curie, *J. Phys. (Paris)*, 1894, **3**, 393.
9. F. E. Neumann, *Poggendorff Ann. Phys.*, 1833, **27**, 240.
10. R. Dubos, *Louis Pasteur: Free Lance of Science*, Da Capo Press, New York, 1986.
11. A. Richards and R. McCague, *Chem. Ind.*, 1997, June 2, 422. 12 J. B. S. 12. 7.
12. J. B. S. Haldane, *Nature*, 1960, **185**, 87.
13. T. D. Lee and C. N. Yang, Phys. Rev., 1956, 104, 254
14. L. Pasteur, *C. R. Acad. Sci.,* Paris, June 1, 1874.
15. M. Hargittai, 'Val L. Fitch', in: M. Hargittai and I. Hargittai, *Candid Science IV: Conversations with Famous Physicists*, Imperial College Press, London, in press.
16. T. D. Lee, *Nature*, 1997, **386**, 334.

Mountain goats in the Budapest Zoo (photograph by Istvan Hargittai) displaying gradual size and age changes. They can be considered to be a segment of an "infinite" succession of *similarity symmetry* (see in Fig. 16a in the following article).

The Universality of the Symmetry Concept[a]

István Hargittai and Magdolna Hargittai

Abstract

The notion of symmetry brings together beauty and usefulness, science and economy, mathematics and human relations. This presentation demonstrates the breadth and versatility of the symmetry concept. There are no symmetries specific to various disciplines, yet there are differences in emphasis in applications of the concept. The sciences, humanities and arts have gradually drifted apart; symmetry can provide a connecting link among them. The symmetry concept may be broadened to include harmony and proportion, constituents of symmetry often present in architectural composition. The symmetries considered here are point group, chiral, space group, and translational. While mathematical symmetry is exact and rigorous, the symmetry we encounter in everyday life is much more relaxed. The broad interpretation of the symmetry concept, coming close to blending fact and fantasy, may help scientists recognize trends, changes, and patterns.

Introduction

The notion of symmetry brings together beauty and usefulness, science and economy, mathematics and music, architecture and human relations, and much more, as has been shown recently with many examples [1–4]. There is a lot of symmetry, for example, in Béla Bartók's music. It is not known, however, whether he consciously applied symmetry or was simply led intuitively to the golden ratio so often present in his music. Bartók himself always refused to discuss the technicalities of his composing and stated merely "We create after Nature." Another unanswerable question is how these symmetries contribute to the appeal of Bartók's music, and how much of this appeal originates from our innate sensitivity to symmetry. This question might be equally asked of symmetries in architectural composition.

The present chapter takes a broad view of the symmetry concept. It demonstrates its breadth and versatility. There are no distinctly different specific symmetries in various disciplines, yet there are discernible differences in emphasis of the application of this concept in different fields. This emphasis changes with time as well. For example, there is a marked emphasis on the presence of symmetry in chemistry, in contrast to physics where the importance of broken symmetries has been stressed during the past decades. Generally though the symmetry concept unites rather than divides the different branches of science, and even helps bridge the gap between what C.P. Snow called "two cultures." Sciences, the humanities, and the arts have all drifted apart over the years and symmetry can provide a connecting link among them. Its benefits are available to us if we free ourselves from the confinements of *geometrical symmetry*.

Everything is rigorous in geometrical symmetry. According to one definition, "symmetry is the property of geometrical figures to repeat their parts" [5]. Another definition says that "a figure is symmetrical if there is a congruent transformation which leaves it unchanged as a whole, merely permuting its component elements" [6]. In the geometrical sense, symmetry is either present or it is absent. Any question regarding symmetry has a restricted *yes/no* alternative. For the real, material world, however, degrees of symmetry and even gradual symmetry is feasible and applicable. Beyond geometrical definitions there is another, broader meaning to symmetry—one that relates to *harmony* and *proportion*, and ultimately to *beauty*. This aspect involves feeling and

[a]K. Williams and M.J. Ostwald (eds.), Architecture and Mathematics from Antiquity to the Future, DOI 10.1007/978-3-319-00137-1_40, © Springer International Publishing Switzerland 2015

First published as: István Hargittai and Magdolna Hargittai, "The Universality of the Symmetry Concept", pp. 81–95 in Nexus I: *Architecture and Mathematics*, ed. Kim Williams, Fucecchio (Florence): Edizioni dell'Erba, 1996.

I. Hargittai (✉) · M. Hargittai
Materials Structure and Modeling Research Group, Budapest University of Technology and Economics, Budapest, Hungary
e-mail: istvan.hargittai@gmail.com; hargittaim@mail.bme.hu

Springer

subjective judgment and, as a result, is especially difficult to describe in technical terms.

Simple considerations are indispensable in classifying different kinds of symmetry. There are two large classes of symmetry, *point groups* and *space groups*. For point group symmetries there is at least one special point in the object or pattern that differs from all the others. In contrast to this, in space groups, there is no such special point. There are also some terms that are useful in the description of different types of symmetry. Thus, the action that characterizes a particular type of symmetry is called a *symmetry operation*. The tool whereby the operation is performed is called a *symmetry element*.

Point Group Symmetry

The simplest kind of point-group symmetry is *bilateral symmetry*. Bilateral symmetry is present when two halves of the whole are each other's mirror images (Fig. 1). This is the most common symmetry and the every-day usage of the term "symmetry" refers to this meaning. The symmetry element is a *mirror plane*, also called a *symmetry plane* or a *reflection plane*. The symmetry operation is *reflection*. Applying a mirror plane to either of the two halves of an object with bilateral symmetry recreates the whole object. Bilateral symmetry is probably the most common symmetry in architecture as well, from simple buildings to larger assemblies (Fig. 2a, b).

Another kind of point-group symmetry is *rotational symmetry* (Fig. 3). It is present when, by rotating an object around its axis, it appears in the same position two or more times during a full revolution. *Rotation* is the symmetry operation

Fig. 2 (a) The whole assembly of the Blue Mosque in Istanbul, Turkey, has bilateral symmetry. (b) The design of St. Peter's Square in Vatican City also shows bilateral symmetry. Photo: authors.

Fig. 1 The orchid has bilateral symmetry. Photo: authors.

Fig. 3 This hubcap has sevenfold rotational symmetry. Photo: authors.

and the *axis of rotation* is the symmetry element. Rotational symmetry may be twofold, threefold, fourfold, etc. It is common that reflection and rotation appear together. The presence of some symmetry elements may generate others

Fig. 4 The Eiffel Tower from below. It shows both reflections and rotational symmetry. Photo: authors.

Fig. 5 The cupola of the Hungarian Parliament with both reflectional and rotational symmetry. Photo: authors.

and vice versa. If we look at the Eiffel tower from below (Fig. 4) we have twice two orthogonal reflection planes which generate a fourfold rotation. The cupolas of many state capitols and other important buildings have reflectional and rotational symmetry together (Fig. 5).

The regular polygons, so basic in architectural design, also have both rotational and reflectional symmetry. Best seen when viewed from above, many buildings have outlines of a regular polygon (Fig. 6). The regular polyhedra, also called Platonic solids, all have equal regular polygons as their faces. As H.S.M. Coxeter, professor of mathematics at the University of Toronto, remarked, "the chief reason for studying regular polyhedra is still the same as in the times of the Pythagoreans." Namely, that their symmetrical shapes appeal to one's artistic sense. There are other highly symmetrical polyhedra, called Archimedian polyhedra, whose faces are also regular polygons but not identical ones. Buckminster Fuller's geodesic dome is composed of lightweight bars forming regular polygons. His geodesic dome at the Montreal expo (Fig. 7) inspired some chemists who saw that the structure of a newly discovered substance may be the truncated icosahedron. This molecule, C_{60}, called buckminsterfullerene (Fig. 8) is characterized, among others, by six axes of fivefold rotation [4: 100–101]. Experimentally discovered in 1985, its great relative stability was predicted already in 1970, based solely on symmetry considerations.

Chirality

A special kind of symmetry relationship is when two objects are related by mirror reflection and the two objects cannot be superposed. Our hands are an excellent example, and the term *chiral* derives from the Greek word for hand. Chiral objects have senses and following the hand analogy they are left-handed (L) and right-handed (D). The simplest chiral molecule is a methane derivative in which three of the four hydrogens are replaced by three different atoms, such as, for example, fluorine (F), chlorine (Cl), and bromine (Br). There may then be a left-handed C(HFClBr) and a right-handed C(HFClBr) molecule which will be each other's mirror images but won't be superposable (Fig. 9). A chiral object and its mirror image are called each other's *enantiomorphs*.

The two chiral molecules look the same in every detail; only their senses are different. The distinctions between the twins of a chiral pair have literally vital significance. Only l-amino acids are present in natural proteins and only d-nucleotides are present in natural nucleic acids. This happens in spite of the fact that the energy of both enantiomers is equal and their formation has equal probability in an achiral environment. However, only one of the two occurs in nature, and the particular enantiomers involved in

Fig. 6 The outline of the Pentagon in Washington, D.C. with its regular *pentagonal shape*. Photo: authors.

Fig. 7 Buckminster Fuller's Geodesic Dome at the Montreal Expo. Photo: authors.

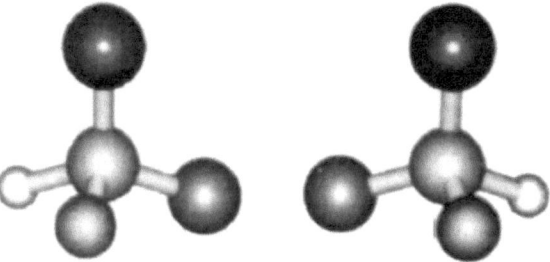

Fig. 9 A chiral pair of molecules. Image: authors.

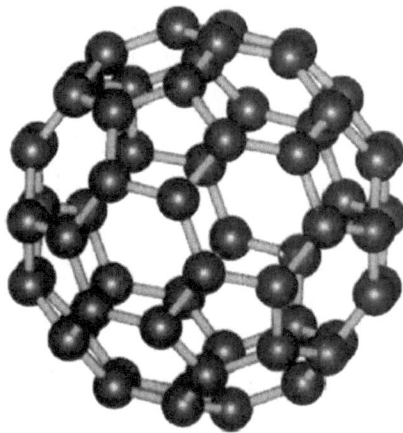

Fig. 8 C_{60}, the buckminsterfullerene molecule. Image: authors.

life processes are the same in humans, animals, plants, and microorganisms. The origin of this phenomenon is a great puzzle.

Once a chiral molecule happens to be in a chiral environment, the two chiral isomers will be behaving differently. This different behaviour is manifested sometimes in very dramatic ways. In some cases one isomer is sweet, the other is bitter. In some other cases the drug molecule has an "evil twin." A tragic example was the thalidomide case in the 1950s in Europe, in which the right-handed molecule cured morning sickness and the left-handed one caused birth defects. Other examples include one enantiomer of ethambutol fighting tuberculosis with its evil twin causing blindness, and one enantiomer of naproxen reducing arthritic inflammation with its evil twin poisoning the liver. Ibuprofen is a lucky case in which the twin of the chiral form that provides the curing is converted to the beneficial version by the body.

Even when the twin is harmless, it represents waste and a potential pollutant. Thus, a lot of efforts are directed toward producing enantiomerically pure drugs and pesticides. One of the fascinating possibilities is to produce sweets from chiral sugars of the enantiomer that would not be capable of

Fig. 10 Chiral rosettes on a building in Bern, Switzerland. Photo: authors.

contributing to obesity yet would retain the taste of the other enantiomer.

Chiral symmetry is also frequently found in architectural design either in two- or in three dimensions, as illustrated by Fig. 10.

Space Group Symmetry

A different kind of symmetry can be created by simple *repetition* of a basic motif leading us to *space-group symmetries*. The most economical growth and expansion patterns are described by space-groups symmetries. There are three basic cases of space groups, depending on whether the basic motif extends periodically in one direction only, or in two, or finally, in three. These three cases are described by the so-called *one-dimensional, two-dimensional,* and *three-dimensional* space groups.

Border decorations are examples of one-dimensional space groups. In border decorations a pattern can be generated simply by repeating a motif at equal intervals. This is *translational* symmetry. The symmetry element is *constant translation;* the operation is the *translation* itself. The resulting pattern shows periodicity in one direction. Repetition can be achieved by a simple shift in one direction as can be seen very often in the rows of columns of grandiose buildings (Fig. 11) or in the ancient aqueducts of the Romans. Fences are typical examples of one-dimensional space groups (Fig. 12), the ease and economy of using the same elements repeatedly makes this obvious. Repetition can also be achieved in other ways, such as by reflection, rotation (Fig. 13), or *glide- reflection.* Glide-reflection is another new element that does not occur in point-group symmetries. It means the consecutive application of translation and horizontal reflection. When we walk in wet sand along a straight line we leave behind a pattern of footprints whose symmetry is described by glide-reflection. There is a total of seven possibilities for generating one-dimensional space-group symmetries.

Helices and spirals have also one-dimensional space-group symmetries although their bodies may extend to three

Fig. 11 Colonnade on St. Peter's square in Vatican City. Photo: authors.

Fig. 12 Repeating pattern of a fence in the Topkapi Palace in Istanbul, Turkey. Photo: authors.

dimensions [7]. *Helical symmetry* is created by a constant amount of translation accompanied by a constant amount of rotation. In *spiral symmetry*, again, translation is accompanied by rotation but the amount of translation and

Fig. 13 Another illustration for one-dimensional space groups: the units turn 90° at every translation in this chain. Photo: authors.

rotation changes gradually and regularly. An extended spiral staircase has helical symmetry. Well-ordered biological macromolecules also have helical symmetry. Helices are always three-dimensional whereas there are spirals that extend in two dimensions only. Occurrences of spirals may be as diverse as chemical waves and galaxies and snails. Spirals and helices have also been used in various ways in architecture, from ancient times to the present, as Trajan's column in the Forum Romanum (Fig. 14) and the spiral ramp of Frank Lloyd Wright's Guggenheim Museum in New York indicate.

Another beautiful example of spiral symmetry is the scattered leaf arrangement around the stems of plants, called *phyllotaxis*. Numbers of the Fibonacci series (1, 1, 2, 3, 5, 8, 13, 21, . . .—each new element is the sum of the two previous elements) characterize the ratios defining the occurrence of every consecutive new leaf in scattered leaf arrangements. Thus, for example, there is a new leaf at each 3/8 parts of the circumference of the stem as we move along the stem in one of the characteristic cases. The pineapple (Fig. 15) displays a pattern of spirals that can be thought of as if it were a result of compressed phyllotaxis. Such ratios when involving very large numbers approximate an important irrational number, 0.381966..., expressing the so-called *golden ratio*. The golden ratio is created by the golden section in which a given length is divided such that the ratio of the longer part to the whole is the same as the ratio of the shorter part to the longer part. If the whole is 1.00, the lengths of the longer and shorter parts will be 0.618 and 0.382, respectively. This may be the single most important proportion in architecture and in artistic expression. Its relationship to phyllotaxis may have inspired Leonardo da Vinci's description of the scattered leaf arrangement as "more beautiful, more simple, or more direct" than anything humans could devise (Leonardo da [8]).

Spiral symmetry can also be considered as belonging to the broad concept of *similarity symmetry*. Here pattern generation always involves an increment of a characteristic property (Fig. 16).

Fig. 14 Spiral symmetry of Trajan's column in the Forum Romanum in ancient Rome. Photo: authors.

Fig. 17 Two-dimensional space group: decoration from the Alhambra Granada, Spain. Photo: authors.

Fig. 15 The pineapple displays a pattern of spirals that can be thought of as if it were a result of compressed phyllotaxis. Photo: authors.

With two-dimensional space-groups, there is a total of *17* ways to generate different patterns. It is a special case when the planar network covers the plane without gaps and

overlaps. Of the regular polygons, only the equilateral triangle, the square, and the regular hexagon are capable of covering the plane without gaps and overlaps. For arbitrary shapes though, there are infinite possibilities. M.C. Escher's periodic drawings and the wall decorations in the Alhambra of Granada, Spain (Fig. 17) are famous examples. The façades of buildings, especially those of modern skyscrapers often display symmetries in two dimensions (Fig. 18).

Space utilization by periodic arrangements seems to be the underlying principle of the occurrence of three-dimensional space-group symmetries. This is a common arrangement of the building elements in *crystals*. The packing of spheres was first considered as the key to the internal structure of crystals by Johannes Kepler. As he was looking at the exquisitely beautiful hexagonal snowflakes, he made drawings of sphere packing, similar to a pyramid of canon balls (Fig. 19).

There are restrictions for the regular and periodic structures, such as the nonavailability of fivefold symmetry in generating them. This can be understood easily when we find it impossible to cover the plane without gaps or overlaps with equal-size regular pentagons.

Fig. 16 (a) Similarity symmetry, the increments being the change in size or the change in age. (b) An architectural example of similarity symmetry where the increment is the change in size of the units of the church-tower in London, England. Photo: authors.

Fig. 18 The façades of modern skyscrapers are typical examples of repetitions in two dimensions. Photo: authors.

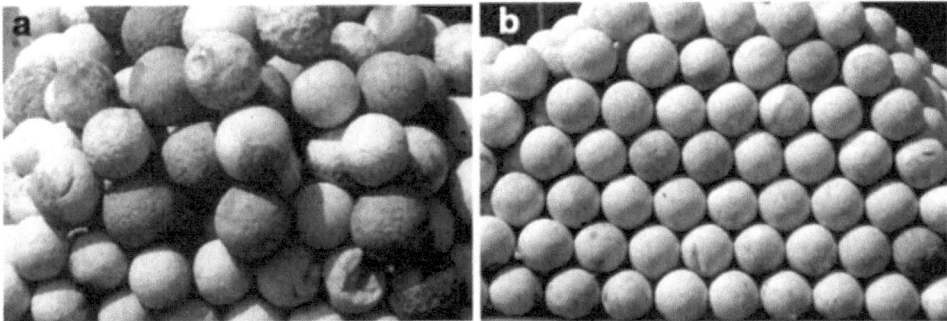

Fig. 19 Random arrangement of canon balls provides much poorer space utilization than their regular arrangement. Photo: authors.

Crystals are advantageous for the determination of the structure of molecules. The great success of X-ray crystallography may have diverted attention from structures of lesser symmetry though of not necessarily lesser importance. The discovery of quasiperiodic crystals (in short, quasicrystals [9]) by the Israeli scientist Dan Shechtman in 1982 has by now persuaded many scientists that their view of crystals is unnecessary narrow. David Mermin compared abandoning the traditional classification scheme of crystallography, based on periodicity, to abandoning the Ptolemaic view in astronomy, and likened changing it to a new foundation to astronomy's adopting the Copernican view [10].

Recently, even such descriptive fields of biology as zoology have displayed a growing activity in symmetry matters. Not surprisingly, the role of external symmetry is being recognized as decisive in mate selection. Empirical evidence supports the notion relating "animal beauty" to the symmetry of outlook. The degree of left-and-right correspondence of the wings seems to correlate with hormone and pheromone production [11: C1].

In view of the fundamental importance of the symmetry concept, it is surprising that even very basic discoveries about it were left to be made in this century. When P.A.M. Dirac was asked about Einstein's most important contributions to physics, he singled out Einstein's "introduction of the concept that space and time are symmetrical" [12: 11]. An important step was Emmy Noether's recognition that symmetry and conservation are connected. Indeed, the idea that the great conservation laws of physics, like the conservation of energy and momentum, are related to symmetry opened up a wholly new way of thinking for scientists. Realizing that Nature included continuous symmetry in her design physicists started to look for new connections.

It was Dirac who had the prescience to write already in 1949, that "I do not believe that there is any need for physical laws to be invariant under reflections" [13]. Yet, even most physicists were surprised by the discovery of the nonconservation of parity in 1957 that brought the Nobel prize in physics to T.D. Lee and C.N. Yang. C.P. Snow called this discovery one of the most astonishing in the whole history of science. Since then broken symmetries have been receiving increasing attention.

There seems to be a difference in approach and emphasis between physicists and chemists in viewing symmetry. It may even be related to the ancient Greek philosophers, stressing the importance of continuum by Aristotle, and of the discreet, by Lucretius and Democritos. From the point of view of continuum, even the ideal crystal may be discussed in terms of broken symmetries. On the other hand, the chemist's approach is succinctly symbolized by Democritos' statement: "Nothing exists except atoms and empty space; everything else is opinion."

Of course, the way symmetry is looked at can vary a great deal. While mathematical symmetry is exact and rigorous, the symmetry we encounter in everyday life is much more relaxed. The vague and fuzzy interpretation of the symmetry concept may also aid scientists to recognize trends, characteristic changes, and patterns. This is getting close to blending fact and fantasy. As Arthur Koestler expressed it, "artists treat facts as stimuli for the imagination, while scientists use their imagination to coordinate facts" [14].

References

1. HARGITTAI, I. ed. 1986. *Symmetry: Unifying Human Understanding 1 and 2*. New York and Oxford: Pergamon Press.
2. HARGITTAI, I. ed. 1989. *Symmetry: Unifying Human Understanding 1 and 2*. New York and Oxford: Pergamon Press.
3. HARGITTAI, I. and M. Hargittai. 1995. *Symmetry through the Eyes of a Chemist*. 2nd edn. New York: Plenum Press.
4. HARGITTAI, I. and M. Hargittai. 1994. *Symmetry: A Unifying Concept*. Bolinas, CA: Shelter Publications. Rpt. New York, Random House, 1996.
5. SHUBNIKOV, A.V. 1951. *Simmetriya I Antisimmetriya Konechnykh Figure*. Izd. Akad. Nauk SSSR: Moscow.
6. COXETER, H.S.M. 1973. *Regular Polytopes*. 3rd edn. Dover Publications: New York.
7. HARGITTAI, I. and C. A. Pickover, eds. 1992. *Spiral Symmetry*. Singapore: World Scientific.
8. LEONARDO DA VINCI. 1939. *The Notebooks. 1508–1518*. Jean Paul Richter trans. Oxford: Oxford University Press.
9. HARGITTAI, I. ed. 1990. *Quasicrystals, Networks, and Molecules of Fivefold Symmetry*. New York: VCH.
10. MERMIN, N.D. 1992. Copernican Crystallography. *Phys. Rev. Lett* **68**, 1172 (1992).
11. ANGIER, N. 1994. *The New York Times*, February 8, 1994.
12. YANG, C.N. 1991. *The Oscar Klein Memorial Lectures*, Vol 1, G. Ekspong ed. World Scientific: Singapore.
13. DIRAC, P.A.M. 1949. Forms of Relativistic Dynamics. *Rev. Mod. Phys.* **21**, 392.
14. KOESTLER, A. 1949. *Insight and Outlook*. Macmillan: London.

István Hargittai, Ph.D., D.Sc., is Professor Emeritus (active) of the Budapest University of Technology and Economics. He is a member of the Hungarian Academy of Sciences and the Academia Europaea (London) and foreign member of the Norwegian Academy of Science and Letters. He is Dr.h.c. of Moscow State University, the University of North Carolina, and the Russian Academy of Sciences. His recent books include: *Buried Glory: Portraits of Soviet Scientists—Makers of a Superpower* (Oxford University Press, 2013); *Drive and Curiosity: What Fuels the Passion for Science* (Prometheus, 2011); *Judging Edward Teller: A Closer Look at One of the Most Influential Scientists of the Twentieth Century* (Prometheus 2010); *Martians of Science: Five Physicists Who Changed the Twentieth Century* (Oxford University, Press 2006; 2008); *The Road to Stockholm: Nobel Prizes, Science, and Scientists* (Oxford University Press, 2002; 2003).

Magdolna Hargittai, Ph.D., D.Sc., Research Professor at the Budapest University of Technology and Economics. She is a member of the Hungarian Academy of Sciences and the Academia Europaea (London). She is Dr.h.c. of the University of North Carolina. Her recent books (with István Hargittai) include: *Symmetry through the Eyes of a Chemist*, 3rd ed. (Springer, 2009, 2010); *Visual Symmetry* (World Scientific, 2009); *Candid Science I-VI* (Imperial College Press, 2000–2006); *In Our Own Image: Personal Symmetry in Discovery* (Kluwer/Plenum, 2000).

"Lonely Discoverer" (courtesy of Anatoly Fomenko). The mathematician Anatoly Fomenko graduated from Moscow University and is head of department of higher geometry and topology at his Alma Mater. Father noticed his striking graphics already in the 1980s in Moscow where street vendors were selling them. The images were by an ostensibly unknown author as abstract art was very much frowned upon in the Soviet Union. Fomenko was not aware of these sales. Had he been accused of practicing abstract art, he could always claim that he did not engage in creating abstract art; he was merely illustrating rigorously defined mathematical ideas to facilitate their understanding. Thus, for example, he produced chapters for Istvan's multi-disciplinary volumes on symmetry and illustrated them copiously with his intriguing drawings and provided a dry description for each in the most mathematical language. Fomenko's stated goal was to discover hidden symmetries in our surroundings that would lead to the understanding of hidden laws governing events. A closer look could always discover some human beings in miniature in his drawings. All his drawings were made by hand without computer assistance. Father gave the title "Lonely Discoverer" to the image presented here, with Fomenko's consent, to be sure. Originally, this was yet another illustration of a mathematical concept.

Introduction

I have always shared Father's interest in history, all kinds of history, but especially the histories of scientific discoveries. Nonetheless, Father has always protested when somebody labelled him as a science historian. It is a profession to which he had not been trained—he argues. Along with the stories of discoveries he has also been interested in the stories of the discoverers. But, then, as Ralph Waldo Emerson said, there is no history, only biography. Especially biographies that are about scientists who had overcome odds and adversities to reach the summit have fascinated Father. The real talent—he says—is not that shines under ideal conditions but that shines under challenging conditions. Father had initiated the broad interviews program of our parents which eventually I also joined to a modest extent. Bona fide interviews are not among the selection I am presenting in this Section, but his writings about scientists and discoveries have been influenced by the hundreds of interviews he had recorded and published. Most of the interviews appeared in the *Chemical Intelligencer* magazine and in the six-volume *Candid Science* book series. In addition, we selected excerpts from 111 interviews and published them in a separate volume.[1]

The title of the first entry in Section 2, "The loneliness of the scientific discoverer," is a telling expression of Father's thinking about scientific discoveries and discoverers. For a moment or for a while, the discoverer is alone with some knowledge that is his or hers alone. It is an uplifting experience and a heavy burden at the same time. It may also be tormenting if the discovery has a hard time getting accepted and especially if it is very much ahead of its time. Michael Polanyi's discoveries concerning adsorption were premature and their negative reception caused Polanyi much aggravation. Polanyi always chose the road less travelled[2] and no wonder he often ended up with new insights in the fields he had recently entered. Eventually, Polanyi left science and moved to epistemology where he produced a milestone book, *Personal Knowledge*. Father's other remarkable encounter in 1969 at the University of Texas was with Michael Polanyi (the first was with Wigner).

Father was interested in discoveries that destroyed rock-solid dogmas or shattered believes about the impossibility of some material or phenomenon. Such were the production of the first noble-gas compounds, the uncovering of the DNA structure, the observation of quasicrystals, or disproving Phoebus Levene's tetranucleotide hypothesis. Most of the entries in Section 2 refer to significant discoverers and other players with whom Father was personally acquainted. They included Neil Bartlett, Linus Pauling, Otto Bastiansen, Erwin Chargaff, Francis Crick, Edward Teller, Jerome Karle, George A. Olah, Michael Polanyi, Glenn T. Seaborg, Paul Lauterbur, Paul von Ragué Schleyer, James D. Watson, Marshall Nirenberg, Frank H. Westheimer, and others. Some of the entries of Section 1 are also relevant for Section 2. Most notably is relevant the discovery of quasicrystals and its long and hard road to acceptance. Father was one of the early proponents of Dan Shechtman's highest recognition. Not long after the announcement of Shechtman's Nobel Prize in 2011, Father and I visited the National Institute of Science and Technology in Gaithersburg, Maryland, the venue of Shechtman's original breathtaking observation. Father gave a talk about the story of the quasicrystal discovery and its afterlife. From the discussion and other conversations at NIST an interesting impression formed. Prior to the Nobel Prize NIST was not quite aware of the significance of what Shechtman and his NIST colleagues achieved and the NIST contribution had also appeared as if being underplayed in the various descriptions of the discovery.

In some of the entries of Section 2, a political message is discernible although always relevant to science. Father has been interested in the achievements of Soviet science (see, e.g., his recent book, *Buried Glory: Portraits of Soviet Scientists*). He has been concerned with the

---

[1] B. Hargittai, M. Hargittai, I. Hargittai, *Great Minds: Reflections of 111 Top Scientists* (Oxford University Press, 2014).

[2] I am paraphrasing here from Robert Frost's "The Road Not Taken."

fate of scientists operating under totalitarianism. Furthermore, he is fascinated by the survival of creativity and the determination of creating even under the most adversarial circumstances. Knowing his personal history what he has to say in these matters is especially relevant. He speaks out not only because of his interest, but because of his deep feeling of obligation of facing these, often neglected, aspects of science history.

Statue of Arthur Koestler (1905–1983) by Imre Varga (2009) on Lövölde Square, District VI in Budapest, close to Koestler's boyhood home (photograph by Istvan Hargittai). The Hungarian-British Koestler was a writer and journalist. He wrote about the nature of scientific discovery in his book *The Act of Creation* (New York: Macmillan, 1964). In his refreshing discussion, he stressed the fruitfulness of bringing different planes of thinking together in creating something new.

The Loneliness of the Scientific Discoverer[a]

István Hargittai

There is a story about a biology class in school when the teacher asks the students to look into the microscope and observe some microbes. So, the students one by one look into the eyepiece, record their observations in their diaries, and yield to the next student. Then, one of the students, the ninth in the row, tells the teacher that he sees only darkness. The teacher is first annoyed, but finds that he had forgotten to remove the cap over the objective lens. It took some courage from the ninth student to differ from the preceding eight. I wonder if I would have had the courage to do so. The story came from a regimented European school rather than from a liberally minded American one.

The scientist who makes a truly seminal discovery often must break with generally accepted previous notions. This may happen to earthshaking discoveries as well as to lesser achievements. When Jacobus Henricus van 't Hoff and Joseph Achille Le Bel came out with the suggestion of the tetrahedral bond configuration of the carbon atom in 1873–1874, it was met not only with disbelief, but also with ridicule. A famous organic chemist [1] of the time accused them of invoking "fanciful nonsense" and using "supernatural explanations" when their knowledge failed them. Today, it seems hard to believe, but then, to many chemists it was utterly alien to extend structural considerations into the third dimension.

Max Planck himself found it difficult to accept his own discovery, when he came to the conclusion of the quantum. However, he had the strength to overcome his own reservations and the courage to come out with his shattering new theory. Perhaps, judging by his reaction to his own discovery, Planck was not very optimistic with respect to others accepting his theory. He noted: "An important scientific innovation rarely makes its way by gradually winning over and converting its opponents. . ... What does happen is that its opponents gradually die out, and that the growing generation is familiarized with the ideas from the beginning" [2].

When Oswald Avery and his two young coworkers reported that, as they called it, the "transforming principle" is DNA [3], very few scientists took notice. There were others, rather vocal ones, who managed to plant the seeds of doubt in changing the generally and for long accepted notion that proteins are the substance of heredity. Avery had every confidence in the correctness of their careful experiments, but he was not prone to publicity and he believed the results should speak for themselves. They did—eventually.

If one looks at the numbers in Erwin Chargaff's chromatographic analyses of DNA from different organisms, and sees the scatter of measurements, the notion is that he needed determination and self-confidence to pronounce the observation about the ratio of 1:1 of purine and pyrimidine bases. The scatter was around 10%, and it took Chargaff some foresight to see the pattern beyond those numbers [4].

[a]Originally published in *Structural Chemistry* 2007, 18, 1–3.

I. Hargittai (✉)
Department of Inorganic and Analytical Chemistry, Budapest University of Technology and Economics, Budapest, Hungary
e-mail: istvan.hargittai@gmail.com

Max Planck (courtesy of the Oesper Collection and William Jensen, University of Cincinnati).

F. Sherwood Rowlamd (photo by the author).

Oswald Avery (courtesy of the late Maclyn McCarty, New York City).

Dan Shechtman (photo by the author).

Erwin Chargaff (courtesy of Tom Chargaff, Surry, Maine).

Alan G. MacDiarmid (photo by the author).

When F. Sherwood Rowland and his postdoctoral associate, Mario Molina first determined and communicated that chlorofluorocarbons may be responsible for the depletion of the ozone layer, they were met with more than ridicule. There were accusations by the industries involved in the production and application of chlorofluorocarbons, and there was virtual boycott by some of Rowland's colleagues finding expression in noninvitations for giving seminars and not sending him graduate students [5].

Being a student of the solid-state, Dan Shechtman had to prove in one of his exams that fivefold symmetry is forbidden in crystallography. It was an irony of fate then that he was the first who observed and reported extended regular structures with fivefold symmetry and no periodicity. Thus, he became the discoverer of the so-called quasicrystals. People warned him that if he insisted on his claims, he would risk his scientific reputation, and Linus Pauling remained their vocal opponent to the end of his life. Shechtman remained alone with his discovery for two painful years [6].

A physicist colleague of Alan McDiarmid also warned him not to risk his scientific reputation when McDiarmid suggested cooperation in understanding the electronic structure of conducting polymers. To the luck of McDiarmid and a whole new future industry, he succeeded in finding another physicist to help him [7]. But the road was bumpy because the idea of such substances was entirely new.

Today, research is often carried out in large groups and sometimes a dozen or more and in certain areas even hundreds of coauthors share the glory and the responsibility represented by research reports. Seminal ideas and observations, however, still belong to individuals. They, more often than not, have to bear the weight of bringing down dogmas or opening entirely new vistas in science. At some point, they must be feeling very lonely out there.

Acknowledgement Our research is supported by the Hungarian Scientific Research Fund (OTKA K 60365).

References

1. For Reference to Hermann Kolbe, see in Ramsay B (1972) Stereochemistry. Longman, London, p 93
2. See in Mackay AL (1991) A dictionary of scientific quotations. Adam Hilger, Bristol, UK, p 195
3. Avery OT, MacLeod C, McCarty M (1944) Studies on the chemical nature of the substance inducing transformation of pneumococcal types: Induction of transformation by a desoxyribonucleic acid fraction isolated from pneumococcus type III. J Exp Med 79:137–158
4. Chargaff E (1950) Chemical specificity of nucleic acids and mechanism of their enzymatic degradation. Experientia 6:201–209
5. Hargittai I (2000) F. Sherwood Rowland. In: Hargittai I (ed) Candid science: conversations with famous chemists. Imperial College Press, London, pp 448–465
6. Hargittai B, Hargittai I (2005) Dan Shechtman. In: Hargittai B, Hargittai I (eds) Candid science V: conversations with famous scientists. Imperial College Press, London, pp 76–93
7. Hargittai B, Hargittai I (2005) Alan G. MacDiarmid. In: Hargittai B, Hargittai I (eds) Candid science V: conversations with famous scientists. Imperial College Press, London, pp 400–409

О ХИМИЧЕСКОМЪ СТРОЕНІИ ВЕЩЕСТВЪ *)

ПРОФ. ХИМІИ А. БУТЛЕРОВÁ.

Нынѣ, послѣ открытія массы неожиданныхъ и важныхъ фактовъ, почти всѣ сознаютъ, что теоретическая сторона химіи не соотвѣтствуетъ ея фактическому развитію. Теорія типовъ, принятая теперь большинствомъ, начинаетъ оказываться недостаточною; несмотря на то, что она возникла еще недавно и много сдѣлала для развитія химіи, нѣкоторые изъ фактовъ, открытыхъ въ новѣйшее время, подтверждаютъ даже справедливость прежнихъ воззрѣній: образованіе оксиэѳильныхъ щелочей открытыхъ Вюрцомъ, говоритъ въ пользу взгляда Берцеліуса, принимавшаго алкалоиды за парныя соединенія амміака, а эѳиленная теорія эѳильныхъ соединеній является справедливою до извѣстной степени, если принять во вниманіе образованіе алкоголя изъ эѳилена и воды, іодистаго эѳила изъ эѳилена и іодоводорода и проч.

Дѣло въ томъ, что большинство старыхъ воззрѣній, также какъ и новыя справедливы лишь для опредѣленнаго круга фактовъ, и преимущественно для тѣхъ, которые легли въ ихъ основаніе.

---

*) Статья эта была уже напечатана въ Zeitschrift f. Chemie und Pharmacie 1861, p. 549. (Einiges über die chemische Structur der Körper.), но такъ, какъ она, вмѣстѣ съ слѣдующей статьей, составляетъ приложеніе къ отчету о заграничной поѣздкѣ. проф. Бутлерова, помѣщенному во II-мъ отдѣлѣ этаго выпуска и при томъ обѣ эти статьи имѣютъ неоспоримую солидарность, то редакція и сочла необходимымъ съ одобрѣнія факультета помѣстить въ Ученыхъ Запискахъ статью уже напечатанную въ заграничномъ изданіи.

The facsimile of the first page of the Russian version [Ref. 3] of the paper shown in the German original [Ref. 2] in the following article. Butlerov wrote about the chemical structure of matter for the first time in this communication.

Aleksandr Mikhailovich Butlerov and Chemical Structure: Tribute to a Scientist and to a 150-Year Old Concept[a]

István Hargittai

Abstract

In a pioneering move, 150 years ago the Russian organic chemist Aleksandr Butlerov (1828–1886) coined the term "chemical structure." He called for basing our understanding of the chemical composition of substances on the concepts of atomicity and structure.

In the 1850s and 1860s, excellent chemists worked on elucidating the composition of organic substances, such as August Kekulé, Hermann Kolbe, Archibald Couper, and others. Concepts such as valence in general and the tetravalence of the carbon atom in particular were born. It was felt intuitively that there should be a correlation between the chemical properties of organic compounds and the distribution of bonds between their atoms. However, it was not yet possible to understand and, accordingly, to depict this correlation properly.

From this background is it only possible to appreciate the importance of Aleksandr Butlerov's (Fig. 1) presentation 150 years ago this year at the meeting of the chemistry division of the 36th congress of the German physicians and scientists. The title of his presentation on September 19, 1861, was "Einiges über die chemische Structur der Körper," which was then printed in the German journal Zeitschrift für Chemie und Pharmacie [2] (Fig. 2). Note the spelling of "Structur" which is "Struktur" in today's German. Butlerov at the time was professor at Kazan University and his paper was reprinted next year in Russian translation in the journal of Kazan University [3]. Figure 3 shows the first page of this paper in which the asterisked footnote refers to the original German publication.

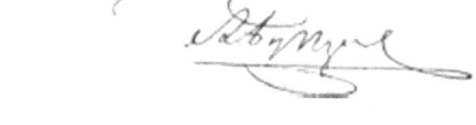

Fig. 1 Aleksandr Mikhailovich Butlerov (1828–1886). The portrait is from Ref. [1].

The term "khimicheskoe stroenie" in Russian and "chemische Struktur" in German means chemical structure and was probably used for the first time. It did not go into general usage very soon, and the term stereochemistry became more popular after it had been coined in 1890 by

[a]Originally published in *Structural Chemistry* 2011, 22, 243–246

I. Hargittai (✉)
Department of Inorganic and Analytical Chemistry, Budapest University of Technology and Economics, Budapest, Hungary
e-mail: istvan.hargittai@gmail.com

(a)　**Einiges über die chemische Structur der Körper.**

Von Prof. Dr. A. Butlerow.

(Vorgetragen in der chemischen Section der 36. Versammlung deutscher Naturforscher und Aerzte zu Speyer am 19. Septbr.)

Bei dem gegenwärtigen Zustande der Chemie, wo wir in den Besitz einer Masse ebenso unerwarteter als interessanter Thatsachen gekommen sind, lässt es sich ziemlich allgemein fühlen, dass die theoretische Seite unserer Wissenschaft ihrer thatsächlichen Entwickelung nicht genug entspricht.

In der That, die jetzt fast allgemein angenommene typische Betrachtungsweise, obgleich sie erst vor wenigen Jahren entstanden, und für die Entwickelung der Chemie ungemein fruchtbar gewesen ist, genügt uns doch kaum.

Es sind sogar in der neuesten Zeit einige Thatsachen entdeckt worden, welche viel mehr für die Wahrheit mancher älteren Ansichten sprechen. In der That spricht die Bildung der von Wurtz neu entdeckten Oxyäthylbasen zu Gunsten der Ansicht von Berzelius, welcher die Alkaloide als copulirte Ammoniake betrachtete, und die Aethylentheorie der Aethylverbindungen erscheint bis zu einem gewissen Grade richtig, wenn man der Bildung des Alkohol's aus Aethylen und Wasser, der Bildung des Jodäthyl's aus Aethylen und Jodwasserstoff u. s. w. gedenkt.

Die Sache ist so, dass die Mehrzahl der älteren und, ebenso die neue Ansicht, nur einem gewissen Kreis von Thatsachen angemessen sind, und zwar denen, auf welche sie sich hauptsächlich stützen.

Dieser Kreis ist natürlicherweise viel grösser für die neue An-

Zeitschrift f. Chemie. 1861. 　　　　　　　36

(b)

560　　Butlerow, Einiges über die chemische Structur der Körper.

da man aber gewohnt ist, unter denselben diese Letztere zu verstehen und da ihre vollkommene Entwickelung nur auf Kosten von sehr vielem Raume möglich ist, so erscheinen sie kaum passend.

Ich bin weit entfernt von dem Gedanken, dass ich hier eine neue Theorie vorschlage, vielmehr glaube ich, solche Ideen auszudrücken, welche sehr vielen Chemikern gehören. Ich muss sogar bemerken, dass der Anschauung und den Formeln von Couper, dessen zu absolute und zu exclusive Schlüsse ich zur Zeit bestritt, ein ähnlicher aber nicht hinreichend klar aufgefasster und ausgedrückter Gedanke zu Grunde lag. Im Vorstehenden wollte ich nur aussprechen, dass es Zeit wäre, die Idee der Atomigkeit und der chemischen Structur in allen Fällen und ganz frei von der typischen Anschauung, als Grundlage für die Betrachtung der chemischen Constitution anzuwenden und dass dieselbe ein Mittel, der jetzigen unbehaglichen Lage der Chemie abzuhelfen, an die Hand zu geben scheint.

Fig. 2 Opening (a) and closing (b) pages of Butlerov's 1861 paper in the German journal [2].

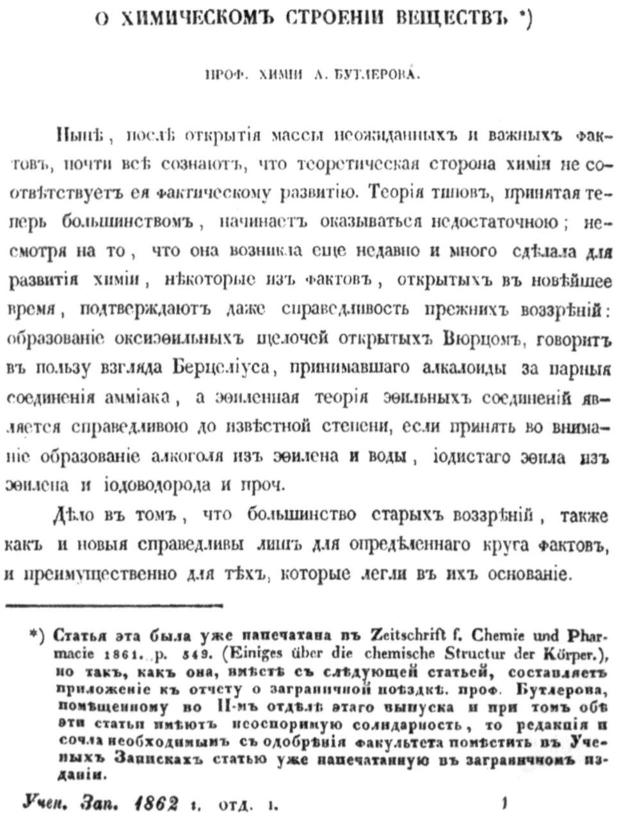

О ХИМИЧЕСКОМЪ СТРОЕНІИ ВЕЩЕСТВЪ *)

ПРОФ. ХИМІИ А. БУТЛЕРОВА.

Нынѣ, послѣ открытія массы неожиданныхъ и важныхъ фактовъ, почти всѣ сознаютъ, что теоретическая сторона химіи не соотвѣтствуетъ ея фактическому развитію. Теорія типовъ, принятая теперь большинствомъ, начинаетъ оказываться недостаточною; несмотря на то, что она возникла еще недавно и много сдѣлала для
развитія химіи, нѣкоторые изъ фактовъ, открытыхъ въ новѣйшее
время, подтверждаютъ даже справедливость прежнихъ воззрѣній:
образованіе оксиэтильныхъ щелочей открытыхъ Вюрцомъ, говоритъ
въ пользу взгляда Берцеліуса, принимавшаго алкалоиды за парныя
соединенія амміака, а эѳильная теорія эѳильныхъ соединеній является справедливою до извѣстной степени, если принять во вниманіе образованіе алкоголя изъ эѳилена и воды, іодистаго эѳила изъ
эѳилена и іодоводорода и проч.

Дѣло въ томъ, что большинство старыхъ воззрѣній, также
какъ и новыя справедливы лишь для опредѣленнаго круга фактовъ,
и преимущественно для тѣхъ, которые легли въ ихъ основаніе.

---
*) Статья эта была уже напечатана въ Zeitschrift f. Chemie und Phar-
macie 1861. p. 549. (Einiges über die chemische Structur der Körper.),
но такъ, какъ она, вмѣстѣ съ слѣдующей статьей, составляетъ
приложеніе къ отчету о заграничной поѣздкѣ. проф. Бутлерова,
помѣщенному во ІІ-мъ отдѣлѣ этаго выпуска и при томъ обѣ
эти статьи имѣютъ неоспоримую солидарность, то редакція из-
сочла необходимымъ съ одобренія факультета помѣстить въ Ученыхъ Запискахъ статью уже напечатанную въ заграничномъ изданіи.

Учен. Зап. 1862 г. отд. 1.   1

Fig. 3 Opening page of the Russian version of Butlerov's German paper in which he introduced the term "chemical structure" [3].

Victor Meyer to describe the relative three-dimensional positions of the atoms in the molecule [4]. Even in Butlerov's native Russian, "struktura" and "stroenie" are used alternatively, and the word "stroenie" has a connotation of being related with the building industry. This connotation though subtracts nothing from its being appropriate for chemical nomenclature since, knowingly or not, Butlerov introduced a term that was supposed to stress the three-dimensional nature of chemical entities. Incidentally, the bonds in Butlerov's paper linked groups of atoms rather than atoms. In this connection we note that even in the 1930s, when electronic theory of bonding was already in the vogue, the arrows in chemical texts, supposed to be indicating electron movements in chemical transformations, were often placed wherever it served the printers' convenience rather than expressing meaningful chemistry.

The tetrahedral arrangement of chemical structures goes back to Louis Pasteur who had to suppose it in order to account for dissymmetry and the optical activity of substances. There was another, little known pioneer of the tetrahedral bond arrangement of carbon, Emanuel Paternò. He published his ideas in an obscure Sicilian journal in 1869 [5]. From what he wrote though it is possible to derive what we call today conformational isomerism [6].

The year of the birth of stereochemistry, however, is considered to be 1874. The basic concepts were proposed by J. H. van 't Hoff and J. A. Le Bel, and in 1875 van 't Hoff published a booklet La Chimie dans l'Espace (Chemistry in Space). Van 't Hoff's and Le Bel's ideas were not readily accepted. The most vocal of their opponents was Hermann Kolbe whose vitriolic words illustrate the barrier the concept of three-dimensional chemistry had to overcome [Ref. 4, p. 93]:

> ...A Dr. J. H. Van 't Hoff, of the Veterinarian College, Utrecht, appears to have no taste for exact chemical research. He finds it a less arduous task to mount his Pegasus (evidently borrowed from the Veterinary College) and to soar to his Chemical Parnassus, there to reveal in his La Chimie dans l'Espace how he finds the atoms situated in the world's space.
>
> It is not possible, even cursorily, to criticize this paper, since its fanciful nonsense carefully avoids any basis of fact, and is quite unintelligible, to the calm investigator...

Butlerov did not participate in the controversy. When he gave his lecture in 1861, he did not even claim credit for any new thoughts. Rather, in the conclusion of his paper he stressed that he was expressing ideas that had occurred to many of his colleagues, and he mentioned in particular Couper. Butlerov added that the ideas he was presenting

Fig. 4 Butlerov's statue in Kazan, courtesy of Boris Solomonov, Kazan.

had not yet been expressed with sufficient clarity. In the last sentence of his paper he stresses that "it is time to base our understanding about the chemical composition of substances on the concepts of atomicity and chemical structure…" [translated from the Russian original, Ref. 1, p. 74].

Recently, David E. Lewis reviewed the significance of Butlerov's contribution to the science of organic structures accompanied by a brief description of his career [7]. Butlerov's complete works have appeared in a monumental series of four volumes in Russian, the first volume being referred to in our Ref. [1]. In the subsequent volumes, his organic chemistry text book was reproduced (Vol. 2), along with his science-popularizing works, correspondence, reviews, and other writings (Vol. 3), and, finally, treatises concerning agriculture and other studies not related to chemistry (Vol. 4).

Butlerov's activities in organic chemistry beyond the structural aspects were also significant (see e.g., [8]), but the evaluation of his oeuvre has been clouded with

Fig. 6 Soviet stamp issued in 1951, at the height of the Soviet resonance controversy, and it says, "A. M. Butlerov, great Russian chemist creator of the theory of chemical structure of organic compounds".

Fig. 5 Butlerov's statue in front of the Chemistry Department, Moscow State University (photograph by the author).

interference from politics. In the Soviet Union, he was at times considered as the founder of the science of organic chemistry, and his memory was kept alive (Figs. 4, 5, 6). In contrast, in the West, his contributions were often underestimated. The blatant actions to politicize his chemistry occurred most conspicuously during the big controversy about the theory of resonance in the Soviet Union in the early 1950s. Butlerov explicitly stated that each compound had one chemical structure and only one. The critics of the theory of resonance used Butlerov's teaching to discard the possibility of resonance structures as it would allow two or more structures to coexist. Had this been part of a sober scientific discussion it could have been considered a reasonable argument.

Unfortunately, in the early 1950s, this grew into an ideological and even nationalistic controversy with grave consequences for the proponents of the theory of resonance. The critics of the theory of resonance contrasted Butlerov's true Russian values with the cosmopolitan views of those who had bowed slavishly to Western values, etc. The proponents of the theory of resonance had to exercise humiliating self-criticism and lost their jobs [9]. The minutes of a meeting in Moscow on June 11–14, 1951, were published in a 440-page hardbound volume [10]. Four hundred and fifty chemists, physicists, and philosophers attended the meeting, including the top chemists from all over the Soviet Union. There was a report on "The status of chemical structure theory in organic chemistry" compiled by a special commission of the Chemistry Division of the Soviet Academy of Sciences. It was followed by 43 oral contributions. The report consisted of eight chapters and the first was titled "Butlerov's teachings and their role in the development of chemistry."

Linus Pauling was among the Western scientists attacked in the Soviet resonance controversy, and he seemed rather puzzled by these attacks and even after many years appeared as if he had misunderstood the situation in Soviet Union in the 1950s [11]. Today, we should not let the unprincipled past misuse of Butlerov's teachings mask the values of his pivotal contributions to organic chemistry as well as to structural chemistry.

Acknowledgments I thank Professor David E. Lewis (University of Wisconsin-Eau Claire) for a copy of his treatise referred to in Ref. [7] and Professor Boris N. Solomonov (Kazan State University, A. Butlerov Institute of Chemistry) for the image of Butlerov's statue in Kazan.

References

1. Butlerov AM (1953) Sochineniya Tom I Teoreticheskie i eksperimental'nie raboti po khimii (in Russian, Collected Works Volume 1 Theoretical and experimental works in chemistry), Izdatel'stvo Akademii nauk SSSR, Moscow
2. Butlerow A (1861) Einiges über die chemische Structur der Körper. Zeitschrift für Chemie und Pharmacie 4:549–560
3. Butlerov AM (1862) O khimicheskom stroenii veshchestv. Uchenie zapiski Kazanskogo universiteta 1:1–11
4. Ramsay OB (1981) Stereochemistry. Heyden, London
5. Paternò E (1869) Giornale di Scienze Naturali ed Economiche 6:115–122
6. Hargittai I, Hargittai M (2000) In our own image: Personal symmetry in discovery. Kluwer/Plenum, New York, pp 66–69
7. Lewis DE (2010) 150 Years of organic structures. In: Giunta CJ (ed) Atoms in Chemistry: From Dalton's Predecessors to Complex Atoms and Beyond. ACS Symposium Series, vol 1044, Chapter 4, pp 35–57
8. Berson JA (2003) Chemical discovery and the logicians' program: A problematic pairing. Wiley-VCH, Weinheim, pp 104–107
9. Hargittai I (2000) The great Soviet resonance controversy. In: Hargittai M (ed) Candid science: conversations with famous chemists. Imperial College Press, London, pp 8–13
10. (1952) Sostoyanie teorii khimicheskogo stroeniya v organicheskoi khimii (in Russian, The state of affairs of the theory of chemical structure in organic chemistry), Publishing House of the Soviet Academy of Sciences, Moscow
11. Hargittai I (2010) Struct Chem 21:1–7

Bust of Aleksandr N. Nesmeyanov (1899–1980; bust by N.I. Komov and photograph by Istvan Hargittai) in front of the Nesmeyanov Institute of Element-Organic Compounds of the Russian Academy of Sciences, 28 Vavilov Street, Moscow. The brilliant organic chemist Nesmeyanov was the President of the Academy at the time of the resonance controversy that the following paper is about. Nesmeyanov served the Stalinist regime unconditionally, but he could also manipulate it within limits. He could not prevent the attacks on theoretical chemistry by zealous ideologists, but he could mitigate the consequences of the attacks.

When Resonance Made Waves[a]

István Hargittai

I was appointed University Professor in 1991. The Diploma of University Professor was signed by the President of Hungary, and it was handed to me by the Secretary of State for Education and Culture. The brief formal ceremony was followed by an informal reception.

It was obvious that the Secretary, himself a former college professor, had read my curriculum vitae. He knew that I had done years of study in Moscow and later spent years at various U.S. universities as well. He asked me whether it was depressing to look back to the darkness of my Moscow years as compared with the freedom I must have experienced in the American laboratories. It was a leading question. In 1991, so soon after the great political changes of 1989–1990, many people tried to dissociate themselves from past Soviet connections. This made me feel a little defiant. I told the Secretary that, regardless of the external political system, inside the laboratory I did not feel any difference. In fact, my 4 years at Moscow State University, 1961–1965, were during Khrushchev's "thaw." That period carried the promise of change. Also, Moscow State University was a great school with excellent teachers and students and with often vibrant discussions in the laboratory. Besides, I was in my early twenties at the time.

All that I told the Secretary was true. Eventually, however, I have felt that my response to his question was not quite complete. Even during the sixties, ideology was penetrating the laboratory. A case in point was the nonacceptance of the concept of electronegativity. It was nothing like what had happened with regard to the theory of resonance, though, during the early fifties in the Soviet Union.

An Episode

Some time in the late eighties during a molecular structure meeting in Austin, Texas, we were having a late-night conversation, and the theory of resonance came up. A Soviet colleague and I quite surprised our Western colleagues when we mentioned the ostensibly grave ideological implications of this theory. In the fifties this theory was heresy in the eyes of Soviet officialdom. For me it was only history, but I remembered that I had read a whole volume about it, containing the minutes of a big meeting against resonance. I mentioned this book and painted a rather gloomy picture of the whole affair. The Soviet colleague tried to soften our Western colleagues' surprise. He told us that none of the proponents of the resonance theory had lost his or her life for advocating this theory. At most, some may have lost their jobs. The atmosphere froze immediately in that warm Texas night.

A little while later, I received two letters about that Texas encounter. An American friend wrote that he found the joke cruel and not funny at all. Obviously, he did not think that the story was true. The Soviet colleague also sent a letter. He apologized for having sounded as if he were belittling the situation, but, he continued, surely I must have known that ideological differences did cost lives in other branches of science, so chemistry did not fare so badly, after all.

There seems to be a widespread ignorance of the resonance controversy in Soviet chemistry among most in the West and among the younger generation of chemists in Russia as well. I don't even know whether the story can teach us anything today. They say, however, that "those who cannot remember the past are condemned to repeat it"

[a]B. Hargittai and I. Hargittai (eds.), *Culture of Chemistry: The Best Articles on the Human Side of 20th-Century Chemistry from the Archives of the Chemical Intelligencer*, DOI: 10.1007/978-1-4899-7565-2_30, © Springer Science + Business Media New York 2015. First appeared in *Chemical Intelligencer* 1995(1), 34–37.

I. Hargittai (✉)
Department of Inorganic and Analytical Chemistry, Budapest University of Technology and Economics, Budapest, Hungary
e-mail: istvan.hargittai@gmail.com

[1]. Having this in mind and in view of the inherent historical interest of the story, I thought it worthwhile to bring it up in these pages. Hence, I will present here a brief review of the minutes of that ominous meeting I had referred to in our Texas conversation. I have inserted only a few of my own comments in braces.

The Minutes of the Meeting

The minutes of the meeting were published in *Sostoyanie Teorii Khimicheskogo Stroeniya v Organicheskoi Khimii (The State of Affairs of Chemical Structure Theory in Organic Chemistry)*. Izdatel'stvo Akademii Nauk SSSR (Publishing House of the Soviet Academy of Sciences): Moscow, 1952; p. 440. This volume is not a light booklet. It is a heavy, hardbound volume of densely printed pages (it makes heavy reading, too). The title page is shown in Fig. 1.

The Chemistry Division of the U.S.S.R. Academy of Sciences held a 4-day, all-union conference between June 11 and June 14, 1951, in Moscow. The subject of the meeting was the structure theories of organic chemistry. Four hundred and fifty chemists, physicists, and philosophers attended. They represented major centers of scientific research and

АКАДЕМИЯ НАУК СССР
ОТДЕЛЕНИЕ ХИМИЧЕСКИХ НАУК

СОСТОЯНИЕ ТЕОРИИ
ХИМИЧЕСКОГО СТРОЕНИЯ
В ОРГАНИЧЕСКОЙ
ХИМИИ

ВСЕСОЮЗНОЕ СОВЕЩАНИЕ
11-14 ИЮНЯ 1951г

✳

СТЕНОГРАФИЧЕСКИЙ
ОТЧЕТ

ИЗДАТЕЛЬСТВО АКАДЕМИИ НАУК СССР
Москва—1952

Fig. 1 Title page of the hardcover volume published in 1952 containing the minutes of the 1951 meeting.

higher education all over the Soviet Union. A report entitled "The State of Affairs of Chemical Structure Theory in Organic Chemistry," compiled by a special commission of the Chemistry Division, was presented, followed by 43 oral contributions. An additional 12 contributions were submitted in writing. The conference adopted a resolution and sent a letter to I.V. Stalin.

The letter to Stalin expressed self-criticism for past deficiencies in appreciating the role of theory and theoretical generalizations in chemical research. This has resulted, the letter added, in the spreading of the foreign concept of "resonance" among some Soviet scientists. This concept was an attempt to liquidate the materialistic foundations of structure theory. However, the letter continued, Soviet chemists have already started their struggle against the ideological concepts of bourgeois science. They have unmasked the falseness of the so-called "theory of resonance" and would cleanse Soviet chemical sciences from the remnants of this concept, the letter concluded.

During the meeting, there were repeated references to Stalin's teachings on the importance of the struggle between differing opinions and of the freedom of criticism [George Orwell's doublespeak of *1984* pales by comparison].

The report of the Chemical Division was submitted to the meeting by Academician A.N. Terenin on behalf of the special commission. [The names of the members of this commission and many of the names of the speakers participating in the subsequent discussions read like a who's who of Soviet chemistry. There were many academicians among them and also many future academicians]. The report consisted of the following chapters:

1. Butlerov's teaching and its role in the development of chemistry (Fig. 2)
2. The development of structure theory during the second half of the nineteenth century and the first half of the twentieth century
3. Advances of Soviet organic chemistry
4. Quantum chemistry and structure theory
5. About the so-called "theory of resonance"
6. On the mistakes of some Soviet chemists
7. Current state of Butlerov–Markovnikov's teaching on the intramolecular interactions of atoms and on reactivity
8. Perspectives on further development of structure theory

These are telling titles and I mention a few points from Chap. 6 only. Here we learn that Professor G.V. Chelintsev had criticized actively the concept of "resonance" in the press. It was mainly owing to him that Soviet scientific society had turned to this question. It was noted, however, that the basis of his criticism was his own "new structure theory," which completely contradicted the modern theory of

Fig. 2 Statue of A.M. Butlerov (1828–1886) in front of the Chemistry Department of Moscow State University. (Photo courtesy of Dr. A.A. Ivanov, Moscow) The term "chemical structure" was introduced by Butlerov in 1861 (cf. O. Bertrand Ramsay, Stereochemistry. Heyden: London, 1981; pp. 55–57).

chemical structure and was contrary to the experimental facts and theoretical foundations of quantum physics.

Y.K. Syrkin and M.E. Dyatkina were named as the main culprits in disseminating the theory of resonance in the Soviet Union. They were accused of having even further developed the erroneous concepts of Pauling and Wheland, of ignoring the works of Soviet and Russian scientists, of idolizing foreign authorities, and of quoting even works of secondary importance by American and English authors.

Others were also mentioned, among them the organic chemist A.N. Nesmeyanov (Fig. 3). He, along with R.H. Freidlina, interpreted the diverse reactivity of chlorovinyl compounds of mercury and other quasicomplex compounds in terms of the "resonance" between their covalent and ionic structures. These lesser sinners, along with

Fig. 3 A.N. Nesmeyanov (1899–1980) on a Soviet stamp. He was Rector of Moscow State University, 1948–1951, and President of the Academy of Sciences of the U.S.S.R., 1951–1961.

many others, however, had eventually repented and had themselves become critics of the application of "resonance."

The report, as well as the subsequent contributions, was followed by questions and answers. Perhaps the most important question was about the idealism of the concept of resonance. The answer to this question started with a quotation from V.I. Lenin, according to which philosophical idealism was a one-sided exaggeration of an insignificant feature of the cognitive process. Such a feature was then detached from the matter and from nature and made into something absolute. The answer then turned to the concept of resonance, where the insignificant features of the cognitive process were specified to be the individual components of the approximate computational techniques employed in the calculation of the molecular wave function. They were made into something of primary importance, as if objectively existing in the molecule, and as if determining a priori the molecular properties. In reality, it was further explained, the resonance structures and their resonance was torn from the matter, and the theory of resonance became an absolute above the matter.

[If it sounds complicated, it is complicated. The sentences were formulated very carefully as the matter was extremely sensitive. The atmosphere was like that of a trial rather than that of a scientific discussion. The problem was not that the theory of resonance was criticized. There are chemists who do not like the description of a molecular structure by a series of resonance structures. What is frightening and mind-boggling is that such a dislike was made into an official dogma with philosophical justification].

Only a small, though very vocal, group of chemists attacked blindly the theory of resonance and even more those whom they declared to be its proponents. They also attacked quantum chemistry and all of the science of the West. They advocated a return to historical Russian results and suggested that their own theories be used. These theories, however, had been shown to be worthless nonsense by many. However, all those present painstakingly dissociated themselves from the theory of resonance. At times the self-criticism of some excellent scientists was humiliating in the extreme.

It was characteristic of the atmosphere of the meeting that a philosopher (B.M. Kedrov) declared Schrödinger to be a representative of modern "physical" idealism. This made Schrödinger a relative of Pauling's. Furthermore, he stated that Dirac's superposition principle was as idealistic as Heisenberg's complementarity principle and even more idealistic than Pauling's theory of resonance.

Another speaker, a writer (V.E. L'vov), criticized the report for a serious political error, namely, that the protagonists of the theory of resonance were equated with the greatest Soviet scientists. These protagonists had been unmasked as spokesmen of the Anglo-American bourgeois pseudoscience by the press and Soviet society. According to L'vov, the report was vague about the main thrust of the ideological struggle taking place in theoretical chemistry. He also quoted, as a positive example, the criticism of Mendel by T.D. Lysenko, who proved that Mendel's work had nothing to do with the science of biology. Furthermore, L'vov attacked fiercely the theories of Heisenberg as well as those of Heitler and London. He protested the report's view of quantum mechanics as a development of Butlerov's teaching. The most important political task of Soviet chemistry, he declared, was the isolation and capitulation of the insignificant group of unrepenting proponents of the ideology of resonance.

[To me it is a great puzzle why a concept as innocent as the resonance of chemical structures triggered a reaction of such enormous proportions. I cannot offer any rational explanation. An important contributing factor, though, may have been the fear of foreign ideas. The story of resonance should not be viewed in isolation from the rest of Soviet life in the early fifties, the last years of Stalin's reign. To me a question that is most telling was that asked of M.E. Dyatkina: "How do you explain that you are so conspicuously familiar with the teachings of foreign scientists? May it be that you, along with Professor Syrkin, are intentionally bowing to foreign scientists?"

Reading this accusation brings back some personal memories. In 1965, as a Master's student at Moscow State University, I traveled every week to the Institute of Inorganic Chemistry of the Soviet Academy, where Professor Dyatkina (Fig. 4) was giving a not-for-credit course, something like Structural Inorganic Chemistry. The large auditorium was always packed with research workers and graduate students. Professor Dyatkina would basically tell us about her readings of the previous week and her interpretation of recent literature. She was not a colorful lecturer, yet she held our attention

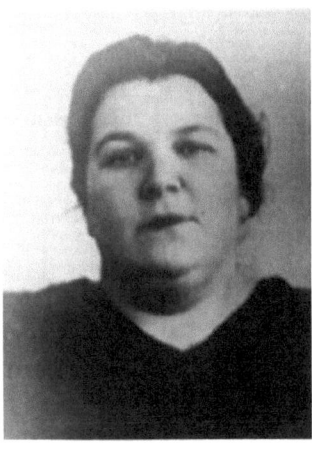

Fig. 4 Photo of the late Professor M.E. Dyatkina (1914–1972) by an unknown photographer. I am grateful to professor L.V. Vilkov (Moscow) for acquiring this photo.

fully. She was also our living library, library facilities being scarce].

The last entry in the minutes of the meeting is a dissenting opinion in the form of a short letter by E.A. Shilov, member of the Ukrainian Academy of Sciences. He was critical of the report and the resolution of the meeting for looking so much backward rather than forward. He suggested concentrating on new results and new teachings instead of conducting scholastic debates about questions such as where does resonance end and mesomerism begin and how does the "healthy" mesomerism of Soviet authors differ from Ingold's "erroneous" mesomerism and how can ideal structures be considered real at the same time. The result of ending such debates would be, Shilov added, that tire efforts and time of Soviet organic chemists could be devoted to valid and productive work. This contribution was not delivered as an oral presentation during the meeting.

A Final Note

It is an irony of history that in 1952 the already world-famous Linus Pauling was denied a passport by the U.S. State Department, based on his leftist politics [2]. This happened exactly at the time when the theory of resonance, associated primarily with Pauling's name, was vehemently attacked in the Soviet Union. The attackers might have asked Professor Pauling himself about the ideological implications of his chemical resonance theory. He could have explained that "the several structures that are used in the description of a molecule such as benzene by application of the theory of resonance are idealizations, and do not have existence in reality" [3]. Such an interpretation would have taken the ostensibly alien and harmful edge off the theory of resonance, had the accusers been interested in his explanation.

According to a later Soviet evaluation of the affair, "The minutes of the meeting is one of the most shameful documents ever created by a collective effort of scientists, and God only knows when it will be possible to wash off this disgrace" [4].

Acknowledgment Professor J.E. Boggs's (Austin, Texas) comments on this manuscript were very helpful.

References

1. *Santayana, Life of Reason.* Quoted in *The Pocket Book of Quotations.* Davidoff, H., Ed.; Pocket Books: New York, 1951.
2. Serafini, A. *Linus Pauling. A Man and His Science.* Paragon House: New York, 1989.
3. Pauling, L. *The Nature of the Chemical Bond,* 3rd ed.; Cornell University Press: Ithaca, NY, 1960; p 287.
4. Okhlobystin, O. Yu. *Zhizn'i Smert' khimicheskikh idei (Life and Death of Chemical Ideas)*; Nauka: Moscow, 1989; p 184.

Bust of Vladimir Engelhardt (courtesy of Larissa Zasurskaya) in the entrance lobby of the Engelhardt Institute of Molecular Biology of the Russian Academy of Sciences, 32 Vavilov Street, Moscow.

Early Molecular Biology in Moscow[a]

Balazs Hargittai and István Hargittai

From left to right, (**1**) Nikolai I. Vavilov, 1977; (**2**) Nikolai I. Vavilov, 1987; (**3**) Sergei I. Vavilov, 1961.

At one time in his life, Nikolai Ivanovich Vavilov (1887–1943) was the most distinguished Soviet biologist. Vavilov's field was plant genetics and, in particular, the immunology of plants. He was a member of the Soviet Academy of Sciences (SAS), the President of the All-Union Academy of Agricultural Sciences (VASKhNIL), and the Director of the Institute of Genetics of SAS. He was also member of academies and scientific societies of Germany, Great Britain, the United States, and other countries. Sadly, he was induced in 1932 to denounce the theoretical pursuit of genetics practiced in the West. He was forced to accept

planned science to serve economic needs, as declared by a Conference on Planning Genetics Selection Research in Leningrad.

Also, sadly, he gave valuable initial support to the uneducated and dogmatic T.D. Lysenko (1898–1976), who was appointed, rather than elected to be an academician. Lysenko rigidly denied any possibility of heredity and made the boldest and never-to-be-fulfilled promises to subsequent Soviet administrations about sharply increasing agricultural production if his "scientific" methods were applied.

The second half of the 1930s was an era of arrests, false accusations, showcase trials, and harsh sentences in the Soviet Union, and the scientific community was not immune against these actions, either [1]. From the late 1930s, whole branches of biology were paralyzed in the Soviet Union until well after Stalin's death in 1953. Nikolai Vavilov's adherence to party policy did not save him. He was dismissed from office in 1939, arrested in 1940, and sentenced to death in 1941. He spent almost a year waiting for his execution, which was then stayed, and he died from hunger in prison in 1943.

[a]Originally published in *The Chemical Intelligencer* 1999, 5(1):62–64.

B. Hargittai (✉)
Saint Francis University, Loretto, PA, USA
e-mail: bhargittai@francis.edu

I. Hargittai
Department of Inorganic and Analytical Chemistry, Budapest University of Technology and Economics, Budapest, Hungary
e-mail: istvan.hargittai@gmail.com

Decades later, Nikolai Vavilov was "rehabilitated" and, subsequently, two Soviet stamps were issued in his honor. The first (**1**) in 1977, somewhat oddly, commemorated his 90th birthday, and the second (**2**) in 1987, marked his centennial.

N.I. Vavilov had a brother, Sergei Ivanovich Vavilov (1891–1951), a leading Soviet physicist, also an academician, Director of the Institute of Physics of the SAS, who also held many other lesser functions. From 1945 until his death he was President of the SAS. Sadly, he appears to have been one of Stalin's instruments in Soviet science [2]. A Soviet stamp (**3**) was issued to mark the seventieth anniversary of his birth and the tenth anniversary of his death, in 1961. That was also the year of the Fifth International Congress of Biochemistry in Moscow.

Unbeknownst to most foreign participants of the Moscow congress, Lysenko was back in power in 1961 following the temporary demise of his influence in the late 1950s. His anti-science actions culminated in 1948 when, with Stalin's direct help, he crushed Soviet genetics. Incredibly, at the time of the Moscow congress, one of his many positions was the directorship of the Institute of Genetics. He was also President of VASKhNIL. However, 1961 was no longer the era of absolute rigidity and, apparently, he had to tolerate the presence of a huge group of foreign scientists and the flow of many new ideas, converging in Moscow.

One of the signs of greater flexibility was the interaction of some of the Soviet scientists with their colleagues in the West in preparation for the meeting. A leading Soviet biochemist, Vladimir Alexandrovich Engelhardt (1894–1984) was in charge of organizing the section called "biological structure and function at the molecular level," which was the code name for molecular biology, still anathema to the Soviet authorities. This is how Engelhardt remembered this two decades later [3]: "We intended to name it simply 'molecular biology section.' But when I proposed this at a meeting of the Moscow group of our organizing committee, strong objections were raised by the more orthodox members: 'What is molecular biology? We do not know such a science or branch.' To satisfy my opponents (I was in the minority of one!) I invented a euphemistic name, 'biological functions at the molecular level.' This was accepted."

Engelhardt turned to Max Perutz in England for help in organizing the section and was able to give him complete freedom as to the program and invited speakers. Engelhardt was one of the initiators of molecular biology in the Soviet Union, albeit without using that name at the beginning. He founded a research institute in 1959, which became the Institute of molecular biology of the Soviet (today, Russian) Academy of Sciences in 1964. Some of Engelhardt's early works was done jointly with Alexei Nikolaevich Bach (1857–1946), another famous biochemist [4]. They placed proteins on insoluble carriers and investigated their interactions with other substances. Bach founded what eventually became the Institute of Biochemistry of the Soviet (today, Russian) Academy of Sciences and served as its first director. The centennial of his birth was commemorated on a Soviet stamp in 1957 (**4**).

Apparently Perutz did not anticipate that a breakthrough would be reported at the congress, and it almost did not happen [5]. A little-known associate of the then little respected National Institutes of Health of the United States, Marshall W. Nirenberg, was scheduled a brief presentation, which he duly gave to a handful of attendees. Those few

(**4**) Alexei N. Bakh, 1957; (**5**) The Fifth International Congress of Biochemistry, Moscow, 1961.

people present, however, immediately recognized the extraordinary importance of Nirenberg's announcement. Nirenberg demonstrated "the first word to be identified in the genetic code" 'One or more uridylic acid residues appear to be the code for phenylalanine' [6]. Walter Gilbert, who attended Nirenberg's original presentation, seems to feel to this day the electrifying effect of Nirenberg's discovery [7].[1] An unprecedented thing happened then in Moscow: Nirenberg was asked to repeat the presentation, this time to a large audience, many of whom rushed home after his lecture to give a new direction and push to their own experiments.

The 1961 Moscow congress, commemorated on a Soviet stamp (5), helped molecular biology get established in the Soviet Union and invigorated this branch of science internationally. In 1968, Marshall W. Nirenberg shared the Nobel Prize in physiology or medicine with Robert W. Holley and H. Gobind Khorana "for their discoveries concerning the interpretation of the genetic code and its function in protein synthesis."

References

1. *Tragicheskie sud'bi: repressirovannie uchenie Akademii nauk SSSR* (in Russian) [*Tragic Fates: Oppressed Scientists of the Soviet Academy of Sciences*]. Collection of articles compiled by the Division of History, Archives of the Russian Academy of Sciences, Kumanev, V. A., Ed.; Nauka: Moscow, 1995.
2. Shnoll, S.E., *Geroi i zlodei rossiiskoi nauki* (in Russian) [*Heroes and Villains of Science in Russia*]. Kron-Press: Moscow, 1997.
3. Engelhardt, W.A., "Life and Science," *Annu. Rev. Biochem.* **1982**, *51*, 1–19.
4. Volkov, V.A.; Vonskii, E.V.; Kuznetsova, G.I., *Vydayushchiesya khimiki mira* (in Russian) [*Outstanding Chemists of the World*]; Vysshaya Shkola: Moscow, 1991.
5. Judson, H.F., The Eighth Day of Creation: Makers of the Revolution in Biology; Simon and Schuster: New York, 1979.
6. Nirenberg, M.W.; Matthaei, H., "The Dependence of Cell-Free Protein Synthesis in *E. Coli* upon Naturally Occurring or Synthetic Template RNA." *Proceedings of the Fifth International Congress of Biochemistry, Moscow, 10–16 August 1961. Volume I: Biological Structure and Function at the Molecular Level.* V.A. Engelhardt, Ed.; Macmillan: New York, 1963; pp. 184–189.
7. Hargittai, I., "Interview with Walter Gilbert." *The Chemical Intelligencer* to be published.

---

[1] See, I. Hargittai, Candid Science II: Conversations with Famous Biomedical Scientists (Ed. M. Hargittai). Chapter 7, "Walter Gilbert" (London: Imperial College Press, 2002), pp. 98–113.

Abram M. Blokh at his typewriter in his office in 2004 in Moscow (photograph by Istvan Hargittai). Blokh has collected relentlessly all available information about Soviet and Russian Nobel laureates and potential laureates, and about all the events that have surrounded nominations for the Nobel Prize and the lacks thereof. The 900-page English translation of his opus magnum has appeared, Abram M. Blokh, *The Soviet Union in the Context of the Nobel Prize* (Singapore: World Scientific). The second Russian edition appeared in 2005 and the first in 2001. The following paper by Istvan appeared in 1999, and one of Istvan's sources was a Blokh article under Reference No. [1].

A Curious Case of Soviet Nobel Aspirations[a]

István Hargittai

To the recent observer it may seem that the importance of the Nobel Prize is going out of proportion. In our shrinking world, it is not only the single most important science prize, it is the only one recognized by the general public internationally. Some tend to measure the success of nations by the number of Nobel prizes earned by their citizens. Looking back over the 100-year history of the prize, however, it is apparent that the Nobel prize had reached improbable heights of importance already decades ago. This is evidenced by examples of concern and involvement of national governments in matters of the Nobel Prize.

(1) B. L. Pasternak

(2) A.D. Sakharov

A conspicuous case in point was when Nazi Germany banned its citizens from accepting Nobel Prizes. This happened after the Norwegian Parliament had awarded the pacifist journalist and concentration camp inmate Carl von Ossietzky the Nobel Prize for Peace (for 1935) in 1936. Then the Soviet government did practically the same with the Russian writer Boris Pasternak (1), who won the prize in literature in 1958. He was warned by the Soviet authorities that if he were to travel to Stockholm to receive the prize, he might not be let back home. So Pasternak declined the prize. When nuclear physicist Andrei Sakharov (2) won the peace prize in 1975, he was not allowed to go to Oslo to receive it, so instead his wife, Elena Bonner, went. I happened to be visiting Oslo University at the time and vividly remember the solemn torch-light demonstration in downtown Oslo on the night of the ceremony, honoring Sakharov.

I mention here two other much milder examples of how much sentiment is attached to the Nobel Prize at a national level. One is a headline I noticed in a British daily after the fullerene prize in 1996, announcing that Harry Kroto scored one for Britain. The other is what Nobel laureate Leo Esaki (physics, 1973) said, commemorating the recently deceased Kenichi Fukui (chemistry, 1981): "Japan should nurture better scientific researchers so they receive more Nobel prizes and thus allow Fukui's soul to rest in peace" [2].

The story I am about to tell is from 1955–1956 and concerns the Nobel Prize in chemistry awarded to Nikolai Semenov. Recently, there was a long article in a Moscow weekly, whose name means "search" in English [1], in which Abram Blokh gave a detailed account of some events and activities preceding the first Soviet Nobel Prize. Much of the present note was compiled on the basis of Abram Blokh's article.

[a]Originally published in *The Chemical Intelligencer* 1999, 5(3), 61–64.

I. Hargittai (✉)
Department of Inorganic and Analytical Chemistry, University of Technology and Economics, Budapest, Hungary
e-mail: istvan.hargittai@gmail.com

(3) D.I. Mendeleev   (4) L.N. Tolstoi   (5) A. P. Chekhov   (6) M. Gorky

Below is an excerpt from Resolution No. 19 adopted on November 1, 1955, by the leadership of the Division of the Physical-Mathematical Sciences of the Soviet Academy of Sciences. The resolution was titled "On the Nomination of Soviet Scientists for the Nobel Prize": "The divisional leadership does not find it advisable to nominate Soviet scientists for the Nobel Prize since this prize cannot be considered international as demonstrated by the lack of Nobel awards to outstanding individuals of science and culture of our country (D.I. Mendeleev, L.N. Tolstoi, A.P. Chekhov, M. Gorkii). Mendeleev and the three writers mentioned in this quote are depicted on stamps (**3–6**)".

The resolution was signed by, among others, such well-known scientists as the mathematician M.A. Lavrent'ev (**7**), head of the division at that time, and the physicists L.A. Artsimovich (**8**) and P.L. Kapitsa (**9**). Present also, as a guest, was a so-called instructor from the science and education section of the Central Committee of the Communist Party of the Soviet Union. (Today he is a member of the Russian Academy of Sciences).

The Soviet Academy of Sciences was structured in divisions; the mathematicians and physicists shared the same division, whereas the chemists and biologists each had their own division, among many other divisions. Lavrent'ev was a

(7) M.A. Lavrent'ev   (8) L.A. Artsimovich   (9) P.L. Kapitsa

noted mathematician, at one time Vice President of the Soviet Academy of Sciences. He initiated the creation of the new and gigantic scientific center in Siberia and served as its first president. Incidentally, while my wife and I were on our honeymoon in 1967, we met Lavrent'ev in Visegrad, Hungary. Artsimovich was a famous physicist, most noted for his work on an electromagnetic method of isotope separation and later on controlled thermonuclear reactions.

"knew" Russian. I have found evidence of this in his [Sillén's] printed lecture to the Sixth Welch Conference on Chemical Research, on November 28, 1962. The title of his presentation was "Aqueous Hydrolytic Species" [3]. When he enumerated his four important principles in equilibrium analysis, he showed his Houston audience these four principles in Russian, stressing also "that I am telling the same things to chemists in Moscow, Leningrad and

(10) N.N. Semenov   (11) I.P. Pavlov   (12) I.I. Mechnikov

The timing of Resolution No. 19 is significant. In October 1955, it was expected in Moscow that Nikolai Semenov (**10**) would receive the Nobel Prize in chemistry but this did not happen. The Nobel Prize in chemistry for 1955 was awarded to the American Vincent du Vigneaud "for his work on biochemically important sulphur compounds, especially for the first synthesis of a polypeptide hormone."

Up until that time, there was indeed no Soviet Nobel prizewinner. The two Russian Nobel laureates had been awarded their prizes back in czarist times, both in physiology or medicine. One was Ivan Pavlov (**11**), who received the Prize in 1904, "in recognition of his work on the physiology of digestion, through which knowledge of vital aspects of the subject has been transformed and enlarged." The other was Ilya Mechnikov (**12**), who shared the Prize in 1908 with Paul Erlich, "in recognition of their work on immunity." There was a Nobel Prize in literature to a Russian writer, I.A. Bunin (**13**), in 1933, but this did not count, or if it did, it made the situation worse, since Buninwas an emigrant, living in Paris.

Lars Gunnar Sillén, Professor of the Royal Swedish Institute of Technology in Stockholm had been trying to get documentation on the scientific activities of Semenov since 1952. He renewed his efforts in 1954 through the Soviet embassy in Stockholm. Blokh notes that Sillén

Kharkov as I am telling you here in Houston." Somewhat later in his lecture, he quoted a Russian proverb in Russian, as if to leave no doubt among his audience about his knowledge of Russian.

Left: Lars G. Sillén (courtesy of Julius Glaser).   Right: Vincent Du Vigneaud (courtesy of Miklós Bodánszky)

Left: Lars G. Sillén (courtesy of Julius Glaser). Right: Vincent Du Vigneaud (courtesy of Miklós Bodánszky).

In 1954, the secretary of the Biology Division of the Soviet Academy of Sciences, the biochemist A. Oparin (1894–1980), famous for his theory of the origin of life, attended a meeting in Stockholm. Upon his return to Moscow, he reported to A.N. Nesmeyanov (**14**), the organic chemist, who was then the President of the Academy, on the discussion he had had in Stockholm with two members of the Nobel Committee of Chemistry. According to Oparin, Professors A. Tiselius and A. Fredga told him that they would like to see Soviet scientists participate in the Nobel movement. They asked Oparin to discuss this with the Soviet authorities in Moscow. Nesmeyanov and Oparin decided to inform the Central Committee of the Communist Party about Oparin's experience. Apparently as a result of all these events, a detailed documentation of Semenov's scientific activities was sent to Stockholm at the end of 1954. The documents arrived too late to play a role in the selection for the 1955 prize, but they were duly taken into account in the selection process for the 1956 prize. Cyril N. Hinshelwood (1897–1967) of Oxford and Semenov would share the chemistry prize in 1956 "for their researches into the mechanism of chemical reactions."

It is not difficult to imagine that back in October 1955 the Moscow forces against the Nobel connections must have rejoiced when the 1955 chemistry prize went to an American scientist rather than to Semenov. Blokh painstakingly analyzes the text of Resolution No. 19 and the circumstances of its initiation and afterlife. He comes to the conclusion that it must have been inserted among the resolutions adopted on November 1, 1955, *after* the actual meeting and notes an active involvement of the Interior Ministry and the Communist party in the affairs of the Academy. In addition to this alleged falsification, he also sees a pattern of similar attempts since the resolution by the physicists and mathematicians was not unique.

The physicists and the mathematicians did not have much at stake in Nobel considerations at that time. The physicists did not think that any Soviet physicist was a potential nominee, and there is no Nobel prize in mathematics. The physicists though may have been mistaken because the Nobel Prize in physics was to be awarded very soon, in 1958, to P.A. Cherenkov (**15**), I.M. Frank, and I.Y. Tamm "for the discovery and the interpretation of the Cherenkov effect." This was not the first time either that Soviet physicists had been considered for the Nobel Prize. When preparing the nominations for the physics prize of 1939, Niels Bohr, one of the nominators, was considering a joint nomination of the American Ernest Lawrence and the Russian Pyotr Kapitsa, but in the end, he dropped Kapitsa [4]. Lawrence went on to receive the Nobel Prize in 1939 and Kapitsa had to wait for his until 1978 ("for his basic inventions and discoveries in the area of low-temperature physics"). By then, he was 84.

In December 1955 the Chemistry Division of the Soviet Academy decided, too, not to nominate anybody for the 1956 Nobel Prize. This was a safe decision though, because the Nobel Committee had already received the documentation supporting Semenov's nomination. The chemists' resolution appears to have been an exercise in political expediency.

That politics was involved is obvious from many episodes. Thus, for example, the secretary of the Chemistry Division of the Soviet Academy of Sciences asked the Soviet embassy in Stockholm back in 1954 to reassure Professor Sillén that "we consider participation in the contest for the Nobel Prize to be a great honor for any scientist." Such a request could not have been transmitted without the support of the state security services.

I am not familiar with past practices regarding the nominating procedure for the Nobel Prize. Currently, however, every invitation to nominate candidates is accompanied

(13) I.A. Bunin   (14) A.N. Nesmeyanov   (15) P.A. Cherenkov

by stern warnings to discourage the discussion of possible candidates in academies or other collective bodies. In the above story, such bodies and their politics obviously were deeply involved, and it is equally obvious that the Nobel organization could not have been unaware of such involvement. Since the materials of the Nobel archives become available for research after 50 years, it will be possible to find out more about the circumstances of Semenov's nomination on the Swedish side in 2006.

There must have been a lot of discussion about the Nobel Prize in Moscow behind the scenes. The years between Stalin's death in 1953 and the 20th party congress in 1956 were full of pendulum-like swings in policy. Blokh anticipates that the full truth will come out only when all the documents of the Central Committee of the Communist Party of the Soviet Union, stored in the Central Archives of Modern Documentation, finally become available for research.

I recorded a conversation with Nikolai Semenov for Radio Budapest in September 1965. It was a good introduction for me as it was my first ever interview of a famous scientist. He spoke about the origin of his interest in chemistry and physics, about the very beginnings of Soviet science in the early 1920s, and about the branched chain reactions, his Nobel Prize-winning research. He was even willing to make predictions about the future, which can be compared today with reality. The English version of our conversations appears in my first volume of interviews [5].[1]

Acknowledgments I am grateful to Professor Eiji Osawa (Toyohashi University of Technology) for the clipping from *Asahi Evening News* (Tokyo) and to Professor Lev V. Vilkov (Moscow State University) for the clipping from *Poisk* (Moscow).

References

1. Blokh, A., *Poisk* Nos. 31–32, July 25–August 7, 1998, p 12.
2. *Asahi Evening News*, January 10, 1998.
3. Sillén, L.G., In *Proceedings of the Robert A. Welch Foundation Conferences on Chemical Research. VI. Topics in Modern Inorganic Chemistry*; Milligan, W.O., Ed.; The Robert A. Welch Foundation: Houston, Texas, 1963, pp 187–234.
4. Heilbron, J.L.; Seidel, R.W., *Lawrence and His Laboratory: A History of the Lawrence Berkeley Laboratory*, Vol. 1; University of California Press: Berkeley, 1989, p 491.
5. Hargittai, I., *Candid Science: Conversations with Famous Chemists*; Imperial College Press: London, 2000 (in press).

---

[1] This was the first volume of the six-volume *Candid Science* series, B. Hargittai, I. Hargittai, and M. Hargittai, *Candid Science: Conversations with Famous Scientists*; Imperial College Press: London, 2000–2006.

Central part of the mural at the Mendeleevskaya metro station in Moscow (photograph by Istvan Hargittai). The mural is at the end of the huge waiting hall and it consists of Mendeleev's portrait and parts of his periodic table of the elements. The station was opened in 1988.

Dmitri I. Mendeleev: A Centennial[a]

Balazs Hargittai and István Hargittai

(*Left*) D. I. Mendeleev in 1869 (*Chronicle*) at the time of his seminal discovery. (*Right*) The Periodic Table of the Elements as a fresco at what is today the Mendeleev Institute of Metrology in St. Petersburg and was the Board of Weights and Measures in Mendeleev's time (courtesy of Alexander V. Belyakov, St. Petersburg).

[a]Originally published in *Structural Chemistry* 2007, 18:253–255.

B. Hargittai (✉)
Saint Francis University, Loretto, PA, USA
e-mail: bhargittai@francis.edu

I. Hargittai
Department of Inorganic and Analytical Chemistry, Budapest
University of Technology and Economics, Budapest, Hungary
e-mail: istvan.hargittai@gmail.com

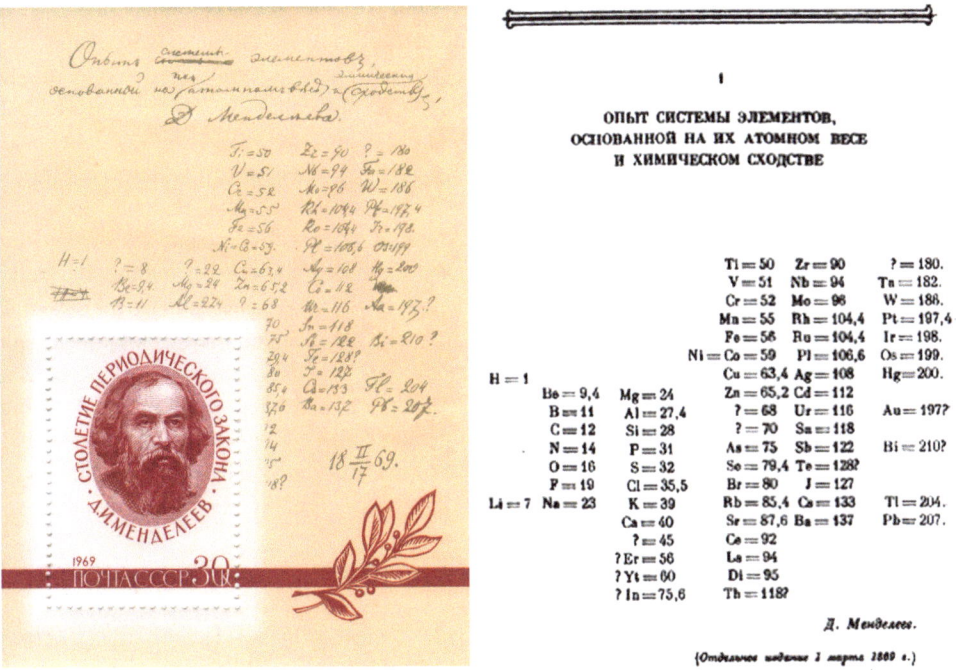

The first version of Mendeleev's Periodic Table of the Elements on a Soviet postage stamp of 1969 and in its original printed form of 1869 (see D. I. Mendeleev, p. 9).

Dmitri Ivanovich Mendeleev (1834–1907) is one of the most famous chemists of all time [1–4]. His Periodic Table of the Elements hangs on the walls of most chemistry classrooms and is printed in many chemistry texts. He died 100 years ago when the importance of his discovery was widely appreciated, though its physical foundation was not yet understood. Although his fame originated mainly from his recognition abroad, he never received the Nobel Prize—the ultimate accolade internationally. In this editorial, we discuss briefly his monumental discovery in 1869 and circumstances of the decision of the Nobel Committee of Chemistry in 1906.

When he was a Professor of Chemistry at St. Petersburg University, Mendeleev discovered the Periodic Table of the Elements while working on his text *Foundations of Chemistry (Osnovi Khimii)*. This text was based on his lectures and the lecture notes prepared by his students. He was much concerned with presenting the plethora of facts in a systematic way. He compiled the first version of his table on February 17, 1869, using the properties and the atomic masses of the elements as governing principles. On the same day he sent his table to the printers and when it was printed on March 1, 1869, he sent out copies to Russian and foreign colleagues.

From the beginning, Mendeleev noticed periodicity in the properties of the elements; made predictions for heretofore unknown elements; and suggested corrections for atomic masses of certain elements. Subsequently, his predictions showed forcefully the value and validity of his discovery.

As early as August 1869, Mendeleev gave a talk about his discovery at the Second Congress of Russian Scientists and in October of the same year, he reported his findings to the meeting of the Russian Chemical Society. He kept refining his table in subsequent months and its complete version appeared in the first edition of his *Foundations of Chemistry*, in 1871. In the same year, he made an extensive trip to Western Europe and popularized his discovery among the international community, in talks as well as in papers.

Soviet postage stamp of 1969 showing the correctness of Mendeleev's prediction of the existence of the element that has come to be known as Gallium.

Mendeleev made contributions to chemistry in other fields and interacted with international chemists in other areas of theoretical and practical importance, including American chemists in oil exploration. However, his lasting contribution is his Periodic Table of the Elements. It was for this discovery that he was nominated for the Nobel Prize in 1905, 1906, and 1907. He was considered in 1905 and 1906, but died before deliberations could have begun for the 1907 prize. All his nominations originated from foreign colleagues, and none from his compatriots. Although by 1905, three and a half decades had passed since his original discovery, recent advances in chemistry and in particular, the discovery of noble gas elements that were quickly reconciled with Mendeleev's system, added timeliness to it and again turned attention to his achievement.

Of the three chemists that made the short list of the Nobel Committee of Chemistry in 1905, viz., the German organic chemist Adolf von Baeyer (1835–1917), the French inorganic chemist Henri Moissan (1852–1907), and Mendeleev, von Baeyer was selected by the Committee unanimously and was awarded the Nobel Prize by the Royal Swedish Academy of Sciences. In 1906, the Nobel Committee of Chemistry recommended Mendeleev to the Academy by a majority vote of 4 to 1, noting that the minority candidate Moissan would also be acceptable. The Academy awarded the prize to Moissan rather than to Mendeleev, by a majority vote. This would have been the last opportunity for Mendeleev to receive the Nobel Prize. Incidentally, Moissan also died a few weeks after Mendeleev.

Mendeleev was never much recognized in his home country. Although he was elected as the corresponding member of the Russian Academy of Sciences, he never made it to full member, which is a much higher distinction in the two-tier system of the Russian Academy. As a Professor, he sympathized with student movements for improving their conditions, and for this, at some point, he was removed from his university position. From 1893, he continued as Controller of the Board for Weights and Measurements (today, the Mendeleev Institute of Metrology), where he was no longer in contact with students. He was much appreciated abroad; his many distinctions included membership of the Royal Society (London), the National Academy of Sciences (USA), and the Royal Swedish Academy of Sciences. He received the Copley Medal—the highest award of the Royal Society—in 1905 "for his contributions to chemical and physical science." Perhaps, as the highest distinction, the radioactive element of atomic number 101, Mendelevium, Md, produced artificially in 1955, carries his name.

References

1. Kedrov BM (ed) (1958) D. I. Mendeleev: Periodic law (in Russian, D. I. Mendeleev: Periodicheskii zakon). Izd. Akad. Nauk SSSR, Moscow
2. Dobrotin RB, Karpilo NG, Kerova LS, Trifonov DN (1984) Chronicle of D. I. Mendeleev's life and work (in Russian, Letopis' zhiznii i deyatel'nosti D. I. Mendeleeva). Nauka, Leningrad
3. Blokh AM (2005) The Soviet Union in the dealings of the Nobel prizes: Facts; Documents; Considerations; Commentaries (in Russian, Sovyetskii Soyuz v inter'yere nobelevskikh premii. Fakti. Dokumenti. Razmishleniya. Kommentarii), 2nd revised and augmented edn. Fizmatlit, Moscow
4. Hargittai I (2002) The road to Stockholm: Nobel prizes, science, and scientists. Oxford University Press, Oxford

Dmitry I. Mendeleev's grave in the museum section, called "Literary Bridges," of the Volkovskoe Cemetery in St. Petersburg (photograph by Istvan Hargittai).

Ulf Lagerkvist and His Nobel Histories[a]

István Hargittai

Abstract

Upon retirement from his professorship in medical biochemistry at Gothenburg University, Ulf Lagerkvist (1926–2010) wrote several books about great discoveries in chemistry and biology and examined both their awarded and missing Nobel Prizes. He combined a deep interest in historical development, intimate knowledge in his subject matters, and a natural gift of an engaging writing style. His latest, posthumous book was about the missing Nobel Prize for Dmitrii Mendeleev's Periodic Table of the Elements.

I met Ulf Lagerkvist first through his book, DNA Pioneers and Their Legacy [1]. Eventually we started corresponding and at one point, in September 2003, I spent a couple of days in the Lagerkvists' home in Gothenburg. Our conversations with him and his wife gave much intellectual pleasure. Only later did I learn that his father was a Nobel laureate author. Ulf did not take up writing science history until his retirement from his professorship of medical biochemistry at Gothenburg University. His book on the DNA scientists was followed by another, Pioneers of Microbiology and the Nobel Prize [2], and a third one, The Enigma of Ferment: From the Philosopher's Stone to the First Biochemical Nobel Prize [3]. In his books he gave a concise introduction to the science he discussed and then narrated the story of the respective Nobel Prizes without shying away from mentioning controversies if they were noteworthy. His style was truly instructive without being pedagogical; all essential information was given yet he managed to maintain a readable style. Another characteristic of his books was economy and the generous application of photographs of fairly large size. All this is present in his last, posthumous book, The Periodic

Table and a Missed Nobel Prize (its cover is shown in Fig. 1) [4]. When Ulf died in June 2010, unexpectedly, the manuscript was ready for publication except the illustrations. Erling Norrby whose book Nobel Prizes and Life Sciences was recently reviewed in this Journal [5], completed the collection of the photographs for the book and saw the manuscript through the production process.

The title of Lagerkvist's last book is too modest, because there is much more covered in it than the Periodic Table and Mendeleev's missing Nobel award, in spite of its brevity. The book consists of three parts, "Elements, Atoms and Molecules," "Atomic Weights and Their Relation to Chemical Properties of Elements," and "The Elusive Nobel Prize." It is almost puzzling how Lagerkvist managed to cram so much information, once again, into this relatively small book of xii + 122 pages of which one fourth of the pages is taken up by illustrations. The style is, again, unhurried and the book certainly reads more like a novel than a scientific monograph, yet it is accurate in all aspects. The highlights of the book are the narratives about the development of atomic weight determination, Lothar Meyer's achievements in discovering the system of the elements and correlating them with their properties, and, of course, Mendeleev's discovery. The quest for the missing elements that Mendeleev had predicted on the basis of his periodic table is retold that made Mendeleev's accomplishment stand out.

By the time the institution of the Nobel Prize had been established, Mendeleev's discovery had become a deed of the past and the rule prescribing recency appeared excluding it from considerations for the award. Here, reference is made to the expression "during the preceding year" in Alfred Nobel's Will. However, already early on the awarders of the prize realized that adhering strictly to the letter of the Will might not correspond to its spirit, and soon introduced a qualifying stipulation. The Statutes of the Nobel Prize determine many of the details that had not been unambiguously clarified in the Will. Its §2 instructs that older works can also be considered "if their significance has not become apparent until recently." For Mendeleev's Nobel Prize this amendment should have

[a]Originally published in Structural Chemistry 2012, 23:1663–1666.

I. Hargittai (✉)
Department of Inorganic and Analytical Chemistry, Budapest University of Technology and Economics, Budapest, Hungary
e-mail: istvan.hargittai@gmail.com

Fig. 1 Cover of Ulf Lagerkvist's last book.

solved all difficulties, because recent discoveries kept enhancing the significance of his discovery. In particular, the discovery of inert gases and their seamless fit to Mendeleev's Table brought the justification of Mendeleev's award into the forefront. In 1904, both the physics and the chemistry Nobel Prizes went to the discoverers of the inter gases. The physics prize went to Lord Rayleigh (John William Strutt) "for his investigation of the densities of the most important gases and *for his discovery of argon* in connection with these studies" (emphasis by me). The chemistry prize went to William Ramsay "in recognition of his services in the discovery of the inert gaseous elements in air, and *his determination of their place in the periodic system*" (emphasis by me).

This is where we pick up the story and follow Lagerkvist's discussion in somewhat more detail. The discovery of the inert gases prompted Otto Pettersson to nominate Mendeleev for the 1905 chemistry prize; Mendeleev had not been nominated before. In 1905, however, the organic chemist Adolf von Baeyer was awarded the Nobel Prize. Incidentally, in his case, it was also §2 of the Statutes that helped justifying his selection. Along with von Baeyer, Mendeleev and Henri Moissan were also seriously considered for the 1905 award. Moissan's merits

were in the discovery of fluorine and the construction of an electric oven, which made it possible to conduct high-temperature experiments. The chemistry Nobel committee though gave preference to Mendeleev over Moissan. The latter had received nominations before and in 1905 the number of his nominations far exceeded that for Mendeleev.

For the 1906 chemistry prize, both Mendeleev and Moissan received nominations, along with other scientists. The deliberations of the chemistry Nobel committee led to the recommendation that Mendeleev should be awarded the prize. There was one dissent; Peter Klason, a professor at the Royal Institute of Technology in Stockholm argued tirelessly for Moissan and submitted a separate report in favor of Moissan over Mendeleev. His main point against Mendeleev's award was that the periodic law had been a fundamental part of chemistry for decades, and he did not see justification in applying §2 to Mendeleev's case. Besides, he argued that without accurate atomic weights Mendeleev could have not discovered his system and had Mendeleev been selected for the award, this could not be done without selecting Stanislao Cannizaro (for the accurate atomic weights) as well. According to the Statutes, however, only people who had been nominated for the Nobel Prize could be considered, and Cannizaro had not been nominated for the 1906 award. The chemistry Nobel committee could not arrive at a unanimous decision; Klason introduced sufficient doubts about Mendeleev's selection, on the one hand, while he kept forcefully arguing for Moissan's selection. Klason appears to having been a shrewd politician. He never belittled Mendeleev's merits and achievements; he only argued that objective considerations made it impossible to give the 1906 Nobel Prize to Mendeleev. Also, Klason was not alone in placing the feasibility of Moissan's selection over Mendeleev's. Moissan did solid experimental work whereas Mendeleev had been engaged in more loosely appearing theoretical considerations that some considered too speculative to be sound. Finally, a narrow majority decided in favor of Henri Moissan. In 1906, there was no chemistry lecture delivered; Moissan never gave his Nobel lecture and died in 1907. Mendeleev also died in 1907; thus one of the greatest discoveries in chemistry remained without Nobel recognition. The Nobel Prize is a great institution, but in this case, its prestige suffered and not Mendeleev's, and his missing prize remains one of the most conspicuous blunders of the Nobel Prize in its history. The Nobel institution's omission of Mendeleev could be compared with what the Academie Française once wrote of Molière: "Rien ne manquait à sa gloire, il manquait à la nôtre" (His crown of glory was complete; ours lacked only him) [6].

Ulf Lagerkvist gives a lively account of the events of more than a 100 years ago. His last book is an important contribution to science history and a beautiful memento of his oeuvre.

In August 2012, my wife and I spent a week in St. Petersburg and visited both the University and the Institute of Technology; Mendeleev used to work at both. We are sharing a few images from our visit at the Institute of Technology, its Mendeleev

Fig. 2 Dmitrii Mendeleev's bust at the Mendeleev Museum of the Institute of Technology in St. Petersburg (photo by I. Hargittai).

Fig. 4 Mendeleev's periodic table of the elements as fresco on the wall of the Central Bureau of Weights and Measures in St. Petersburg (photo by Magdolna Hargittai).

Museum, and at the building of the Central Bureau of Weights and Measures in St. Petersburg (Figs. 2, 3, 4, 5). This is only a minute fraction of how the world of chemistry and the world at

Fig. 3 Mendeleev's bust in the yard of the Institute of Technology in St. Petersburg (photo by Magdolna Hargittai).

Fig. 5 Mendeleev's statue in front of the fresco (photo by Magdolna Hargittai).

large remember Dmitrii Mendeleev. His name is commemorated also in the Periodic Table of Elements as Element 101, mendelevium, Md, just before the one commemorating Alfred Nobel, Element 102, nobelium, No.

References

1. Lagerkvist Ulf (1998) DNA pioneers and their legacy. Yale University Press, New Haven and London
2. Lagerkvist Ulf (2003) Pioneers of microbiology and the Nobel Prize. World Scientific, Singapore
3. Lagerkvist Ulf (2003) The enigma of ferment: from the philosopher's stone to the first biochemical Nobel Prize. World Scientific, Singapore
4. Lagerkvist Ulf (2012, edited by Erling Norrby) The periodic table and a missed Nobel Prize. World Scientific, Singapore
5. Hargittai I (2011) Struct Chem 22:483–487
6. In this, I followed R. Dubos who wrote Oswald Avery's biography and applied such a comparison to Avery's missing Nobel Prize. Avery's discovery that DNA is the substance of heredity was left without Nobel recognition constituting another most conspicuous blunder of the Nobel Prize institution. See, Dubos RJ (1976) The professor, the institute, and DNA. Rockefeller University Press, New York, p 159; The English translation of the French sentence in our case was by Alan L. Mackay, FRS, London, private communication, 2001

Andrei Sakharov's statue on Sakharov Square in St. Petersburg with university buildings in the background (photograph by Istvan Hargittai). Sakharov's hands are tied behind his back.

Los Alamos and "Los Arzamas"[a]

István Hargittai

Abstract

Marking the seventieth anniversary of the Los Alamos Laboratory provides an opportunity for comparison with its Soviet counterpart, Arzamas-16 (nicknamed "Los Arzamas"). There were similarities and differences, but in their principal motivations and treatments of their scientists, they diverged irrevocably. This Editorial is based on an invited presentation on June 12, 2013, at the Norris E. Bradbury Science Museum, Los Alamos National Laboratory, in Los Alamos, New Mexico.

This year, the Los Alamos Laboratory celebrates its seventieth anniversary. It came to life in 1943 as the concluding segment of the Manhattan Project to produce the atomic bombs for the US Army. In August 1945, these bombs were dropped over Hiroshima and Nagasaki. Apart from the devastation and human tragedies they caused, their immediate consequences included the surrender of Japan and the conclusion of World War II (Fig. 1). The Los Alamos Laboratory had importance well beyond World War II and the scientists working for the Soviet nuclear program at the secret Soviet installation, Arzamas-16, nicknamed their laboratory "Los Arzamas." This note focuses on some similarities and differences between Los Alamos and Arzamas-16.

The two laboratories had a one-way direct connection through espionage due to which the first Soviet atomic bomb was a copy of the American plutonium bomb. Only the leadership of the Soviet project was aware of the source of information, the scientists were merely given the tasks of what solutions to work out. It proved to be a good exercise for them, but a frustrating experience since they could not bring in their own ideas. For the hydrogen bomb, with less intelligence, the Soviet physicists could utilize their innovative talents. Even later, the shadow of Los Alamos over Arzamas-16 did not disappear entirely. In 1983, the long time scientific director Yulii Khariton wondered loud in a critical moment whether they could guess what the Americans might do in a similar situation. His colleagues sometimes called Khariton the Soviet Oppenheimer [1].

J. [Julius] Robert Oppenheimer (1904–1967), the first scientific director of Los Alamos, filled the post for only 2 years yet his name has become synonymous with Los Alamos (Fig. 2, top). Oppenheimer trained for physicist, but his early achievements included seminal discoveries in chemical physics; suffice it to mention the Born–Oppenheimer approximation worked out jointly with his Göttingen mentor, Max Born. Oppenheimer was an unlikely choice for the post of Los Alamos scientific director, but the military commander of the Manhattan Project, General Leslie Groves, had the right instinct in making it. Oppenheimer had the intellectual capacity to oversee a complex project; possessed the talent in theoretical physics to wield authority over his colleagues, and was eager to prove himself. His past involvement in leftist politics made him feel insecure, and Groves probably sensed that this made Oppenheimer pliable. The physicist came from an upper-middle-class nonreligious Jewish family. His youth was at the time when anti-Jewish discrimination was still widespread in American academia. According to the renowned physicist Isidor I. Rabi, "Oppenheimer was Jewish, but he wished he weren't and tried to pretend that he wasn't" [2]. This must have contributed to Oppenheimer's feeling vulnerable. Oppenheimer performed impeccably against all odds and in spite of the harassments he suffered from the security services that did not trust him.

There was a conspicuous concentration of Jewish refugee scientists from Europe at Los Alamos. By the time the laboratory came to life, most other scientists had already been engaged in war-related projects. The refugees were latecomers in becoming US citizens to allow them

[a]Originally published in *Structural Chemistry* 2013, 24, 1397–1400.

I. Hargittai (✉)
Department of Inorganic and Analytical Chemistry, Budapest University of Technology and Economics, Budapest, Hungary
e-mail: istvan.hargittai@gmail.com

Fig. 1 "Whose son will die in the last minute? Minutes Count!" The poster refers to the deployment of the atomic bombs in anticipation of the expected huge sacrifices of the invasion of Japan in 1945. Photograph of the legendary Ed Westcott; courtesy of Oak Ridge National Laboratory.

participation in other classified projects. The atomic bomb project was a latecomer among war-related research projects. The refugees had been kicked out of their home countries, and in the US, they were welcome and were found needed. The physics of nuclear weapons was challenging, and the refugee scientists were dedicated to the fight against Germany. The anti-Nazi Jewish resistance expressed itself not only in the uprisings of the Warsaw Ghetto and the Vilna Ghetto, but in the Manhattan Project as well [3]. The Hungarian refugee Eugene P. Wigner, later, Nobel laureate, reasoned that if he could come to the US, surely, so could Hitler. The scientists in the US were not unique in recognizing the potentials of the new nuclear physics for defense. Their German, Japanese, and Soviet colleagues came to similar conclusions, but their circumstances were different.

The Soviet nuclear weapons project had its roots at the time right after the discovery of nuclear fission in December 1938 in Berlin. Just as in the US, a few scientists began a project before it could have been sanctioned and financed by the government. Yakov Zeldovich and Yulii Khariton were the principal protagonists and they worked at Nikolai Semenov's Institute of Chemical Physics in Leningrad. Khariton started his career in chemical physics and he and Semenov had co-discovered the branched chain reactions in chemistry in the early 1920s (for which Semenov would receive the Nobel Prize in Chemistry in 1956). In 1933, the Hungarian refugee Leo Szilard in London came to the idea of the analogous nuclear chain reaction; he patented it in 1934, and deposited his patent with the British Admiralty.

Even the small-scale Soviet attempts came to a halt between 1941 and 1943 when the scientists had to work on improving traditional weapons, among them the famous Katyushas. The German troops were still fighting on Soviet territory, however, when the nuclear program, by now as a state project, resumed. Soon after the war ended, the Soviet government established the secret nuclear installation, Arzamas-16, some 240 miles east of Moscow.

Many of the most prominent Soviet physicists happened to be Jewish and some joined Arzamas-16. The nuclear weapons project protected the physicists during the difficult period of 1948–1953 when Stalin's paranoia developed into active anti-science as well as anti-Semitic persecution. When Zeldovich got into trouble in Moscow, he found refuge at Arzamas. Another Jewish physicist, Ovsei Leipunskii, found shelter at the even more distant Semipalatinsk Proving Ground in Eastern Kazakhstan, to ride out a crisis. Under Stalin, as well as under subsequent Soviet leaders, if there was a project deemed exceptionally important, it was exempted to observe quotas or even complete ban on hiring Jewish scientists.

Yulii Khariton (1904–1996), the long-time director of Arzamas-16, himself was a conspicuous exception (Fig. 2, bottom). His year of birth and his first name were not the only similarities with Oppenheimer (Yulii being the Russian equivalent of Julius). They both spent years in Western Europe for postgraduate studies. For both, this included Ernest Rutherford's Cavendish Laboratory in Cambridge, England. There, each had a future Nobel laureate for mentor;

Fig. 2 Top: Upper part of the statue of J. Robert Oppenheimer in Los Alamos, New Mexico (photograph by I. Hargittai). Bottom: Yulii B. Khariton on Russian postage stamp, 2004.

Patrick Blackett for Khariton and, a little later, James Chadwick for Oppenheimer. Khariton blended well into the Cavendish program and he earned his doctorate there whereas Oppenheimer did not, and left for Göttingen. Later as scientific director of Arzamas, Khariton tried to emulate what he experienced at the Cavendish—without success—but at least that was on his mind.

Like Oppenheimer, Khariton was Jewish, a life-threatening condition under Stalin and a definite disadvantage under the subsequent Soviet leaders. There was substantial difference between American anti-Semitism in academia—while it existed—and anti-Semitism in the Soviet Union. In the US, it was discrimination; in the Soviet Union it often developed into persecution. Khariton's situation was especially difficult. His mother lived in Palestine and his father had been kicked out of the Soviet Union and lived in a Baltic state. When in 1940, the Soviet Union annexed the

Baltics, he was arrested and directed to the Gulag. Every time Khariton had to submit an autobiography, he painstakingly described his family background—known to the authorities in more detail than to him—lest he be accused of hiding it.

It was for Khariton's exceptional talent and abilities that in spite of his circumstances he was made, and retained for 46 years, the scientific leader of the nuclear weapons installation. It was 46 years of luxurious isolation, a "golden cage," with his private railway car for travel and other perks and the highest decorations.

There was a superficial similarity in the motivations of the American and Soviet programs. With few exceptions, the Soviet scientists were dedicated to their nuclear weapons program, at least initially. They were past a bloody war called with good reason the Great Patriotic War, in which their nation literally fought for survival. In the early 1950s, they were taught that a yet more dangerous foreign enemy might attempt their annihilation. This is why even the future fearless human rights fighter Andrei Sakharov could propose murderous schemes to destroy densely populated foreign ports with Soviet thermonuclear devices.

Gradually, however, the Soviet scientists came to the realization that placing nuclear weapons into the hands of a dictator could have led to unforeseeable tragedies. Clashes between Sakharov and the Soviet leader Nikita Khrushchev demonstrated the blatant recklessness of the Soviet leadership in connection with the nuclear arms race. When during the 1967 war between Israel and its neighbors, Zeldovich heard about the consideration of dropping a nuclear bomb over Israel, he deposited a suicide note in secure hands (he knew the authorities would destroy such a note if they found it) and decided to kill himself if the bombing happened. Fortunately, it did not. The nature of relationship of the scientists toward the Soviet nuclear program changed. Los Alamos and "Los Arzamas" diverged irrevocably.

Khariton, on his part, never expressed dissidence. However, when in 1990, amid the great political changes in the Soviet Union, the octogenarian Khariton greeted the first US visitors at Arzamas-16, he told them: "I was waiting for this day for 40 years" [4].

References

1. Hargittai I (2013) Buried Glory: Portraits of Soviet scientists. Oxford University Press, New York
2. Cassidy DC (2005) J. Robert Oppenheimer and the American Century. Pi Press, New York, p 32
3. Hargittai M, Hargittai I (2004) "Yuval Ne'eman" in Candid Science IV: Conversations with famous physicists. Imperial College Press, London, pp 32–63
4. Khariton Yu B (2005) Put' dlinoyu v vek (in Russian, Century-long journey). Nauka, Moscow, p 428

Neil Bartlett and Magdolna Hargittai in Bartlett's office at the University of California at Berkeley, on May 11, 1999 (photograph by Istvan Hargittai). It was the last day of operation of Bartlett's laboratory at Berkeley.

Neil Bartlett and the First Noble-Gas Compound[a]

István Hargittai

Abstract

Neil Bartlett (1932–2008) was the first to produce non-transient amounts of a compound of a noble gas. By describing some of the circumstances of his seminal work we gain a glimpse into the process of his extraordinary experimental discovery. It happened in the spring of 1962 and was followed by an avalanche of other discoveries in noble-gas chemistry leading—over the years—to many new substances with the most interesting bonding properties and structures.

Neil Bartlett (at the 14th International Symposium on Fluorine Chemistry in Yokohama, Japan, 1994, photograph by I. Hargittai).

[a]Originally published in *Structural Chemistry* 2009, 20:953–959.

I. Hargittai (✉)
Department of Inorganic and Analytical Chemistry, Budapest University of Technology and Economics, Budapest, Hungary
e-mail: istvan.hargittai@gmail.com

In the June 1962 issue of *Proceedings of the Chemical Society* Neil Bartlett, then of the University of British Columbia, published one of the shortest—less than 250 words—seminal papers in science history announcing the discovery of a 1:1 compound between xenon and hexafluoroplatinate, $XePtF_6$, a mustard-yellow solid, whose formula could also be written as $Xe^+[PtF_6]^-$ [1]. Bartlett first sent his report on $XePtF_6$ to *Nature*, on April 2, 1962. It seemed to him that his manuscript might have been lost when he did not hear from the journal, so after a few weeks time, he withdrew his report. As he learned later *Nature* did acknowledge the receipt of his sending, but it mailed the acknowledgment by surface mail and it arrived in Vancouver only in June. By then Bartlett had sent off his note to the *Proceedings*, which was the forerunner of *Chemical Communications*. He was fully aware of the importance of bringing out the report about his discovery as quickly as possible. His manuscript arrived on May 4 at the *Proceedings*, was quickly accepted, and was printed in the next available issue, in June 1962 (Fig. 1).

Several ingredients came together for Bartlett's discovery in a fortunate way and it is of interest to discuss at least some of the circumstances of the discovery, because it is usually only the dry facts directly related to successful experiments that are described in the printed reports. The first interesting question is how the idea of trying to make a noble-gas compound came to Bartlett and how he came to the idea of trying the particular reaction between xenon and PtF_6. Here, our narrative is based on what Bartlett actually told me in an interview on May 11, 1999 [2]. The interview was followed up by e-mail messages. Bartlett was unequivocal in stating that nobody else suggested to him to try this particular reaction; it was his own idea [Bartlett N (June 13, 2000) Private communication by e-mail] (Figs. 2, 3).

He did not start thinking about this until he had identified O_2PtF_6. Bernard Weinstock and his group at the Argonne National Laboratory had recently published their preparation of platinum hexafluoride. Bartlett did not find it easy to make his own PtF_6, which he needed for some other experiments;

Reprinted from Proceedings of the Chemical Society, June, 1962, page 218

Xenon Hexafluoroplatinate(v) Xe⁺[PtF₆]⁻

By Neil Bartlett

(Department of Chemistry, The University of British Columbia,
Vancouver 8, B.C., Canada)

A recent Communication[1] described the compound dioxygenyl hexafluoroplatinate(v), $O_2^+PtF_6^-$, which is formed when molecular oxygen is oxidised by platinum hexafluoride vapour. Since the first ionisation potential of molecular oxygen,[2] 12·2 ev, is comparable with that of xenon,[2] 12·13 ev, it appeared that xenon might also be oxidised by the hexafluoride.

Tensimetric titration of xenon (AIRCO "Reagent Grade") with platinum hexafluoride has proved the existence of a 1:1 compound, $XePtF_6$. This is an orange-yellow solid, which is insoluble in carbon tetrachloride, and has a negligible vapour pressure at room temperature. It sublimes in a vacuum when heated and the sublimate, when treated with water vapour, rapidly hydrolyses, xenon and oxygen being evolved and hydrated platinum dioxide deposited:

$$2XePtF_6 + 6H_2O \rightarrow 2Xe + O_2 + 2PtO_2 + 12HF$$

The composition of the evolved gas was established by mass-spectrometric analysis.

Although inert-gas clathrates have been described, this compound is believed to be the first xenon charge-transfer compound which is stable at room temperatures. Lattice-energy calculations for the xenon compound, by means of Kapustinskii's equation,[3] give a value ∼ 110 kcal. mole⁻¹, which is only 10 kcal. mole⁻¹ smaller than that calculated for the dioxygenyl compound. These values indicate that if the compounds are ionic the electron affinity of the platinum hexafluoride must have a minimum value of 170 kcal. mole⁻¹.

The author thanks Dr. David Frost for mass spectrometric analyses and the National Research Council, Ottawa, and the Research Corporation for financial support.	(*Received, May 4th, 1962.*)

[1] Bartlett and Lohmann, *Proc. Chem. Soc.*, 1962, 115.
[2] Field and Franklin, "Electron Impact Phenomena," Academic Press, Inc., New York, 1957, pp. 114—116.
[3] Kapustinskii, *Quart. Rev.*, 1956, 10, 284.

Fig. 1 The first report of a noble-gas compound in June 1962 with Bartlett's dedication to the author in 1999.

his trials, together with his student, Derek Lohman, however, led to an accidental preparation of a new substance of the

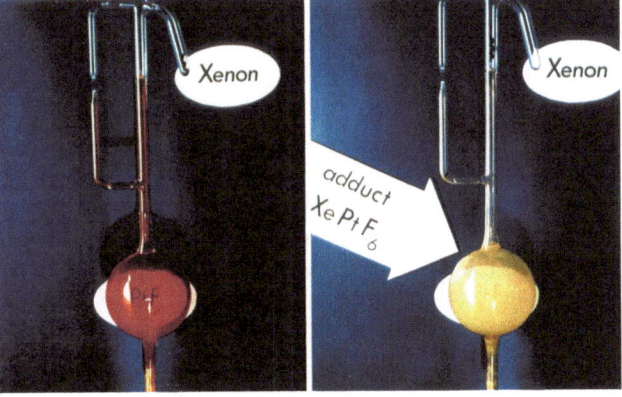

Fig. 2 The historic experiment producing the first noble-gas compound as it was recreated at the Lawrence Berkeley National Laboratory (courtesy of the late Neil Bartlett).

composition of PtO_2F_6 [3]. It took then a lot of experiments, thinking, and convincing doubting colleagues before Bartlett came to the conclusion that the substance could be looked at as being $O_2^+PtF_6^-$. When he finally convinced not only his peers but also himself of the correctness of this characterization, he realized that in PtF_6 he possessed the most potent oxidizing agent ever discovered. This made him think about what might be even more difficult to oxidize than molecular oxygen.

It was another fortunate circumstance that at this time Bartlett was preparing for his undergraduate chemistry course, and studied the plot of ionization potential as a function of atomic number, which is a common illustration of periodicity. He noticed that the first ionization potential of molecular oxygen is comparable with the ionization potential of xenon. From this he concluded that it should be possible to oxidize xenon by platinum hexafluoride, just as it was possible to oxidize molecular oxygen. When he realized that the way to go was reacting xenon with PtF_6, he ordered some

Fig. 3 Neil Bartlett at the time of the production of the first noble-gas compound (courtesy of the late Neil Bartlett).

xenon. His means were limited and he had to restrict his order to 250 cc. of the noble gas. Although he already had two graduate students, they were inexperienced, and Bartlett had to do even the glassblowing for the apparatus he designed for the reaction.

The crucial experiment happened on March 23, 1962; it was a Friday, and the apparatus was brought together by 7 p.m. His two graduate students both had left at this hour as they lived in a student residency where dinner was served at 6:30 p.m. Bartlett was all alone in the lab. He broke the seal between PtF$_6$ and xenon, and there was an immediate reaction. Bartlett desperately wanted to tell somebody—anybody—about it; he went out into the corridor, but there was not a soul in the whole building. So he went back to the lab and suddenly he was flooded with doubts, "Maybe xenon was impure, maybe there was some oxygen present, maybe I'm just fooling myself." Both the urge to communicate about his experiment and the doubts seem to be typical in such moments of discovery.

Bartlett then sublimed the reaction product by heating it under vacuum, then condensed water onto it and collected the evolving gas in one part of his apparatus. He then sealed this part of the apparatus and the next day he gave it to his mass spectrometrist colleague, David Frost. The mass spectrometric analyses during the next couple of days provided

reassuring proof to Bartlett as to the composition of his product.

It is interesting that the world of chemistry appeared ready for Bartlett's discovery because within a couple of months a group of scientists at Argonne National Laboratory, Howard Claasen, Henry Selig, and John Malm reported the synthesis of the first binary xenon compound [4]. Then, before the end of 1962, the preparations of further noble-gas compounds were announced by Rudolf Hoppe and his group at the University of Münster, Germany [5], Jozef Slivnik and his group at the Jozef Stefan Nuclear Institute in Ljubljana (then Yugoslavia, now Slovenia) [6], and Paul Fields and his associates at the Argonne National Laboratory [7].

Word was also spreading fast about Bartlett's discovery even prior to its publication. Parallel to his initial submission to *Nature*, Bartlett informed four people whom he trusted about the manuscript; R. S. Nyholm at University College London, A. G. Sharpe and H. J. Emelius at Cambridge, UK, and R. D. Peacock at Imperial College, London [Bartlett N (June 13, 2000) Private communication by e-mail]. Peacock later told Bartlett that he had shared this confidential information with others whom he trusted at a meeting of the Chemical Society. The news traveled fast. This was important for a later dispute that developed concerning the priority of the discovery of xenon tetrafluoride between the German group and the Argonne laboratory. Bartlett referred to this development as follows: "Certainly the Argonne discovery was a direct consequence of mine. I had after all strayed into their territory when I discovered the nature of $O_2^+[PtF_6]^-$. They were all set to repeat my experiment and to extend it to the other transition series hexafluorides. It was their experiments with xenon and RuF$_6$, which held the key (they recognized RuF$_5$ among the products). Professor Hoppe has never acknowledged that his efforts were stimulated by my publication of XePtF$_6$, although he has acknowledged my priority" [Bartlett N (June 13, 2000) Private communication by e-mail]. In contrast with other priority disputes in connection with some of the follow-up discoveries, Bartlett's achievement has remained recognized as the pioneering one.

Bartlett was by far not the first who tried to prepare noble-gas compounds. The earliest attempts go back to the time of the isolation and identification of the noble gases and to Henri Moissan and William Ramsay at the end of the nineteenth century. The first noble gas discovered by Ramsay was argon and the discovery of fluorine by Moissan preceded it by 9 years only. When Ramsay observed the inertness of argon, he sent a sample to Moissan who mixed it with fluorine and passed the reaction mixture through an electric discharge. He analyzed the effluent very carefully and could not detect any sign of any combination. He reported this observation to Ramsay and this early experiment had a strong impact in how chemists viewed the noble gases for a long time.

The most noteworthy of later attempts to prepare noble-gas compounds were by a professor at the Gates Chemical Laboratory of the California Institute of Technology, Don M. Yost, and his graduate student Albert L. Kaye in 1933 [8]. They acted upon Linus Pauling's suggestion and used a sample of xenon from Fredrick John Allen. Allen was Pauling's old chemistry teacher who was by then at Purdue University. Yost and Kaye's attempt did not succeed, but was found important enough to describe it in the *Journal of the American Chemical Society*. They concluded that "It cannot be said that definite evidence for compound formation was found." Even more significantly, they stated that "It does not follow, of course, that xenon fluoride is incapable of existing" [8].

According to Bartlett, it was unfortunate that Yost and Kaye used a quartz bulb for carrying out their reaction. When they passed an electric discharge through the bulb all they must have gotten was an attack on the walls of the quartz container. Bartlett supposed that they got silicon tetrafluoride in addition to oxygen liberated from the container material. Bartlett ascribed their main mistake to the fact that they did not take their reaction mixture out into the California sunlight [2].

Three decades later, after the discovery of the first noble-gas compounds, Yost summarized their efforts and the lessons that could be learned from them [9]. He described the rudimentary conditions of their experiments and noted that "only visionary scholars" could have dreamed of the conditions among which decades later similar experiments could be performed. He put it succinctly, "Fluorine chemistry was then carried out in the days of wooden ships and iron men." Yost did not in any way though belittle the significance of the achievements of the discoverers. On the contrary, he stated that "chemistry had reached a stationary state, during which no profoundly fundamental discoveries were reported or even deemed possible.... This state of affairs has now been changed.... So long as man shows any interest whatever in chemistry, the discovery of xenon and other noble-gas fluorides will not be forgotten." Further, Yost advanced his vision of the future: "One can envision a whole sombrero full of studies that can be made on noble-gas compounds and their derivatives, which will enrich our knowledge of nature and her laws." He also predicted that applications will come forward.

Incidentally, Bartlett used Pyrex glass in his experiments. In this connection, he learned important lessons from the Manhattan Project where the people involved in making UF_6 could work with fluorine more easily if they scrubbed the traces of HF out of their fluorine sample. This was possible to do by passing the fluorine over sodium fluoride, forming sodium bifluoride. The fluorine thus treated was much easier to contain in dry glass. Water content, on the other hand, would destroy the container regardless whether it was glass or quartz.

Actually, Bartlett considered it a great advantage for his experiment that he worked with glass and could always follow what was happening to his reaction mixture. In contrast, the Argonne people used metal lines and metal valves and they did not have the possibility to observe everything during the reaction. Bartlett stressed though that he had learned a lot from the Argonne work on platinum hexafluoride and he praised the beauty and sophistication of their work, especially their spectroscopy [2]. The Argonne group was also very strong in understanding the physics of the new compounds, including the Jahn–Teller features of their molecular structures [10].

On his part, Bartlett stayed in preparative inorganic chemistry throughout his career and produced exciting new substances that enriched our field enormously. When asked if there was ever a compound that he would have liked to prepare but never succeeded, he named gold hexafluoride. He said in 1999 as he was closing down his laboratory at the Berkeley campus, "Somebody else will now have to find the effective route to it. It should exist, if made at low temperature and kept cold" [2].

Neil Bartlett was a gracious man who could be happy at seeing other people succeeding as well. He could hardly contain his joy when he learned about the discovery of the new substance of HArF by a group of chemists in Helsinki [11]. He heard about it over the public radio at dinner time on August 24, 2000. The next day he sent an e-mail message to his fluorine chemist colleagues describing his ideas about what he supposed might have been the bonding properties of HArF. He also wrote to congratulate the group in Finland and subsequently "had an interesting and pleasant exchange of letters with them" [Bartlett N (January 11, 2001) Private communication by e-mail].

The molecular structures of noble-gas compounds have provided beautiful and instructive examples of various models of molecular geometry and chemical bonding. In particular, xenon hexafluoride provided an early proof for the validity and usefulness of the valence shell electron pair repulsion (VSEPR) model [12]. Its structure had been predicted to be regular octahedral on the basis of some rudimentary molecular orbital considerations. However, the valence shell of xenon contains a lone pair of electrons in addition to six bonding pairs in XeF_6 and thus the shape of a distorted monocapped octahedron would be more consistent with the VSEPR model than a regular octahedral shape. Reliable experimental determinations then provided evidence for the structure suggested by the VSEPR model. We have to add though that the last word has not been said about the molecular shape of xenon hexafluoride as there is no clear-cut decision between a regular octahedral shape and a distorted one. There are competing effects and the outcome of even the most sophisticated calculations depends on the extent these various effects are being taken into account.

Toward the end of his career as an experimentalist, Bartlett and his group at Berkeley returned to the topic of his original discovery and performed an extensive investigation concerning the nature of $XePtF_6$ [13]. It is of interest to look at the authors' list of this paper. Two of them were Bartlett's associates at Berkeley. Then there was Narendra K. Jha who had been Bartlett's doctoral student at UBC in the early 1960s and received his Ph.D. degree there in 1965, but by the time of this more recent paper he had long before left UBC. Nonetheless, he was assigned UBC affiliation, which is the more conspicuous since no other affiliation was given for him. In an almost romantic gesture, Bartlett added UBC to his own Berkeley affiliation.

In this review they used experimental results from the early 1960s and added new ones. They confirmed that when xenon was in large excess over platinum hexafluoride in the reaction mixture, the stoichiometry of the product approached 1:1. However, when the platinum hexafluoride exceeded the xenon, a different product formed, which was a sticky solid of deep red color. This product approached the composition of $Xe(PtF_6)_2$. This could correspond also to the product obtained when XeF_2 and PtF_5 were brought into interaction. This work, performed in 1999, investigated in great detail the nature and physical characterization of all possible products in related reactions. Here we note only that Bartlett never felt satisfied until he uncovered all possible angles of a problem. His new findings did not subtract anything from the significance of his original discovery, on the contrary, further enhanced the weight of his pioneering work. He gave a full account of his 40 years in fluorine chemistry as a chapter in a monograph in the year 2000 [14].

At this point it may be illuminating to have a few words about Neil Bartlett's life and career [2]. He was born in 1932 in Newcastle-upon-Tyne, England. His parents started their married life at the time of the Great Depression and they ran a corner shop. His father had served in World War I and was severely gassed. The father died early and left his widow with three children, but she managed to cope, and encouraged her children to study. She was even very tolerant of Neil's running chemical experiments at home accompanied with foul smells and frightening bangs. Neil and his brother augmented their pocket money by making and selling ice cream on weekends. Neil used the proceeds to buy chemicals and books.

Bartlett's maternal grandfather had the German-sounding name Vock, Friedrich Wilhelm Heinrich Johannes Vock in full, and he was from the small archipelago in the North Sea, Heligoland, which was British at the time. He migrated from Heligoland to Newcastle. When Queen Victoria and the German Kaiser exchanged Heligoland and Zanzibar in 1890, the grandfather became German. Bartlett's mother changed her name from Vock to Voak during World War I to make it sound more English.

Bartlett went to school in Newcastle and received his B.Sc. degree in 1954 and his Ph.D. degree in 1958, both from King's College, University of Durham, at Newcastle, which is today the University of Newcastle. In 1958, he moved to Vancouver, became a faculty member of the University of British Columbia, and stayed there until 1966. It was during this period that he discovered the first noble-gas compound.

Bartlett's stay at UBC was successful also for his rapid rise from lecturer to full professor in 5 years. On the other hand, he sensed some personal animosities—at least this is what he alluded to when looking back to this period in our conversation in 1999. In spite of his rapid advance in rank, he felt that UBC should have done better. At the same time, his contacts with the Argonne National Laboratory showed him the contrast in opportunities that he lacked in Vancouver. Then there was a laboratory accident on January 27, 1963, which somehow became associated with Vancouver in Bartlett's mind. He and one of his students were looking at an experiment and each was wearing a plastic visor. Although Bartlett had very good eyesight, the visor barred him from having a good look at the surface of the crystal that was being formed, so he put his visor up and so did his student. At that very moment the sample they were looking at blew up. They both were injured; Bartlett more heavily than the student. They were carted off to hospital where they stayed for over 4 weeks. Bartlett had some glass in his eye that kept bothering him for the next 27 years, and came out only in 1990.

There were other unfavorable conditions at UBC at the time, so when an attractive offer came from Princeton, he accepted. Eventually he realized that he felt more comfortable at UBC than at Princeton, at least in hindsight, that he preferred a smaller campus to a big one, and that he was a West Coast person rather than an East Coast one. When he had an invitation from Berkeley, he accepted it and moved to California in 1969. There he was Professor of Chemistry at the University and served also as Principal Investigator at the Lawrence Berkeley National Laboratory. When my wife and I visited him on May 11, 1999, he did not strike us as a happy person. The date was significant, because it was the day when he was closing down his laboratory for good.

Neil Bartlett was an internationally renowned scientist. Among many other distinctions, in 1973, he was elected Fellow of the Royal Society (London); he was a Foreign Associate of the National Academy of Sciences of the USA and of the French Academy of Sciences. In 1976, he received the Robert A. Welch Award in Chemistry. In 2006, the American Chemical Society and the Canadian Society for Chemistry honored him with an International Historical Chemical Landmark plaque on the wall of the building where his seminal discovery took place at UBC in 1962. It reads, both in English and in French.

Neil Bartlett and reactive noble gases

In this building in 1962 Neil Bartlett demonstrated the first reaction of a noble gas. The noble gas family of elements—helium, neon, argon, krypton, xenon, and radon—had previously been regarded as inert. By combining xenon with a platinum fluoride, Bartlett created the first noble gas compound. This reaction began the field of noble gas chemistry, which became fundamental to the scientific understanding of the chemical bond. Noble gas compounds have helped create anti-tumor agents and have been used in lasers.

The first page of the first chapter in the Italian chemist turned famous writer Primo Levi's book, *The Periodic Table* says the following, "As late as 1962 a diligent chemist after long and ingenious efforts succeeded in forcing the Alien (xenon) to combine fleetingly with extremely avid and lively fluorine, and the feat seemed so extraordinary that he was given a Nobel prize" [15]. Bartlett never received the Nobel Prize, but this mark of distinction from Primo Levi reinforces the fact that the name Bartlett is immortalized as a result of his beautiful experiment in the annals of science.

Acknowledgments Our research is being supported in part by the Hungarian Scientific Research Foundation (OTKA Nos. K60365 and T046183). I thank Professors Emeritus William R Cullen and Robert C Thompson, both of the Chemistry Department, University of British Columbia, for advice.

References

1. Bartlett N (1962) Proceedings of the Chemical Society, p 218
2. Hargittai I (2003) Candid science III: more conversations with famous chemists, edited by M. Hargittai (interview with Neil Bartlett), pp 28–47
3. Bartlett N, Lohmann DH (1962) Proceedings of the Chemical Society, p 115
4. Claassen HH, Selig H, Malm JG (1962) J Am Chem Soc 84:3593
5. Hoppe R, Dähne W, Mattauch H, Rödder KM (1962) Angew Chem 74:903
6. Slivnik J, Brcic B, Volavsek B, Marsel J, Vrscaj V, Smalc A, Frlec B, Zemljic Z (1962) Croat Chem Acta 34:253
7. Fields PR, Stein L, Zirin MH (1962) J Am Chem Soc 84:4164–4165
8. Yost DM, Kaye AL (1933) J Am Chem Soc 55:3890–3892
9. Yost DM (1963) A new epoch in chemistry. In: Hyman HH (ed) - Noble-gas compounds. The University of Chicago Press, Chicago and London, pp 21–22
10. Hargittai M, Hargittai I (2009) Struct Chem 20:537–540
11. Khriachtchev L, Pettersson M, Runeberg N, Lundell J, Räsänen M (2000) Nature 406:874–876
12. Hargittai I (2009) Struct Chem 20:155–159
13. Graham L, Graudejus O, Jha NK, Bartlett N (2000) Coord Chem Rev 197:321–334
14. Bartlett N (2000) Forty years of fluorine chemistry. In: Banks RE (ed) Fluorine chemistry at the millennium. Elsevier, Amsterdam, pp 29–55
15. Levi P (1984) The periodic table (translated from the Italian by Raymond Rosenthal). Schocken Books, New York, p 4

British physicist David Shoenberg (1911–2004) in 2000 in Cambridge, UK (photograph by Istvan Hargittai). Istvan dedicated the following Editorial about Graphene to David Shoenberg who was a pioneer in the application of strong magnetic fields, including levitation of Type I superconductors (see, Refs. [13 and 14] in the following paper). Shoenberg told Istvan about his interesting life and about his experiences with Petr Kapitsa and Lev Landau.

Graphene 2010[a]

István Hargittai

Science does not need to be boring to be good.
Andre Geim in 2006 [1]

Abstract

The first production of graphene was awarded the 2010 Nobel Prize in Physics. This discovery has implications for chemistry and within it for structural chemistry as well.

Andre Geim (courtesy of Andre Geim).

Konstantin Novoselov (courtesy of Konstantin Novoselov).

On October 5, 2010, the Royal Swedish Academy of Sciences in Stockholm awarded the 2010 Nobel Prize in Physics jointly to Andre Geim and Konstantin Novoselov "for groundbreaking experiments regarding the two-dimensional material graphene" [2]. By happy coincidence, earlier on the same day, our editorial office had accepted a manuscript for publication, "Remarkable diversity of carbon–carbon bonds: Structures and properties of fullerenes, carbon nanotubes, and graphene," in which T. C. Dinadayalane and Jerzy Leszczynski reviewed, among others, the results of computational studies on graphene [3]. The manuscript had been submitted in July, however, due to the summer vacations, its reviewing process took a little longer than it normally would have.

The Nobel Prize in Physics in 2010 was among those rather rare cases when the Nobel recognition followed relatively quickly the discovery. Geim and his co-authors communicated the production of graphene in 2004 for the first time. Graphene is a single sheet of carbon atoms. It was a fortunate circumstance that they gave the easy and appealing name of graphene to the new material. Coining a name—and an easy and attractive one at that—is arguably one of the most important, if not the most important component of a scientific discovery [4]. The new material is extremely strong along with many other favorable properties promising a plethora of applications. It may well be that the speedy Nobel recognition was a consequence of the anticipation of an avalanche of studies in the foreseeable future in this area of research after which it might be more difficult to single out up to three awardees. Here graphene is symbolized by an obviously strong window fence in the Topkapi Sarayi, Istanbul, which could be taken as a model of graphene (Fig. 1).

[a]Originally published in *Structural Chemistry* 2010, 21:1151–1154.

I. Hargittai (✉)
Department of Inorganic and Analytical Chemistry, Budapest University of Technology and Economics, Budapest, Hungary
e-mail: istvan.hargittai@gmail.com

Fig. 1 "Graphene model" as a window fence at Topkapi Sarayi in Istanbul (photograph and © by the author).

Fig. 2 "Nanotube model" as detail of a decoration at the Royal Palace in Bangkok (photograph and © by the author).

The principal discoverer, Andre Geim was born in 1958 in Sochi, Russia, which is on the Eastern shore of the Black Sea. When he was 6 years old, the family moved to Nalchik, further East, on the Northern slopes of the Caucasian Mountains, approximately midway between the Black Sea and The Caspian Sea. There he graduated from a high school specializing in the English language. He continued his studies in Moscow and did his Diploma work (Master's degree equivalent) in 1982 at Moscow Institute of Physics and Technology. He earned his Candidate's degree (PhD equivalent) in 1987 at the Institute of Solid State Physics of the Soviet Academy of Sciences (now Russian Academy of Sciences). Geim came from a family of Jewish-German origin and as being Jewish was considered to be a nationality his identity documents carried this designation causing barriers in his receiving higher education. He always felt he had to outperform others to survive in the Soviet system. Following his doctorate, Geim did research at the Institute of Microelectronics Technology of the Soviet (then, Russian) Academy of Sciences in Chernogolovka, near Moscow. With the crumbling and then collapse of the Soviet Union, travel became easier and Geim continued as postdoctoral fellow at the University of Nottingham, University of Bath, and University of Copenhagen. Finally, he got his appointment as associate professor at the Radboud University in Nijmegen, Holland. In 2001, he became

Langworthy Professor of Physics at the University of Manchester and he has directed its Center for Mesoscience and Nanotechnology. He is now a Dutch citizen.

Fig. 3 "Slightly irregular fullerene model" under the paw of a dragon at Forbidden City outside Beijing (photograph and © by the author).

Konstantin S. Novoselov was born in 1974 in Nizhny Tagil, Russia, in the Southern region of the Ural Mountains. He did his Diploma work at the Moscow Institute of Physics and Technology and in the early 1990s he moved to Nijmegen where he started his PhD work with Geim as his mentor. Novoselov moved to Manchester along with Geim in 2001. He holds both Russian and British citizenships.

Geim (while still in Nijmegen) and his colleagues communicated a photograph in April 1997 displaying a levitating frog [5]. It was taken as an April Fool's joke. In reality, however, Geim and his colleagues suspended the frog, among many other similar experiments with non-magnetic objects, by creating an upward magnetic force from a powerful magnet and thus they succeeded in compensating for the effect of gravity. Soon a British scientist Michael Berry developed a theory to interpret the phenomenon. When it was realized that Geim's experiment was not meant to be a joke; rather, the frog was truly levitated in Geim's experiment, this piece of research was deemed so outrageous that he and Berry were awarded the 2000 Ig Nobel Prize "for using magnets to levitate a frog."

The Ig Nobel Prize is a joke—taken seriously—which had been created to recognize scientific contributions that should not have been made. Ridiculing the experiment of frog levitation was a severe misunderstanding on the part of the organizers of the very popular Ig Nobel award. Apparently, they did not realize that using a frog was merely a device to attract attention, but the science behind it was serious pioneering achievement. The initiation of the Ig Nobel Prize was a sign of a great sense of humor and it was ironic that the awarders of the Ig Nobel Prize did not recognize humor when others practiced it. Thus, Geim has become the only scientist so far who has received both an Ig Nobel and a real Nobel Prize. Incidentally, the Ig Nobel Prize did not diminish Geim's affinity for joking. The following year he published another serious result, this time about the detection of earth rotation using a diamagnetically levitating gyroscope, and he listed as his co-author H. A. M. S. ter Tisha, which is supposed to be his favorite hamster by the name of Tisha [6].

The discovery of graphene and the award for it make one think about the history of the fullerene discovery. The surprise in Geim's and Novoselov's discovery in 2004 was that they were able to isolate and stabilize a single-atom-sheet carbon [7]. Prior to their report, many believed that such a two-dimensional atomic crystal simply could not exist. In contrast, back in 1966, David Jones wondered about the possibility of graphite sheets curling up and forming huge balls [8]. When Eiji Osawa described the C_{60} molecule of truncated-icosahedron shape in his Japanese-language

Fig. 4 "Buckminsterfullerene model" as an entrance decoration at Topkapi Sarayi in Istanbul (photograph and © by the author).

publication, he based his suggestion on symmetry consideration and did not follow it up neither by computation nor by experiment [9]. This was followed by a computation-based prediction of the stability of truncated icosahedron-shaped C_{60} molecule by DA Bochvar and EG Galpern [10]. This work was not followed up either. Both these reports disappeared in oblivion to be discovered again only after the actual observation and production of buckminsterfullerene in 1985 [11] and 1990 [12], respectively.

The experimental observation of C_{60} was eventually awarded a chemistry Nobel Prize in 1996 to Robert Curl, Harold Kroto, and Richard Smalley. Two physicists pioneered the production of C_{60}, Donald Huffman and Wolfgang Krätschmer, and they might have been recognized by a similar distinction, but they were not. Graphene could be considered at least in principle to be the initial material of all carbon nanotubes and fullerenes. The nanotubes are represented here by a detail of the decorations of the Bangkok Royal Palace (Fig. 2) and the fullerenes by a slightly irregular model under the paw of a dragon in the Forbidden City outside Beijing (Fig. 3) and a buckminsterfullerene shape from the Topkapi Sarayi in Istanbul (Fig. 4).

Geim's and Novoselov's discovery and their Nobel Prize in Physics might also be viewed as the closing act of this beautiful round of discoveries and their ultimate recognition. By this it is not meant that the discoveries might also end; on the contrary, these latest events will undoubtedly contribute to further invigoration of the field and its researchers.

Dedication The present Editorial is dedicated to the memory of a late friend, David Shoenberg (1911–2004) who was a pioneer in the application of strong magnetic fields, including levitation of type I superconductors [13, 14].

References

1. Geim A (2006) in a Profile: A physicist of many talents. Physics World February, pp 8–9
2. The motivation is quoted after the Nobel Prize announcement on October 5, 2010. The motivations for all Nobel Prizes are communicated in the Nobel Foundation Directory, which is published periodically by the Nobel Foundation
3. Dinadayalane TC, Leszczynski J (2010) Struct Chem 21: to be completed
4. Hargittai I (2002, 2003) The Road to Stockholm: Nobel Prizes, Science, and Scientists. Oxford University Press, Oxford and New York, pp 184–191
5. Berry MV, Geim AK (1997) Of flying frogs and levitrons. Eur J Phys 18:307–313
6. Geim AK, ter Tisha HAMS (2001) Detection of earth rotation with a diamagnetically levitating gyroscope. Physica B 294–295:736–739
7. Novoselov KS, Geim AK, Morozov SV, Jiang D, Zhang Y, Dubonos SV, Grigorieva IV, Firsov AA (2004) Electric field effect in atomically thin carbon films. Science 306:666–669
8. Jones DEH (1966) New Scientist 32:245
9. Osawa E (1970) Superaromaticity (in Japanese). Kagaku 25:854–863
10. Bochvar DA, Galpern EG (1973) Hypothetical systems: Carbododecahedron, s-Icosahedron, and Carbo-s-Icosahedron (in Russian). Dokl Akad Nauk SSSR 209:610–612
11. Kroto HW, Heath JR, O'Brien SC, Curl RF, Smalley RE (1985) C_{60}: Buckminsterfullerene. Nature 318:162–163
12. Krätschmer W, Lamb LD, Fostiropoulos K, Huffman DR (1990) Solid C_{60}: a new form of carbon. Nature 347:354–358
13. Shoenberg D (1952) Superconductivity. Cambridge University Press, Cambridge, UK
14. Hargittai I (2004) David Shoenberg. In: Hargittai M, Hargittai I (eds) Candid Science IV: Conversations with Famous Physicists. Imperial College Press, London, pp 688–697

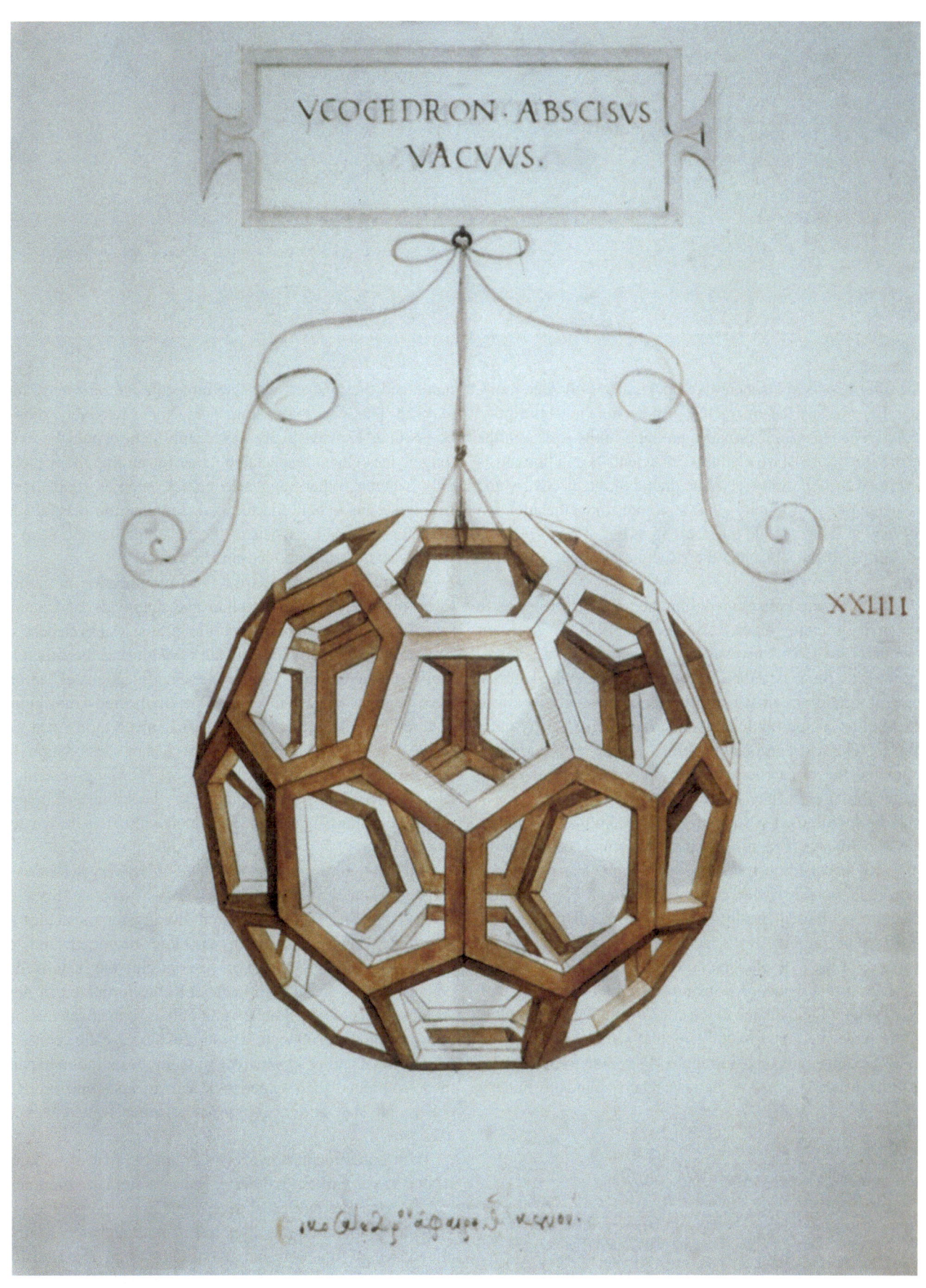

YCOCEDRON · ABSCISVS
VACVVS.

XXIIII

Leonardo da Vinci's truncated icosahedron, drawn for Luca Pacioli's *De Divina Proportione*. In the original article the Editorial Office of the magazine Leonardo erroneously replaced this image by Leonardo da Vinci's drawing of the pentagonal dodecahedron.

A Fuller Bridge[a]

István Hargittai

The discovery of the sphere-like buckminsterfullerene molecule, followed by the emergence of the whole new discipline of *fullerene chemistry,* provides an opportunity to lessen the separation of the "Two Cultures" of scientists and humanists described by C.P. Snow [1]. The drama of the discovery and the magnificent simplicity of the structure attract the attention not only of chemists and other scientists but of non-scientists as well. Both the story and the structure are rich in cultural implications.

During an experiment involving the use of laser beams to evaporate graphite in early September 1985, a group of scientists at Rice University in Houston, Texas, Harry Kroto, Rick Smalley, Bob Curl, and their students identified a set of conditions in which the C_{60} species could be produced in an incredibly high abundance relative to any other cluster. The extraordinary stability of the C_{60} molecule prompted the researchers to look for the structural reason for its formation. They first came to the conclusion that it must be a closed-cage structure. Having known this much, they should have recognized that its shape must be that of the truncated icosahedron, one of the 13 Archimedean polyhedra. Instead, the scientists were merely searching for a sphere-like structure composed mostly of same-size regular hexagons, based on the graphite sheets. They remembered, however, the structure of the U.S. pavilion at Montreal's Expo '67, which led them to the works of Buckminster Fuller. Working with models, they finally came to the conclusion that the structure of the molecule consists of 12 regular same-size pentagons and 20 regular same-size hexagons. The route to the discovery thus was connected in the researchers' minds to Fuller's

name and they named the new molecule *buckminsterfullerene* [2]. This is a rather long name for a relatively simple compound. However, any systematic name would be even longer. It is also a respectable name for an important molecule, whereas other suggested names, such as footballene, soccerene, buckyball and the like, sounded too playful (the official soccer ball consists of the same number and form of patches as the truncated icosahedron).

Buckminsterfullerene is the third modification of carbon to be discovered (after graphite and diamond), and nature seems to have kept it secret for a long time. An avalanche of similar, all-carbon molecules, all belonging to the fullerene family, and technically as many new modifications of carbon, have become known to exist. One of the most intriguing features of fullerene chemistry is that metal atoms can get inside the C_{60} ball, requiring that a new designation be devised to describe this mode of forming chemical associations. Thus, for example, the buckminsterfullerene molecule containing a lanthanum atom within it is designated as $La@C_{60}$.

Another interesting feature of this discovery was that it resulted from a lucky crossing of two separate lines of research. In one, Kroto had been looking for molecules of interstellar space. For him, the laser-beam method of evaporating graphite served to mimic the interstellar conditions that are thought to lead to the formation of new species. In the other, Smalley had built a sophisticated apparatus in which loosely bound groups of atoms, called clusters, were formed and observed. The graphite-evaporation experiment combined their experience and interests and brought cluster physics and astrophysics together in a chemical exercise.

Their experiment, however, was not the first of its kind. About a year earlier, another group in a similar experiment had detected and published their research on the products of graphite evaporation by laser beam. Although the relative abundance of C_{60} was not as striking as in the Houston

[a]Originally published in *Leonardo* 1996, 29(1):1–3.

I. Hargittai (✉)
Department of Inorganic and Analytical Chemistry, Budapest University of Technology and Economics, Budapest, Hungary
e-mail: istvan.hargittai@gmail.com

experiment, again, in hindsight, it should have been noticed. Not only was it not noticed by the researchers who produced the data, but the readers of the prestigious journal where the report had appeared did not notice it either [3].

When the report of the Houston group was published, it generated interest, but the real landslide of a new chemistry started when another team, led by Wolfgang Kratschmer of Heidelberg and Donald Huffman of Tucson, Arizona, found a simple way to produce the buckminsterfullerene in measurable quantities [4]. This enabled any chemist to experiment with the new substance.

The discovery of buckminsterfullerene, although serendipitous, was not made possible through luck alone but through hard work, training, experience and curiosity. As Louis Pasteur stated, "In the field of observation, chance only favors those minds which have been prepared" [5]. Curiously, several suggestions, unknown to the discoverers, had preceded the discovery, all pointing to the feasibility of the substance that today we call buckminsterfullerene. In 1970, Eiji Osawa of Japan suggested the existence of C_{60} with a truncated icosahedral shape, based purely on symmetry considerations. In 1973, D.A. Bochvar and Elena G. Gal'pern of Moscow carried out some theoretical calculations that led them to postulate the great relative stability of a C_{60} molecule with a truncated icosahedral shape. Even before, in 1966, David Jones of Britain mused in print about the possibility of graphite sheets curling up into hollow ball-like molecules.

Thus, there are many threads of the buckminsterfullerene story, and Fuller's involvement in the picture had more than symbolic significance. Fuller was not only the creator of geodesic domes but also an advocate of a physical geometry in which the dodecahedron and the icosahedron play an important role. His writings may be controversial but it is his influence, exerted over a broad range of disciplines, that not only survives him but also appears to provide fruitful stimuli in different fields. The physical importance of the icosahedron and the relevance of Fuller's teachings were stressed by the discoverers of virus structures, Donald Caspar and Aaron Klug, who stated in the early 1960s:

> The solution we have found … was, in fact, inspired by the geometrical principles applied by Buckminster Fuller in the construction of geodesic domes…. The resemblance of the design of geodesic domes … to icosahedral viruses had attracted our attention at the time of the poliovirus work …. Fuller has pioneered in the development of a physically orientated geometry based on the principles of efficient design [6].

Alas, it seems that the influence on this important microbiological research has not spilled over to other fields.

The situation may be different with the buckminsterfullerene story. A whole new field is evolving, one that is new not only because of the unique shape of the C_{60} molecule but also because of the size range involved. Fullerenes appear not only as isolated molecules but also in a great variety of sheets, tubes and molecular wires. Potential applications range from superconductivity to anti-AIDS agents. Fullerene chemistry is becoming an important part of a new area of science, often called nanochemistry, which refers to a size-range based on a great, though finite, number of molecules.

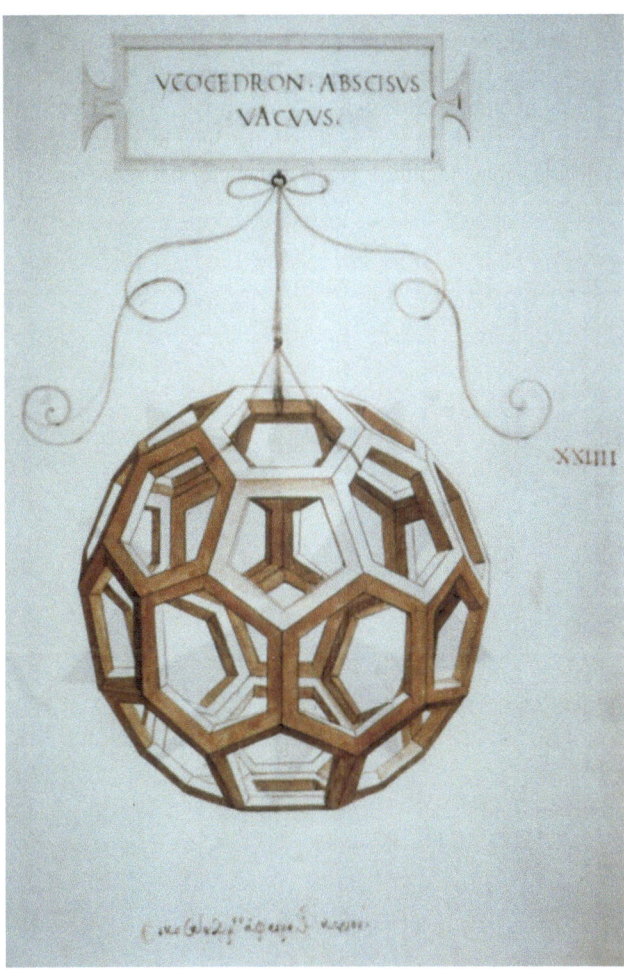

Leonardo da Vinci's truncated icosahedron, drawn for Luca Pacioli's De Divina Proportione.

This new chemistry has an added attraction in the accessible shape of the buckminsterfullerene molecule. The truncated icosahedron is not as common as the cube, yet it is not so complicated that it is difficult for non-scientists to understand and recognize. It is special enough to catch the eye and is shown in a conspicuously beautiful drawing by Leonardo da Vinci. When this shape is recognized outside the discipline of chemistry, it is pleasing to interested laypersons and recognizable as a children's climber, a lamp or the soccerball itself.

Had the original discoverers known their geometry better, or had Euler's formula come to mind as soon as they started looking for the shape of a cage consisting of 60 carbon atoms, they might have had no reason to reach out to Buckminster Fuller. In that case the synergistic impact of their discovery would almost certainly have been much less significant. I am not praising the lack of being versed in geometry, however. On the contrary, one of the side benefits of this outstanding chemical discovery may be an enhanced interest in three-dimensional geometry and a strengthened commitment towards geometry education in our schools. This is but one aspect in which the smooth-rolling buckminsterfullerene molecules may facilitate closing the gap between our two cultures.

References

1. C.P. Snow, *Two Cultures and a Second Look* (Cambridge, U.K.: Cambridge University Press, 1964). Snow's lecture, *Two Cultures*, was presented in 1950.
2. H.W. Kroto, J.R. Heath, S.C. O'Brien, R.F. Curl, and R.E. Smalley, "C60: Buckminsterfullerene," *Nature 318* (1985) pp. 162–163.
3. E.A. Rohlfing, D.M. Cox, and A. Kaldor, "Production and Characterization of Supersonic Carbon Cluster Beams," *Chem. Phys. 81* (1984) pp. 3322–3330.
4. W. Kratschmer, L.D. Lamb, K. Fostiropoulos, and D.R. Huffman, "Solid C: A New Form of Carbon," *Nature 347* (1990) pp. 354–358.
5. "Dans les champs de l'observation le hasard ne favorise que les esprits préparés." *Encyclopaedia Britannica*, 11th Ed., Vol. 20 (1911), quoted here after A.L. Mackay, *A Dictionary of Scientific Quotations* (Bristol, U.K.: Adam Hilger, 1992).
6. D.L.D. Caspar and A. Klug, "Physical Principles in the Construction of Regular Viruses," *Cold Springs Harbor Symposia on Quantitative Biology 27* (1962) pp. 1–24.

Linus Pauling visiting Moscow State University in 1984; courtesy of Larissa Zasurskaya.

Linus Pauling's Quest for the Structure of Proteins[a]

István Hargittai

Abstract

Linus Pauling, arguably the greatest chemist of the twentieth century, never publicly admitted that there was a race for the determination of the structure of the most important biopolymers. But according to his competitors there was a race, in fact, there were two, and Pauling won one and lost the other. He had a tremendous amount of ideas, many of them worthless, but a few were spectacular. Not only did he make seminal discoveries, he was also a master of announcing them in a most dramatic way. Eventually, Pauling shifted toward politics and controversial issues, but his science ensured him his place among the greats. Here, we follow Pauling's route to the discovery of the alpha-helix; the defeat of the star-studded British team in the same quest; and a seemingly unrelated story about the fate of the theory of resonance that assured Pauling's victory yet at the same time it was excommunicated in the Soviet Union.

For most of the first half of the twentieth century a large number of scientists were not even sure biopolymers existed. The view survived for long that the principal components of living matter were in a colloidal state, that is, conglomerates of smaller molecules. It was only in 1953 when Hermann Staudinger was awarded the Nobel Prize in Chemistry for his discoveries about macromolecules that the existence of polymers was irrevocable accepted. By then, though, a lot about the structures of the biologically important macromolecules had been discovered. Thus, during the first half of the twentieth century, the efforts to establish the nature of biological substances and to uncover their structures went in parallel.

The British father and son team, W. H. Bragg and W. L. Bragg, pioneered the technique of X-ray diffraction crystallography in 1913, with the son playing the leading role. When the two Braggs were awarded the Nobel Prize in 1915, the son became the youngest ever Nobel laureate and has stayed the youngest to this date. After a hiatus due to World War I, this field took off spectacularly in the realm of small molecular systems. As early as the 1920s, fibrous materials were subjected to X-ray diffraction for the first time by Michael Polanyi in Herman Mark's laboratory at the Kaiser Wilhelm Institute in Berlin. Polanyi's experiments on cellulose indicated the presence of crystallites in cellulose and they were oriented in the direction of the fiber axis. He could not have performed a full structure analysis at that time, but Mark and Polanyi observed characteristic changes on stretching the cellulose fibers. Mark was to become one of the century's foremost polymer chemists. When he was forced out of Germany, he moved to his native Vienna where he helped one of his students, Max Perutz to be accepted as a doctoral student in Cambridge, UK, in 1935. Perutz would become a key player in the quest for the structure of proteins. But he was only slowly moving to the area of his ultimate success when Linus Pauling was already a major force in the field.

Pauling came from a humble background, but he was ambitious. He lost his father when he was 9 years old and his mother found it difficult to cope with her obligations. She certainly did not appreciate her son's intentions to stay in school even when it was no longer mandated for him to continue his studies. Pauling's schoolings were not at top places, and when in 1922 he went to the California Institute of Technology (as it was later; Caltech in short) it was far from the preeminent research-oriented institution into which it would develop. But the school was as ambitious as its new student, and there were visionary movers of it who were set to making Caltech a top-notch institution of higher education and research. They were smart enough not only looking into the distant future and only for big names to recruit from

[a]Originally published in *Structural Chemistry* 2010, 21:1–7.

I. Hargittai (✉)
Department of Inorganic and Analytical Chemistry, Budapest University of Technology and Economics, Budapest, Hungary
e-mail: istvan.hargittai@gmail.com

Springer

faraway places, but recognized in Pauling the potentials of a star scientist who would even challenge British preeminence in the science of chemical structures.

When Pauling started his doctoral studies with Roscoe Dickenson, a fresh home-grown PhD in X-ray crystallography at Caltech, this field was less than a decade old. Pauling became engaged in the determination of the structure of many inorganic and organic molecules and amassed a large amount of information about them during the ensuing decade. What kind of information was that? It was about the geometrical arrangement of the atoms in the molecules and the arrangement of the molecules in the crystals.

Not all the modern knowledge was to be had at Caltech at the time, and not even in other laboratories in the United States. The leading country of science was Germany and a few other places in Europe, and Pauling—like many other aspiring American scientists—paid pilgrimage to a series of European research centers in order to learn from the likes of Arnold Sommerfeld in Munich and Erwin Schrödinger in Zurich. They were both physicists, but Pauling's aim was not to transform himself into a physicist. Rather, his goal was to apply the latest discoveries in physics, and above all the new quantum mechanics, to solving a wealth of problems in chemistry in which he proved to be unique.

The most intriguing question in chemistry at that time was about the forces that keep the atoms together in a molecule, that is, about the nature of the chemical bond. If there is anything truly associated with Pauling's name, it is the understanding the nature of the chemical bond. He used the achievements of modern physics, the experimental information about the geometry of molecules and his thinking, to put together a theory. He then kept refining it in accordance with the emergence of the latest experimental information. The science of chemistry has a great deal of intuitive approach in it, very often stemming from a desire to represent on paper what the chemists experience in the laboratory. Thus, for example, they started using a straight line connecting the symbols of two elements to represent their bonding without really understanding anything about what that straight line represented. Nowadays when we know so much about what it means, we still find this straight line an excellent representation of the chemical bond. Lewis's description of the covalent bond in 1916 was not much less intuitive than this; nonetheless he made a big step forward. He introduced the idea of the shared electron pair, meaning the covalent bond between two atoms. During the late 1920s two physicists, Walter Heitler and Fritz London used the new quantum mechanics and their sophisticated mathematical apparatus to rigorously describe this covalent bond. It was so rigorous that it was too sophisticated for most chemists to understand it let alone to apply it to solving their problems that were usually more complex than the hydrogen molecule for which Heitler and London had worked out their theory.

Linus Pauling bridged this gap in a series of brilliant articles in the *Journal of the American Chemical Society*. Eventually he developed his ideas and his repository of structural information into a bestseller *The Nature of the Chemical Bond* [1]. Its last, third edition appeared in 1960 and many of the later stars of chemistry benefited from it by getting their introduction to the intricacies of this branch of science. A new book would be timely, but nobody seems brave enough to try filling Pauling's shoes in producing a new comprehensive monograph about the chemical bond.

Had Pauling produced his series of articles about the chemical bond, and nothing else, he would have already written his name into the annals of the history of chemistry. However, he did not limit his interest to theoretical studies. He utilized X-ray crystallography broadly and was constantly on the lookout for new techniques. While in Europe, he visited Herman Mark's laboratory in Ludwigshafen, Germany (where he was at the time), and Mark introduced a new experimental technique to his visitor for the determination of molecular structure, gas-phase electron diffraction. It was similar to X-ray crystallography, but there were two major differences. It used electrons rather than X-rays and the target was not a crystal but a gaseous sample in which the molecules had no well-defined order in their mutual arrangements.

One of the great advantages of using electrons was the very high intensity of the interaction between electrons and molecules. Thus, the duration of the required interaction was measured in minutes rather than many hours as with X-rays. The other important advantage was that in the gaseous sample the molecules were by themselves and their structures were not impacted by the closeness of their neighbors. For the X-ray technique, the molecules were required to be able to form a crystal in the first place, and there was no such requirement for using the electron diffraction technique. The structures determined by the new technique depended only by the molecule itself and not by the way they were arranged relative to each other as was the case in the crystal. Other limitations of the new technique, however, have restricted it from becoming so widely used as X-ray crystallography, which truly has been the preeminent tool for uncovering the structures of biopolymers.

Mark's industrial laboratory was not the proper environment to expand the studies of molecular structures and he happily offered Pauling to take the new technique with him to Caltech. Mark even supplied him with the blueprints of his apparatus. Pauling not only introduced the gas-phase electron diffraction technique quickly in the United States, but he and his student, Lawrence Brockway further developed it. They added a mathematical step to handling the experimental data that made it possible to extract structural information in a graphically direct and attractive way from the probability density distribution of the internuclear distances in the molecule (usually it is referred to as the radial distribution curve,

Fig. 1 Ava and Linus Pauling (photograph by and courtesy of Karl Maramorosch, Scarsdale, NY).

which is a misnomer). From the experiment to reading off the curves directly the distances between atoms in simple molecules took only a few days' work.

Pauling (Fig. 1) established relationships among various experimental facts and made predictions about structures not yet investigated. He then worked out a theoretical technique based on quantum mechanics, but simple enough for a broad circle of chemists, to describe molecular structures. It was called the valence-bond or VB theory and it was one of the two major theoretical approaches developed over the decades. The other is the molecular orbital or MO theory. The VB theory builds the molecules from individual atoms linked by electron-pair bonds. For chemists, the VB theory appealed as more straightforward, alas, it did not stand well the test of time. The MO theory has proved more amenable to computations, which itself has become a major thrust in modern structural chemistry. However, for a long time the VB theory dominated the field.

An important feature of the VB theory was that a molecular structure could be described by a set of "resonating" structures. This did not mean that each structure in such a set would be considered as present individually, but that the sum of these resonating structures represented the emerging structure better than any other description at the time. It needs to be stressed that what the resonance theory provides is merely a model, an approach, rather than a unique reflection of reality. There were proponents and opponents of the theory as is the case with most theories. Yet the resonance theory proved to be eminently useful for Linus Pauling—who was

one of its initiators—in his quest for the protein structure. It happened so that this theory showed him the way and brought him a resounding victory over his competitors who lacked this tool and could not arrive at the right solution.

Pauling was advancing in a systematic manner in his quest for building up structural chemistry. First, he busied himself with inorganic substances and after the first 10 years he moved to organic substances. Among the organic molecules he often observed structures in which the lengths of the bonds between atoms were intermediate between single bonds and double bonds, so the theory of resonance came in handy in their understanding and description. Today, chemists no longer tend to think in terms of purely single bonds and double bonds, or triple bonds for that matter, and, accordingly, the utility of the resonance theory has largely disappeared, but in the 1930s it was considered to be of great help.

As Pauling was learning more and more about the structures of relatively simple molecules, in the mid-1930s, it occurred to him that he might as well make an attempt to learn about larger systems. He was aware of the importance of biopolymers and that the understanding of their structures might be a step toward understanding biological processes. Proteins were an obvious choice, because they were the most important biopolymers. At that time nucleic acids were already known, and their building blocks, the nucleotides, had been identified, but the nucleic acids were not considered to be of great significance. There was a hypothesis by Phoebus Levene about the tetranucleotide structure that was based on an erroneous observation that the four nucleotides in nucleic acid were present in equal amounts [2]. Hence, the nucleic acids were thought to be dull, uninteresting molecules, not capable of carrying any great amount of information.

When Pauling started thinking about protein structures, the first protein to attract his attention was hemoglobin, which is the vehicle of carrying oxygen in our organism. Incidentally, the British group engaged in protein structure studies had also selected hemoglobin for their target; their choice was independent of Pauling's interest. At the end of the 1920s, Gilbert Adair in Cambridge, UK, showed that the hemoglobin molecule consists of four units each with an iron atom, and each iron could bind an oxygen atom. Pauling formulated a theory about the oxygen uptake of hemoglobin and the structural features of this molecule related to its function of disposing of and taking up oxygen.

His interest in protein structures was further whetted when a visiting scientist and protein specialist, Alfred Mirsky of the Rockefeller Institute, spent the academic year 1935–1936 in his laboratory. They jointly studied the phenomenon of denaturation of proteins by heat or chemical substances, and formulated a theory about it. In this theory, they described the native protein as having a regularly folded structure in which hydrogen bonds provided the stability of the structure. Hydrogen bonding was a recently discovered phenomenon; it was becoming recognized as a crucial mode of interaction in chemical structures and especially in those of biological importance. In retrospect, it was a pivotal discovery, but its significance emerged only gradually over the years. For many biological molecules it is the hydrogen bonds that keep their different parts together.

Pauling postulated that the subsequent amino acid units are linked to each other in the folded protein molecule not only by the normal peptide bond but also by hydrogen bonding that is facilitated by the folding of the protein, which brings the participating atoms sufficiently close to each other for such interactions. In Pauling's and Mirsky's conclusion, when the protein molecule is denatured it undergoes complete or partial unfolding accompanied by breaking the hydrogen bonds. This was a hypothesis, because they knew practically nothing about the nature of folding; finding more about it occupied Pauling's mind for the next 15 years.

By the time Pauling became engaged in this research it had been established from rudimentary X-ray diffraction patterns that there might be two principal types of protein structure. Keratin fibers, such as hair, horn, porcupine quill, and fingernail belonged to one, and silk to the other. The foremost British crystallographer of fibers, William T. Astbury showed in the early 1930s that the diffraction pattern of hair underwent changes when it was stretched. He called the one producing the normal pattern alpha keratin and the other, which was similar to the pattern from silk, beta keratin. In 1937, Pauling set out to determine the structure of alpha keratin. He did not just want to rely on a single source of information. He planned to use all his accumulated knowledge in structural chemistry and find the best model that would make sense on this background and would be compatible with the X-ray diffraction pattern.

There was one piece of information from X-ray diffraction that seemed to be a good point of reference and that was the structural unit—whatever it would be—along the axis of the protein molecules repeated at the distance of 5.1 angstrom. He also knew the dimensions of the peptide group, that is, the characteristic sizes of the group linking the amino acids to each other in the protein chain. The C–N bond in the peptide linkage was not simply a single bond, but it was not a purely double bond either. Pauling's involvement with the resonance theory taught him that the emerging structure could be represented by two resonating structures.

Hence, the resonance theory suggested that the C–N bond in the peptide linkage had a partial double bond character. From the accumulated structural information he also knew that the bonds around a double bond are all in the same

plane. This was a very important piece of information because rather than taking into account all kinds of rotational forms with respect to the peptide bond, he could assume that it was a planar configuration. This assumption greatly reduced the number of possible models he had to consider for describing the structure of alpha keratin. Nonetheless, at this time Pauling was unable to find a model that would fit the X-ray diffraction pattern and he postponed further study on protein structures.

During the ensuing years Pauling and his newly arrived associate, Robert Corey, an expert in X-ray crystallography, carried out a large amount of experimental work determining the structures of individual amino acids and simple peptides. At some time every doctoral student in Pauling's laboratory was supposed to determine the structure of an amino acid for his PhD dissertation. The study was interrupted by World War II, but continued vigorously upon its conclusion. Pauling returned to the question of the structure of alpha keratin in 1948 while he was a visiting professor at Oxford University in England.

Not only had the amount of experimental information in the meantime expanded considerably, but Pauling could take a more detached view of the problem in his renewed efforts. When he was looking for the solution more than a decade before, he was bothered by the knowledge that his model was supposed to accommodate the possible presence of 20 different amino acids in the protein chain. At this time, in 1948, he decided to ignore their differences and assumed them to be equivalent for the purpose of his model. This was yet another example of Pauling's ability to distinguish between essential features and those that could be ignored in building his models.

Pauling remembered a theorem in mathematics he learned about at Caltech a quarter of a century before. It stated that the most general operation to convert an asymmetric object into an equivalent asymmetric object is a rotation–translation and that repeated application of this operation produces a helix. Here the asymmetric objects are the amino acids constituting the protein chain; the rotation should take place about the molecular axis of the protein; and the translation is the movement ahead along the chain. The amount of rotation was such that took the chain from one amino acid to the next while the peptide group was kept planar, and this operation was being repeated and repeated all the time. An additional restriction was keeping the adjacent peptide groups apart at a distance that corresponded to hydrogen bonding. In Pauling's model the turn of the protein chain did not involve an integral number of amino acids—he did not consider this a requirement whereas his British counterparts did. This was yet another relaxed feature of the structure that served him well in finding the best model whereas it served as an unnecessary restriction for his competitors.

Pauling—ever the model builder—sketched a protein chain on a piece of paper and folded the paper while looking for structures that would satisfy the assumptions he had made (Fig. 2). He found two and called one the alpha helix and the

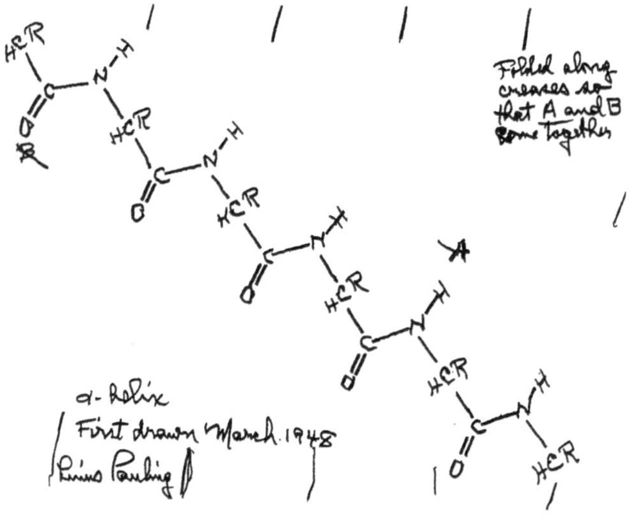

Fig. 2 Linus Pauling's sketch of the polypeptide chain in 1948. When he folded the paper along the creases, the alpha-helix appeared [3] (Fig. 3).

other the gamma helix, the latter being much less probable than the former. He determined the distance between repeating units in the protein chain and noticed a marked difference between his estimation from the model and the experimental value from the diffraction pattern. This was disappointing but the model was so attractive and so sensible that Pauling had little doubt in its correctness. Nonetheless, he decided to wait

Fig. 3 Model of alpha-helix with 3.7 amino acid residues per pitch after [4].

with its publication until the discrepancy would be understood. His confidence was enhanced when he visited the British group involved also in the structure elucidation of proteins and Max Perutz showed him his diffraction patters. From the X-ray diagrams it was obvious to Pauling—though not yet to Perutz—that the structure was alpha helix. Pauling did not say anything to Perutz.

When Pauling returned to Pasadena, he and his associates double checked all his calculations and found no errors in them. In the meantime, after about a year, Bragg, Perutz, and John Kendrew of Cambridge, UK, published a big article about protein structures and communicated about 20 models, none of which contained a planar peptide group and none of which described alpha keratin satisfactorily [5]. Finally, Pauling decided to ignore the discrepancy of the repeat distance between his model and the experimental observation and he and his associates published the alpha helix.

Eventually, the origin of the discrepancy was understood; it was caused by the alpha helices twisting together into ropes. This interaction between the chains caused a change in the experimental data as compared to what it would be for a single chain for which the model had been constructed. Thus, Pauling's alpha helix was confirmed even in this detail. The alpha helix has proved to be a great discovery because it is a conspicuously frequent structural feature of proteins.

Pauling's approach to solving this complex problem was exemplary in focusing on what was essential and ignoring what had little consequence. When it turned out that the turn about the chain did not correspond to an integer number of amino acids, hinting at less than perfect symmetry, he did not let himself bothered by this. He thus expanded the realm of crystallography toward structures that were not part of classical crystallography yet included literally vital substances. It was also noteworthy that he could skip a decade in working on this most important discovery without much danger of others scooping him. They almost did, but only in their timing and not in knowledge, because his knowledge proved to be superior to anyone in his field at that time.

Pauling must have sensed the precarious nature of the situation and restrained himself from revealing crucial information to Perutz during his visit to Cambridge (vide supra). The Cambridge X-ray diffraction pattern showed the helical nature but Perutz did not think about it and thus did not notice it whereas for Pauling it provided additional evidence of the correctness of his model. This episode showed both his competitive spirit and his self-discipline. Finally, Pauling was sure enough in himself and his model that he went ahead with publishing the alpha helix without having yet resolved the remaining (apparent) discrepancy between his model and the available experimental evidence. First they published a short note [6], followed by a longer article [7] and soon they wrote seven more papers to report their findings.

Pauling was a master in creating publicity for his discoveries. When he prepared for announcing the discovery of the alpha helix it was to be in a big lecture hall at Caltech. The model stood on the rostrum, but it was under a cover, waiting to be unveiled, just as a sculpture would be, and it came toward the end of Pauling's lecture. When it was finally unveiled, the effect was dramatic and the audience was stunned by its beauty. I myself experienced the mesmerizing effect of Pauling's lecturing at the University of Oslo in 1982. He covered the board with complicated formulas and from time to time he looked at the audience as if checking whether we were duly impressed. Otherwise, the formulas were not at all necessary for us to understand the points he was making. He was already an octogenarian, but watching him gave an impression of a young assistant professor who came for interview and was presenting his research with the usual arrogance of such scenes. During the lunch following the lecture he was more vigorous than the rest, led the discussion, and fired away questions, mostly answering them himself.

In research publications there is no place for the human sides of the discoveries and Pauling wrote up the story of his alpha helix discovery separately, but it never appeared while he was alive. It was published 2 years after he died when I was running a chemical magazine and his former secretary of his last 20 years, Dorothy Monro, suggested to bring it out there. Research papers usually lack the human element and the blind alleys in research, so this paper by Pauling was especially valuable for our understanding how this particular discovery happened [3].

The Cambridge group suffered a defeat in this case, which was especially heavy for W. Lawrence Bragg to bear, because he was the pioneer of X-ray crystallography and the American group came out on top in their undeclared race. It was not possible to pinpoint a single reason for this defeat, but it was a crucial difference that Pauling could limit the number of possible models because of his superior knowledge of structural chemistry. The Cambridge group had no such guideline although it could have. It turned out that Lord Todd the soon to be Nobel laureate organic chemist who worked in the next building to Perutz's and Kendrew's laboratory had told Bragg that the peptide bond had some double-bond character. Bragg, however, could not from this piece of information make any conclusion about the configuration of the peptide bond, namely, that it was planar.

Years after this fiasco, Perutz complained about their lack of knowledge of the planarity of the peptide group. He blamed the Medical Research Council (MRC) for having him denied the use of a Rockefeller Fellowship for travel to America in 1948. The Secretary of the MRC thought that rather than going to learn from the Americans, the Americans should come and learn from the British. In hindsight, Perutz thought that he could have learned about the peptide bond planarity from Pauling had he been allowed to travel [8]. Of course, he could have just walked across the street to visit Lord Todd for the same information.

It is not at all sure whether had Perutz visited Pauling he would have learned from Pauling as much as he might have supposed in retrospect. We have seen Pauling withholding his observation from Perutz that he had noticed the evidence of helical structure on Perutz's X-ray diffraction diagram. During his Oxford sojourn, Pauling wrote to Corey back to Pasadena that he felt uncomfortable about the English competition. In their turn, the British considered protein crystallography their own territory. It was not only that the Braggs discovered X-ray crystallography and that Astbury was a pioneer in taking X-ray pictures of proteins. It was also J. Desmond Bernal who had prepared the first ever X-ray diffraction diagrams of a single-crystal protein—a pepsin single crystal—that clearly showed the possibility of deducing atomic positions from it. This was in 1934. In the future Nobel laureate X-ray crystallographer Dorothy Hodgkin's description, "that night, Bernal, full of excitement, wandered about the streets of Cambridge, thinking of the future and how much it might be possible to know about the structure of proteins if the photographs he had just taken could be interpreted in every detail" [9]. The British self-confidence in dominating this field reached such proportion that Astbury and Bernal divided it by a gentlemen's agreement between the two of them. They decided that Bernal would take up the investigation of the crystalline substances and Astbury the fibrous ones [10].

Perutz on his part, for their failure blamed Astbury's X-ray diffraction picture, which showed a discrepancy between the repeat distances as compared with reasonable structures, a discrepancy—as we have seen—Pauling daringly disregarded. Perutz was disheartened when he found Pauling's paper about the alpha helix model. He devised an additional X-ray experiment that gave further evidence for the correctness of Pauling's result, something that Pauling had missed. When Perutz reported his finding to Bragg, Bragg asked him, "How did you think of that?" Perutz's response was that it was because he was so angry that he hadn't thought of the structure himself. To which Bragg replied coldly, "I wish I'd made you angry earlier" [11]. Perutz told me this story in 1997, and he used this phrase as the title of his next book. Perutz might have thought that Pauling would be pleased that he provided additional evidence for alpha helix, but was disappointed by Pauling's reaction, which was clearly dismissing.

Pauling's fascination with proteins served him well in his focusing his attention to their structures at a crucial period in twentieth century science. However, he continued his protein bias even when the next big task appeared before structural chemistry that was the structure of nucleic acids. Pauling entered that race too, but there is ample evidence that Pauling did not concentrate on it with the intensity and dedication as he had done for the protein structures. In case of the quest for the structure of nucleic acids he was defeated by the British teams. Pauling published an erroneous triple helix and he was not in possession of the best X-ray diffraction patterns of nucleic acids that were available at the early 1950s either. As is well known, those patterns were produced at King's College in London and the winning double helix model came out from the Cavendish Laboratory in Cambridge, UK, but this is a different story.

We can add a footnote about Pauling's theory of chemical resonance, which served him so well in the above story. At about the same time, this theory was in the center of attack by rabid ideologists in the Soviet Union [12]. The culmination was a 4-day conference in Moscow in 1951, organized by the Soviet Academy of Sciences. Leading Soviet chemists, physicists, philosophers, and others attended the meeting. A small but vocal group of chemists attacked the theory of resonance as an ideological aberration and together with it quantum theory and the science of the West. They insisted on returning to traditional Russian values and offered their own worthless theories. Excellent scientists suffered ruthless criticism for having applied the theory of resonance in their work, and they, in turn, offered humiliating self-criticism.

The affair has been referred to as the great Soviet resonance controversy and it was a chapter in the anti-science events following World War II that touched biology even more severely. Physics was spared in the last minute due to its decisive role in producing nuclear weapons. Stalin's terror did everything to protect his empire from even the slightest influence by the West, the purest sciences included. There was irony in this story in that Pauling was a friend of the Soviet Union and suffered persecution in the McCarthy era, but this was not yet known in the Soviet Union. In 1993, I asked Pauling for his comments about this affair. He appeared as if he misunderstood it or did not want to understand it. He wrote that it took years "for the chemists in the Soviet Union to get a proper understanding of the resonance theory" ([12], p 5). In reality, they understood it well enough and applied it with great success, that is, until 1951, when the main proponents of the theory lost their jobs. If it was a consolation, their lives were spared in contrast with some of their biologist colleagues in a similar ideological controversy.

Acknowledgment Our research is being supported in part by the Hungarian Scientific Research Foundation (OTKA Nos. K60365 and T046183).

References

1. Pauling L (1960) The nature of the chemical bond and the structure of molecules and crystals: an introduction to modern structural chemistry (Third edition, First edition, 1939). Cornell University Press, Ithaca, NY
2. Hargittai I (2009) Struct Chem 20:753–756
3. Pauling L (1996) Chem Intell 2(1):32–38
4. Corradini P (1982) In: Carra S, Parisi F, Pasquon I, Pino P (eds) Giulio Natta: present significance of his scientific contribution. Editrice di Chimica, Milano, p 134
5. Bragg WL, Kendrew JC, Perutz MF (1950) Proc R Soc 203A:321–357
6. Pauling L, Corey RB (1950) J Am Chem Soc 72:5349
7. Pauling L, Corey RB, Branson HR (1951) Proc Natl Acad Sci USA 37:205–211
8. Olby R (1994) The path to the double helix: the discovery of DNA. Dover Publications, New York, p 291
9. Hodgkin DC, Riley DP (1968) In: Rich A, Davidson N (eds) Structural chemistry and molecular biology. WH Freeman, San Francisco and London, pp 15–28
10. Bernal JD (1968) Labour Monthly, pp 323–326
11. Hargittai I (2002) Candid science II: conversations with famous biomedical scientists (edited by Magdolna Hargittai). Imperial College Press, London, p 288
12. Hargittai I (2000) Candid science: conversations with famous chemists (edited by Magdolna Hargittai). Imperial College Press, London, pp 8–13

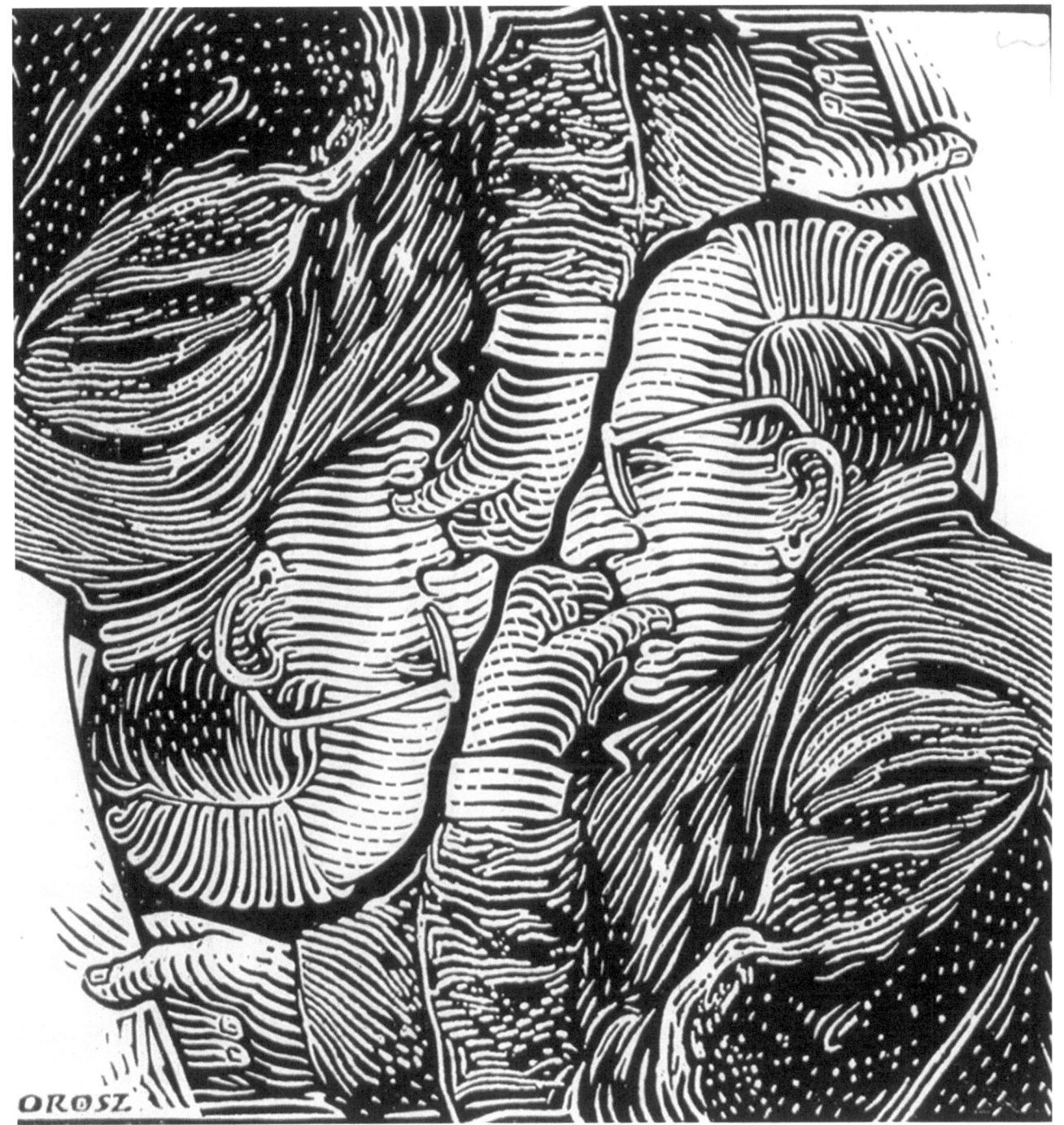

"Complementary Kitaigorodsky" by the graphic artist Istvan Orosz (Budapest). The great Russian crystallographer predicted early on that the most frequent symmetry will be a complementary arrangement of molecules in molecular crystals. Orosz prepared this drawing, along with some others, at my Parents' request for their book, Istvan Hargittai and Magdolna Hargittai, *In Our Own Image: Personal Symmetry in Discovery* (New York: Kluwer/Plenum, 2000).

Crystallography in *Structural Chemistry*[a]

István Hargittai

Abstract

This anniversary article has three functions: It marks Volume 25 of our journal; it honors 2014, the International Year of Crystallography; and it celebrates the centennial from the birth of a great crystallographer, Aleksandr I. Kitaigorodskii.

Introduction

Our journal, *Structural Chemistry*, is completing its 25th volume this year; it seems recent that we marked the twentieth anniversary of this publication [1]. The goal of our journal has not changed since its inception; viz., to provide a venue for high-quality research reports and overviews in diverse areas of the ever expanding discipline of structural chemistry. The year 2014 is the International Year of Crystallography, and we are happy to note that crystallography has constituted a substantial portion of the papers we have published. Lately, the emphasis of our crystallographic contributions has changed. We have moved away from publishing mere structure reports toward more comprehensive papers, including such that deal with the foundations of crystallography and the various approaches to the determination of crystal and molecular structures. During the past quarter century the foundations of crystallography have broadened and it has increasingly become the science of structures. Following a brief historical introduction, I am singling out a few contributions from these 25 volumes for illustration, but this sampling is indeed a sampling only and is far from a comprehensive coverage.

Aleksandr I. Kitaigorodskii in his laboratory at the Institute of Element-organic Compounds (INEOS) of the Soviet (now Russian) Academy of Sciences, Moscow. Photograph courtesy of the late Erlen Fedin. (Kitaigorodskii's name is spelled in a variety of ways in the scientific literature: Kitaigorodskii, Kitaigorodsky, Kitaigorodski, Kitaigorodskij, depending on the approach to transliterating his name from the Russian original).

[a]Originally published in *Structural Chemistry* 2014, 25:1321–1326.

Dedication: This Editorial is dedicated to the memory of Aleksandr I. Kitaigorodskii (1914–1985). He was a unique thinker, an original scientist, and a versatile human being. He uncovered a general principle of how molecules build up crystal structures, and he predicted correctly the relative frequencies of occurrence of the various space groups among crystal structures at the time when relatively few organic crystal structures had been yet determined. His teachings live on in his papers and monographs and in the scientific and pedagogical activities of his pupils and their pupils.

I. Hargittai (✉)
Department of Inorganic and Analytical Chemistry, Budapest University of Technology and Economics, Budapest, Hungary
e-mail: istvan.hargittai@gmail.com

X-Ray Crystallography

In the November 14, 1912, issue of *Nature*, the British scientist A. E. H. Tutton reported [2]: "During a visit to Munich at the beginning of August last the writer was deeply interested in some extraordinary photographs which were shown to him by Prof. von Groth, the *doyen* of the crystallographic world, and professor of mineralogy at the university of that city. They had been obtained by Dr. M. Laue, assisted in the experiments by Herren W. Friedrich and P. Knipping, in the laboratory of Prof. A. Sommerfeld in Munich, by passing a narrow cylindrical beam of Röntgen rays through a crystal of zinc blende, the cubic form of naturally occurring sulphide of zinc, and receiving the transmitted rays upon a photographic plate. They consisted of black spots arranged in a geometrical pattern, in which a square predominated, exactly in accordance with the holohedral cubic symmetry of the space-lattice attributed by crystallographers to zinc blende." It seems that Tutton recognized the broader significance of what he saw, "Crystallography thus affords to its sister science Chemistry the first visible proof of the accuracy of Dalton's atomic theory, and now enters into a new sphere of still greater usefulness.... Crystallography has thus become an exact science leading us to a practical knowledge of the hitherto mysterious world where Dalton's atoms and molecules reign supreme."

I hasten to note that crystallography had existed as a science long before the Munich experiments. Two hundred years before Dalton and 300 years before the Munich experiment, Johannes Kepler discussed the shape and inner structure—in today's terms, the atomic arrangement—of crystals. He did this in his Latin-language treatise, Strena, seu De Nive Sexangula of 1611, which was published in English translation in 1966 [3]. Kepler presented arrangements of closely packed spheres. Incidentally, Dalton invoked the image of close packing of spheres in his work on gas absorption [4]. The history of classical crystallography from Kepler to Laue is a shining page in science history.

Laue had initiated the experiments because P. P. Ewald, a doctoral student in physics, had raised the possibility of X-ray scattering by crystals in his dissertation. Ewald completed his thesis earlier in 1912 and turned to Laue with his query. Ewald's genius connected the propagation of electromagnetic radiation and the supposed internal structure of crystals. He assumed that if the crystal is looked at as a regular arrangement of resonators, and the distance between these resonators would be commensurate with the wavelength of the radiation, there should be a diffraction phenomenon, and it should be possible to observe it. He consulted Max Laue (as he was then), and the experiment mentioned above followed. The significance of the experiment was immediately recognized in Max von Laue's Nobel Prize in 1914 (by then, "von," because his father received hereditary nobility in 1913).

Direct Methods

The Nobel laureate mathematician turned crystallographer Herbert Hauptman (1917–2011) described in the pages of *Structural Chemistry* [5] the development of X-ray crystallography to which he and Jerome Karle (1918–2013) [6] contributed in a seminal way by proposing and working out the direct methods of structure analysis. Hauptman and the physical chemist Karle shared the Nobel Prize in Chemistry in 1985. They solved the so-called phase problem using mathematical techniques [7–9]. Their discovery and the discoveries of others made it possible for X-ray crystallography to expand toward larger systems than before and to increase the accuracy of structure determination. David Sayre (1924–1912) was another outstanding contributor to the solution of the phase problem and he gave a detailed critical analysis in this journal of how it happened [10].

Isabella Karle was one of the pioneers in applying the direct methods for actual structure determinations. She and Jerome had utilized their vast experience in modernizing a less well-known technique of structure determination, gas-phase (often, simply, gas) electron diffraction (GED) [11]. Some of the intricacies of the structure analysis of GED were helpful in developing the direct methods. At this point, it is proper to stress that the term crystallography has become synonymous with the science of structure; thus, it embraces structural studies in gases and liquids as well. The development of the gas-electron-diffraction technique has also received exposure in our periodical *Structural Chemistry* [12, 13]. Beside gases, the electron diffraction technique has also been used extensively for the analysis of solid structures. An outstanding contribution to electron crystallography is mentioned here as it was applied for a broad range of materials [14]. The development of holographic methods was among the recent technical innovations in crystallography, on the road toward structure determination with atomic resolution [15].

Biological Macromolecules

The crystallographic investigation and structure determination of biological macromolecules was one of the most spectacular scientific achievements of the twentieth century in which X-ray crystallography played a pivotal role. Linus Pauling's triumph in discovering the alpha-helix structure of proteins served as useful example in a host of research projects [16]. Pauling's accumulation and utilization of structural data was a key element in his success. The discovery of the double helix structure of DNA was a spectacular application of Pauling's model-building approach [17]. Pauling's attitude, however, was less than welcoming toward the cyclol hypothesis, which came up in the quest for protein structures. Further support for the rejection of this hypothesis appeared recently from theoretical calculations [18]. The ultimate utilization of the quest for biomolecular structures will be the fascinating approach of personalized medicine toward which the mapping of the human genome constituted a great stride [19].

Generalized Crystallography and Quasicrystals

Our journal has paid much attention to generalized crystallography—a term much cultivated by Alan L. Mackay—expressing the structure of science [20]. In particular, this interpretation of crystallography went beyond the classical system of 230 three-dimensional space groups. Here we quote only a small set of contributions that paid homage to Mackay and his concepts and extended the realm of structure considerations to non-classical constructions [21–23]. This area of crystallography overlaps with nanoscience and nanotechnology, molecular biology, and condensed state physics.

A specific area of non-classical crystallography is quasicrystals. The history of their discovery and the barriers its concept had to overcome before acceptance by some of the leading scientists was most instructive [24–26]. In particular, Ref. [25] anticipated the high recognition for the quasicrystal discovery just a few months before Dan Shechtman's Nobel Prize was announced in October 2011.

The structures of quasicrystals have remained a puzzle in many aspects, but the accomplishments in the area have also been most impressive. The time has come to discuss quasicrystal structures at the atomic level [27]. Steurer and Deloudi took up the challenge of describing quasicrystals consisting of clusters, and found packing principles for them [28].

Fundamental Concepts

Finally, a sampler of papers follow to represent studies of rather general character that belong to the expanded interpretation of crystallography. They cover a wide range of topics, but all examine questions related to fundamental features or concepts of structure. Above we mentioned a paper about quasicrystal clusters. Ilyushin communicated a comprehensive study of clusters in general, their self-organization, and described geometrical modeling of nanocluster precursors, building up a hierarchical system [29]. Malenkov investigated the possibility of understanding the regularities of non-crystalline substances on the level of their inherent structures. The concept of inherent structures and the history of the development of this concept are exposed [30]. Shevchenko carries a similar question to philosophical depths. He follows the process of structures building up from fundamental configurations to clusters and to the whole structures. He finds unity in the basic principles of hierarchical construction regardless whether the final structure is periodic or aperiodic [31]. Efforts to find the concept universal optimum and general principles of "inorganic genes" directed Shevchenko and Krivovichev to investigate paulingite-related zeolites and minerals [32].

Meyer investigated the notions of size and shape separately and in combination; he showed the advantage in considering them "wedded" [33]. He involved molecular volume, surface area, packing densities and other properties in the discussion, and used the example of aromatic organic compounds for the application of his conclusions. One of the conditions influencing molecular size is temperature. For molecules with high degree of deformation motion, the size may expand considerably at elevated temperatures. Varga et al. examined the extent of such expansion as a function of temperature [34]. For this, intricacies of intramolecular motion and its anharmonicity have to be taken into account. The concerted use of experimental data and computational results yielded noteworthy conclusions.

There have been valuable attempts to uncover regularities and trends in the variations of various properties of related substances. Slovokhotov, Batsanov, and Howard analyzed the trends in melting temperatures and boiling temperatures of organic compounds [35]. Their observations supported the notion of molecular van der Waals symmetry developed earlier by the authors. They thus seemed to transfer successfully information about molecular properties to information about bulk properties.

In 2007, a rather unusual contribution was published in *Structural Chemistry* by an unusual scientist. The late John E. Scott was professor of chemical morphology at the Department of Chemical Morphology, University of Manchester, UK. His research field was the structure and function of various polysaccharides that constitute building blocks in our body. As he taught, "Our shape is defined and maintained by the connective tissues (skin, tendons, cartilages, blood vessels, etc.) or more precisely by their extracellular matrices. These highly ordered supramolecular organisations are modules of protein fibrils held together by elastic carbohydrate strings" [36]. His review covered a broad range of structural studies on a variety of polysaccharides involving various experimental techniques as well as molecular modeling and computer simulations. Scott's studies served as one of the inspirations for us to look more into the structural intricacies of a particular polysaccharide, hyaluronic acid, called also hyaluronan [37]. Lately, this substance has gained great visibility and fame for its presence in relatively great concentrations in some specific areas of the human body (such as, for example, the vitreous, the umbilical cord, the joints, and in the skin); for its most efficient clinical use; and for its popularity as an anti-aging agent in cosmetics. It has interesting structural features, including a double-helix configuration with intramolecular hydrogen bonding, not unlike the double helix of DNA.

In this overview, already a few studies have been mentioned in connection with developing a systems approach to inorganic structures. David Brown's work on chemical topology belongs to this domain of inquiries. He singled out three components in the description of structures [38]. They are the properties of atoms participating in the structure; the three-dimensional space hosting the structure; and the topology describing the structure. He analyzed the relevant topologies and concluded that this analysis together with electrostatic theory and augmented with empirical observation led to a helpful model comprising of localized chemical bonding.

Geometry and models are most useful ingredients of structural chemistry and in particular, structural inorganic chemistry [39]. The geometrical model has been utilized in the description of molecular systems although its utility has limitations; the more rigid the system the better it works. Geometrical modeling has helped uncovering molecular structures from the simplest systems to the most intricate biological macromolecules. In this connection, the areas surveyed included: the technique of gas-phase electron diffraction; the notion of "experimental error" in quantum chemical calculations; precision and accuracy; the application of qualitative models, such as the VSEPR model; the investigation of isomerism; chirality; and molecular packing in organic substances.

Al Kitaigorodskii and Molecular Packing

The degree of understanding of molecular packing has been an evolving measure in our understanding how molecules, atoms, and ions build crystal structures. Aleksandr Kitaigorodskii made outstanding contributions to this question. Although he could not fully solve it; he was among the first who asked this pivotal question. He was an extraordinary scientist as well as human being [40]. He pioneered the observation that the distances between molecules showed a characteristic constancy in diverse classes of organic substances. This observation helped him formulate the concept of molecular shape from which it was natural to pose the question about molecular packing in crystals. This fundamental question arose simultaneously in the minds of more than one scientist, which is not a rare occurrence in science history. In 1940, Linus Pauling and Max Delbrück published a note about interactions between molecules assigning precedence for interactions between parts that are complementary to each other rather than parts that are identical with each other [41]. This notion could be applied directly to considerations about molecular packing although Kitaigorodskii developed his ideas independently from Pauling and Delbrück.

Kitaigorodskii communicated his ambitious research program in a brief paper in 1945 in a then still existing English-language Soviet journal [42]. Kitaigorodskii declared in this paper that in a molecular crystal, "the mutual location of molecules is determined by the requirements of the most close-packing." The packing of molecules in organic crystals remained his leitmotif throughout his scientific career and he opened up a new area in crystal chemistry. With painstaking and systematic work, Kitaigorodskii determined the frequency distribution of molecular crystals among the 230 three-dimensional space groups. Later observations on hundreds of thousands experimentally determined crystal structures confirmed the correctness of his predictions. The importance of complementary arrangements has proved a basic governing factor in the structure of molecular crystals.

Kitaigorodskii had a spectacular initial career; he was successful in original research, in popularizing science, in building a research center, in developing a great school of pupils, and in gaining international recognition. He was less successful in gaining official recognition in his home country where his free spirit and irreverence toward authority gained him enemies and generated jealousy. We remember Aleksandr Kitaigorodskii as a great contributor to the science of chemical structures, a devoted teacher, and a unique human being.

We at *Structural Chemistry* recognized the importance of Kitaigorodskii's oeuvre from the beginning of our publication. Soon after the inception of this journal, in 1992, we initiated a special issue to honor his memory. An excellent collection of papers came together; of them, a few remembered Kitaigorodskii the scientist and the man, and most were outstanding research contributions. Unfortunately, by the time the collection was ready for publication, a crisis had developed about our journal and the survival of the journal was uncertain for some time before a change in publishers could be arranged. In the meantime, as we wanted to be sure that the special collection of papers honoring Kitaigorodskii would not be lost we had to find an alternate venue for bringing it out. It happened to be *Acta Chimica Hungarica—Models in Chemistry* of the Hungarian Academy of Sciences. This venue saved the collection, but it also meant a reduced visibility, the more so, because *Acta Chimica Hungarica* soon ceased to exist. This is also, why I mention this special collection with emphasis in this Editorial [43].

Trends

Recently we have reviewed the Nobel Prizes awarded for discoveries in the domains of structural chemistry [44, 45]. It needs to be stressed that science history cannot be compiled on the basis of Nobel Prizes as they are often accidental and sometimes the award givers succumb to demands of fashion. Nonetheless, it is noteworthy that structural chemistry, directly or indirectly, figures conspicuously often in the award-winning achievements. Table 1 in Ref. [45] listed 19 Nobel Prizes related to structural chemistry, awarded through 2011. If extended to two more years available as of July 2014 when this account is being written, the 2013 Nobel Prize in Chemistry "for the development of multi-scale models for complex chemical systems" should be added, because the application of this approach is aimed at solving structural problems. Thus, during the last six decades, 20 Nobel Prizes were related to structural chemistry. Even a superficial browsing of these 20 awards shows that at least half of them were related to crystallography if it is taken in the modern sense as the structure of science.

Crystallography is as old as science itself; this is so if we consider the appearance of "scientific" crystallography from Kepler's treatise on the snowflakes. Moreover, the "science" of crystals could be dated at the first moment when a crystalline substance was distinguished from an amorphous body. Alan Mackay used this telling example from *Kama Sutra* of Vatsayana that in the India of the sixth century, the courtesans had to learn some basics of mineralogy in order to distinguish real crystals from paste [20].

Contemplating about the development of crystallography during the past decades, a shift is noteworthy from ordered structures toward less ordered ones. The initial success of X-ray crystallography originated to great extent from the enormous amount of routine determinations of ordered structures. However, at certain point this also became a barrier to further development, because many scientists were hesitant to extend the realm of their inquiry toward less ordered structures. Between the two world wars, it was still possible to divide large research areas among a few scientists. Thus, for example, the two outstanding British crystallographers J. Desmond Bernal and William Astbury decided to delineate their research areas. In Bernal words: "I took the crystalline substances and he [Astbury] the amorphous or messy ones. At first it seemed that I must have the best of it but it was to prove otherwise. . . . It may be paradoxal that the more information-carrying methods should be deemed the less useful to examine a really complex molecule but this is so as a matter of analytical strategy rather than accuracy" [46].

Acknowledgments At this point of the publication of the 25th volume of our journal, I am grateful for a most fruitful and pleasant cooperation in all matters of editing this journal to my friend, Editor Jerzy Leszczynski; to Senior Publishing Editor at Springer-Verlag, Sonia Ojo; to the Production Editor in India, Ms. Muthulakshmi and her associates; to the members of our Editorial Board; to our reviewers; and most significantly, to the authors and the users of our journal, worldwide.

References

1. Hargittai I, Kovács A (2009) The twentieth year in *Structural Chemistry*. Struct Chem 20:1–10
2. Tutton AEH (1912) The crystal space-lattice revealed by Röntgen rays. Nature 90:306–309
3. Kepler J (1611) Strena, seu De Nive Sexangula; English translation by L. L. Whyte, The Six-cornered Snowflake. Clarendon Press, Oxford, 1966
4. Dalton J (1805) Memoirs and Proceedings of the Manchester Literary and Philosophical Society. Manchester, Vol 6, p 271; Alembic Club Reprints (1961), Edinburgh, no 2, p 15
5. Hauptman HA (1990) History of X-ray crystallography. Struct Chem 1:617–620
6. Hargittai I, Hargittai M (2013) Jerome Karle (1918–2013)—Nobel laureate; Charter member of the Editorial Board of Structural Chemistry. Struct Chem 24:2219–2222
7. Karle J, Hauptman H (1950) Acta Crystallogr 3:181
8. Hauptman H, Karle J (1950) Phys Rev 80:244
9. Hauptman H, Karle J (1953) Solution of the phase problem I. The centrosymmetric crystal. American Crystallographic Association Monograph No 3. Polycrystal Service, Dayton, Ohio
10. Sayre D (2002) X-ray crystallography: the past and present of the phase problem. Struct Chem 13:81–96
11. Karle I, Karle J (2005) Gas electron diffraction and its influence on the solution of the phase problem in crystal structure determination. Struct Chem 16:5–16
12. Hedberg K (2005) Fifty years of gas-phase electron diffraction structure research: a personal retrospective. Struct Chem 16:93–109
13. Hargittai I (2005) Looking back and ahead: gas-phase electron diffraction at 75. Struct Chem 16:1–3
14. Dorset DL (2002) From waxes to polymers—crystallography of polydisperse chain assemblies. Struct Chem 13:329–337
15. Faigel G, Tegze M (2003) X-ray holography. Struct Chem 14:15–21
16. Hargittai I (2010) Linus Pauling's quest for the structure of proteins. Struct Chem 21:1–7
17. Hargittai I (2004) Francis Crick (1916–2004). Struct Chem 15:545–546
18. Alkorta I, Sánchez-Sanz G, Trujillo C, Azofra LM, Elguero J (2012) A theoretical reappraisal of the cyclol hypothesis. Struct Chem 23:873–877
19. See, e.g., Hargittai I (2010) The Human Genome Project—A triumph (also) of structural chemistry: On Victor McElheny's new book, *Drawing the Map of Life*. Struct Chem 21:667–671
20. Mackay AL (2002) Generalized crystallography. Struct Chem 13:215–220
21. Kuo KH (2002) Mackay, anti-Mackay, double-Mackay, pseudo-Mackay, and related icosahedral shell clusters. Struct Chem 13:221–230
22. Ogawa T, Ogawa T (2002) Proportional representation system as generalized crystallography and science on form. Struct Chem 13:297–303
23. Shevchenko VY, Madison AE, Mackay AL (2007) A generalized model for the shell structure of icosahedral viruses. Struct Chem 18:343–346
24. Hargittai B, Hargittai I (2012) Quasicrystal discovery—from NBS/NIST to Stockholm. Struct Chem 23:301–306

25. Hargittai I (2011) "There is no such animal (כזו חיה אין)"—lessons of a discovery. Struct Chem 22:745–748

26. Hargittai I (2007) Quasicrystals: 25 years. Struct Chem 18:533–534

27. De Boissieu M (2012) Atomic structure of quasicrystals. Struct Chem 23:965–976

28. Steurer W, Deloudi S (2012) Cluster packing from a higher dimensional perspective. Struct Chem 23:1115–1120

29. Ilyushin GD (2012) Theory of cluster self-organization of crystal-forming systems: geometrical-topological modeling of nanocluster precursors with a hierarchical structure. Struct Chem 23:997–1043

30. Malenkov GG (2007) Inherent structures of condensed phases. Struct Chem 18:429–436

31. Shevchenko VY (2012) What is a chemical substance and how is it formed? Struct Chem 23:1089–1101

32. Shevchenko VY, Krivovichev SV (2008) Where are genes in paulingite? Mathematical principles of formation of inorganic materials on the atomic level. Struct Chem 19:571–577

33. Meyer AY (1990) More on the size of molecules. Struct Chem 1:265–279

34. Varga Z, Hargittai M, Bartell LS (2011) On the thermal expansion of molecules. Struct Chem 22:111–121

35. Slovokhotov YL, Batsanov AS, Howard JAK (2007) Molecular van der Waals symmetry affecting bulk properties of condensed phases: melting and boiling points. Struct Chem 18:477–491

36. Scott JE (2007) Chemical morphology: the chemistry of our shape, in vivo and in vitro. Struct Chem 18:257–265

37. Hargittai I, Hargittai M (2008) Molecular structure of hyaluronan: an introduction. Struct Chem 19:697–717

38. Brown ID (2002) Topology and chemistry. Struct Chem 13:339–355

39. Hargittai I (2011) Geometry and models in chemistry. Struct Chem 22:3–10

40. See, e.g., Hargittai I (2013) Aleksandr Kitaigorodskii: Soviet maverick (Chap. 11). In: Buried glory: portraits of Soviet scientists. Oxford University Press, New York, pp 250–266

41. Pauling L, Delbrück M (1940) The nature of the intermolecular forces operative in biological processes. Science 92:77–79

42. Kitaigorodskii AI (1945) The close-packing of molecules in crystals of organic compounds. J Phys (USSR) 9:351–352

43. Hargittai I, Kálmán A, Guest Editors (1993) A. I. Kitaigorodskii Memorial Issue, Parts 1 and 2. Acta Chimica Hungarica—Models in Chemistry Volume 130, Number 2, pp 151–298 and Numbers 3–4, pp 301–555

44. Hargittai B, Hargittai I (2011) Nobel Prize and structural chemistry I. Struct Chem 22:961–964

45. Hargittai B, Hargittai I (2012) Nobel Prize and structural chemistry II. Struct Chem 23:1–5

46. Bernal JD (1968) The material theory of life. Labour Monthly July, pp 323–326, actual quotation, p 324

Otto Bastiansen (left) and Istvan Hargittai (right) in spring 1969 on the campus of the University of Texas at Austin. Bastiansen was a visiting professor and my father a visiting research associate, both for 1 year and both at the Department of Physics. In fact, Bastiansen's appointment gave him the possibility to bring with him two associates; one was Jon Brunvoll from Trondheim and the other was my father.

The Bastiansen Phenomenon[a]

István Hargittai

Otto Bastiansen in 1969 in Austin, Texas (photograph by I. Hargittai).

It was a grey November day in 1967 in Budapest. We were expecting the famous Norwegian scientist, Otto Bastiansen. On the eve of his arrival I had broken my leg and for the first 24 h I could not move around at all. My wife Magdi went to the airport to meet Professor Bastiansen. We just got married a couple of months before. She was a third year student and I was a junior research associate. I had described him to her as a very mobile and smiling person, and she had no difficulty identifying him.

When I started building my gas electron diffraction laboratory in Budapest in the fall of 1965, after having learned the technique in 1964/65 with Lev Vilkov in Moscow, I sent out a letter to what I thought were the four most important laboratories in the world, one Norwegian, one Japanese, and two in the United States. I had studied all the literature I could find but to build up this experiment I had to ask for any additional information they might send me. There were three responses. The American scientist explained painstakingly why it was hopeless for me to initiate such an experiment in Budapest, and, in hindsight, he was almost right. From Tokyo I received four lengthy reports in Japanese, full of detailed information truly unavailable from publications. I had immediately organized their translations into Hungarian and they proved to be very useful indeed. The letter from Bastiansen was not only accompanied by useful documentation but it also offered encouragement and further help.

I could describe Bastiansen to Magdi because I had seen him already in the summer of 1966 at the open session of the electron diffraction commission during the International Union of Crystallography Meeting in Moscow. It was my first visit to Moscow since I had finished my studies there. I went to the meeting with my Russian friends. It was in a large amphitheater-type auditorium. The commission meeting was down in the center and we were sitting in the upper back rows. We were guessing who was who as we knew we had met some of the names in the literature. We assigned Bastiansen's name to the oldest and most respectable looking person. In fact, he turned out to be an old French professor. The real Bastiansen turned out to be one of the younger persons whose behavior we found a little annoying as he seemed to be dominating the scene, not letting the real great persons, such as our designated 'Bastiansen', speak more.

As it turned out, Otto's visit and lecture in Budapest were a superb public relations move for electron diffraction. Also, he must have realized that I needed more training and brought me to Oslo the following year for 3 months. On April 2, 1968, I arrived for my first visit in the West. I was assigned to Reidar Stølevik. I saw very little of Oslo on that first visit, but learned and did a lot. I had a special permit to spend the

[a]Originally published in *Journal of Molecular Structure* 2998, 445:xiii–xviii, © 1998 Elsevier, reproduced with permission. Here we communicate fewer images than in the original.

I. Hargittai (✉)
Department of Inorganic and Analytical Chemistry, Budapest University of Technology and Economics, Budapest, Hungary
e-mail: istvan.hargittai@gmail.com

nights in the computation center. Otto was following my progress, lent me his bicycle for transportation, and it was the beginning of a special relationship for the next 25 years. During my stay in Oslo, Otto was already preparing for his visiting professorship at the University of Texas the following fall. The University of Texas had offered him to take two young associates with him. He had asked Jon Brunvoll and Reidar Stølevik, and when Reidar bowed out, he asked me to go with him to Austin. I had difficulties in getting my passport in Hungary and this caused a 4-month delay in my arrival, but finally I could join Otto and Jon in Austin. For 6 months, the three of us were sharing an office. Otto was full of ideas and he was interested in everything and everybody. Mr. Pemberton, the janitor, who happened to be black, and happened to have a B.A. degree in English literature, would often come to have a chat with Otto. Jon and I were also sharing Otto's all-embracing attention. He took us along whenever something interesting, like an outing with Ilya Prigogine, was happening. I got my first driver's license in Austin, and so did Jon, and Otto was our instructor. I still remember and pass on his extremely useful advice about safe driving.

When our son, Balazs, was born in 1970, Otto became his godfather, an institution we otherwise would not have cared to consider but it came naturally in our relationship. When my passport was revoked after my brother had left Hungary 'illegally' in 1978, Otto gave a talk in Hungary, praising me out of proportion, hoping that the authorities would return my passport; alas nobody paid attention. When the ban on my travel was lifted, my first trip was, again, to Oslo and, again, on Otto's initiative. It was the 1981 Odd Hassel Lectureship. Earlier in the morning of the day the Hassel Lecture was scheduled for, Odd Hassel had died. Otto was especially distressed by this sad event. He was very close to Hassel although their personalities could not have been more different. He was also concerned about the Hassel Lecture. He remembered an event under the German occupation in Norway, when Hassel had been arrested by the Norwegian Nazis and was turned over to the Gestapo and the university people were quite at a loss what to do. Some even contemplated a strike, until word came from the jail, from Hassel himself, "The lectures must go on." So Otto contacted the Hassel family, and the decision was, "The lecture must go on."

Otto Bastiansen and Istvan Hargittai in 1982 in Norway (by unknown photographer).

In 1988 Otto was the President of the Norwegian Academy of Sciences and I was elected foreign member. The invocation, in the presence of the Honorary President, King Olav V, was followed by a reception. At one point word came that His Majesty wanted to see me. He was sitting in a corner with Otto, two old men, immersed in a conversation, obviously having known each other for a long time.

The last time I saw Otto at the University of Oslo was after he had retired from all his various functions and he was visibly happy to become active again, during my visit, if only for a few days. He was rather distressed that his office was quite removed from the rest of the Group and that his photograph was already displayed among the deceased rather than getting an oil painting among the past Professors on the appropriate wall in the Department. The worst thing about it was that the healthy Otto would have been the first person to joke about this.

The last time I saw Otto was not long before his passing away, in the nursing home. His health had deteriorated by then but this was not an unhappy meeting. Hans Seip and I went together. Otto was very fond of both Ragnhild and Hans Seip. It pained him when he felt it prudent to remove himself from the voting for the Professor of Theoretical Chemistry many years before, for Hans was among the candidates and he felt Hans to be so close to him. There were many lucid moments during this last visit. He was obviously enjoying it and when we thought we should be leaving he would not let us. When the departure finally came, he hugged me, something he had never done before. He asked me to carry his best wishes to our families, our friends, our colleagues. He was thinking, as always, of everybody else.

Otto Bastiansen was born on September 5, 1918, in Balsfjord, northern Norway, and died on October 2, 1995, in Oslo. He attended the University of Oslo and was a student of Odd Hassel. It was at the time of the German occupation. He must have participated in some anti-German activities but nobody seems to know in what and to what extent. He never talked specifically about it. However, he had started learning Russian about that time, and he prided himself in later years in having read Dostoevskii's *Brothers Karamazov* in the original form.

He worked with Odd Hassel at the time Hassel discovered conformational equilibria in his gas-phase electron diffraction studies of cyclohexane derivatives. Hassel published his results in Norwegian in an obscure Norwegian chemical journal in 1943. He was not allowed to publish in English and he refused to publish in German. This paper by Hassel has been much cited but little read. It contained essentially everything for which Hassel got the Nobel Prize in Chemistry in 1969, shared with Derek Barton. Twenty-five years after the publication of the Norwegian article, the paper was re-published in the same Norwegian journal in English.

Bastiansen was instrumental in developing gas-phase electron diffraction conformational analysis. This was also the topic of his doctoral dissertation. Curiously, the details of his working together with Hassel was another of those few topics Otto would not like to talk about much although he was not given to shyness. In the mid-1970s when I was spending 3-month periods in Oslo, I had some interesting conversations with Sven Furberg (1920–1983), Professor of Chemistry, an X-ray crystallographer. Furberg wanted me to know that Otto's role was more instrumental in the seminal discovery of conformational equilibria than I might have guessed. Apparently, Otto's contribution was the simple, in hindsight, notion that when neither individual conformers of some cyclohexane derivatives would account for the experimental electron diffraction data alone, he suggested to try their mixtures. Furberg told me I ought to know these details of the story because it was useful and instructive to get to know more details about scientific discoveries. I did not know then about the crucial role Sven Furberg had played in the research leading to the discovery of the double helix. Although his work was cited in the first announcement of the structure in 1953 by James Watson and Francis Crick, J. Desmond Bernal was moved to state the following, in his review of Watson's *Double Helix*: "I do think that, for historic justice, in the light of the importance of his work, Furberg's contribution has been grossly overlooked." Recently, out of curiosity, I looked up Hassel's Nobel lecture to see the references and acknowledgments. Alas, Hassel's Nobel lecture was not about conformational analysis but about his work on charge transfer complexes. It is not unprecedented but rare that the topic of the Nobel lecture is different from the research for which the Prize had been awarded.

Bastiansen first worked at the University of Oslo and then got a professorship at the (then) Norwegian Technical University (1955–1962). He subsequently returned to the University of Oslo, where he was Professor of Theoretical Chemistry and later, as Hassel's successor, Professor of Physical Chemistry until his retirement in 1988. He held all the high offices there were for him to hold in Norway, including Chair of the Norwegian Research Council, Chair of the Norwegian Chemical Society, Rector of the University of Oslo, and president of the Norwegian Academy of Science and Letters. He received awards and prizes, the highest of which was the Commander's Cross of the Order of St. Olav's.

He was a truly public figure in Norway for some time. Everybody seemed to know about him. He was also well known internationally. He did postdoctoral work in Linus Pauling's Caltech laboratory in Pasadena and returned to the United States several times as a coveted visiting professor. He was not indifferent to anything and he touched the lives of many both around him and at a distance. He was a great educator but not by words. He was an educator to us in the best way, possibly the only way, by his example. He was friend to the janitor in Austin, never condescending, and to the King of Norway.

Otto was always ready to discuss science. His love was conformational analysis, and he also advocated the need to reinvestigate the structures of fundamental substances, again and again, as the techniques, and, in particular, gas electron diffraction, were being perfected. He was deeply concerned about the concerted consideration of geometry and motion in structure analysis and interpretation. He gave a course in Austin about the foundations of gas electron diffraction and its applications in structure analysis. He favored classical treatment and the presentation did not always appear well organized. He would periodically interrupt his line of thought to demonstrate the principle or equation under discussion on practical examples, mostly from his own experience. In hindsight, and browsing my lecture notes nowadays, it was a well-organized and very didactic course.

One of Otto's main concerns was the joint application of various techniques for better structure determination. He had started advocating this much before it became fashionable. It was not to a small extent due to his efforts that by the early 1980s this truly became a much practised approach and is continuing in today's research. In 1981 he gave a talk at the Hungarian Academy of Sciences entitled "Joint Application of Several Methods in the Investigation of Molecular Structure and Molecular Dynamics." This talk was translated and published in Hungarian [*Kémiai Közlemények* 59 (1983) 109–115].

On the occasion of his 1981 visit in Budapest, he gave a second talk for a broader audience at the Academy, "Fundamental Research and Modern Society," which also appeared in Hungarian [*Magyar Tudomány* (1982) 258–266]. His talk was recorded and here I would like to share a few excerpts as transcribed from the recording, and polished somewhat, taking the Hungarian publication into account. I conclude my sketch of what I call the Bastiansen Phenomenon with these excerpts. I think we can still learn from what he has to say. Besides, I still don't find it easy to write in past tense about Otto Bastiansen, so I am asking him to rescue me, and take it over.

On his interest in science history:

"I have dealt a lot with the value of history in understanding our present. The past, present, and future lie on one line yet it is very difficult to extrapolate, nonetheless, we have to try. I find this so important that I had even written a little book about it and it is presently being published by the Government."

On the uses and costs of research:

"A Norwegian example of the usefulness and costs of scientific research. We are very proud of our Viking ships. Some years ago when I was Rector of Oslo University, they discovered another Viking ship. Everybody was very enthusiastic. We will excavate it, dig it up, bring it up, prepare it, build a nice building for it, and exhibit it. Everybody agreed, this was a very important project. But how can we get money for it? It was a terribly expensive affair. What I could do, as Rector, I freed some people from their work for this project. Then we begged money from industry, especially the chemical industry. Finally we had everything ready, we had the Museum, and the exhibition opened. We thought God has helped us to get all this money. Nevertheless, we sat down to calculate how expensive it was. Very near the Museum a huge new highway was being built, and the excavation and all the auxiliary expenses were equivalent to two and a half meters of that highway. We did more calculations and found that all our expenses were equivalent to the portion of eight minutes of the Norwegian military budget."

On the so-called university dreams:

"I was the chairman of the Research Council when the oil in the North Sea, off the Norwegian shores entered our life. I often ask people in Norway, When do you think the first article was printed in plain Norwegian that there was oil in the North Sea? Some people say, it must have been in the sixties, others said, in the fifties. Somebody said, in 52, and that was correct. But it was not 1952, not even 1852. It was in 1752. Bishop Erich Pontoppidan [*Norges Naturlige Historie*, Berlingske Arvingers Bogtrykkeri, Copenhagen, 1752, p. 116] wrote a book about it in which he tells in detail about the oil in the North Sea. In 1874 a group of professors of Oslo University got together and produced a study of the continental shelf of Norway providing useful data and maps even for today's explorers. Fridtjof Nansen, the famous polar explorer, who was also an oceanographer, also enriched greatly our knowledge on the continental shelf. Already some 30 years ago there was a doctoral dissertation about how to drill oil in the North Sea. In the 1960s the Norwegian Research Council, under my chairmanship, provided substantial support for oil research. Our Research Council was supposed to support fundamental research, such as medicine, theology, history, etc. Yet we initiated this investigation and carried it on for several years.

At some point I started feeling uncomfortable and visited the Chairman of the Applied Research Council and suggested that they take over. I told him that we anticipated oil and, accordingly, income from oil for Norway, but he said, no. I repeated my visit half a year later and he called my proposal a university dream. I got very upset but I soon calmed down. I think he was right, it was exactly what he said it was. Everything that is important starts as a university dream before it develops into systematic research.

Another story dates back to 1958. There were negotiations in Geneva about the rights of countries with respect to the continental shelf off their shores. Our Ministry of Foreign Affairs turned for advice to the Norwegian Geological Survey. The question was whether we could expect useful minerals in the continental shelf. The answer was an unqualified no. Accordingly the Norwegian negotiators suggested strong limitations of the rights for the countries adjacent to the continental shelf. Fortunately, Norway was voted down. Had the meeting accepted the Norwegian recommendation, we would have no Norwegian oil today. Luck was better than brains as the Norwegian saying goes. The Foreign Ministry should have turned to one of our universities for advice."

On the recognition of research results: "Fundamental research goes in a certain sequence. We have a great enemy in Norway, and it comes as a small beetle. Its Latin name is *fps typographus* and it makes interesting drawings on trees but kills them ruthlessly. We have won the war with this enemy and it has disappeared from our forests. An outstanding colleague of mine, Professor Lars Skattebøl, investigated the pheromone component of the beetle secretion; he isolated it, and worked out its synthesis. Eventually, he developed a substance to eliminate this beetle. When this was done, some other people came in and distributed big tubes dispersing this substance in all Norwegian forests, and got tremendous publicity from the media. Skattebøl's name was never mentioned, although he had done 95% of the research. The last 5% got all the recognition. Eventually the balance was reestablished but it is rather typical that the last small step gets much more recognition than the long, painstaking research preceding it. This is usually true for financial support as well."

On the question whether small countries should be involved in fundamental research?

"Sometimes I hear suggestions that a small country like Norway should give up fundamental research. They suggest building good institutions which will be receiver of the results of international research and adapt them to our technologies and health services. However, it never works. It would never work because it is impossible to import knowledge without producing knowledge. I consider all the universities and all the research institutes as part of an international un-bureaucratic system of linkages, Universitas Mundi, the university of the world. We have an international bank of knowledge, but you can draw from this bank only if you also contribute to it."

Erwin Chargaff in 1994 in his home in the Upper West Side of Manhattan (photograph by Istvan Hargittai).

Chargaff Centennial: Erwin Chargaff (1905–2002)[a]

István Hargittai

Erwin Chargaff.

Erwin Chargaff (1905–2002) wrote himself into the annals of science by his seminal discoveries related to DNA and as a thinker. He started working on nucleic acids in 1945. For some time the nucleic acids had been considered to be built up uniformly from four nucleotides. This is why they had been thought to be uninformative substances. Researchers were looking for the secret of life in proteins rather than in nucleic acids.

There was, however, a seminal publication in 1944 by Oswald T. Avery and his two associates, Colin MacLeod and Maclyn McCarty, which provided experimental evidence that DNA was the substance of heredity [1]. Chargaff was one of the few scientists who immediately recognized the validity and importance of Avery *et al.*'s findings. He was 40 years old, worked at Columbia University, and had been a very successful biochemist. However, in an extraordinary move, he cleared his desk and from that point he devoted his efforts and those of his laboratory to studying nucleic acids.

Chargaff's discoveries in the next few years can be summarized in the following two points. He showed that the composition of DNA is organism-specific by the four nucleotides and their relative proportions differing considerably in different organisms. He also showed that the amount of adenine always equaled the amount of thymine, and the amount of guanine always equaled the amount of cytosine. This latter discovery has become especially famous for it provided the foundation for establishing the DNA base pairs that then greatly facilitated the establishment of the double helix structure of DNA. The pair-wise equality of bases is commonplace today, and it is hard to imagine how daring it must have been to first report it. The measurements scattered and the recognition of equality needed sharp eyes and even recklessness because it meant a break with the previous tetranucleotide dogma. In his 1950 *Experientia* paper, Chargaff let his readers into his hesitations over the validity of the observation (quoted here after *Heraclitean Fire*, p. 93). "It is, however, noteworthy—whether this is more than accidental, cannot yet be said—that in all desoxypentose nucleic acids examined thus far the molar ratios of total purines to total pyrimidines, and also of adenine to thymine and of guanine to cytosine, were not far from one."

Erwin Chargaff was born in 1905 in Czernowitz, Austria-Hungary and died in 2002 in New York City. He was educated in Austria and received his doctorate in chemistry from the University of Vienna. He spent some postdoctoral years at Yale University, then returned to Europe and worked for a few years in Berlin, then in Paris when he became a refugee from the Nazis. His mother was killed by the Germans in a concentration camp in 1943. Chargaff spent 40 years at the Department of Columbia University until his retirement in 1975.

[a]Originally published in *Structural Chemistry* 2005, 16:455–456.

I. Hargittai (✉)
Department of Inorganic and Analytical Chemistry, Budapest University of Technology and Economics, Budapest, Hungary
e-mail: istvan.hargittai@gmail.com

Chargaff was a much decorated scientist; among others, by the Pasteur Medal (Paris, 1949); the Charles Leopold Mayer Prize (French Academy of Sciences, 1963); the H. P. Heineken Prize (Royal Netherlands Academy of Sciences, 1973), the National Medal of Science (1975), and many others. He was a member of the National Academy of Sciences of the USA and many other learned societies. In spite of the recognition, he was known to be a bitter person, probably because of his natural demeanor, and also because he felt having been slighted in connection with his contribution to the discovery of the double helix structure.

By the time I had my first personal meeting with him in 1994, I had read a lot of his works. By then he had mellowed and my encounters with him were pleasant. He was lonely after what he considered too early retirement from Columbia University and especially after his wife's death.

There is no space here to analyze his philosophical–historical writings. Instead we give a small sample of quotable Chargaff from three of his books [2–4]:

Science is wonderfully equipped to answer the question "How?" But it gets terribly confused when you ask it the question "Why?" [*Voices*, p. 8]

It is true of every scientific discovery that the road means more than the goal. [*Voices*, p. 17]

To the scientist nature is like a mirror that breaks every thirty years; and who cares about the broken glass of past times? [*Voices*, p. 24]

The kind of questions we ask is conditioned by the kind of answers we expect. [*Voices*, p. 43]

The DNA may well predetermine the shape and composition of the "biopiano," but not the music that is being played on it. [*Voices*, p. 72]

A classic in science is a man who no longer has to be quoted. For the pickpocket, the man with the widest pockets is a classic. [*Voices*, p. 99]

A good teacher can only have dissident pupils. [*Heraclitean*, p. 7]

Never before has science become so alienated from the common man, and he, in turn, so suspicious of science. [*Heraclitean*, p. 158]

Models—in contrast to those who sat for Renoir—improve with age. [*Heraclitean*, p. 171]

A balance that does not tremble cannot weigh. [*Heraclitean*, p. 179]

Books last longer than empires. [*Questions*, p. 21]

I prefer the search for the truth to its possession. [*Questions*, p. 111]

Rome did not have to scream that it had to be Number One: it was. [*Questions*, p. 118]

Bibliography

1. Avery, O. T.; MacLeod, C.; McCarty, M. *J. Exp. Med.* **1944,** *79,* 137–158.
2. Chargaff, E. *Voices in the Labyrinth: Nature, Man and Science*; The Seabury Press, New York, 1977.
3. Chargaff, E. *Heraclitean Fire: Sketches from a Life before Nature*; The Rockefeller University Press, New York, 1978.
4. Chargaff, E. *Serious Questions: An ABC of Skeptical Reflections*; Birkhäuser, Basel, 1986.

Odile and Francis Crick in February 2004 in the Cricks' home in La Jolla (photograph by Istvan Hargittai). My Parents were visiting in Pasadena and went to see the Cricks at their invitation. The visit was uncertain until the very last moment because of Crick's illness. The visit turned out to be a great success. My Parents learned a great deal and it appeared that the Cricks were enjoying their presence. Francis Crick died in July 2004.

Francis Crick (1916–2004)[a]

István Hargittai

Francis Crick in February 2004 in the Cricks' home in La Jolla (photograph by Istvan Hargittai).

Francis Crick, the physicist turned biologist and one of the most influential, if not the most influential scientists of the second half of the twentieth century, died on July 28, 2004. He was co-discoverer—with James D. Watson—of the double helix structure of DNA in Cambridge, England. This was not just the solution of a heretofore-unsolvable problem but it was also the starting point of a whole new area of research. The consequences have reverberated for the next 50 years and are expected to strongly impact science and our lives for many years to come. Crick continued his research in molecular biology for decades after the double helix discovery, again, with outstanding achievements, including pivotal

findings for the genetic code. For the last decades of his long productive life he moved geographically to La Jolla, California, and thematically to understanding consciousness. In one of the numerous obituaries about him, Charles F. Stevens stated that had Crick been "20 or 30 years younger when he started in neurobiology, he might have found gold in the study of consciousness too." Crick was a Nobel laureate, but he was one of the very few whose being a laureate was more important to the institution of the Nobel Prize than the prize was to him.

The discovery of the double helix was a seminal event of structural chemistry even though chemists were a little slow in recognizing the importance of biological macromolecules for our science. It was not only a beautiful construction, but its architecture uniquely suggested its function. The structure had two strikingly novel features [1]. One was that it had two helical chains, each coiling around the same axis but having opposite direction, thus complementing each other. This is a simple consequence of the twofold symmetry of the double helix with the symmetry axis being perpendicular to the axis of the double helix. The other novel feature was how the two chains are held together by the purine and pyrimidine bases with hydrogen bonds connecting these bases. The consequence was that "if the sequence of bases on one chain is given, then the sequence on the other chain is automatically determined" [1]. The authors referred to the functionality of this structure in the last sentence of their brief paper, "It has not escaped our notice that the specific pairing we have postulated immediately suggests a possible copying mechanism for the genetic material" [1].

Shortly after their first paper, Watson and Crick published a somewhat longer note [2] elaborating on the genetic implications of the double helix structure of DNA. In this note, they describe the recipe for self-duplication and give more details of the structure, with emphasis on the hydrogen bonds in the base pairs. There were four antecedents to the double helix discovery although that is not obvious from

[a]Originally published in *Structural Chemistry* 2004, 15:545–546.

I. Hargittai (✉)
Department of Inorganic and Analytical Chemistry, Budapest University of Technology and Economics, Budapest, Hungary
e-mail: istvan.hargittai@gmail.com

these papers by Watson and Crick. One was the discovery of Oswald Avery and his two co-workers that DNA is the substance of heredity [3]. The second was Erwin Chargaff's observation that the purine and pyrimidine bases occurred in a 1:1 ratio in the DNA of all organisms [4], which came to be expressed as base-pairing in the Watson–Crick model. The third was Linus Pauling's discovery of the alpha-helix [5] of protein structure in which he showed the helicity of a biological macromolecule. The fourth was Rosalind Franklin's careful X-ray diffraction measurements of the DNA [6], which, among others, pointed to the C_2 symmetry of the structure.

The importance of the knowledge of C_2 symmetry, which came from the X-ray patterns, was not explicitly acknowledged in Watson and Crick's communications, which is the more curious because the double helix structure could have been described in a simplified way stressing the presence of this symmetry. This is also a point that is of special importance from a structural chemistry point of view and which has had further lessons. Of Watson and Crick, Watson gave much less importance to C_2 symmetry than Crick, perhaps because Watson had no background in crystallography. Furthermore, knowing that Watson had very little faith in the importance of Chargaff's observations, it is remarkable how far he advanced in his intuitive model-building. Crick, on the other hand understood the implications of C_2 symmetry and when they arrived at the concept of base-pairing, he pointed out that the base pair implied a twofold axis in the planes of the bases, perpendicular to the axis of the helices. In other words, the two chains were anti-parallel (Crick, F., Private communication, 2003). Incidentally, Crick's background prompted him—with two associates—to work out the theory of X-ray diffraction of helices, that is, that of a single helix [7]. Apparently, Crick gave so much importance to his work on a single helix that when he decided to erect a symbolic structure over the entrance of their home in Cambridge, it was a single rather than a double helix.

Over the years, I have recorded conversations with many famous scientists and his name—from among the living—came up most often in answers to questions about the heroes of these scientists (from among the deceased, it was Einstein's name) [8]. This is the more remarkable because Crick did not have anybody whom he would have called his pupils; he did not take graduate students either (Crick, F.,

Private communication, 2001). On the other hand, he had close collaborators, one at a time, like James Watson, Sydney Brenner, and most recently, Christof Koch. To lesser extent there were Aaron Klug, Beatrice Magdoff, Leslie Orgel, and Graeme Mitchison. Crick also had interactions with many scientists and, earlier, he gave many lectures. He was a superb lecturer and at meetings no one liked to have a lecture after him.

During the past few years I exchanged several letters with Crick; I turned to him with my questions when I had exhausted all other sources; I knew I should not bother him with trivialities. We corresponded about Sydney Brenner's then missing Nobel Prize, about what is success in science, George Gamow, J. Desmond Bernal, Max Perutz, religion, Rosalind Franklin, some details of the double helix discovery, and other topics. He was already gravely ill and underwent a major operation, but he was patient and helpful and most generous in our correspondence. Then, when my wife and I was to spend a few days at the California Institute of Technology in Pasadena, a mere 3-h drive from La Jolla, the Cricks invited us for lunch and we spent a most memorable visit with them on February 7, 2004, less than 6 months before his death. Crick was lively and joyful during this visit. He was interested in everything and especially some gossiping, our conversation covered even more topics than our correspondence, and he displayed his great sense of humor and his legendary laughs in spite of his severe illness. It was our first and, sadly, last meeting in person, and it was a beautiful gift for us from fate and from Odile and Francis Crick.

References

1. Watson, J. D.; Crick, F. H. C. *Nature* **1953,** *171*, 737.
2. Watson, J. D.; Crick, F. H. C. *Nature* **1953,** *171*, 964.
3. Avery, O. T.; MacLeod, C.; McCarty, M. *J. Exp. Med.* **1944,** *79*, 137.
4. Chargaff, E. *Experientia* **1950,** *6*, 201.
5. Pauling, L.; Corey, R. B.; Branson, H. R. *Proc. Natl Acad. Sci. USA* **1951,** *37*, 205.
6. Franklin, R. E.; Gosling, R. G. *Nature* **1953,** *171*, 740.
7. Cochran, W.; Crick, F. H. C.; Vand, V. *Acta Crystallogr.* **1952,** *5*, 581.
8. Hargittai, I. *Candid Science II: Conversations with Famous Scientists*; Imperial College Press: London, 2002, and other volumes in the series.

Edward Teller performing an experiment in preparation for a lecture (courtesy of Wendy Teller). The following communication is about the history of the Jahn–Teller and the Renner–Teller effects. Theory was Teller's primary interest, but he liked experimentation and liked to enliven his presentations with experimental demonstrations.

Hermann Jahn and Rudolf Renner of the Jahn–Teller and Renner–Teller Effects[a]

Magdolna Hargittai and István Hargittai

Abstract

Vibronic interactions have received increasing attention in modern structural chemistry. Edward Teller played a pioneering role in understanding and describing them during the "molecular physics" period of his scientific career. Very little is known about the two scientists who contributed significantly to our knowledge about these effects and whose names have become associated with Teller's. This Editorial is devoted to Hermann Jahn and Rudolf Renner and attempts to lift them out of oblivion by paying them tribute for their contributions.

Left: Hermann A. Jahn (1907–1979; in the early 1930s; courtesy of Michael Jahn and Margaret May, Jahn's son and daughter, London) and *right*: Rudolf Renner (1909–1991; courtesy of Beate Bauer-Renner, Rudolf Renner's daughter-in-law, Dorum, Germany).

[a]Originally published in *Structural Chemistry* 2009, 20:537–540.

M. Hargittai · I. Hargittai (✉)
Department of Inorganic and Analytical Chemistry and Materials Structure and Modeling Research Group of the Hungarian Academy of Sciences, Budapest University of Technology and Economics, Budapest, Hungary
e-mail: hargittaim@mail.bme.hu; istvan.hargittai@gmail.com

During the 62 years since its publication [1] the importance of the Jahn–Teller effect has grown substantially; partly due to its importance in high-T_c superconductivity [2]. According to the original formulation of the Jahn–Teller effect, a non-linear symmetrical molecule with a partially filled set of degenerate orbitals is unstable and gets distorted, thereby removing the electronic degeneracy until a non-degenerate ground state is achieved. In simpler terms, in such a molecule there is a mismatch between the high symmetry of the ensemble of the atomic nuclei and the lower symmetry of the electron density distribution and this causes some of the nuclei to move from their original positions into ones that match the symmetry of the electron density distribution and, as a consequence, the symmetry of the molecule decreases. Since the Jahn–Teller effect entails the coupling of the electronic and vibrational motions of a molecule, it represents a breakdown of the well-known Born–Oppenheimer approximation, according to which the electronic and vibrational motions of a molecule can be separated because the relatively heavy nuclei move much more slowly than do the electrons. Thus, it can be assumed that the latter move about a fixed nuclear arrangement [3].

That the Jahn–Teller effect does not apply to linear molecules is simply due to an inconsistency between the symmetry of the degenerate electronic state and the symmetry of the vibration that can take the molecule out of the linear configuration, e.g., the doubly degenerate bending vibration of a linear triatomic molecule, AB_2. There is though still a possibility for such molecules to become unstable in their linear shape and that is described by the Renner–Teller effect (vide infra). The Jahn–Teller and Renner–Teller effects belong to the so-called "vibronic interactions." Although the Jahn–Teller effect is the best known among them, historically it was preceded by the discovery of the Renner–Teller effect and both of them were preceded by some related work of Edward Teller and Gerhard Herzberg.

Teller met with Herzberg, the future Nobel laureate spectroscopist, in the early 1930s in Darmstadt, Germany. More than 50 years later, Herzberg remembered vividly his interactions with Teller:

Our discussions at that meeting later led to our collaboration in the paper on the vibrational structure of electronic transitions in polyatomic molecules, which was written during visits of mine to Göttingen and visits by Teller to Darmstadt. My function was that of a midwife: Teller had the ideas, which I tried to get out of him by describing the experimental results to him and by drafting a tentative form of the paper, which he then corrected. Teller had an extraordinary reservoir of ideas in this field (as well as in other fields) and was always ready to share his knowledge. Working with him was an experience that I shall never forget. Although the ideas came from him, he insisted that on the title page we follow the alphabetical order of the authors [4].

Their still often cited paper, written on the above topic, was published in 1933 [5] and it discussed the coupling of the electronic and vibrational motions of molecules during electronic transitions. This is generally referred to as the Herzberg–Teller effect. It can certainly be considered as a forerunner of the Jahn–Teller effect. Chronologically, the paper by Renner follows (published in 1934) [6], with the description of the vibronic interaction in the first electronic excited state of CO_2. Although, due to the above-mentioned symmetry mismatch, there is no linear coupling in this molecule, there still can be considerable quadratic coupling that can cause the molecule to bend and thus the electronic degeneracy is removed. Interestingly, it took a quarter of a century to find the first experimental evidence of the Renner–Teller effect, in the electronic absorption spectrum of the NH_2 radical [7]. The NH_2 radical has one electron on a π orbital and thus a Π electronic state in its ground state, where it is bent; while it is linear in the excited nondegenerate state.

Edward Teller and Lev Landau used to have discussions on the possibility and result of interactions between the electronic and vibrational wave functions of molecules and, according to Teller, they argued about this question. The way he explained Landau's contribution to this effect is interesting but also somewhat confusing. We are quoting here the relevant part of his description [8, 9]:

> This effect had something to do with Lev Landau. I had a German student in Göttingen, R. Renner, and he wrote a paper on degenerate electronic states in the linear carbon dioxide molecule, assuming that the excited, degenerate state of carbon dioxide is linear.

> In the year 1934 both Landau and I were in Niels Bohr's Institute in Copenhagen and we had many discussions. He disagreed with Renner's paper, he disliked it. He said that if the molecule is in a degenerate electronic state then its symmetry will be destroyed and the molecule will no longer be linear. Landau was wrong. I managed to convince him and he agreed with me. This was probably the only case when I won an argument with Landau.

What is confusing in this story is that Teller assumes that the degenerate excited state of CO_2 is linear—but, in fact, Renner's paper suggested it to be bent. What makes the difference is whether the quadratic (or higher order) terms of the interaction matrix are taken into account or not. If higher order terms could be ignored, then the above quotation from Teller would be valid and, indeed, the CO_2 molecule should be rigorously linear even in its degenerate electronic state. Let us

quote from the beginning of Renner's paper here, in which he refers to earlier results about the coupling of the electronic and rotational motions in a diatomic molecule. Renner explains why the rotations and not the vibrations are coupled with the electronic motion there: "The reason for this is the high (cylindrical) symmetry with respect to the line connecting the two nuclei. This symmetry is, of course, conserved for the vibration and so inhibits a coupling between the vibrational and electronic frequencies [10]."[1] The same is true for a symmetric linear triatomic molecule like CO_2, for which the Σ term is nondegenerate but the Π and Δ terms are degenerate. No splitting is expected for the Σ state—in case of CO_2 this is the ground state and, indeed, the CO_2 molecule is linear in its ground electronic state. For the Π term the splitting will depend quadratically on the coordinate of the bending vibration, while for the Δ term the splitting will be proportional to the fourth power of the displacement coordinate—a very small quantity—and can mostly be ignored [10].

There might be different explanations of the quote from Teller from his late years [8, 9]. Perhaps in his earlier discussions with Landau, Teller did not take into account the possibility that the quadratic terms might also cause distortion and Landau might have thought of this possibility but did not elaborate on it [11]. It is also possible that Landau made a mistake as the story implies—but, again, as Teller himself told us, "it bothered me that he was usually not wrong, so maybe he was always right with the exception of linear molecules [8, 9]." And that made Teller ask Jahn to do the computations for all different symmetries and that eventually led to their famous paper and the emergence of the Jahn–Teller effect [1]. Of course, a trivial explanation for Teller's late remark might be simply that he did not remember correctly the events having taken place over 60 years prior to our conversation in 1996 quoted above [8, 9].

Rudolf Renner

Rudolf Renner was born in 1909, in Schweidnitz in Silesia (then Germany, now Swidnica in Poland). He studied first in Hannover, then, from 1929, in Göttingen, where he was awarded his doctor degree in 1934. His former supervisor, Max Born, and his consultant, Teller, by then had left Germany as a consequence of the anti-Jewish legislation following the Nazi accession to power in Germany in 1933. This might have been the reason why Renner published his paper under his name alone, but it might also have been that Born and Teller would have urged him to publish alone in any case.

---

[1] The quotation is from the English translation of Renner's paper [6]: "On the theory of interaction between electronic and nuclear motion for three-atomic, bar-shaped molecules".

In preparation for his doctoral examinations, Renner had to submit an autobiography, which was dated November 1, 1933, and has been preserved in the Archives of Göttingen University. The roughly one-page handwritten document profusely thanks Born not only for scientific guidance but also for financial assistance without which Renner could not have completed his studies. For his substantial assistance in molecular vibrational spectroscopy Renner acknowledges Teller who was staying a little longer in Göttingen after Born had left. Renner notes that Professor Eucken made it possible for him to have his doctoral exams, which is a hint that his original supervisor was no longer available [12].

Teller did not know anything about Renner for a long time, until 1980, when he received a letter from his former pupil.[2] As the physics program in Göttingen disintegrated, Renner entered a 2-year teacher's training program, and received a teaching certificate, but could not find employment. At this time the German Reich Weather Service was looking for natural scientists and Renner entered this service in 1936 and worked there until the end of the war. He married a pharmacist's daughter from Dorum, Lower Saxony, in 1942. His wife's two brothers were killed in the war and there was no one who could be the successor of his father-in-law in the pharmacy. Hence Renner acquired training as a pharmacist and took over the family business in 1950. In 1955, his wife died of cancer and later he married a pharmacist from Dorum. Renner worked in the pharmacy for 30 years and retired in 1980.

Teller expressed his sorrow in his response that Renner could not continue in physics and noted that Renner's work was the forerunner of the Jahn–Teller effect ("Ihre Dissertation war der Vorläufer des sogenannten 'Jahn–Teller Effekts' von dem immer noch viel gesprochen wird").[3] He called his life "quite active," which was certainly an understatement. He mentioned that apart from physics he also worked for Peace. It is very unlikely that Renner and Teller had any further interaction. Renner asked Teller to visit him in Dorum, but in his response Teller noted that Dorum was situated quite outside of the main thoroughfares. Apart from this single exchange of letters with Teller, Renner lived his life without ever telling his family about his "first career." Renner died in 1991.[4]

Hermann A. Jahn

Teller and Jahn worked together at University College London on the famous paper that eventually brought vibronic interactions into the limelight. London was Teller's last stop

before he left Europe and moved permanently to the United States. In contrast to Teller, the other eponym of the Jahn–Teller effect is hardly known. The best source of information about him is the obituary by one of his former colleagues, P. T. Landsberg [13]. Jahn was of German descent. His father left Germany and settled in England in the 1890s. Jahn was born in Colchester in 1907, grew up in Lincoln and received his BSc degree in chemistry at University College London in 1928. He became interested in quantum mechanics and continued his studies in Leipzig under the supervision of Werner Heisenberg and the Dutch mathematician Bartel L. van der Waerden. He completed his doctorate in 1935 and returned to England. He worked at the Royal Institution and during the war at the Royal Aircraft Establishment. In 1943 he married Karoline Schüler, a teacher who was a Jewish refugee from Frankfurt am Main. For 2 years from 1946 he worked at the physics department of Birmingham University under Rudolf Peierls's guidance. He was appointed Professor of Applied Mathematics at the University of Southampton in 1949 and stayed there until he retired in 1972. He died in 1979.

Jahn and Teller interacted when Teller was at University College London and Jahn at the Royal Institution. The paper they did jointly was followed by a paper by Jahn alone. Jahn published a number of other studies, mostly on the structure of molecules, especially treating molecular vibrations. He utilized his acumen in mathematics for solving problems in the theory of X-ray scattering and applied group theory to nuclear structures. Vibrations stayed with him in his war-time projects when he studied the vibrations of larger structures, including airplanes. This may have been the reason that he never wanted to fly and never used air travel. When after the war he was interviewing for a job at Harwell, which might have involved classified work in the British nuclear program, he remarked on his German ancestry and expressed the hope that it would not prevent him from getting clearance. The irony of the situation was that the interviewer was Klaus Fuchs, a German who had participated in the Manhattan Project and turned out to be a spy for the Soviet Union.

Jahn was a quiet and modest man, very different from Edward Teller. He had a wide range of intellectual interests, including European literature, and was noted for his patient encouragement of his postgraduate students.

Afterlife

Today we can speak about Jahn–Teller "effects," rather than just one effect, as they include some derivatives of the originally postulated effect, such as the pseudo-Jahn–Teller effect. There are monographs published about the Jahn–Teller effect and international conferences held regularly about it. The Jahn–Teller effect served as a starting point in the quest for high-temperature superconductors by J. Georg Bednorz and K. Alex Müller at an IBM research laboratory in Switzerland. They

[2] Letter of November 11, 1980, from Rudolf Renner to Edward Teller; Hoover Institution Archives, Stanford University.

[3] Letter (undated, presumably in November 1980) from Edward Teller to Rudolf Renner; Hoover Institution Archives, Stanford University.

[4] Letter of April 30 (2009) from Beate Bauer-Renner, Renner's daughter-in-law.

made their discovery in 1986 and were awarded the Nobel Prize in Physics for 1987, one of the fastest awards ever made after a discovery in the history of the Nobel Prizes. The laureates paid ample tribute to the Jahn–Teller effect in their joint Nobel lecture and illustrated it in two attractive figures [2].

Acknowledgment Our research is being supported in part by the Hungarian Scientific Research Foundation (OTKA Nos. K60365 and T046183). We are grateful to Michael Jahn and Margaret May (London) and Beate Bauer-Renner (Dorum) for their kindness in sending us the photographs of Hermann Jahn and Rudolf Renner, respectively. The assistance of University Archives Göttingen (Germany), is gratefully acknowledged and so is the kind help by the associates of the Hoover Institution Archives at Stanford University. Our research at the Hoover Institution was generously supported by a grant to one of us (IH) from the Alfred P. Sloan Foundation (New York) for a book project in progress, *Judging Edward Teller: A Closer Look at One of the Most Influential Scientists of the Twentieth Century* (Amherst, New York, Prometheus, 2010).

References

1. Jahn HA, Teller E (1937) Proc Roy Soc Lond A 161:220–235
2. Bednorz JG, Müller KA (1993) Nobel lectures physics 1981–1990. World Scientific, Singapore, pp 424–457
3. Born M, Oppenheimer R (1927) Annalen der Physik 84:457–484
4. Herzberg G (1985) Annu Rev Phys Chem 36:1–30, 10
5. Herzberg G, Teller E (1933) Z Phys Chem 21:410
6. Renner R (1934) Z Phys 92:172
7. Dressler K, Ramsay DA (1959) Phil Trans Roy Soc Lond 251A:553–602
8. Hargittai I, Hargittai M (2008) Struct Chem 19:181–184
9. Hargittai M, Hargittai I (2009) Symmetry through the eyes of a chemist, 3rd edn. Springer, USA, pp 305–306
10. Hettema H (2000) Quantum chemistry: classic scientific papers. World Scientific, Singapore, pp 61–62
11. Hargittai M (2009) Struct Chem 20:21–30
12. Renner R (1933) "Lebenslauf des cand phys," Göttingen, 1. XI. 1933, in the collection of Universitätsarchiv Göttingen
13. Landsberg PT (1980) Bull Lond Math Soc 12:383–386

From left to right, Istvan and Magdolna Hargittai and Isabella and Jerome Karle at an international crystallography meeting in 1978 in Pécs, Hungay (by unknown photographer).

JEROME KARLE (1918–2013)—Nobel Laureate; Charter Member of the Editorial Board of *Structural Chemistry*[a]

István Hargittai and Magdolna Hargittai

Abstract

Jerome Karle (1918–2013) had to overcome adversities before he acquired the education he strived to get. He did pioneering work in modernizing the gas-phase electron diffraction technique of molecular structure determination. Drawing on this experience, he and Herbert Hauptman jointly came to a seminal discovery that solved the phase problem in crystal structure determination by X-ray diffraction. All along, he enjoyed close cooperation with his wife, the most distinguished scientist Isabella Karle. Jerome Karle was a revered colleague and a faithful friend.

Jerome Karle (Fig. 1) was born in 1918 in Coney Island, a part of Brooklyn, New York. When in 2009 he retired from the US Naval Research Laboratory in Washington, DC, he

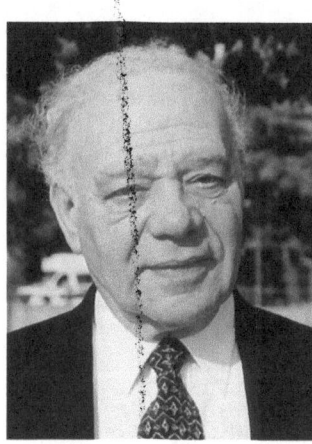

Fig. 1 Jerome Karle in 2000 in Washington, DC (photo by I. Hargittai).

[a]Originally published in *Structural Chemistry* 2013, 24:2219–2222.

I. Hargittai (✉) · M. Hargittai
Department of Inorganic and Analytical Chemistry, Budapest University of Technology and Economics, Budapest, Hungary
e-mail: istvan.hargittai@gmail.com; hargittaim@mail.bme.hu

was Chief Scientist of the Laboratory for the Structure of Matter. He received his M.A. at Harvard University (in biology) in 1937 and his M.S. (in physical chemistry) at the University of Michigan in Ann Arbor in 1941, and finished his work there for the Ph.D. degree in 1943. In between his studies at Harvard and graduate work at the University of Michigan, he worked at the New York State Health Department where he developed, as one of his chores, the standard method used for testing the amount of fluoridation that was applied to water supplies. He worked on the Manhattan Project at the University of Chicago in 1943 and in 1944. He returned to the University of Michigan to work on a project for the Naval Research Laboratory (NRL). In 1946, he and his equally distinguished wife, Isabella Karle, became members of the NRL. Among many other distinctions, he was a member of the National Academy of Sciences of the U.S.A., served as President of the International Union of Crystallography, and received the Nobel Prize in Chemistry in 1985, jointly with Herbert Hauptman, "for their outstanding achievements in the development of direct methods for the determination of crystal structures." In 1989, he was one of the charter members of the new periodical Structural Chemistry, and the fledgling journal greatly benefited from his active support.

We were in friendly and collegial interactions with Jerome and Isabella Karle for decades (Fig. 2); from the early 1970s, we enjoyed our get-togethers at meetings and visited each other in our homes. We learned a great deal about the Karles over the years; then on two separate occasions, we recorded more formal conversations with them, in 2000, with Isabella, and in 2003, with Jerome. Jerome was not very eager to share some of his experiences of his youth, though he found them important so that others might find them instructive. Some of these experiences are quoted below that have been largely absent in the obituaries and reminiscences we have seen about Jerome.

Jerome's paternal grandfather was a painter in the artistic sense. He specialized in decorating ceilings both with

Fig. 2 The Karles and the Hargittais (the two Karles in the middle) in 1978 in Pécs, Hungary.

paintings and with reliefs. Almost everybody on Jerome's father's side of the family had artistic inclinations, except Jerome's father and one of his sisters. Jerome's father was born in 1882 in Poland, and he came to the United States as a small child. On Jerome's mother's side, there was not much artistic inclination. Jerome's mother was the fourth child in her family. Her father had to face a difficult situation because his wife, Jerome's maternal grandmother, died, leaving Jerome's grandfather alone with his four children. He felt that he could take care of three, but not of all four. So the smallest child—she would become Jerome's mother—was adopted by members of her extended family, by the name of Kun, who had come to the US from Budapest. Jerome's mother was very talented and she was an excellent pianist. Jerome also learned to play.

Jerome's immediate family environment was not conducive for him to become interested in science, but when he was 7 or 8 years old, his mother took him to a science museum. This hands-on museum captivated Jerome's attention and imagination. The museum did not exist for long, but long enough to set Jerome on his course for science. He read popularized science books and he wrote book reports that caught his teachers' attention, and when he was 9 years old he qualified for pushing ahead. Jerome graduated from the eighth grade before he was 11 years old. Then, he went to Abraham Lincoln High School, followed by City College of New York, from which he graduated when he was 19 years old.

So far, Jerome's path was smooth sailing, but rough times were ahead. He was Jewish and it proved impossible for him to get into medical school. This was a painful story for Jerome to remember, and he was hesitant whether it was proper for him to speak about it, but he decided that he should.

"I went to Harvard and spent a year there obtaining a master's degree in biology. I had the illusion that being a good student was all that was necessary to get admitted to medical school. I applied to Harvard and some other places and, of course, I was turned down. I wanted to try again early the next year and I was allowed to have a conversation with the Dean of the [Harvard] Medical School. The only thing I got from him was a harangue. He said 'We have enough Jews in Massachusetts; we don't need any from New York City'. He was not at all interested in my record as a student. For example, when I graduated from City College I received the first award given at graduation for 'excellence in the natural sciences'" [1].

Jerome did not give up.

"I had applied to various graduate schools just to do graduate work and I was turned down by all of them. So I wasn't doing anything. Then there was just a stroke of luck. In the summer of 1938 I was working in Coney Island and a good friend of mine, with whom I still communicate, told me that exams were forthcoming for civil service jobs in the New York State Health Department. I took the exam and I had the highest grade among those they accepted. There was a rule that after a certain period, perhaps 3 months, the Health Department could not dismiss anyone without an explanation. I stayed for about 2 years. I learned only later that they had wanted me to leave, along with the rest of the people who arrived when I did, but my boss said that if they tried to dismiss me, he would not accept that and that he needed me for his work. This fine gentleman's name was F. Wellington Gilcrease. He remained my good friend and we kept in touch until the end of his life. I did not know that he saved me from dismissal until I left to go to the University of Michigan. During those 2 years, I was saving up money as I knew I couldn't get any money from graduate school. At that time, someone told me that if I went to the University of Michigan, I would be treated properly. That is why I went there. After the first year, I was funded to continue my education at the University of Michigan. After that I have never experienced any anti-Semitism." [1]

Jerome met his future wife, Isabella, who was a fellow student, at Michigan. Their cooperation evolved during the years. Initially they both did about equal amounts of experimental work, but gradually Jerome moved more and more toward theory. They both did their doctoral work under the supervision of L. O. Brockway. Brockway had been a Linus Pauling disciple who in the 1930s worked out the technique of Fourier-transform radial distribution technique with Pauling for an improved analysis of electron diffraction data. The two Karles advanced greatly the technique of gas-phase electron diffraction during the second half of the 1940s, especially by evaluating the impact of molecular motion on the experimental distributions, both the molecular intensities of electron scattering and the radial distributions (a misnomer), essentially, the probability density distribution of intramolecular internuclear distances. The present authors learned a great deal from the Karles' classic papers at the beginning of their research careers.

At the Naval Research Laboratory, the Karles established one of the world's most active and most innovative gas-phase electron diffraction laboratories, including the construction of new and improved experimental apparatus, and, of course, always up-to-date computational techniques. The innovations and the careful evaluation of the techniques of analysis in electron diffraction proved beneficial much beyond the molecular structure determination by electron diffraction. Jerome was familiar with the X-ray diffraction field and attended meetings where both X-ray diffraction and electron diffraction were discussed.

Jerome soon recognized that because it was supposed that the phases of X-ray diffraction could not be extracted from the experimental data, a large portion of experimental information was lost and the unambiguous determination of crystal molecular structures was hindered. Jerome started interacting with Herbert Hauptman who had also joined NRL and was even more mathematically oriented than Jerome [2]. Their joint research resulted in discovering what has been known as the direct methods of X-ray crystallography that, on the one hand, have greatly expanded the scope of structures that are possible to determine and, on the other, have greatly enhanced the reliability and accuracy of crystal structure determinations. The technique is by now well known, and the readers of Structural Chemistry have had a unique opportunity to be introduced to it directly by its creators, Herbert Hauptman [3] and Isabella and Jerome Karle [4]. Another principal contributor to the technique was David Sayre [5]. Isabella Karle did pivotal work in demonstrating the applicability of the direct methods and this was especially beneficial as crystallographers were rather slow in recognizing the new perspectives that the Karles and Hauptman were offering [6]. At one point, there was a paper in Acta Crystallographica in which two renowned crystallographers posed the question "Have Hauptman and Karle Solved the Phase Problem?" This was the title of the communication and the paper was devoted to showing that the answer to this question was an emphatic "no!" [7]. Not only did the two authors eventually come around; they became two of the most devoted and prolific researchers to employ the direct methods.

One of us (IH) remembers: One October day in 1985, our guest, the well-known Italian crystallographer Aldo Domenicano, and I were sitting in my office in Budapest when the telephone rang and it was a reporter from Radio Budapest. He asked me about the new chemistry Nobel laureates, Herbert Hauptman and Jerome Karle and about their achievement. I must admit that both Aldo and I were surprised by the news. We both knew Herb and Jerry and admired them, but had not thought about a possible Nobel Prize for them. We needed some time to absorb the information, but once it sank in, we had no doubt whatsoever that their award was amply justified. Looking back, perhaps their natural modesty was what made the Nobel announcement appear unexpected rather than any doubt about the importance of their discovery.

Jerome had only one regret about the Nobel Prize—that Isabella was not among the awardees. She would have deserved it, but perhaps others would have just as well. The selection of Hauptman and Jerome Karle was beyond reproach; filling the third slot of the available three places would have become inevitably controversial, whereas nobody has ever criticized the actual choice. For Jerome, among other benefits, the Nobel Prize provided a welcome opportunity to reach out to a broader audience than they could before. Jerome never had a university position and NRL while providing ideal conditions for his research, not being an educational institution did not offer the possibility for a researcher, however distinguished he or she was, to be surrounded all the time by young students. Jerome was open-minded and generous and in subsequent years indulged in enhanced visibility whereas he could always withdraw into the quiet of his NRL laboratory.

As for his personal traits, we can both attest to his having been a most generous and helpful human being; broadly interested in a wide variety of topics, including the arts. IH: I experienced Jerome's friendship already at the time when he was already a world-renowned authority and I was hardly more than a beginner in science. I had known about the works of the Karles, but our first personal meeting took place in the early 1970s. I was on my way to Austin, Texas, and the Karles invited me to their home for 3 days. I learned a great deal of science mainly from Jerome during those 3 days. What is more remarkable though, and I almost took it for granted, that Jerome toured with me all the major museums and galleries in Washington, DC, and spent a great deal of time introducing me to their exhibits.

Jerome Karle was an outstanding scientist, a distinguished contributor to the scientific revolutions of the twentieth century, and a remarkable human being. Not only will the fruits of his discoveries, but also the sense of his personality, be long remembered.

References

1. Hargittai I, Hargittai M (2006) Jerome Karle. In: Candid Science VI: More conversations with famous scientists. Imperial College Press, London, pp 422–437
2. Hargittai I (2003) Herbert A Hauptman. In: Hargittai M (ed) Candid Science III: More conversations with famous chemists. Imperial College Press, London, pp 292–317
3. Hauptman HA (1990) Struct Chem 6:617–620
4. Karle I, Karle J (2005) Struct Chem 16:5–16
5. Sayre D (2002) Struct Chem 13:81–96
6. Hargittai I, Hargittai M (2006) Isabella L. Karle. In: Candid Science VI: More conversations with famous scientists. Imperial College Press, London, pp 402–421
7. Cochran W, Woolfson MM (1954) Acta Crystallogr 7:450–451

Lev D. Landau in 1958; courtesy of the late Boris S. Gorobets. Father told me a story about Landau as he heard it from Gorobets. After the war, Landau was ordered to participate in the nuclear program. Landau obliged, but some of his non-scientist supervisors found his working style unacceptable. Landau worked when he liked and often appeared to be doing nothing—just thinking. This and Landau's high salary upset the general of the secret police who wanted to have Landau fired. The general backed off only when higher-ups explained to him that no matter how Landau appears to be behaving, his contribution was vital for the project.

Lev D. Landau (1908–1968): In Memoriam[a]

István Hargittai and Magdolna Hargittai

Lev Davidovich Landau's name is among the greats of twentieth century science. He was born into a Jewish-Russian family in Baku where his oil-engineer father was sent to in the line of his work. His mother was a gynecologist.

Lev Landau (courtesy of Alexei Abrikosov, from A.A. Abrikosov: *Academician L. D. Landau: Short Biography and Review of his Scientific Work*, Nauka, Moscow, 1965, in Russian).

[a]Originally published in *Structural Chemistry* 2008, 19:181–184.

I. Hargittai (✉) · M. Hargittai (✉)
Department of Inorganic and Analytical Chemistry and Materials Structure and Modeling Research Group of the Hungarian Academy of Sciences, Budapest University of Technology and Economics, Budapest, Hungary
e-mail: istvan.hargittai@gmail.com; hargittaim@mail.bme.hu

Landau was not a child prodigy, but he had a fast start. He completed his studies at 13 years of age, went to university at 14, and published his first paper at 18. He arrived in Leningrad from Baku in 1924 and his later traits manifested themselves in the company of young scientists. They were succinctly characterized by Yevgenia Kanegiesser, the future Mrs. Rudolf Peierls, in a light verse in which Landau also figured [1]:

> To tuneful songs, Landáu the clever
>   Who'll gladly argue anywhere,
> At any time, with whomsoever,
>   Holds a discussion with a chair.

The rest of the time they were playing tennis, went swimming and to the movies watching Hollywood films, but they were mostly doing theoretical physics. His unruly character found some expressions that were not tolerated by the authorities although in the mid-1920s, the repercussions were still relatively mild. When in 1925 the *Encyclopedia Sovietica* carried a long entry on ether, George Gamow, Landau, and others sent a letter to the author ruthlessly making fun of his article. There was a heavy reaction to their joke; one of the consequences was that Landau was dismissed from his teaching job in the Polytechnic Institute although he could continue as researcher in Abram Ioffe's Roentgen Institute (in Leningrad) [1].

Landau was lucky to have had the opportunity to spend some time in Western Europe. In the late 1920s, the Soviet authorities found it meritorious to build closer contacts with the leading research laboratories abroad. Landau spent some time in Leipzig, Berlin, Cambridge, Zurich, and especially in Copenhagen, and worked with Werner Heisenberg, Edward Teller, George Gamow, Wolfgang Pauli, Ernest Rutherford, and especially with Niels Bohr whom he considered his teacher. However, in the early 1930s, the attitude of Soviet officialdom toward scientists started to change, and it was becoming gradually more difficult for Soviet scientists to travel to and even attend meetings in the West. Increasingly,

Landau on an Israeli stamp as part of a large, multi-part block of stamps.

they frowned upon western contacts and started dividing science into capitalist science and proletarian science (in a similar way as science was being divided into Jewish science and Aryan science in Germany). Nothing illustrates this better than the fact that Einstein's theory of relativity was not considered to be compatible with the philosophy of dialectical materialism.

In 1932, Landau moved to Kharkov, where he worked at the Ukrainian Physical Technical Institute (UFTI) as head of the Theoretical Department. UFTI was fast becoming an internationally recognized center of experimental and theoretical physics. Sadly, by about 1938, UFTI was destroyed by the arrest of several of its leading scientists. This was the fate of Landau as well, who by then had transferred to Peter Kapitsa's institute, the Institute for Physical Problems of the Academy of Sciences of the U.S.S.R. in Moscow. Some of the Kharkov physicists, most notably the exceptionally gifted experimentalist Lev V. Shubnikov and others, were executed. Landau was incarcerated for 1 year and was freed because Kapitsa had pledged personal responsibility for Landau's behavior. It is noteworthy that Landau's accusations of anti-Soviet activities were not withdrawn upon his liberation from prison.

He was interested not only in doing research in theoretical physics. Parallel to his position at Kapitsa's institute in Moscow he also taught as a professor of theoretical physics at Moscow State University, just as during his Kharkov period he had taught at Kharkov State University [2]. His work covered all branches of theoretical physics, ranging from fluid mechanics to quantum field theory. A large portion of his papers deal with the theory of the condensed state. These studies started in 1936 with a formulation of a general thermodynamics theory of the phase transitions of the second order. After P. L. Kapitsa's discovery, in 1938, of the super-fluidity of liquid helium, Landau's extensive research led him to the construction of the complete theory of "quantum liquids" at very low temperatures. He devoted a series of papers to the theory of quantum liquids of the "Bose type," referring to the super-fluid liquid helium (the usual isotope ^{4}He). Later, he formulated the theory of quantum liquids of the "Fermi type," referring to liquid helium of isotope ^{3}He.

Landau was awarded an unshared Nobel Prize in Physics in 1962 "for his pioneering theories for condensed matter, especially liquid helium." Sadly he could not truly enjoy and appreciate this distinction. On January 7, 1962, 9 months before the Nobel announcement, he was victim of an automobile accident in which his skull and brain were damaged. Landau was always very fragile; he was never strong physically, and the consequences of the accident were heavier and more extensive than they might have been had his body been stronger. It was a medical miracle that he was brought back to life and the international community provided all the help there was to assist Landau and the medical team fighting for his life. After the accident he lived for 6 more years without fully recovering the legendary capacity of his mind. It was a painful period not only for him but also for everybody who was close to him: his family, his friends, and his colleagues. Incidentally, his laconic biography on the Nobel web site makes no mention of the persecution he suffered for an entire year in 1938–1939, nor of the tragic accident he fell victim to, almost a whole year before the Nobel Prize.

Landau may have owed his life to Kapitsa who bravely protested his incarceration and assumed responsibility for him so that he would be freed. By a strange turn of fate, eventually Landau became indispensable for the Soviet nuclear program and he remained untouchable even when his protector, Kapitsa, fell out of favor after World War II. However, Landau always considered that he was a learned slave who participated in the nuclear program only because he was forced to and because it was a sort of insurance for him.

This was an exceptional attitude and very different from that of most Soviet scientists, who considered their work in the nuclear program a patriotic duty in the defense of their fatherland against a possible foreign aggression. This general attitude was understandable right after World War II, since the Soviet Union suffered terrible losses in the war as a consequence of the attack and invasion by Nazi Germany. After Stalin's death, Landau resigned from the nuclear program.

Also noteworthy is Landau's famous seminar that contributed to the creative atmosphere in Soviet physics for many years. Another of his significant contributions to world science was the textbook series he authored jointly with Evgenii M. Lifshits, titled *Theoretical Physics*. This is a multi-volume series of books that may well be on a par in significance with his discoveries. After Landau's death, Lifshits continued publishing new volumes and revising previous ones. Evgenii Lifshits was an important theoretical physicist in his own right as well. Lifshits was a member of the Soviet Academy of Sciences, and he and Landau were jointly awarded a high state award, the Lenin Prize, for their book series in 1962 [3].

Tombstone over Landau's grave in snow-covered Novodevichi Cemetery, Moscow (photograph by I. Hargittai in March 2006).

Evgenii M. Lifshits and Lev D. Landau in the resort place Borzhomi in 1960 (from the archives of E. M. Lifshits; courtesy of Boris S. Gorobets, the son of Lifshits's late widow's, Z. I. Gorobets-Lifshits, himself a scientist and science historian).

Landau was the doyen of theoretical physics in the Soviet Union while he was in his creative period. Many theoretical physicists considered him their teacher and mentor. Officially though, only 43 of them could pride themselves as having taken his famous, nine-part examination, called Theor-minimum. It is most remarkable how great an impact and influence he had on how physics in the Soviet Union and even on the world scene developed in the middle of the twentieth century. We have personally experienced the awe by other exceptional physicists when Landau's name came up in our discussions [4, 5].

Lev Landau received ample recognition in the Soviet Union. Perhaps the most important among them was his election to full membership of the Soviet Academy of Sciences in 1946. His foreign recognition—in addition to the Nobel Prize—included foreign membership of the Royal Society (London), the National Academy of Sciences of the U.S.A., and other learned societies. In 1961, he received the Max Planck Medal and the Fritz London Prize.

In conclusion, we mention a contribution by Lev Landau to chemistry, which is seldom associated with his name. In this connection, we quote Edward Teller about the discovery of the Jahn–Teller effect [6]:

> This effect had something to do with Lev Landau. I had a German student in Göttingen, R. Renner, and he wrote a paper on degenerate electronic states in the linear carbon dioxide molecule, assuming that the excited, degenerate state of carbon dioxide is linear.
>
> In the year 1934 both Landau and I were in Niels Bohr's Institute in Copenhagen and we had many discussions. He disagreed with Renner's paper, he disliked it. He said that if the molecule is in a degenerate electronic state then its symmetry will be destroyed and the molecule will no longer be linear. Landau was wrong. I managed to convince him and he agreed with me. This was probably the only case when I won an argument with Landau.
>
> A little later I went to London, and met Jahn. I told him about my discussion with Landau, and about the problem in which I was convinced that Landau was wrong. But it bothered me that he was usually not wrong. So maybe he is always right with the exception of linear molecules. Jahn was a good group-theorist, and we wrote this paper, the content of which you know, that if a molecule has an electronic state that is degenerate, then the symmetry of the molecule will be destroyed. That is the Jahn–Teller theorem.

The Jahn–Teller theorem has a footnote: this is always true with the only exception of linear molecules. So the amusing story of the Jahn–Teller effect is that I first worked with my student, Renner, on a paper that presented the only general exception to the Jahn–Teller effect. It really should be the Landau–Jahn–Teller theorem because Landau was the first one who expressed it, unfortunately using the only exception where it was not valid.

References

1. Gamow G (1970) My world line: an informal autobiography. The Viking Press, New York
2. Gorobets B (2006) Krug Landau (in Russian, Landau's Circles). Letnii Sad, Moscow-St. Petersburg. A new, revised and much enlarged edition of this book is due to appear in two volumes in 2008
3. Ginzburg VL (1995) O fizike i asztrofizike (in Russian, About Physics and Astrophysics), 3rd edn. Buro Kvantum, pp 364–382
4. Interview with Alexei A. Abrikosov. In: Hargittai B, Hargittai I (2005) Candid Science V: conversations with famous scientists. Imperial College Press, London, pp 176–197
5. Interview with Vitaly L. Ginzburg. In: Hargittai I, Hargittai M (2006) Candid Science VI: more conversations with famous scientists. Imperial College Press, London, pp 808–837
6. Interview with Edward Teller. In: Hargittai M, Hargittai I (2004) Candid Science IV: conversations with famous physicists. Imperial College Press, London, pp 404–423. See, also, Teller E (1991) Structural Chemistry 2:vii

Istvan Orosz's portrait of George Olah, prepared for Olah's 90th birthday. It appeared on the cover of the special issue of the journal *Structural Chemistry* dedicated to Olah. My father and I jointly edited this issue. The official date of the appearance of the issue was April 2017 and Olah had died in March. However, the issue had been ready well in time for Olah to see it and enjoy it.

Structures and Mechanisms in Chemical Reactions: George A. Olah's Life-long Search of Chemistry[a]

István Hargittai

The realization of the electron donor ability of shared electron pairs could one day rank equal in importance with G. N. Lewis' realization of the electron donor unshared pairs.

George A. Olah [1]

Abstract

The Hungarian-born American chemistry Nobel laureate George A. Olah used superacids to give longer life to carbocations. He resolved a long-standing debate on reaction mechanism in organic chemistry and, more importantly, opened new vistas in hydrocarbon chemistry to produce hosts of new compounds. The concerted utilization of organic synthesis, physical techniques, and computational methods led to spectacular achievements in hydrocarbon chemistry. Olah has always been on the lookout for the practical applications of his discoveries in fundamental chemistry. He continued his research after his Nobel award and has worked out the idea, which he labeled "the methanol economy." Olah's example shows that a great researcher can also be a devoted and caring human being.

Introduction

In 1962, George A. Olah (Fig. 1) delivered an invited talk at the Brookhaven Organic Reaction Mechanism Conference. He had immigrated only 5 years before to North America and was working in a Canadian industrial laboratory. It was at the time of the famous debate about the reaction mechanism: whether the 2-norbornyl ion—an intermediate in the hydrolysis of the 2-norbornyl esters, for which there was significantly higher

Fig. 1 George A. Olah in 1995 in the author's office at the Budapest University of Technology and Economics (photograph by the author).

This contribution is dedicated to George A. Olah in celebration of the forthcoming 90th birthday of this great scientist and wonderful friend.

[a]Originally published in *Structural Chemistry* 2017, 28:259–277.

I. Hargittai (✉)
Department of Inorganic and Analytical Chemistry, Budapest University of Technology and Economics, Budapest, Hungary
e-mail: istvan.hargittai@gmail.com

rate for the 2-exo versus the 2-endo derivatives—had a "non-classical" or a "classical" structure. Saul Winstein suggested that the "non-classical" ion had a bridged structure as a consequence of the sigma participation of the C1–C6 bond leading to electron delocalization. Herbert C. Brown ascribed the observed difference in the rate of hydrolysis to steric hindrance of the endo side causing rapidly equilibrating "classical" trivalent ions. Winstein and Brown were giants of organic chemistry and their public debates were popular spectacles of organic chemistry meetings.

In his lecture, Olah reported to have applied a new method of producing long-lived carbocations by means of superacids. Thus, he gave hope of resolving the long-standing 2-norbornyl ion controversy. The experimental observations concerning the rate difference in the hydrolysis of the 2-exo versus 2-endo-norbornyl esters had never been questioned. They were well established facts. The debate concerned the mechanism of the reaction. Uncovering the mechanism of a chemical reaction has been compared to uncovering Hamlet's story between the opening and closing acts of Shakespeare's drama [2]. Often, only the identities of the reactants and the products are known and the mechanism of the reaction leading from the reactants to the products need to be understood. There was solid evidence about the presence of cationic species in the reaction of hydrolysis of 2-norbornyl esters, but they were short-lived, "elusive," hence their nature and structure could not be determined. This is why Olah's claim of giving longer lives to such ions was so stirring. The two protagonists of the debate, Winstein and Brown each, separately, told Olah to be careful with his claim, citing the ease in which unsubstantiated claims could ruin a young chemist's promising career. Winstein and Bown also told Olah that should his claims prove true they expected him to come up with evidence supporting the "non-classical" (Winstein) and the "classical" (Brown) nature of the 2-norbornyl cation.

Eventually, and with the help of NMR spectroscopy and theoretical calculations, Olah provided unequivocal evidence in favor of the "non-classical" nature of the 2-norbornyl cation. The resolution of the famous debate was by itself not a pivotal achievement, but it enhanced Olah's visibility among his peers. Its real significance was in using superacids to produce long-lived, "persistent" carbocations. It pointed to the creation of a whole new chemistry involving hardly reactive hydrocarbons. The development of Olah's new chemistry happened in stages rather than in outbursts of earthshaking discoveries. Olah took his growing fame with attractive humility. He must have felt enormous inner satisfaction though when looking back to the road leading to this exalted status in his science. That road was anything but easy and uneventful.

The Beginnings

George A. Olah was born (as György Oláh, on May 22, 1927) into an upper-middle-class intellectual family. His father was a lawyer and the family lived in downtown Budapest. The house (Fig. 2) in which they lived in an apartment stood across from the Budapest Opera House. George attended good schools. In particular, for high school, he attended the Gimnázium of the Piarist Order (Fig. 3). The Catholic Piarist Order has taught in Pest since 1717 (Figs. 4 and 5). The school boasts another chemistry Nobel laureate among its graduates, George de Hevesy (or Georg von Hevesy, depending on the language he was using). Hevesy attended this school between 1895 and 1903. Hevesy was Jewish and his family converted at around the time of his graduation or soon after.

Olah was looking forward to a pleasant and fulfilling life with whatever he would choose for a profession. Everything was given for him, except security as Europe and Hungary were rapidly moving toward World War II. As he was growing up, especially during his upper classes in high school, racial laws of increasing severity were threatening not only his well-being, but eventually even his life. We are circumspect in describing these years of his life in accordance with his own tacit wishes. When in 1957, he, his wife, and their first child immigrated to North America, he and his wife felt they could leave behind all the unpleasantness and horrors of their lives in Hungary. He expressed this in a letter

Fig. 2 The Olah family lived at 13–15 Hajós Street, District VI (photograph by the author), where George was born. It is just across the street from the Budapest Opera House.

Fig. 3 George A. Olah as a high school student (courtesy of George A. Olah).

to a friend in Budapest in 2003, "... My life is a life of an American of Hungarian origin, and I am no longer living in the shadow of the [anti-Semitic] Nuremberg Laws" [3]. In his autobiography, he devotes a single, though poignant, sentence to this period: "I do not want to relive here in any detail some of my very difficult, even horrifying, experiences of this period, hiding out the last months of the war in Budapest" ([4], p 45).

In 2003, Olah received an award from the University of Szeged, the Klebelsberg Prize, honoring the memory of the long-time minister of religion and public education. Kuno Klebelsberg had a broad vision for the dominance of Hungary in the region through Hungarian 'cultural superiority,' which was an expression of blatant nationalism. This included a desire to regain territories referred to as Greater Hungary, racism and in particular anti-Semitism.

He aimed at bringing back some of the talent that had left Hungary, but he did not include the Jewish expatriates in the circle of those he wanted to return to Hungary. There was an irony in Olah's receiving the Klebelsberg Prize. It is highly doubtful whether Olah could have had a career in academia under Klebelsberg's reign of culture and education in the anti-Semitic Horthy regime that lasted in Hungary for 25 years, between 1920 and 1944.

In their tolerance, the Piarists built on their liberal traditions. At the time of the Holocaust, the school meant to exclude persecution within its walls and had its Jewish students remove

Fig. 4 and Fig. 5 The two principal buildings of the recently renovated Gimnázium of the Piarist Order at the Pest bridge-head of Erzsébet [Elizabeth] Bridge (photographs by the author).

the yellow star from their clothing. However, eventually, Olah had to seek refuge outside the school and Olah's above quoted sentence referred to this last period.

During the Hungarian Holocaust, representatives of a number of nations distinguished themselves in saving lives, and there were a number of Hungarian saviors as well. Gábor Sztehlo was a Lutheran minister who responded to the Lutheran Bishop Sándor Raffay's call to save persecuted Jewish children who had converted, and Sztehlo organized protective homes for them (Figs. 6 and 7). The high school student Olah was among his charges. Soon, Sztehlo extended his efforts to all Jewish children and eventually to all children that he found abandoned as he continued his activities after

Fig. 6 and Fig. 7 Two views of the Gábor Sztehlo statue (erected in 2009) on Deák Ferenc Square, District V, by Tamás Vigh and Barnabás Winkler. Photographs by the author.

liberation. In 1972, Yad Vashem granted Sztehlo the title "Righteous among the Nations" [5].

George's brother, Peter, 3 years his senior, did not survive the war. He had also been a student of the Piarists, which he attended between 1934 and 1942. His name is listed among the martyrs of the school of the period 1938–1958 on a memorial plaque in the lobby of the school.

When the Hungarian legislation had ordered the Hungarian high schools to restrict the number of Jewish pupils for the academic year 1943/1944, the Piarists ignored these restrictions except for the first-graders. They did not send away any of the upper-class pupils. The Arrow-Cross (Hungarian Nazi movement) took over the government on October 15, 1944, and the school closed for a few days. The instruction stopped entirely from October 25. It resumed on March 12, 1945. There were some new teachers and some classes were combined. The yearbook of the school lists the names of the pupils that never returned. George completed his studies at the Gimnázium of the Piarist Order in the spring of 1945 (Fig. 8). The school, keeping with its academic standards and regardless of the immediate post-war conditions, instituted a demanding final examination.

The sentence quoted above from Olah's autobiography about placing the painful experience of his Hungarian period behind him should not be interpreted as indifference to political and other developments in Hungary. He has remained conscious and proud of his Hungarian roots. He has observed keenly and critically the recent political developments in Hungary that include the restoration of much of the spirit of the Horthy regime between the two world wars. He finds it especially painful that the Hungarian responsibility for past tragedies has still not been faced [6].

Start of a Career

Up to his graduation from high school, Olah had been especially interested in literature and history and he had not planned a career in the hard sciences. When in 1996, soon after his Nobel award, the periodical *Chemistry & Industry* asked him, "If you hadn't become what you are, what else would you most likely to have been?," Olah's response was, "Writer, historian" [7]

His experiences and the post-war conditions in Hungary, however, prompted him to rethink the direction he was going to take. One generation before him, other future great scientists had to face similar dilemmas. Eugene P. Wigner, for example, was interested in becoming a physicist and John von Neumann, a mathematician. However, parental advice directed each of them to earn a diploma in chemical engineering first as it was offering a more secure future than physics or mathematics at the time as far as jobs were concerned. Olah chose chemical engineering rather than history or literature, and once he became engaged to chemistry, he never left it (Figs. 9 and 10).

The Budapest Technical University (today, Budapest University of Technology and Economics) had an academically very strong Faculty of Chemical Engineering (today, Faculty of Chemical Technology and Biotechnology). There was and has been as much emphasis on learning basic chemistry as on the subjects more directly related to technology. By all

Fig. 8 Graduates of Olah's Class 1945; Olah's portrait is in the lower left corner (courtesy of the Gimnázium of the Piarist Order).

available information, Olah enjoyed his studies and valued the direct interactions with his teachers. A reviewer of his autobiography noted: "Lectures, albeit compulsory, by active professors so inspired him that he continues to advocate a historical perspective in teaching and, despite the accessibility of electronic communication, direct teacher-student interaction in informal lectures" [8].

It took 4 years—eight semesters—of structured and intensive studies to earn the Diploma of Chemical Engineer (Fig. 11). Immediately upon graduation, in June 1949, Olah

Fig. 9 The middle section of the central building "K" of the Budapest University of Technology and Economics, 1–3 Műegyetem Quay, District XI (the building was inaugurated in 1909). Photograph by the author.

Fig. 10 The chemistry building "CH" of the Budapest University of Technology and Economics, 4 St. Gellért Square, District XI (it was built in 1902 and houses about half of the chemistry faculty) Photograph by the author.

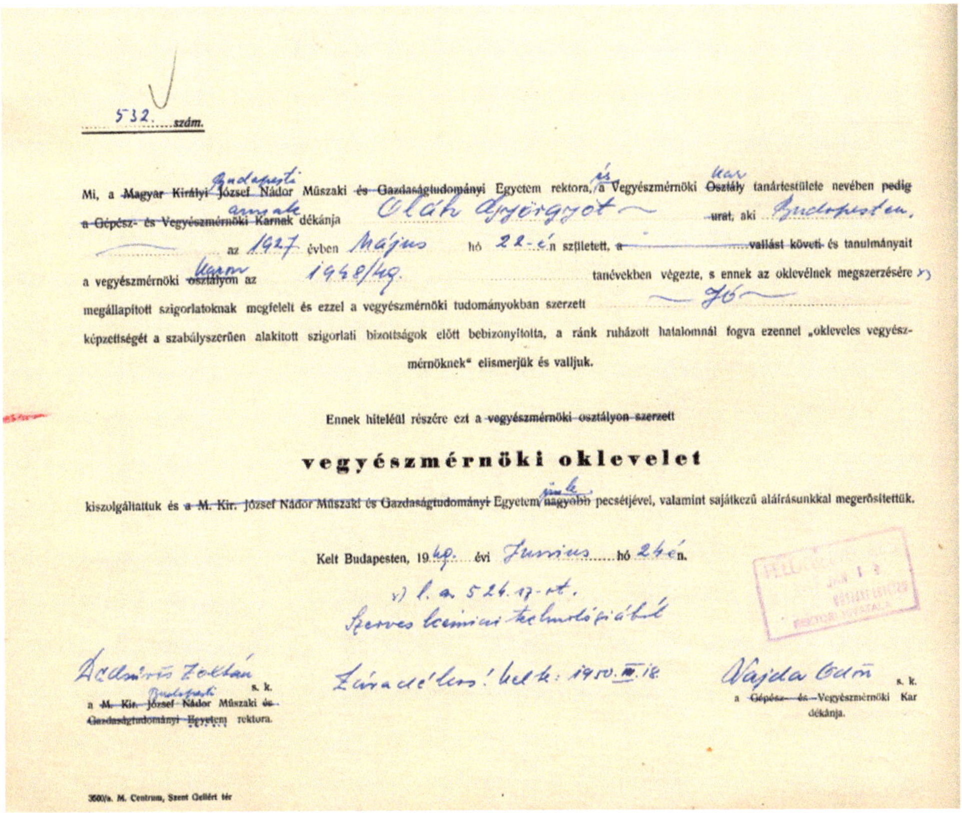
532. szám.

Mi, a ~~Magyar Királyi~~ Budapesti ~~József Nádor~~ Műszaki ~~és Gazdaságtudományi~~ Egyetem rektora, és a Vegyészmérnöki ~~Osztály~~ Kar tanártestülete nevében ~~pedig~~ ~~a Gépész- és Vegyészmérnöki Karnak~~ annak dékánja Oláh Györgyöt ~~urat~~, aki Budapesten, az 1927 évben Május hó 22-én született, ~~a ... vallást követi~~ és tanulmányait a vegyészmérnöki ~~osztályon~~ karon az 1948/49. tanévekben végezte, s ennek az oklevélnek megszerzésére x) megállapitott szigorlatoknak megfelelt és ezzel a vegyészmérnöki tudományokban szerzett — jó — képzettségét a szabályszerűen alakitott szigorlati bizottságok előtt bebizonyította, a ránk ruházott hatalomnál fogva ezennel „okleveles vegyészmérnöknek" elismerjük és valljuk.

Ennek hiteléül részére ezt a ~~vegyészmérnöki osztályon szerzett~~

vegyészmérnöki oklevelet

kiszolgáltattuk és ~~a M. Kir. József Nádor Műszaki és Gazdaságtudományi~~ Egyetem ~~nagyobb~~ pecsétjével, valamint sajátkezű aláírásunkkal megerősitettük.

Kelt Budapesten, 1949. évi Junius hó 24-én.

x) l. a. 524.17-et. Szerves kémiai technológiából

Zemplén Zoltán s. k. a ~~M. Kir. József Nádor~~ Budapesti Műszaki ~~és Gazdaságtudományi Egyetem~~ rektora.

Vajda Ödön s. k. a ~~Gépész- és~~ Vegyészmérnöki Kar dékánja.

3600/a. M. Centrum, Szent Gellért tér

Fig. 11 Copy of George A. Olah's (György Oláh) Diploma of Chemical Engineer dated June 24, 1949; the grade is "good," the final examination was in organic chemistry technology. The many amendments of the printed form were due to the fact that in 1949, the school was still using pre-war printed forms (courtesy of the Library of the Budapest University of Technology and Economics).

was appointed assistant professor at the Institute of Organic Chemistry of the Technical University. Géza Zemplén (1883–1956, Fig. 12), a former disciple of the great German organic chemist Emil Fischer in Berlin did postdoctoral studies (using today's term) with Emil Fischer in 1907 and 1908–1910. Zemplén was Professor of Organic Chemistry at the Budapest Technical University from 1913 until his death, and from 1950, he was the head of the Institute of Organic Chemistry of the University. He was the principal figure in organic chemistry in Hungary. His main interest was carbohydrate chemistry and did a great deal of work for pharmaceutical companies as well. Zemplén proved to be a good mentor who could serve as a knowledgeable example but who let Olah go his own way when Olah wanted to develop an independent research line in fluoro-organic chemistry.

Olah excelled from the start of his research career and in several aspects. He published papers that caught the attention of foreign researchers; he, with a colleague, compiled the index to an organic chemistry text; embarked on writing a book on theoretical organic chemistry; and did his share of teaching. In the years 1950 and 1951 Olah's primary research focus was in carbohydrate chemistry—the area of Professor Zemplén's studies. From 1951, Olah developed his independent line of research.

Soon after graduation at the Technical University, Olah applied for and was granted a scholarship for doing his postgraduate work in a structured framework. This led to the scientific degree, which used to be called "Candidate of Science" following the Soviet example. For all practical purpose, it was equivalent to a PhD degree in a good western university, but it was not granted by a university; rather it was granted by a special degree-granting institution of the Hungarian Academy of Sciences. Olah did all his research at the Technical University, submitted his dissertation in 1953, and defended it in 1954. The dissertation was about the chemistry of organic fluorine compounds; it is in Hungarian; and a copy of it is stored in the Manuscript Collection of the Hungarian Academy of Sciences, available for inspection. It consists of 186 pages with a vast amount of hand-drawn formulae and reaction equations, and it reports about a tremendous amount of innovative synthetic work. The working conditions were

Fig. 12 Bust of Géza Zemplén in the aula of the central building "K" of the Budapest University of Technology and Economics (photograph by the author).

poor, the reactants that elsewhere might have been readily available often had to be prepared from scratch, but the work is overwhelmingly impressive.

In his thesis work, Olah applied techniques and procedures of organic fluorine chemistry practiced already elsewhere and invented new techniques and procedures as well. He constructed what he called a Freon reactor (Fig. 13), a technique for fluorinating carbon tetrachloride and chloroform under ultraviolet irradiation using NaF, KF, and CaF_2. He produced chlorofluorocarbons in continuous operation. Olah listed 16 entries as publications containing the materials of his dissertation. They include a series of 12 papers most of which appeared parallel in Hungarian and in English (the latter in the English-language chemistry journal of the Hungarian Academy of Sciences, *Acta Chimica Hungarica*, which no longer exists). In most of these papers, Attila Pavlath, then Olah's student, much later, President of the American Chemical Society, was a co-author. Two entries referred to Olah's inventions of new techniques and procedures for producing organic fluorine compounds, filed one each in 1952 and in 1953. In his summary, Olah stressed the importance of his inventions for the industrial production of Freon compounds. In addition, two entries among Olah's publications referred to papers co-authored with colleagues at the medical school about the impact of organic fluorine compounds on experimentally induced tumors in animals.

Fig. 13 Olah's "Freon reactor" constructed at the Budapest Technical University in the early 1950s as part of his thesis work (from George A. Olah's PhD-equvivalent dissertation; courtesy of George A. Olah and the Archives of the Hungarian Academy of Sciences).

One of these two papers appeared in a German tumor research journal, the *Archiv für Geschwulstforschung*. Olah defended his dissertation and was granted the PhD-equivalent Candidate of Science degree in June 1954.

Olah's interactions with his colleagues in the medical school were anything but superficial. He signed up and completed the first three years of the subjects in the medical school, passed the examinations and fulfilled other requirements. All this, he was doing in the years 1951–1953. Also in this period he studied the Russian language and passed the exam for the PhD candidates with flying colors. He had to take also the obligatory political subjects prescribed for the PhD candidates during this same period. Olah listed his working engagements as 64 hours weekly, which included 14 contact hours with students. In reality his engagement was most probably more than 64 hours per week.

In 1954, Olah was appointed deputy director of the newly organized Central Research Institute of Chemistry of the Hungarian Academy of Sciences. There, Olah found another worthy mentor in the director of the Institute, the physical chemist Géza Schay (1900–1991, Fig. 14). He was a former disciple (postdoc) of the great Hungarian-born physician-turned physical chemist (and later turned philosopher) Michael Polanyi in Berlin, in the years 1926–1928 and 1930. Schay was Professor of Physical Chemistry of the Budapest Technical University (1949–1965) and in 1954, he was appointed director of the new institute. Schay's primary interest in physical chemistry was thermodynamics and reaction kinetics, and in particular, adsorption.

Olah must have been working with improbable high intensity and efficiency. Within 2 years from 1954, having been awarded the Candidate of Science (PhD-equivalent) degree, in 1956 he submitted his dissertation for the Doctor of Science degree. This has no exact equivalent in the American system; it is not the same as the British DSc and it is more than the German habilitation. This was a degree in which substantial scientific research production had to be demonstrated and served as prerequisite for a professorial appointment. Furthermore, only those who possessed this degree could be considered for getting elected to the Science Academy. Olah was not yet 29 years old at the time. It sometimes happens that mathematicians produce such a dissertation at an early age, but I know of no other chemist

Fig. 14 Relief of Géza Schay in building "F" at the Department of Physical Chemistry of the Budapest University of Technology and Economics (photograph by the author).

having completed the Doctor of Science work at that age. Olah's dissertation is available for inspection at the Archives of the Hungarian Academy of Sciences. The Manuscript Collection and the Archives are two separate sections of the Library of the Academy. The Manuscript Collection stores the dissertations that had been defended, whereas the dissertations that have not been are stored in the Archives. Olah left Hungary before he could have completed the process of the defense.

Just like Olah's Candidate of Science dissertation, his Doctor of Science dissertation is also substantial. It is in Hungarian and its title in English translation is "Data for the mechanism of electrophilic reactions of aromatic substitution." Almost all publications on which his previous thesis was based appeared in Hungarian journals (even if they were published in English). In contrast, most of the papers for his DSc thesis appeared in important western periodicals, such as the British *Nature* and the *Journal of the Chemical Society* and the German *Chemische Berichte* and *Naturwissenschaften*. Almost all of these papers were co-authored by Olah and his students, Attila Pavlath and Istvan (later, Steven) Kuhn.

The period between 1949 and 1956—the years during which Olah operated in Hungary after graduation—were busy. He established joint research with colleagues at the medical school in Budapest, attended meetings and visited research laboratories in Switzerland, East Germany as well as West Berlin (as they were then), and the Soviet Union, among others. He met with outstanding scientists, such as the Germans Weygand and Bohlmann, the Czech Wichterle, the Romanian Nenitzescu, and the Russian Reutov, Nesmeyanov, Semenov, and Kitaigorodskii. All these Russian scientists were among the top in Soviet science, and not only scientifically. Nikolai N. Semenov was the founding director of the Institute of Chemical Physics of the Soviet Academy of Sciences and in 1956 he was to receive the Nobel Prize in Chemistry. Aleksandr N. Nesmeyanov was Professor of Organic Chemistry at Moscow State University, the director of the Institute of Organic Chemistry between 1939 and 1954, and from 1954, the director of the Institute of Element-Organic [Heterorganic] Compounds of the Soviet Academy of Sciences. More importantly, he was the President of the Soviet Academy of Sciences between 1951 and 1961. It was at the start of his tenure as president that there was a big conference in Moscow organized to condemn Linus Pauling's resonance theory and condemn its Soviet followers as well. It was part of Stalin's anti-science policies as the theory of resonance was considered as being against Marxist-Leninist dogma. Stalin's anti-science policies were in concurrence with his paranoiac anti-Semitic policies [9]. Olah sensed the impossibility of doing science freely in such an atmosphere and he mentions this in his autobiography.

Olah applied for and was awarded the Dutch van 't Hoff Fellowship (but had no opportunity to utilize it). Reviewing his activities and interactions at the time, his situation may be called exceptionally favorable among his peers. His complaint that "Isolation clearly was a most depressing aspect of pursuing science in Communist-dominated Hungary" ([4], p 62) becomes understandable only if considering the flurry of his later activities under freedom in Canada and the United States.

Toward the Summit

In 1949, Olah married a colleague at the Technical University, Judit Lengyel (born in Budapest in 1929; Judit was later changed to Judith, Fig. 15). She was at the time a secretary at the University, but soon she studied chemical engineering and graduated from the Budapest Technical University. They shared a heavy burden of the recent past. In 1944, the Red Cross helped Judit and her 22-year old sister hide in a convent. On December 17, the Hungarian Nazis took them and others and marched them through the city. In a brave moment Judit escaped from the column, went into hiding, and survived; her sister stayed in the column and perished.

Judith and George had a boy in 1954 and another boy after they immigrated to Canada. The Olahs left Hungary in November, shortly after the Soviet tanks suppressed the Hungarian Revolution of October 23, 1956, but the borders to the West remained open for a short while. After brief stops in Vienna and London—where Olah initiated valuable interactions with colleagues—they moved on to Canada. Olah started looking for a suitable job already in London and in this he was assisted by Ms. Esther Simpson of the Academic Assistance Council (AAC).

The AAC was formed in 1933 to help refugee scientists from Germany and it was initiated by William Beveridge,

Leo Szilard, and a group of internationally renowned British scientists, with Ernest Rutherford as its first president. Ms. Simpson was already working for the organization at that time. The AAC had been renamed to the Society for the Protection of Science and Learning by the time Ms. Simpson was trying to help Olah. Today, the successor of AAC is the Council for At-Risk Academics and it still performs a much needed function.

The Olahs did not intend to stay in England; they were headed to Canada because they had close family connections there. Olah did not find employment in academia and started working in an industrial laboratory of the Dow Chemical Company in Sarnia, Ontario. Years later, when Olah had already become an internationally renowned scientist, a professor of organic chemistry at the University of Toronto apologized to him for opposing Olah's appointment to the University in 1957. Olah was unknown and this professor thought he was not worth the risk of employing him at the University of Toronto.

Olah found an industrial position and he has maintained ever since that "it is good to be challenged" [10]. Beside fulfilling the obligations of his job at the Dow laboratory, Olah continued his fundamental research he had begun in Hungary. In this industrial laboratory, he discovered ways to prolong the lifetime of carbocations and reached results that—again, showing Dow's magnanimity—he was allowed to publish. Olah's publications made an impact and resulted in his invitation to give a talk at the 1962 conference in Brookhaven I referred to in the Introduction. Furthermore, still at the time he was with the Dow Canadian laboratory, the American Chemical Society conferred upon him in 1964 its Award in Petroleum Chemistry.

The broader chemistry community recognized Olah's achievements over the years and the first impulses came through his decisive contribution to the resolution of the Winstein-Brown controversy (Figs. 16 and 17). Gradually, a

Fig. 15 The Olah family in 1962: George, Jr, George, Judith (Judy), and Ronald (Ron). Courtesy of George A. Olah.

Fig. 16 Saul Winstein in 1951 (photograph by and courtesy of J. D. Roberts).

Fig. 17 Herbert C. Brown in 1995 in front of the plaque of the Herbert C. Brown Laboratory of Chemistry at Purdue University (photograph by the author).

whole new chemistry was emerging from Olah's discoveries. His first pivotal results came from his works in the Dow laboratory in Canada, from where he moved to another Dow laboratory in Massachusetts. While in industry, Olah was doing everything to maintain his fundamental research and never lose connection with academia. His colleagues in academia responded to his efforts; they visited Olah's laboratory, invited him to participate in seminars and meetings, read and appreciated his papers, and attended the seminars he organized. Dow was good to him, but within limits. His research director did not recognize the significance of NMR spectroscopy and Olah had to bring or send his samples for NMR recording to university laboratories. However, this was another opportunity to enhance his interactions.

Back in his brief sojourn in London in 1957, he established interactions with the English chemist Ronald J. Gillespie (Fig. 18), one of Christopher Ingold's disciples. Ingold was Gillespie's mentor who helped enormously his associates, which was very good in the beginning of a research career, but became burdensome when it was time for Gillespie to establish his independence. Ingold was not the exploiting type and did not let his name figure on Gillespie's papers even when Ingold did write Gillespie's manuscript on the basis of Gillespie's investigation. However, Gillespie was told by Ingold who his graduate students should be, what apparatus to acquire next time, and so on. When an opportunity arose for a fully independent position at McMaster University in Canada, combined with the possibility of acquiring the most up-to-date equipment, such as a high frequency NMR machine, Gillespie moved [11]. He

then welcomed Olah's technician to run Olah's samples on his NMR equipment. Gillespie's laboratory was not the only one that assisted Olah and his group with NMR spectroscopy during Olah's industrial activities.

Gillespie's and Olah's research interests had an important overlapping area, and that was the superacids. The Harvard professor James B. Conant coined the name superacids for very strong acids as early as 1927, but he did not define their strength. Gillespie did just that in the 1960s, and according to him superacids are protic acids stronger than 100% sulfuric acid. Gillespie did a great deal of pioneering work in the superacid field. Olah recognized this to the extent that according to him, "Had the Nobel Prize been given for

Fig. 18 Ronald J. Gillespie in 1998 in Austin, TX (photograph by the author).

superacids, Ron in my opinion—as he well knows—should have certainly been included" [12]. However, the Nobel Prize was given for Olah's discoveries in carbocation chemistry, and it was, most deservedly, an unshared award.

In hindsight, it was almost inevitable that sooner or later Olah would find his way back to academia. This happened in 1965 when he moved to Cleveland as Professor and Chairman of the Department of Chemistry of Western Reserve University and he stayed in Cleveland for a decade (Fig. 19). He showed his acumen as an organizer, but he never slowed down in his research and fulfilled enthusiastically his teaching duties as well. The five portraits on the wall of his Cleveland office demonstrated his loyalty and his sense for the importance of continuity. When in 1996, he was asked the question: "Who is your biggest influence/hero and why?," his response was: "Hans Meerwein, who never considered himself a 'hyphenated' chemist and contributed much to synthetic, as well as mechanistic chemistry" [7].

Olah strengthened the chemistry department and by far not only through his own activities and those of his group. For example, he invited his fellow émigré chemist Miklos Bodanszky (Fig. 20), well known for his research in peptide chemistry and for his monographs in the field. As an organizer, Olah had an eye for the obvious that is sometimes the most difficult to notice: Across from his department, there was another chemistry department with only a parking lot between them. The other chemistry department belonged to the Case Institute of Technology. Within a couple of years, at

Fig. 20 Miklos Bodanszky in 1999 in Princeton (photograph by and courtesy of Eszter Hargittai).

Olah's initiative, the two departments joined and the merger was so successful that subsequently the two schools joined as well, creating Case Western Reserve University as it is well known today. Olah served as chair for the joined department for a while, but then he let others run it.

Olah's acumen as researcher manifested itself also in bringing together all the techniques that he found necessary for solving the problems he was working on. It was not only a task of finding the right instrumentation but finding the right experts as well. At some stage it became obvious—at least to Olah—that the reliable solution of the carbocation problems could not happen without high-level quantum chemical computations. This is how his life-long cooperation and friendship developed with Paul von Ragué Schleyer (Fig. 21) [13]. Schleyer's computations contributed significantly to Olah's discoveries as the application of physical techniques and computation become jointly much more powerful than the sum of the two approaches when applied independently of each other. Schleyer's commitment to Olah's research lasted his entire research career and he returned to the question about the structure of the 2-norbornyl cation in one of his last papers.

Fig. 19 George A. Olah in 1976 in his office in Cleveland. The pictures on the wall are of Hans Meerwein, Christopher K. Ingold, Saul Winstein, Herbert C. Brown, and Frank Whitmore. Courtesy of George A. Olah.

Fig. 21 Paul von Ragué Schleyer in 1995 in Vicksburg, MS (photograph by the author).

In it, he and his colleagues showed unambiguous X-ray crystallographic evidence, in concert with high-level computations, for the bridged, non-classical geometry of this carbocation [14].

At the Top

In 1972, Olah published a seminal paper in the *Journal of the American Chemical Society* (Fig. 22) in which he described the general concept of carbocations. His discovery of the reactivity of sigma donor single bonds in electrophilic reactions was nothing short of a revolution in hydrocarbon chemistry. The reactivity of these single bonds was "due to their ability to form carbonium ions via electron-pair sharing with the electrophile in two-electron, three-center bond formation." ([15], p. 808) There are a few characteristic drawings to illustrate a few aspects of the production and structure of carbocations after Olah. These are from a book series of the chemistry division of the Hungarian Academy of Sciences in which these drawings were reproduced (Figs. 23, 24, and 25) [16].

In 2015, Olah published the second, updated edition of his autobiographical book, *A Life of Magic Chemistry* [4]. In it, he once again evaluated the significance of the norbornyl

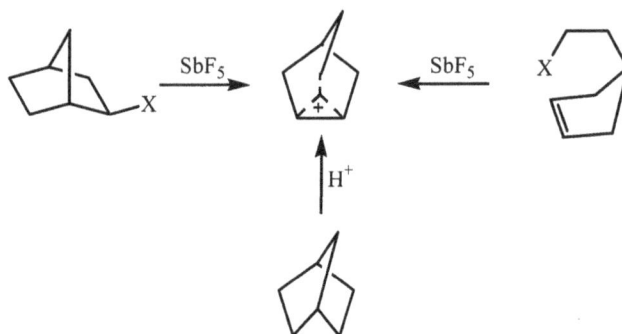

Fig. 23 A variety of routes to the long-lived non-classical 2-norbornyl ion ([16], p 35).

controversy. He did not formulate his views on this anew; rather, he referred to a conversation I recorded with him in 1996 in his office at the University of Southern California. The conversation was printed in full in 2000 ([10], pp 273–274) and the quoted passage was printed again in 2014 [17]. This is what Olah reproduced in 2015:

> I came into it because around 1960 I discovered methods to generate positive organic ions, called now carbocations, as long-lived species, and we were able to take all kinds of spectra and establish their structure, including that of the norbornyl

808

The General Concept and Structure of Carbocations Based on Differentiation of Trivalent ("Classical") Carbenium Ions from Three-Center Bound Penta- or Tetracoordinated ("Nonclassical") Carbonium Ions. The Role of Carbocations in Electrophilic Reactions[1a-e]

George A. Olah

Contribution from the Department of Chemistry, Case Western Reserve University, Cleveland, Ohio 44106. Received February 26, 1971

Abstract: The general concept of carbocations (the suggested generic name for all cations of carbon compounds, in accordance with carbanion for negative ions) is defined based on the differentiation of *trivalent* ("classical") *carbenium ions* from three-center bound *penta-* or *tetracoordinated* ("nonclassical") *carbonium* ions. Carbenium ions usually have a planar or nearly planar sp^2 hybridized electron-deficient carbenium center, although linear vinyl and acyl cations are also known. The carbocation centers in carbonium ions are substantially less electron deficient penta- or tetracoordinated carbon atoms bound by three single bonds and a two-electron three-center bond (either to two additional bonding atoms or involving a carbon atom to which they are also bound by a single bond). Thus, in a carbonium ion the two electrons from the original π or σ bond are delocalized over three centers. Long-lived carbenium and carbonium ions can be experimentally differentiated from each other showing marked differences, for example, in their nmr and core photoelectron spectra. The structures of typical carbenium and carbonium ions, like alkenium ions, alkonium ions, alkenonium ions, cycloproponium ions, norbornonium ions, cyclopropylcarbonium ions, ethenebenzenium ions [bicyclo[2.5]octadienyl cations], are discussed in the context of the general carbocation concept. It is emphasized that division of carbocations into limiting trivalent ("classical") and penta- or tetracoordinated ("nonclassical") categories is frequently arbitrary. In many carbocation systems an intermediate range of delocalization (partial carbonium ion character) must be considered, as is the case in the 2-methylnorbornyl cation. Both carbenium and carbonium ions play important roles in electrophilic reactions involving not only n- and π- but also σ-donor systems. The recently discovered general reactivity of single bonds (σ donors) in electrophilic reactions is due to their ability to form carbonium ions via electron-pair sharing with the electrophile in two-electron, three-center bond formation. Subsequent cleavage to trivalent carbenium ions is followed by typical carbenium ion reactivity.

Fig. 22 Part of the first page of the pivotal paper in the *Journal of the American Chemical Society* in which Olah suggested the carbocation name and described the general concept and structure of carbocations [15] (© 1972 American Chemical Society).

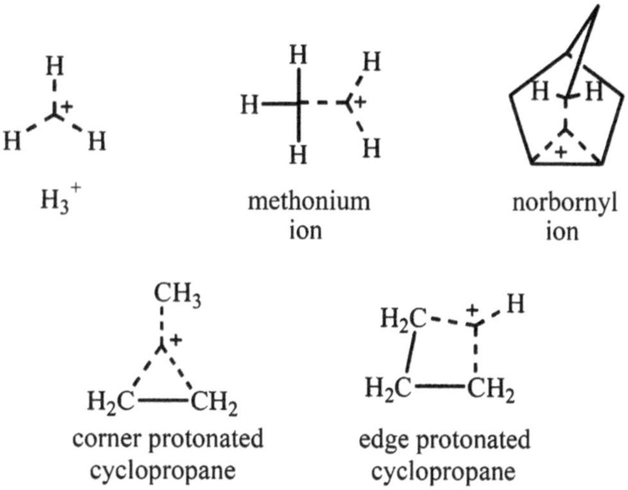

H_3^+

methonium
ion

norbornyl
ion

corner protonated
cyclopropane

edge protonated
cyclopropane

Fig. 24 Characteristic bonding examples in non-classical ions ([16], p 33).

cation. In the course of this work I realized, however, that the problem has much wider implications. In the norbornyl ion the C–C single bond acted as an electron donor nucleophile. In this particular case this happens within the molecule, that is, intramolecularly. This delocalization, which had been originally suggested by Winstein, was indeed there and we were able to see it directly for the first time. Later came, what I thought was a logical idea. The question what I asked myself one day was, if this can happen within the molecule, why can't it happen between the molecules? This led to the discovery of a wide range of electrophilic reactions of saturated hydrocarbons, that is, of C–H or C–C single bonds and the realization that carbon, under some conditions, can indeed bind five or even more neighboring groups ([4], pp 152–153).

Structure of Carbocations: The Case of CH_5^+

The overlapping interests of Olah and Gillespie were manifest also in the application of Gillespie's qualitative model for molecular geometry for testing some of Olah's unusual

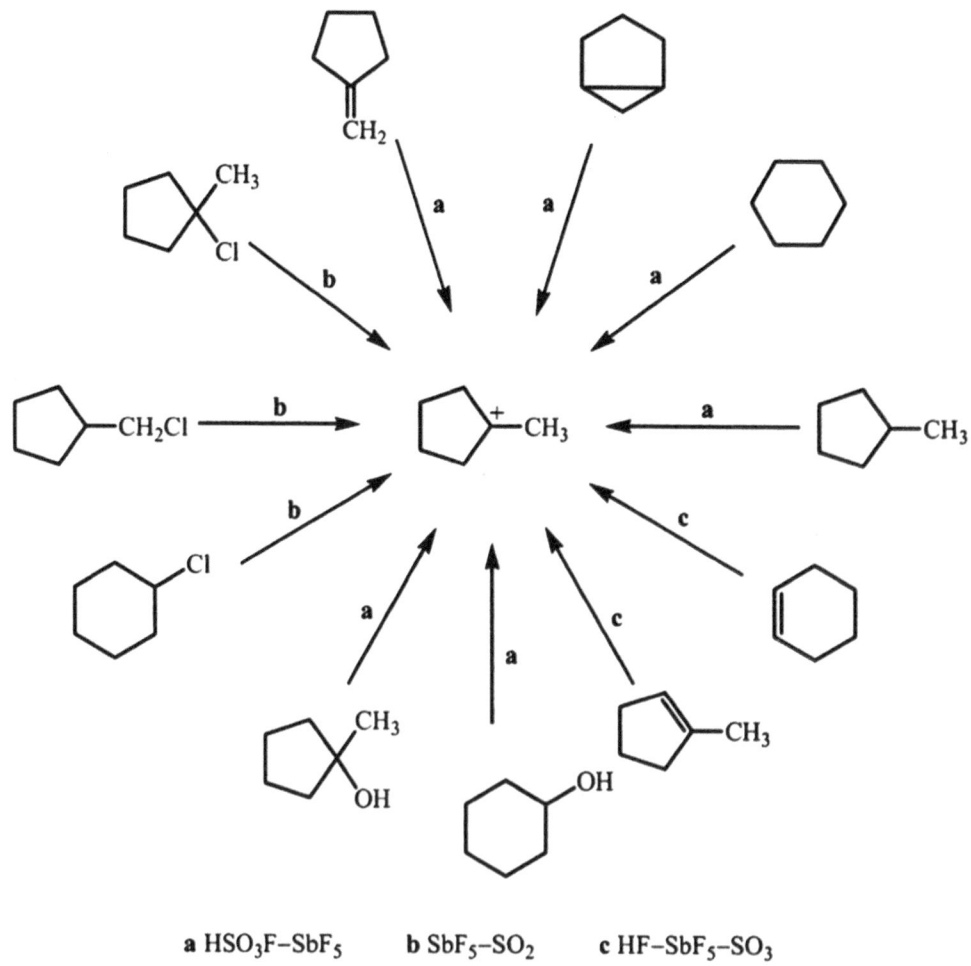

a $HSO_3F–SbF_5$ **b** $SbF_5–SO_2$ **c** $HF–SbF_5–SO_3$

Fig. 25 The utilization of a variety of precursors for the preparation of the methylcyclopentyl carbocation ([16], p 27).

structures. Gillespie's valence shell electron pair repulsion (VSEPR) model or theory predicts the geometry of the molecule on the basis of the number of electron domains (bonding pairs, lone pairs, multiple bonds) in the valence shell of its central atom [18]. The predicted shapes and symmetries depend not only on the general number of electron domains but to various extents also on the nature of those domains, whether they are single bonds, lone pairs or multiple bonds. For the shape of CH_5^+, that is, for five electron domains in the valence shell, Gillespie's model would predict a trigonal bipyramid or a tetragonal pyramid—these two configurations are too close in energy to be distinguished unambiguously.

From the beginning, however, computations predicted a C_s symmetry structure for the CH_5^+, which would correspond to the presence of three two-electron two-center bonds and one two-electron three-center bond (see, eg, [4], p 158). This structure can be viewed either as having a high-degree of localization, or as having a fluxional character by exchanging the positions of the two-electron two-center bonds and the two-electron three-center bond. If the C_s symmetry structure corresponds to a sufficiently deep energy well, it can be observed in experiments, provided that the life-time of this structure is sufficiently long as compared with the interaction time for the physical technique employed. In this respect, the spectroscopic techniques, and NMR spectroscopy especially, are at disadvantage as compared with the diffraction techniques. The interaction times of the former are orders of magnitude longer than those of the latter. Alas, for practical reasons, the structure determination of CH_5^+ by diffraction techniques would not be feasible.

Highly sophisticated high-resolution spectroscopic experiments on CH_5^+, as recent as 2015, have suggested the preeminence of geometries fully consistent with Olah's description of the structure (Fig. 26) [19]. There is a caveat though, because all the spectroscopic evidence point to the highly fluxional character of this carbocation: "the five proton swarm around the central carbon" [20].

Structural studies and considerations for the CH_5^+ carbocation have a rich history (going back much before this nomenclature had been introduced) [21]. Historically,

the structure of CH_5^+, called also the methonium ion, has presented a puzzle ever since it was first observed experimentally in 1952 in a mass spectrometric investigation [22]. Before the first computational studies, it used to be an assumption that the five hydrogens around the carbon would be equivalent or close to equivalent, which means a VSEPR-type geometry. The early computational studies suggested this carbocation consisting of two parts; one, a pyramidal CH_3^+ unit and the other, a hydrogen molecule bound to it. In other words, they were consistent with Olah's model of three two-electron covalent bonds and one two-electron three-center bond.

Thus, the VSEPR model could not predict the geometry of CH_5^+. In contrast, the geometry of monopositively charged carbocation $\{[(C_6H_5)_3PAu]_5C\}^+$ containing five-coordinate carbon has been found to be trigonal bipyramidal in agreement with VSEPR predictions (Fig. 27). As Olah has suggested, this gold complex represents an isolobal analog of CH_5^+, hence the isolobal analogy would favor a trigonal bipyramidal geometry for CH_5^+ as well, which, as we have seen, was not the case. Considering the highly fluxional character of the CH_5^+ carbocation, it means not only the exclusion of VSEPR-type configurations but also a distinct CH_3^+ plus H_2 configuration in which three distinguished hydrogens form two-electron two-center bonds and two hydrogens participate in one two-electron three-center bond.

A discussion similar to the CH_5^+ carbocation could be provided for the CH_6^{2+} carbocation and the $\{[(C_6H_5)_3PAu]_6C\}^{2+}$ carbocation. Six equivalent electron domains would favor a regular octahedral geometry (of O_h symmetry). In the gold complex, indeed, the six bonding directions point to the vertices of a regular octahedron in agreement with the prediction of the VSEPR model. In contrast, for the CH_6^{2+} carbocation, again, the computations have suggested the presence of two two-electron covalent bonds and two two-electron three-center bonds. There is then the CH_7^{3+} carbocation with one two-electron covalent bond and three two-electron three-center bonds.

Here, we are dealing with hypercarbons, though not with hypervalent carbons! The distinction is important. Carbon has no d orbitals available, hence it cannot extend

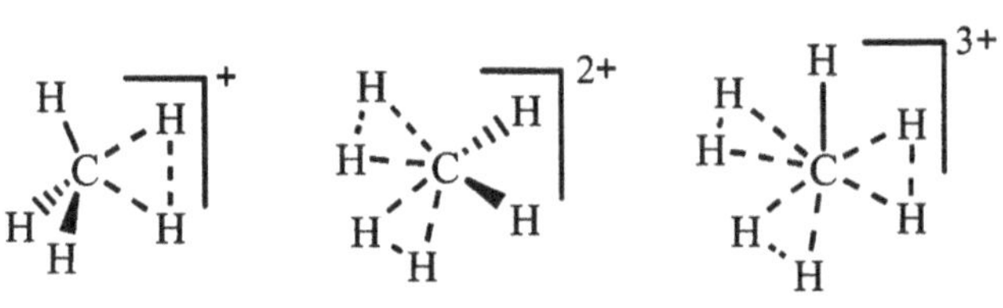

Fig. 26 Two-electron two-center bonds and two-electron three-center bonds in protonated alkanes ([16], p 38).

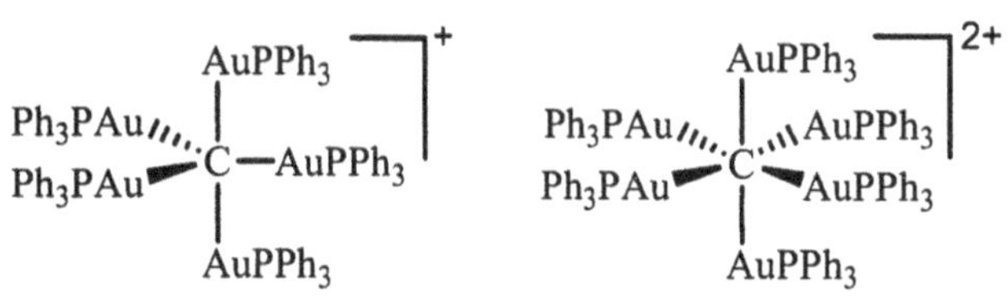

Fig. 27 The trigonal bipyramidal monopositively charged carbocation $\{[(C_6H_5)_3PAu]_5C\}^+$ and the octahedral dipositively charged carbocation $\{[(C_6H_5)_3PAu]_6C\}^{2+}$ ([16], p 38).

its valence shell; it can only extend its coordination. Thus the correct reference to it is hypercoordinated carbon rather than hypervalent carbon. ([4], p 160)

We still do not know the structure of CH_5^+ *in full*, but by now we do not know it at a much higher level of sophistication than before. One of the leading spectroscopists of this and similar species, Takeshi Oka of The Enrico Fermi Institute of the University of Chicago, called CH_5^+ "the 'enfant terrible' of chemical structures." According to Oka, its theoretical understanding "will take at least a few more decades"! [20].

The Nobel Prize

In 1977, Olah and his group moved to the University of Southern California (USC) at Los Angeles. There he developed a research institute known today as the Loker Hydrocarbon Research Institute. It focuses its research efforts in a single well-defined area. This is unusual for the American university setting, but it is not unlike some European research laboratories of science academies. Olah has emerited from the directorship, which G. K. Surya Prakash took over from him, and Olah has continued with the title of Founding Director. Donald P. and Katherine B. Loker were the principal benefactors of the Institute but others have contributed generously to it as well. Olah's Nobel Prize brought a great deal of publicity to the Loker Institute, but the Institute had been operating with great intensity and efficacy for years.

George A. Olah received the Nobel Prize in Chemistry for 1994. It was an unshared award, which is not too common as far as recent science Nobel Prizes go. On the other hand, awarding Olah the Nobel Prize and an unshared one at that met with universal satisfaction, which is also not too common. It was obvious to all that he deserved it and deserved receiving it alone. The official motivation for the prize was terse and even sounded a little noncommittal: "for his contribution to carbocation chemistry." I would not just call his works "contributions;" rather, they were bona fide discoveries.

The presentation speech by Salo Gronowitz, the Chairman of the Nobel Committee for Chemistry at the award ceremony gave a more direct description of why Olah received the award. He said, in part, "Olah's discovery resulted in a complete revolution for scientific studies of carbocations, and his contributions occupy a prominent place in all modern textbooks of organic chemistry." Further, Gronowitz noted: "Olah demolished the dogma that carbon in organic compounds could at most be tetra-coordinated, or bind a maximum of four atoms" [2].

Gronowitz stressed that "Olah found that superacids were so strong that they could donate a proton to simple saturated hydrocarbons, and that these penta-coordinated carbonium ions [in Olah's nomenclature, carbocations] could undergo

further reactions" [2]. In other words, even a carbon-carbon single bond or a carbon-hydrogen single bond could become an electron donor under the impact of an extremely strong proton donor, that is, in the presence of superacids. There is no doubt that Olah recognized the significance of his discovery for the whole of chemistry as he stated what we chose as the motto for this overview: "The realization of the electron donor ability of shared electron pairs could one day rank equal in importance with G. N. Lewis' realization of the electron donor unshared pairs" ([1], first pronounced in Olah's 1972 paper [15]).

The Nobel Prize in Chemistry is supposed to be given specifically "to the person who shall have made the most important chemical discovery or improvement" according to Alfred Nobel's Will (Fig. 28). This stipulation is preceded by a general one applicable to all categories of the prize: the awardees "shall have conferred the greatest benefit on mankind." Of course, even a discovery of purely fundamental nature with no foreseeable practical applications may qualify for conferring great benefits on mankind (the more so that seldom are there even purely fundamental discoveries that would not become eventually the roots of practical applications). However, it is always an advantage for the award selection if a discovery will have already shown direct benefits for improving the lot of humankind by the time of the Nobel Prize. Salo Gronowitz pointed out such features of Olah's discoveries in his presentation speech. Here are two examples from that speech: One that [Olah's] "discoveries have led to the development of methods for the isomerization of straight chain alkanes, which have low octane numbers when used in combustion engines, to produce branched alkanes with high octane numbers." Another is "With superacid catalysis it is also possible to crack heavy oils and liquefy coal under surprisingly mild conditions" [2].

This synergy of fundamental discoveries and their practical applications has continued shaping Olah's activities ever since the Nobel Prize as well. He decided not to sit on his laurels but continue his work and the practical applications may have taken up an even greater share of his activities since the Nobel distinction than before. This was so not necessarily by design, but because he had reached a stage in his research when the application of the vast amount of accumulated fundamental knowledge became almost inevitable. Thus, for example, to utilize the possibilities of "hydrogen economy," and to utilize them safely, Olah proposed storing the hydrogen in the form of methanol—this has become known as his "methanol economy." He recognized the utility of this approach at several levels. It is not only good storage; its production by reduction of the carbon dioxide in the atmosphere helps to counter global warming. The development of the direct methanol fuel cell was a natural outcome of these efforts. Olah's updated autobiographical volume provides

Fig. 28 A group of Nobel laureates at the Award Ceremony in Stockholm on December 10, 2001 (Photo by Hans Mehlin, © The Nobel Foundation, reproduced with permission), Row 7: Klug, X, Huber, X, Varmus (partly hidden), Bishop, Row 6: Blobel, M.S. Brown, Goldstein, H.C. Brown, Neher, Gilman, Row 5: Doherty (partly hidden), Heeger, Pople, Lewis (partly hidden), Nüsslein-Volhard, Wieschaus, Row 4: MacDiarmid, Shirakawa, Olah, Kroto, Schally (partly hidden), Mullis, Row 3: Nirenberg, Crutzen, E.H. Fischer, E.G. Krebs, Watson, Row 2: Benaceraff, Walker, Gilbert, T.N. Wiesel, Rowland, Row 1: Samuelsson, Vane, Edelman, Jacob, Row 0: Two members of the Royal Family: Prince Carl Philip and Princess Lillian.

detailed and readable accounts of the methanol economy and the methanol fuel cell.

Secrets, Not Classified

One wonders if there is any secret that would be the clue to Olah's great success. In fact, there appear to be many such secrets, fortunately, none of them is classified, and here we mention a few. Of course, every scientist must travel his or her own road and there is no generally valid single route to recommend. Olah benefitted from a broad-based education. That his interest in chemistry came relatively late in his youth meant the development of his great interest in reading and in history.

It is instructive to follow the carbocation story in perspective with Olah as a knowledgeable guide. Even in possession of an unshared Nobel Prize he is sufficiently humble to give credit where it belongs and go back to the roots of his science ([4], pp 72–76). As early as 1899, Julius Stieglitz at the University of Chicago raised the question of the possibility of ionic carbon compounds. This was an isolated episode that did not generate any follow up although Stieglitz was an influential chemist. Just as an aside, Herbert C. Brown, who has figured in Olah's story above, attended Stieglitz's lectures years later. Brown referred to his interactions with his professor in this way "That began my acquaintance with Julius Stieglitz, one that changed my life" ([10], p 255).

In the early 1900s, subsequently, several researchers produced and described compounds that could be interpreted as having ionic carbon in them. Hans Meerwein discovered in the 1920s that there are reactions that, while both the reactants and the products are covalent compounds, may have ionic intermediates (carbocations, in today's nomenclature). Still in the 1920s, Ingold, Hughes and their associates discussed further the role of carbocations in reactions.

Olah gives much credit to Frank Whitmore who in the 1930s established the transient role of the ionic intermediates that could not be observed directly, but just had to be there. Sadly, Whitmore's ideas met with so much disbelief that he could not use the trivial notation of cationic carbon species in his papers published in the *Journal of the American Chemical Society*. No wonder, there is loneliness for true discoverers.

In subsequent years, the scenery was changing and broadly recognized chemists took up the problem of cationic carbon species to which they ascribed the presence of transient intermediates in some organic reactions. The famous debate between Winstein and Brown developed, and the conditions had gradually become ripe for Olah's discoveries. With Saul Winstein's untimely death in 1969, Olah had to take up Winstein's role and the discussions continued to 1983 when there was no longer any doubt that there was nothing more to argue about; the idea of the non-classical ion presence had been proved unambiguously.

Olah though did not find the debates superfluous, because they have contributed to a better formulation of his discoveries. Olah appreciated the utility of criticism and he fully embraced what another Hungarian-born American Nobel laureate Georg von Békésy advocated in 1960 about the need of a few selfless enemies: "[One] way of dealing with errors is to have friends who are willing to spend the time necessary to carry out a critical examination of the experimental design beforehand and the results after the experiments have been completed. An even better way is to have an enemy. An enemy is willing to devote a vast amount of time and brain power to ferreting out errors both large and small, and this without any compensation. The trouble is that really capable enemies are scarce; most of them are only ordinary. ... Everyone, not just scientists, needs a few good enemies" [23]. Olah noted that the term adversaries would be a more proper term than enemies in this case. As it happened, nobody could stay long even to be an adversary to Olah, and his former adversaries have become his friends.

"The idea that ionization of alkyl fluorides to stable alkyl cations could be possible with an excess of strong Lewis acid fluoride that also serves as solvent first came to me in the early 1950s while I was still working in Hungary..." ([4], p76) Here Olah magnanimously dates the origin of his road to success back to his tenure at the Budapest Technical University. His direct observation of the long-lived carbocations, called also persistent, happened in the late 1950s at the Sarnia, Ontario, Dow Chemical laboratory. Thus, these two pivotal steps came about in two supposedly unlikely places for important fundamental research. It is not surprising that Olah in his autobiography shares his wondering about the advantages and disadvantages of the famous research universities and the venues he had labored in from Budapest via Sarnia and Cleveland, to Los Angeles. He offers encouraging words to those that, like himself, were not born with a silver spoon as far as research conditions were concerned.

It is a proof of Olah's greatness that if the conditions were not around, he created them. Witness to this the Loker Hydrocarbon Research Institute—a unique institution for an American university setting. He must have been a persuasive individual who could share his enthusiasm and dedication with people of means that were outside of chemistry and outside of science. Hydrocarbon chemistry may have not sounded too exciting even to many chemists and yet Olah could convince people of business that it was. He was right, of course; it is easy to see this in hindsight. Olah has had loyal friends and supporters and he has always been a loyal mentor, teacher, and colleague. When he was escaping from Hungary with his wife and little child, he was also thinking of his associates. I have had limited personal interactions with Olah and even being far outside of the center of his activities I felt his care in the warmest of ways. How privileged it could have been to be closely associated with him. Olah states it unequivocally, "I have always put great importance on loyalty" ([4], p 85).

Olah always gave priority to research over positions and was "never bitten by the bug that makes many people feel important by exercising power." ([4], p 87) He rarely held administrative positions, except for the chair in Cleveland and the directorship of the Loker Institute. He held that the people who do not really want to give up their research and teaching make the best university administrators, because they have the intention and a place to return to their natural calling when their administrative tenure is over. He did not decline invitations to join various committees. However, while in the committee, he freely spoke his mind and this he found to be a foolproof method of never getting asked back to the committee again.

Olah always took teaching duties seriously and in his teaching, he conveyed his personal experience as a researcher. He maintains that good teaching supposes successful research activities. He makes this general comment that monies for education should not be considered expenses; rather, they should be considered investment [24]. He spoke about this at the University of Szeged in Hungary where the sources for education are in the state budget and politicians tend to decrease these sources for education first when the budget has to be cut.

Legacy

Olah has continued his interest and his participation in his science and human affairs, but approaching 90 it is inevitable that the question of legacy comes to mind. Olah is leaving a multifaceted and rich legacy. When reviewing his legacy it becomes clear that his resolving the famous Winstein-Brown dispute is fading away. In contrast, his discoveries that made the resolution of the dispute possible shine in ever increasing intensity. Olah applied the extremely strong superacids to prolong the lifetime of carbocations and his realization of the electron donor ability of shared electron pairs opened a new direction in organic chemistry—one may wonder whether the same realization would not open new vistas in inorganic chemistry as well. Olah's new chemistry led to the creation of countless new compounds and he has enhanced the practical applications of newly synthesized substances. This did not just happen; Olah has always had an eye for and interest in practical applications. The culmination of his efforts in this aspect was the emerging methanol economy for which future developments will be the measure of the scale of its success.

The mention of success brings us to an exchange with Olah I had in 2003 about its meaning for him. This is what he had to say in response to my query:

> Success in science it looks to me means different things to different people. Many judge it in outside recognition of someone's work (prizes, membership in academies, honorary degrees, quotation numbers etc.). These may please the ego, but frankly are only trimmings. What I always felt is important is your inner satisfaction. After all, you should know best, if you are honest about it, whether you had achieved something in your scientific field, which has some lasting importance to our knowledge and understanding. If unexpectedly this can have also some application and benefit to society it adds to the feeling of success. However, most scientists are generally quite selfish and are inquiring because of their personal interest in a topic, which drives them not necessarily because they want to do something for society. Some of course judge success also based on material aspects (i.e. making money) but frankly, this never tempted me [25].

His books have a considerable place in Olah's legacy as this is similarly valid for many scientists. When I asked James D. Watson of the DNA double-helix fame about what he expected to be his longest ranging impact, his response was: "Probably my books" [26]. Watson thought that the DNA discovery "was just waiting to be made," but as for his *Double Helix* book, it "was probably unlikely to have been written by anyone beside myself." The relationship of books and discoveries may be different in Olah's legacy, but his books have undeniable importance. They have closely followed his progress in research. Whenever he completed his work in one research area, a summarizing monograph or an edited volume followed. This makes it possible to compile an approximate progression of his research career on the basis of these books. It started with his treatise on theoretical organic chemistry, on which he worked in Budapest and completed in Canada, to appear then in German in 1960. This was also part of his learning process. A selection of more research oriented volumes follows here without co-authors and co-editors and without full bibliographic references, which can be found elsewhere, for example in Olah's autobiographical volume. These books started appearing right from the start of Olah's research career.

Friedel-Crafts and Related Reactions (edited, in four volumes, 1963–1965)

Carbonium Ions (edited in four volumes, 1968–1973)

Carbocations and Electrophilic Reactions (1973)

Friedel-Crafts Chemistry (1973)

Halonium Ions (1975)

Superacids (1985)

Hypercarbon Chemistry (1987, updated 2011)

Nitration: Methods and Mechanisms (1989)

Cage Hydrocarbons (edited, 1990)

Electron Deficient Boron and Carbon Clusters (edited, 1991)

Chemistry of Energetic Materials (edited, 1991)

Synthetic Fluorine Chemistry (edited, 1992)

Hydrocarbon Chemistry (1994; 2004)

Onium Ions (1998)

A Life of Magic Chemistry (2000, updated 2015)

Across Conventional Lines (edited, selected papers, two volumes, 2003)

Carbocation Chemistry (2004)

Beyond Oil and Gas: The Methanol Economy (2006, updated 2009)

Superelectrophiles and Their Chemistry (2008)

Superacid Chemistry (2009)

Across Conventional Lines (edited, selected papers, third volume, 2014)

The synergy of fundamental science and the applications of its achievements is another important component in Olah's legacy. One might think that hydrocarbon chemistry is such a field that by its nature is close to practical aspects. However, some scientists in purely fundamental areas may also be more interested in practical applications than others. Eugene P. Wigner and John von Neumann, for example, were theoreticians, yet they were eager to find challenges related to applications, especially when they sensed the need for them [27]. For von Neumann, it meant primarily building computers and for Wigner, to use his knowledge of materials to help developing nuclear reactors. I am not suggesting that the shared origin and the shared life experience of Olah, Wigner, and von Neumann played a role in their shared interest in turning their scientific acumen into practical use, but the thought has crossed my mind.

I consider it part of Olah's legacy, the example he has set before others with his human demeanor. He has demonstrated that even a great scientist can stay a caring human being, a loyal friend, and an individual dedicated to assist others. He demonstrated tremendous inner strength when just surviving the last life-threatening months of persecution in 1945, he caught up with his school duties and passed his matriculation examinations at the Gimnázium of the Piarist Order. He then began his studies at the Budapest Technical University without any interruption let alone skipping a year, which could have been understandable. He succeeded in overcoming all barriers in the gradually hardening communist dictatorship of the early 1950s. As a refugee, he used the London sojourn of his family to build interactions with fellow scientists that would soon become useful especially during his tenure at the Dow Chemical industrial laboratory. When it became clear that no academic appointment would be available, he adjusted himself to the conditions of an industrial laboratory; built up his fundamental research after-hours; and in addition to fulfilling his duties, he functioned also as if it were an academic research venue, holding research seminars and attracting world-renowned scientists for visits.

His associates and disciples have always had a special place in Olah's life and activities. It was so during his Budapest years and it was so when he was making his escape with his family and his associates. It was also so from the start of his career in North America, and later in the Loker Hydrocarbon Research Institute. His care for his associates was always present regardless of his difficulties in securing a job, raising a young family, and overcoming all barriers that came his way. He concerned himself with the professional progress of his disciples and about the well-being of his students. Just an atypical example of the use of prize monies for an American professor: In 1979, the American Chemical Society presented Olah with the Award for Creative Work in Synthetic Organic Chemistry. He "used the prize money to send his students and postdoctoral researchers on a vacation to Hawaii" [28].

Wisdom and gaiety helped Olah to overcome difficult situations in his career. Although he states that "Human nature helps to block out memories of hardship and difficulties," ([4], p 294) some crept through even into his autobiography. It is telling that he considered hardship as character-building, but there was a limit to how much of it he wanted to tolerate. He remembers that "personal attacks and criticism which frequently came along were at times not easy to take" ([4], p 268). Early on, Olah received the lesson that being successful, even mildly successful, will generate envy. There were always some who would enjoy seeing him and his group failing. It may have helped him that he had experienced this kind of responses a great deal before his immigration. It is quite telling that the English language has no succinct equivalent to the German word "Schadenfreude" (enjoyment obtained from the troubles of others). As Olah took setbacks and difficulties in stride, he did not allow Schadenfreude to get in the way.

Olah has found great joy in chemistry, and chemistry has remained the focus of his attention throughout. However, this did not prevent him from seeing the beauty of the rest of the world surrounding us. I single out his interest and fascination with the concept of "Symmetry," which is present in both science and the arts, and serves to connect them. This played a role in our personal interactions, which started before his Nobel Prize, a fortunate circumstance, because the number of a Nobel laureate's friends usually grows exponentially after the award, but the friendships that had begun before it usually prove to be stronger.

When in the spring of 1995 George and Judy visited us in Budapest, we talked, among others, about the symmetry concept. My wife and I had already been producing books about symmetry, first about symmetry in chemistry and, eventually, about symmetry everywhere else [29]. George invited me to give a lecture on symmetry at his university in February 1996 (Fig. 29). By the end of December 1995, everything was settled for this event planned for February 20, 1996. At that point, the chair of the USC chemistry

Fig. 29 George A. Olah and Istvan Hargittai on February 20, 1996, the day of the Inaugural George A. Olah Lecture at the University of Southern California (photograph by and courtesy of Magdolna Hargittai).

department informed me that they were setting up a George A. Olah annual lectureship and decided to transform my presentation into its inaugural lecture. This is how it happened that my talk as "The 1st George A. Olah Lecture in Chemistry" on February 20 at USC was about symmetry. Plenty of chemistry found its way into this presentation [30].

Acknowledgments I appreciate the kind assistance I received in the preparation of this Editorial from Anders Bárány, Krisztina Batalka, Sándor Görög, Balazs Hargittai, Magdolna Hargittai, Diána Hay, Bálint Horváth, András Koltai, George A. Olah, Éva Sz. Kovács, Jonna Petterson, Zoltán Varga, Bob Weintraub, and Irwin Weintraub.

References

1. Olah GA (1994) My search for carbocations and their role in chemistry. Nobel Lectures in Chemistry 1991–1995. World Scientific, Singapore, pp 149–176; actual quote and references, p 173
2. Gronowitz S (1997) Presentation speech at the Nobel Award Ceremony (in 1994). In Nobel Lectures in Chemistry 1991–1995. Singapore, World Scientific, pp 139–140
3. Private communication from George A. Olah to Sándor Görög by e-mail, July 16, 2013
4. Olah GA (2015) A Life of Magic Chemistry: Autobiographical Reflections Including Post-Nobel Prize Years and the Methanol Economy. Second updated edition (with Mathew T). John Wiley & Sons, Hoboken, NJ
5. Hargittai I, Hargittai M (2015) Budapest Scientific: A Guidebook. Oxford University Press, Oxford, UK, p 288
6. Private communication from George A. Olah by e-mail, September 10, 2013
7. Chemistry & Industry 15 July 1996, in the "Personal Chemistry" column (p 529)
8. Jones DW (2001) Olah's organic odyssey. Chemistry in Britain August, p 44
9. Hargittai I (2013) Buried Glory: Portraits of Soviet Scientists. Oxford University Press, New York
10. Hargittai I (2000) Candid Science: Conversations with Famous Chemists (edited by Magdolna Hargittai). Imperial College Press, London, p 276
11. Hargittai I (2003) Candid Science III: More Conversations with Famous Chemists (edited by Magdolna Hargittai). Imperial College Press, London, pp 48–57
12. Private communication from George A. Olah in hard-copy letter, January 29, 2002
13. Hargittai I (2015) Paul von Rague Schleyer (1930–2014). Struct Chem 26:1–4
14. Scholz F, Himmel D, Heinemann FW, Schleyer PvR, Meyer K, Krossing I (2013) Crystal structure determination of the nonclassical 2-norbornyl cation. Science 341:62–64
15. Olah GA (1972) The General Concept and Structure of Carbocations Based on Differentiation of Trivalent ("Classical") Carbenium Ions from Three-Center Bound Penta- or Tetracoordinated ("Nonclassical") Carbonium Ions. The Role of Carbocations in Electrophilic Reactions. J Am Chem Soc 94:808–820
16. Oláh Gy (2008) Fél évszázadot felölelő, hagyományos határokon átlépő kutatások. A kémia újabb eredményei, Volume 100, pp 17–59
17. Hargittai B, Hargittai M, Hargittai I (2014) Great Minds: Reflections of 111 Top Scientists. Oxford University Press, New York, p 194
18. See, eg, Gillespie RJ, Hargittai I (2012) The VSEPR Model of Molecular Geometry (Reprint edition; original edition 1991). Dover, Mineola, NY
19. Asvany O, Yamada KMT, Brünken S, Potapov A, Schlemmer S (2015) Experimental ground-state combination differences of CH_5^+. Science 347:1346–1349
20. Oka T (2015) Taming CH_5^+, the "enfant terrible" of chemical structures. Science 347:1313–1314
21. Schreiner PR (2000) Does CH_5^+ Have (a) "Structure?" A Tough Test for Experiment and Theory. Angew Chem. Int. Ed. 39:3239–3241
22. Tal'rose VL, Lyubimova AK (1952) Dokl Akad Nauk SSSR 86:909–912
23. Békésy G von (1960) Experiments in Hearing. McGrawHill, New York, p 8
24. George A. Olah's comments quoted in the newspaper *Délvilág*, June 28, 2003, pp 1 and 3
25. Private communication from George A. Olah by e-mail, August 18, 2003
26. Hargittai I (2007) The DNA Doctor: Candid Conversations with James D. Watson. World Scientific, Singapore, p 37
27. Hargittai B, Hargittai I (2016) Wisdom of the Martians: In Their Own Words with Commentaries. World Scientific, Singapore
28. Wilson E (1997) Olah's 70th birthday draws devoted crowd. Chemical & Engineering News December 22, p 43
29. Hargittai M, Hargittai I (2009 and 2010) Symmetry through the Eyes of a Chemist (Updated Third Edition; previous editions, 1986 and 1995). Springer Science + Business
30. Hargittai I (1996) Symmetry in Science and Art—the Chemical Connection. The First George A. Olah Lecture in Chemistry. University of Southern California, Los Angeles, unpublished presentation

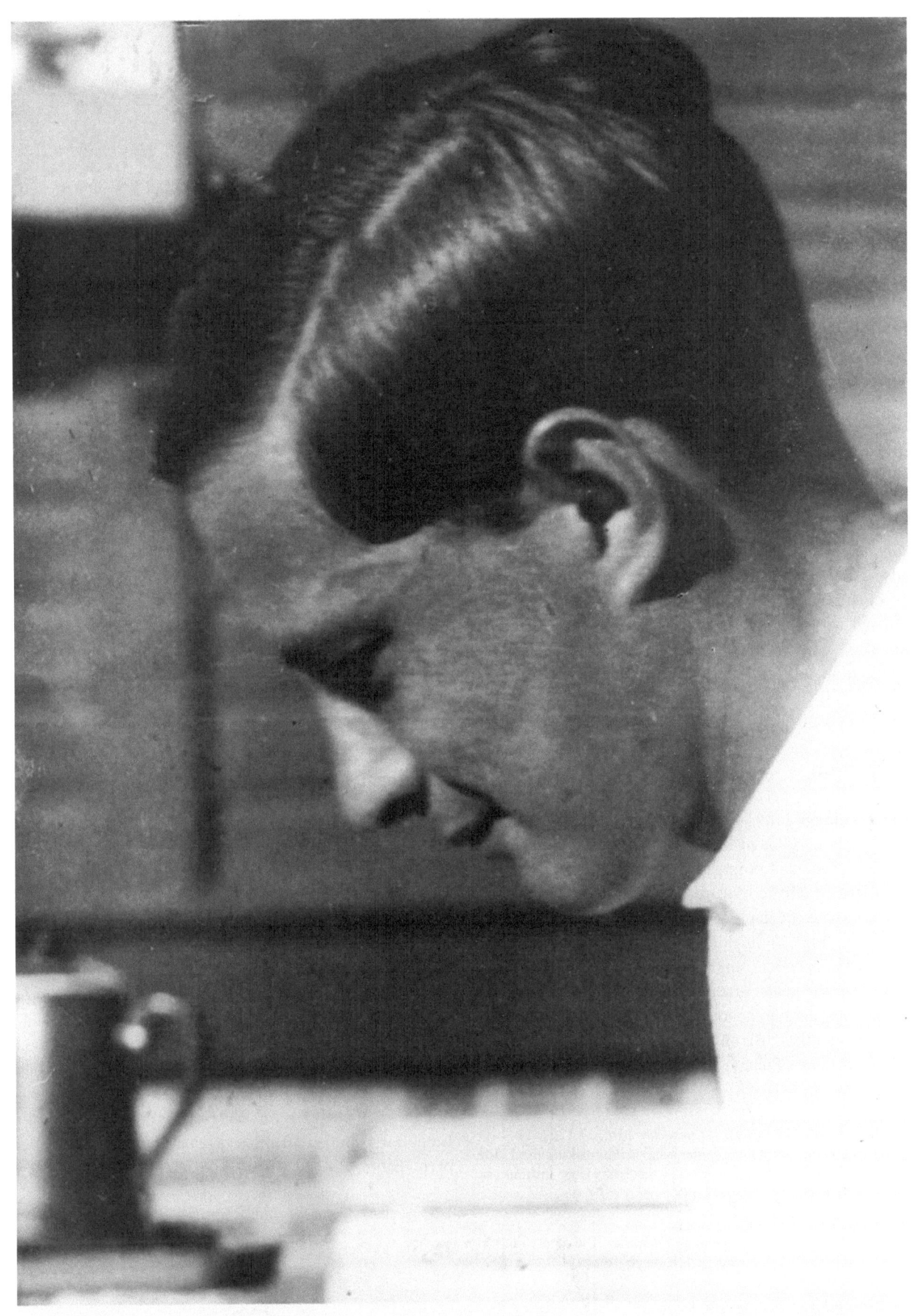

Michael Polanyi in 1931 in Berlin; courtesy of John C. Polanyi.

Michael Polanyi: Pupils and Crossroads—on the 125th Anniversary of His Birth[a]

István Hargittai

Two roads diverged in a wood, and I—I took the one less traveled by
Robert Frost, "The Road Not Taken"

Abstract

Michael Polanyi (1891–1976, Fig. 1) was a Hungarian-born British physician turned physical chemist turned philosopher. His milestone epistemological treatise *Personal Knowledge* followed his substantial discoveries in adsorption studies, X-ray crystallography, materials science, and the mechanism of chemical reactions. Michael Polanyi was one of the last polymaths and his teachings impacted the world views of other outstanding contributors to twentieth century science and culture.

Introduction

Michael Polanyi (1891–1976, Fig. 1) was born into an upper-middle-class Jewish family in Budapest during an era of unprecedented progress in Hungary, which was then part of the dualistic Austro-Hungarian Monarchy. He attended the secular Model High School (Minta Gimnázium, Fig. 2) in downtown Budapest, one of the city's many excellent high schools. The Model High School, over the years, graduated such future luminaries as the American aerodynamicist Theodore von Kármán, the British economists Baron Thomas Balogh and Baron Nicholas Kaldor, Polanyi's economist historian brother

Fig. 1 Michael Polanyi in 1937 in Manchester (courtesy of John C. Polanyi).

Karl Polanyi, the British physicist Nicholas Kurti, the molecular and nuclear physicist Edward Teller, and the American Abel laureate mathematician Peter Lax.

At the time, the high school, called gimnázium, was an important venue for the intellectual development of young boys. Girls were not yet supposed to attend such a school; rather, they went to schools that more directly prepared them for their future tasks in family life. One of Michael Polanyi's siblings (Fig. 3), Laura Polanyi, was exceptional; she attended another famous high school, the Lutheran Gimnázium, as a private student with a special permission.

[a]Originally published in *Structural Chemistry* 2016, 27:1327–1344. Based on an invited contribution to the symposium marking the 125th birthday of Michael Polanyi, organized by the Fritz Haber Institute, at the Technical University in Berlin on October 5, 2016

I. Hargittai (✉)
Department of Inorganic and Analytical Chemistry, University of Technology and Economics, Budapest, Hungary
e-mail: istvan.hargittai@gmail.com

Fig. 2 The Model (Minta, now Trefort) Gimnázium in 2014 (photo by the author).

Fig. 3 The Polanyi siblings: standing from left to right: Sofie, Adolf, Laura, and Karl; sitting, from left to right: Paul and Michael. (Courtesy of László Füstöss).

Fig. 4 Michael Polanyi in 1915 in uniform (courtesy of John C. Polanyi).

Fig. 5 Michael Polanyi and Magda Kemeny on their honeymoon in 1921 (courtesy of John C. Polanyi).

Beginnings

The discoveries, the writings, and the pupils are the true legacy of a scientist. In this account I focus on how some of Polanyi's former pupils remembered him, in particular Eugene P. Wigner and Melvin Calvin. The noted physicist and historian of science, Abraham Pais, opined that Polanyi "decisively marked Wigner's thinking, not just about physics, but also about philosophy and politics." [1] Wigner was referring to Polanyi when he stated that "Man's capacity to think is his most outstanding attribute." [2]

I met Michael Polanyi only briefly (see, below), but that brief meeting gave me an impression how fortunate those were that could spend longer periods of time with him. I have been fascinated not only with Polanyi's science but also with his life. He was hesitant in moving from Germany to England, because he found it difficult to accept that the Nazi madness could take over such a cultured land as Germany. Wigner commented on Michael Polanyi's emigration from Germany, "He moved to Manchester, England, in 1933, when Hitler came to power, a reason *very similar* to that which had originally prompted him to leave Hungary." [2, p. 154] Here, I added emphasis to "very similar," because there are some that do not consider the departure of Wigner, Polanyi, and others from Hungary in the early 1920s, at the time of the anti-Semitic Horthy regime, to be forced emigration. Wigner knew better.

Michael Polanyi (Figs. 4 and 5) graduated from the Model High School in 1908 and obtained his MD degree from the Budapest University in 1913. He served as a physician in the Austro-Hungarian Army in World War I. He had started his scientific research before having completed his medical degree. His professors sent his results in thermodynamics to Albert Einstein who liked Polanyi's paper a great deal. Polanyi received his Ph.D. degree in physical chemistry in Budapest, based on his 1917 dissertation entitled "Gázok (gőzök) adsorptiója, szilárd nem illanó adsorbensen" ("Adsorption of gases (and vapors) on non-volatile solid adsorbent").

Polanyi had important appointments both under the democratic revolution in 1918 and under the communist dictatorship in 1919 in Budapest, but his activities were of purely professional rather than of political nature. Theodore von Kármán occupied an even higher position in the revolutionary governments than Polanyi. Von Kármán, Polanyi and their colleagues saw to it that the best people were appointed at the universities. When the extreme right counter-revolution took over, and the autocratic and anti-Semitic Horthy regime came to power, those appointees became unemployable for the entire quarter-century of the Horthy era. These were tragic consequences of von Kármán's and Polanyi's most benevolent activities. Polanyi understood that in the Horthy regime, a young ambitious scientist, especially if Jewish, had no future in Hungary. This is also why von Kármán, Polanyi, and many others, such as, for example, George de Hevesy, John von Neumann, Leo Szilard, Edward Teller, Eugene P. Wigner, Dennis Gabor, felt compelled to leave.

As forced as Polanyi's departure from Hungary was, it upset him when some time in the 1920s he was accused of denying being Hungarian. I am quoting here, in full, his answer in 1929 to this accusation [3]:

> In 1904, when I was 13, I lost my father. Since then I have supported myself from stipends and my earnings. In the model high school, where I went, my teachers were taking care of me, got stipends and tutoring engagements for me. From the second semester of the university, I have been engaged in Ferenc Tangl's laboratory, who did not cease taking care of me. I graduated in

1913 as Doctor of Medicine. Due to the concern of Ignác Pfeifer, the next year I got to the Technical University of Karlsruhe to study chemistry, as a companion of a rich boy. I was then 22.

> In Germany the professors grab the students' hands, if he is supposed to be gifted. They are like art collectors whose obsession is discovering talent. They educated me and gave me a position where I could address myself to my abilities. They gave me everything and demanded nothing of me. They trust that who gets to know the joy of scientific work, will never leave it as long as he lives.

> Why am I telling you this? Because, looking back, its meaning is exactly what Ady had written about a hundred times, a long time ago, when only a few gray clouds hinted at the upcoming night. Looking back, I see the depth from which I was rescued by helping hands, the lucky one out of many. Looking back, I see other Michael Polanyis bogged half-way down and disappearing, I see them in my good friends, who stayed behind, I see them in unknown poor boys, by the dozen, like me and worthier, cast out of the university, thrown to the ground in front of the barbed wires of *numerus clausus* and other restrictions—onto a hip of invalids.

> Yes, a few words by Ady suffice: On the heap of invalids—In the Gare l'Est—Am I not Hungarian!?—This is what connects me with you, my comrades at home, Endre Ady's spirit. The hope that Ady's nation has not pushed itself away from the West forever, that there will be another Széchenyi and Kazinczy, that there will be new Ferenc Tangls and Ignác Pfeifers at the universities—open doors, helping hands.

> The professors will be looking for talent among the poor, honoring the new manifestation of the spirit for which they have lived. Everybody will be ashamed if his betters are in a lower position than himself, and won't rest until he lifts them into among his colleagues. There will though be unfortunate Official Authorities, but they won't be able to bar the way of the true spirit. I believe what we have here in Germany as the natural foundation of our lives, won't stay a utopia back home forever [4].

Polanyi (Figs. 6 and 7) makes references to the Hungarian poet Endre Ady (1877–1919). Ady published "Am I Not Hungarian?" in 1907 (*Budapesti Napló*). Its original title was "Who Is Hungarian?" It was Ady's response to his accusers who waged a concerted attack on him against his new lyrics. The attack against Polanyi was not dissimilar to the one against Ady, containing accusations of treason and cosmopolitanism, un-Hungarian behavior. There is unison between Ady's poem and Polanyi's response to the question of the editor of the *Pesti Futár*.

I met Michael Polanyi in 1969 in Austin, Texas, at a luncheon in the plush private club, the "Forty Acres," attended by three of us. Polanyi was the guest of honor, the chairman of the Physics Department, Harold P. Hanson, was our host, and I was the third participant. At that time Polanyi was a famous physical chemist for me, and I was not familiar with his works in social sciences, such as his seminal book, *Personal Knowledge* [5]. Polanyi was gentle and unpretentious. Our conversation covered a broad range of topics, from the Turkish and Russian/Slavic words in the Hungarian language to history and philosophy. We also talked about the difficulties of keeping up with the exploding scientific

Fig. 6 Associates of the Kaiser Wilhelm Institute for Physical Chemistry and Electrochemistry in 1931 (courtesy of Éva Gábor). Fritz Haber is second from the left in the upper row sitting and Michael Polanyi is second from the right in the same row standing.

Fig. 7 Michael Polanyi and his research group in Berlin-Dahlem in August 1933, immediately before his move to Manchester. On the back of the photo, there is a dedication of the photo by Polanyi to Andreas Szabo (first from the right, first row). (Courtesy of Éva Gábor).

literature. The aura of our conversation remained more in my memory than the actual topics and I am still under its impression. The quiet and simple way of communicating firm and reliable knowledge has been imprinted in my mind [6].

Polanyi was an excellent pedagogue who recognized the needs of young men (getting higher education at the time was almost exclusively men's business) who turned to him for advice. The future noted low-temperature Oxford physicist Nicholas Kurti (Miklós Kürti, 1908–1998, Fig. 15) had also studied at the Minta Gimnázium. Then, he attended the Sorbonne in Paris and in 1928, he moved to Berlin to study for his doctorate in physics. This is how in 1994 Kurti described what happened and I am quoting Kurti liberally in order to convey the atmosphere in which he found himself following Polanyi's advice [7]:

> ... I had a letter of introduction to Michael Polányi who was at that time in Berlin. Polányi suggested to me to do one year of postgraduate work and then to do a doctorate. The field I chose was low-temperature physics and Professor Franz Simon was my supervisor. He was one of the founders of low-temperature physics in Germany. Those three years, between 1928 and 1931, in Berlin were the most fantastic. As a city to live in, Berlin did not appeal to me. What I missed most was the Quartier Latin of Paris where I used to live. Walking up and down the Boulevard Saint Michel was the best recreation I could ever have. Berlin was different. Compared with Paris, it was a soulless city.

It was all right though because I just wanted to work hard. Still I managed to do a few good things. For example, a few weeks after the premiere of the *Dreigroschen Opera* by Bertold Brecht, I went to see it four times.

The most important thing though was the *Physik Kolloquia*, organized by Max von Laue in the Physics Department. These were not colloquia in the present sense of the word. They were more like the American journal clubs, just one two-hour session every Wednesday. A few people simply reported on recent publications from the literature. It was characteristic that in 1929 or 1930, Max von Laue could have an overview of the whole physics literature by looking at the *Proceedings of the Royal Society, Physical Review*, and *Physikalische Zeitschrift*.

If you went regularly to this colloquium, you could know what was going on in physics. Then you could keep up with everything. Laue would ask the audience about papers as he was looking for volunteers to review them for next time. It was regarded as the thing for graduate students to volunteer. Just think of it, you were reporting about a recent paper by a famous physicist and there was the audience, in the front row, Planck, Schrödinger, von Laue, Gustav Hertz, Haber, Nernst, about 6 or 7 Nobel Laureates or future Nobel Laureates. Behind them were Wigner, Szilárd, and others.

It was a very interesting experience. It was also wonderful to see that every now and then the great men could also make some silly mistakes. I remember when once Schrödinger suddenly stood up in the middle of a discussion of the spectra of triatomic molecules and suggested that the calculations could be simplified if you assumed that the three atoms are in the same plane. There was a silence, followed by laughter.

Wigner: "What a Mentor Michael Polanyi Was!" [8]

Wigner (Fig. 8), with Andrew Szanton's assistance, produced a gentle autobiography in which Wigner narrated his encounters with Polanyi. It is interesting to notice that Wigner spotted Polanyi's early interest in philosophy [8, pp. 76–79]:

> ...there at the Kaiser Wilhelm Institute worked a man who decisively marked my life: Dr. Michael Polanyi. Few people in this century have done such fine work in as many fields as Polanyi. After László Rátz of the Lutheran Gimnázium, Polanyi was my dearest teacher. And he taught me even more than Rátz could, because my mind was far more mature. After Rátz and my parents, Polanyi was my greatest influence as a young man.
>
> The Germans have a tremendous word for fiber chemistry: "Faserstoffchemie." Michael Polanyi had his own laboratory in the Kaiser Wilhelm Institute for Faserstoffchemie. The Mauthner Brothers tannery in Budapest employed a fine chemical engineer named Paul Beer, who somehow knew Polanyi and gave me a strong letter of introduction to him.
>
> So Dr. Polanyi asked me over to his home one evening. A chemist named Herman Mark also came that night. Mark was an energetic, chatty man from Vienna. He was only seven years my senior, but seemed much older.
>
> Mark had fought in the Austrian ski troops during the First World War on both the Russian and Italian fronts and had escaped from an Italian prison camp disguised as an Englishman. He had quickly completed his education at the University of Vienna and taught at the University of Berlin before joining the Kaiser Wilhelm Institute as a research associate.

Fig. 8 Eugene P. Wigner and the author in 1969 in front of the old Physics Department of the University of Texas at Austin (by unknown photographer).

> Polanyi and Mark had a fabulous discussion that evening, just two physical chemists discussing one topic after another. Mark smoked a few cigarettes. I sat by without opening my mouth, amazed at how much physical chemistry they knew. Topics at the farthest edge of my comprehension they discussed with the greatest fluency and ease. They spoke with graceful insightful wit, following each other perfectly.
>
> When Herman Mark finally rose to leave, my involuntary reaction betrayed my great disappointment. Mark put on a little half-smile, sat down again, and revived the conversation. My embarrassment at having kept Mark in the room soon faded in the face of their startling conversation. Listening with all of my limited intelligence, I knew that I was deeply happy.
>
> That was my introduction to Dr. Mark and Dr. Polanyi. Soon I knew Polanyi closely. He told me to call him "Misi" (pronounced "*Mee*-she"), placed me in his laboratory, and asked me to contribute to meetings and colloquia.
>
> About three other students worked for Polanyi. I studied theory: crystal symmetries and the theory of the rates of chemical reaction. I spent just a few hours in the lab and many more hours calculating figures in my room. I also learned a great deal about the life of Michael Polanyi.

Further down, Wigner mentioned their joint work [8, pp. 76–79]:

> Polanyi and I wrote a joint article in 1925, introducing assumptions that seemed drastic then; they later proved quite correct. We wrote another joint paper in 1928. What a pleasure it was to assist a man of such keen mind and deep insight. Polanyi

took an interest in all of his assistants, but I felt that he liked me especially. He freely advised me on various personal matters. In time his generous wife did too. Polanyi even loaned me a bit of money when I needed it. But his finest gift was to encourage my work in physics, and this he did with all of his very great heart. In all my life, I have never known anyone who used encouragement as skillfully as Polanyi. He was truly an artist of praise. And this praise was vital to me because it was often missing at the great afternoon physics colloquia.

Because Polanyi was a decade my senior and held a far higher position, it was not quite proper for him to befriend me as he did. But Polanyi cared nothing for formal questions of age and status. That was part of his great sweetness. Polanyi was concerned instead that young men should love science and labor to understand it. He was concerned that he could never fully share his love and the knowledge he had gathered.

Like me, Polanyi enjoyed asking questions outside the realm of basic science: Why is the world divided into separate nations? Why do all nations have governments? How should a man live his life in a world filled with evil? Polanyi even taught me some poetry. He made learning a great pleasure.

Dr. Polanyi and I did not always see eye to eye. Polanyi found quantum theory too mathematical for his liking. I was the only one in his lab deeply interested in it.

Once I made an observation to Polanyi about the impossibility of an association reaction. He heard my idea without grasping it. I felt sure that I was right and even that my idea had merit. But I was too modest to press it home.

Months later, Polanyi told me one day, "I am quite sorry. This point which you have always made on association reactions: I have just heard it in a paper of [Max] Born and [James] Franck. I told them that you had the same idea, but they have already sent in the article, and nothing can be done." Polanyi paused a moment. "I am quite sorry," he said again, "I don't know why I failed to understand you."

Well, I think I know. Even a man as open-hearted as Polanyi does not easily accept the brash ideas of a modest and untried assistant. What I had told him was radically new, and however open-minded people may seem, very few are prepared to embrace radical ideas.

Wigner worked out a variation of his original idea and published it, but it never made the impact it might have if Wigner had secured his priority in tackling the problem. This is quite a story and it is always a delicate question when the mentored overtakes the mentor even if it is in a single research idea. Both Polanyi and Wigner came out of this story impeccably though.

Wigner did his research for his Diploma work (Master's degree-equivalent) with Herman F. Mark, but opted to do something different for his doctoral work. He decided to investigate the rates of chemical reactions and he signed up for being Polanyi's doctoral student [8, pp. 80–81]:

Polanyi advised my doctoral dissertation at the hochschule. ... I wondered: How do colliding atoms form molecules? We knew that hydrogen and oxygen make water in a container, but how soon? How much depends on pressure and how much on temperature? I pursued such questions with elements far more complex than hydrogen and oxygen.

Polanyi was a wonderful advisor. He understood chemical reaction rates both in theory and practice. He accepted my proposal that angular momentum is quantized and that the atoms collide in a proportion consistent with Planck's constant. This idea is now widely known, but then it was rather brash. And studying chemical reaction rates taught me much about nuclear reaction rates that would be useful in future years.

My thesis paper for the engineering doctorate was submitted, with Polanyi's name attached, in June 1925. We called it "Bildung und Zerfall von Molekülen" ("Formation and Decay of Molecules").

Once Wigner completed his studies in Berlin, he returned to Budapest in 1925 and started working in the tannery directed by his father. He may have not been an enthusiast for tannery work, but he was conscientious in everything he did. He learned whatever there was to learn about the processes involving leather and even visited other tanneries to learn more about the processes he was using. Even decades later, he was proud of his knowledge of the chemistry of leather treatment. Yet he missed physics and subscribed to the *Zeitschrift für Physik* to keep up with the developments in his favorite subject. A year had barely passed when he received an invitation to return to Berlin to work for the crystallographer Karl Weissenberg at the Kaiser Wilhelm Institute (today, we would call this a postdoctoral position). The invitation was the work of Michael Polanyi, who knew that Wigner was destined not for tannery work but for creative science.

Wigner adored Polanyi (Fig. 9), "Michael Polanyi was really the miraculous one [teacher]. Polanyi loved to ask the fundamental question: 'Where does science begin?' He listened to the thoughts of others on this question, but he also had his own well-crafted answer [see below]. ... Polanyi loved and honored the scientific method with great truth and devotion. He managed to keep all of science within his fond gaze and a great deal more besides. What a mentor Michael Polanyi was." [8, pp. 80–81]

Fig. 9 Eugene P. Wigner (on the right) with Michael Polanyi and his son, John C. Polanyi in 1934 in Manchester (courtesy of John C. Polanyi).

When Wigner's Nobel Prize came and he had to give the traditional two-minute speech at the Nobel Banquet, he returned to what he had learned from Polanyi about where science begins: "I do wish to mention the inspiration received from Polanyi. He taught me, among other things, that science begins when a body of phenomena is available which shows some coherence and regularities, that science consists in assimilating these regularities and in creating concepts which permit expressing these regularities in a natural way. He also taught me that it is this method of science rather than the concepts themselves (such as energy) which should be applied to other fields of learning." [9]

Wigner's interactions with Polanyi did not end when both had left Germany and Wigner spent a few precious months with Polanyi in the mid-1930s in Manchester. In his memoirs, Wigner gratefully remembered that Polanyi was still capable of praising Wigner even when Polanyi's faculties were diminishing during Polanyi's terminal illness. One wonders how much Polanyi's example influenced Wigner in Wigner's later years when he was increasingly turning to discuss philosophical questions.

Melvin Calvin About Polanyi's "Curious Mind"

The American Melvin Calvin (1911–1997, Fig. 10) received the Nobel Prize in Chemistry in 1961 "for his research on the carbon dioxide assimilation in plants." Calvin spent two years with Polanyi as postdoctoral fellow for which Polanyi (Figs. 11, 12 and 13) used a grant from the Rockefeller

Fig. 10 Melvin Calvin in 1962 at Berkeley by Berkeley LRL Graphic Arts (courtesy of Marilyn Taylor and Heinz Frei).

Fig. 11 Michael Polanyi (middle) and Alwyn G. Evans (right) in 1940 in Manchester (courtesy of John C. Polanyi).

Foundation. Calvin referred to his time with Polanyi in his Nobel lecture in the following way: "Our own interest in the basic process of solar energy conversion by green plants … began some time in the years between 1935 and 1937, during my postdoctoral studies with Professor Michael Polanyi at Manchester. It was there I first became conscious of the remarkable properties of coordinated metal compounds, particularly metalloporphyns as represented by heme and chlorophyll." [10]

Calvin narrated in detail about these studies in a recorded conversation with Clarence Larson, former Commissioner of the US Atomic Energy Commission. Larson and his wife, Jane Larson, in their retirement recorded conversations with famous scientists and technologists. Melvin Calvin was one of them and their recording took place in 1984 [11]:

Michael Polanyi had been studying reactions of sodium atoms with alkyl halides in a dilute gas. He also had undertaken a study of the reaction of the hydrogen atom with the hydrogen molecule. The way he made that measurement was to use H atoms and D_2 molecules and measured the formation of HD. He was measuring the simplest kinds of reactions, which were susceptible to first principles quantum mechanical calculations, and he succeeded in doing that and in developing what we now know as a transition state theory of reaction kinetics. His more famous pupil was Henry Eyring who preceded me in that work. By the time I got to Polanyi, he had moved to Manchester and by that time the theory of transition state had been sorted out.

Polanyi asked me to study the mechanism of activation of molecular hydrogen on platinum, starting with on polarized platinum. He had the idea that you could study the reaction of hydrogen atoms attached to polarized platinum with hydrogen molecules, which were not attached to platinum. That way you'd be able to affect the activation energy of the atom/molecule reaction, and that's what he put me on. I began to study the effects of polarization on platinum electrodes carrying hydrogen atoms on the rate of exchange between the hydrogen atom and the D_2 or HD molecule. This led to a more general question, which Polanyi now posed.

CHEMISTRY HONOURS, 1936

S.I. Cowton. C.S Walker. C.N.S Farrand. J.B Alcock.. J.W Haworth. F.C Reed. F Thomas

...Hadley. A Atkinson. T.C Mills. E Lofthouse. J.N Haresnape. ..Wincoll. Docter.M Snider. A Ramsden. H Clough.

..G Hey. Mr Herbert. Dr Burt. Prof Polanyi. Prof Heilbron. Dr Campbell. Dr. Sutton. Dr Burkhardt. Dr H.R.Wright

Fig. 12 Faculty and chemistry honors students in 1936 in Manchester (courtesy of Éva Gábor). Michael Polanyi is fourth from the left, first row.

Before that though, you should understand who Polanyi was. He was a refugee both from Hungary and Germany. He was a surgeon in World War I for the Hungarian Army. After the war was over he realized that his interests were in basic science. He went to Berlin and that's where his physical chemistry and his ideas about reaction mechanisms were born and developed, in Berlin-Dahlem. After Hitler came to power in Germany, Polanyi left. He went to England. I went there in 1935 and spent two years with him.

Polanyi's background had some biology in it; he was aware that there were enzymes in living systems that could deal with molecular hydrogen. He thought that those enzymes, and all had metals in them, would probably be important to understand how to activate hydrogen properly. At that time he believed that the active site of hydrogenase, the enzyme, which activates molecular hydrogen and allows it to exchange with water, was an iron-porphyrin-bearing enzyme. The reason, I think, he thought that way, and I have to say, "I think" because he never did tell me, was that most of these enzymes were oxidation and reduction enzymes, enzymes that catalyzed the addition or removal of electrons from substrates. If the enzyme activated molecular hydrogen so it will exchange with the protons of water, presumably the enzyme was oxidizing H_2 to get protons and holding the electrons back somehow. When then the protons would exchange, they would then come back again as molecular hydrogen.

Polanyi had been studying these exchange reactions in various ways. He invented, for example, the micropicnometer to measure the density of water in order to measure the amount of deuterium in it. He would use a few tens of microliters of the water to measure its density. These micropicnometers were little floats. The picnometer would hold a hundred or fifty microliters of water and it was put in through a microcapillary. The top of that picnometer bore a little sphere, a bulb of five millimeters of diameter. That sphere was very thin glass and flat on one side. When the picnometer was dropped in water, it would float with the water-containing part down and the bulb up. The volume of that bulb depends on the pressure. He could measure the density of a hundred microliters of water to five or six or seven places that way. That was the kind of man he was. He invented it, designed it and had it built. We didn't have mass spectrometers in those days. So we were measuring water densities that way and measuring exchange rates that way.

Polanyi had the idea that the enzymes must have some peculiar properties, which are dependent upon the porphyrins because almost all redox systems in biology that he knew about, the hemin of red blood cells, the chlorophyll of the green plants, all were porphyrin type molecules with metal centers. The hemin had an iron center, chlorophyll had a magnesium center. He put me onto that after I had been there a year and a half. He supposed that there must be something very special about this tetrapyrrolic structure which surrounds the metal and which makes it do funny

CHEMISTRY HONOURS 1947

JUNE A.   AILEEN   LOUISE E.
BETON. HESMONDHALGH, I.E.SMITH. E.W.FELTON, R.RANDS.   J.D.SHIMMIN,   HAISELL

J.S.ROBERTS, J.J.CONNELL, D.W.CHADWICK, J.KWILKINSON D.W.MOORE R.P.HANDFORD, S.BEESLEY, J.J.GARNER.   M.V.LOCK   RUTH M.
                                                                                           HAINSWORTH
KATHLEEN
CUNLIFFE.   S.E.ARNOLD. H.SPEDDING.   O.H.GELLNER, T.E.WALKER-SMITH, A.W.CRAIG.   J.R.EMERY,   P.K.BINGHAM,   A.THOMPSON
J.C.WOODS . DR.T.H.QUIBELL. DR.G.N.BURKHARDT, DR.C.CAMPBELL. PROF.E.L.HIRST, PROF.M.POLANYI, DR.F.FAIRBROTHER. MR.J.B.M.HERBERT, T.P.C.MULLHOLLAND

Fig. 13 Faculty and chemistry honors students in 1947 in Manchester (courtesy of Éva Gábor). Michael Polanyi is fourth from the right, first row. (Note the threefold increase in the ratio of female students as compared with 1936.)

things in biology. The biological tetrapyrrols are very unstable compared to the kinds of things he was used to doing.

About that time, in 1934, R.P. Linstead, Professor of Organic Chemistry at Imperial College in London, had discovered phthalocyanine. He was a consultant for ICI. ICI was making phthalonitrile, which is ortho-dicyanobenzene in glass lined kettles. Phthalonitrile crystallizes in beautiful white crystals, but on one occasion it turned into a blue mess. Linstead determined that the glass lining in one of the iron kettles had cracked and phthalonitrile had come in contact with the iron, and this had catalyzed the cyclization of the four phthalonitriles around an iron center. He had iron phthalocyanide. That was the beginning of a new dyestuff, which turned out to be very stable, and became one of the most important organic pigments for a period of 20 or 30 years. It is known as a tetraazaporphyrin. The bridges between

the four pyrrol rings were nitrogen atoms instead of carbons that are the bridges in nature.

Polanyi told me to go down to London, find out how to make that stuff and bring it back. He gave me two weeks to do that. Polanyi then suggested to put different metals in the center and study their catalytic properties for activating hydrogen, like platinum. You could heat it up, cool it, do what you liked. I've spent a lot of time doing that and I enjoyed that very much. In so doing, I became thoroughly aware of the importance of that particular type of structure, always involving the movement of electrons and protons. Of course, the chlorophyll in the green plants, although not the same, is a very close relative of porphyrin. That also involves photochemical oxidation/reduction. That's how I got started on that business. My last experiments with

Polanyi were hydrogen activation on metalphthalocyanines with copper and zinc.

Michael Polanyi was willing to participate in the war efforts in Great Britain. At about the outbreak of World War II, he made inquiries of whether he could participate in the war efforts doing applied research, but was given a negative response. However, his teachings found their way, through Wigner, into the Manhattan Project. As soon as nuclear fission was discovered, the imagination of physicists captured the possibility of the atomic bomb. One of them was John A. Wheeler who helped Niels Bohr in working out the theory of fission, and in this, Wheeler enlisted Wigner's assistance. This is how Wheeler recalled this period in the early 2000s [12]:

> We had to understand this new nuclear phenomenon, fission. It was obvious that the nucleus of such a heavy element as uranium must undergo a considerable deformation before it splits. For that it needs energy. When the uranium is bombarded by neutrons, the neutron can provide this energy; we say that the nucleus is excited. This excitation then could initiate a vibration in the nucleus that could deform it. Our Hungarian friend, Eugene Wigner helped us out. He ate some oyster downtown Princeton and got sick and was in the hospital on the campus. I went to see him at the hospital to get some help. The questions that Bohr and I were dealing with were like a chemical reaction. Uranium breaking up is like carbon monoxide breaking up into carbon and oxygen. I remembered that he [Wigner] had worked in that field with Michael Polanyi. And he helped us and, eventually, getting also ideas from discussions with other colleagues, such as Placzek and Rosenfeld, Bohr and I saw how fission works. Bohr left Princeton in April of that year and during the following months I wrote the paper and we submitted it to *Physical Review* in June. It came out in the September 1, 1939, issue; by strange coincidence the same day when Germany invaded Poland.

John C. Polanyi: Learning Directly and Indirectly

Considering that having a father of the stature of a Michael Polanyi may not only provide a great advantage, but may also be a great burden, John C. has handled it with grace. I am quoting here a few excerpts of our recorded conversation in 1995 (Fig. 14) at the University of Toronto [13]:

Let's speak about your teachers. Was your father your teacher? (Fig. 15)

JCP: Formally he was my teacher for one year. I entered Manchester University in 1946 when I was 17. He lectured to me in the first year. That was the last year he lectured in science. Then he transferred to philosophy. He also taught me a great deal in conversations despite my many absences away from home, first in boarding school and then for three years as an evacuee in Canada.

Fig. 14 John C. Polanyi in 1995 at the University of Toronto (photo by the author).

Fig. 15 John C. Polanyi and Michael Polanyi (courtesy of John C. Polanyi).

Most of what he taught me about physical chemistry I learned at one remove from him. I was a student for six years in the Department that he had shaped in Manchester. My professor Meredith Evans was one of his favorite students and my Ph.D. supervisor Ernest Warhurst was another student of his. What I learned from his students gave me a sense of scientific values—where the field was going, what were the important questions to tackle, and, to a degree, how to tackle them. Without those things I would have been lost. But it happens that I didn't get them directly from him, but from people who owed a lot to him.

When you speak about transition-state spectroscopy, it seems to me to have a close relationship to Michael Polanyi.

JCP: It does, of course, but I don't think that's the closest I got to his interests. He would have thought it far-fetched that one might get light to interact with this subpicosecond entity which is neither reagents nor products. Though it was not first done with lasers, it was the existence of lasers—of which of course, he never dreamed—that got people thinking about "seeing" the transition state.

I find myself now at the age of 66 engaged with great excitement in some novel experiments in which we are trying to look at transition states for sodium-atom reactions. It is this project that brings me eerily close to my father's interests of 1929 and subsequent years.

When I was being conceived (I was born in 1929), my father was establishing himself as the most perceptive interpreter of sodium-atom reactions, which he understood as being in a sense the simplest of all reactions. They are so simple that even a physicist can understand them. The sodium, which is easily ionised, comes up to a molecule with high electron affinity, and an electron jumps across. Then the positive sodium ion is drawn to the negative molecule. Because the electron hops a large distance, my father coined the term "harpooning" for this. It is also called this because the positively charged sodium hauls in its negative catch. This is a uniquely simple reaction. It is different from most reactions which are fascinating because they are *not* sequential events. Harpooning reactions can however be described as sequential. Step 1, reagent approaches; step 2, the harpoon jumps across; step 3, the alkali fisherman pulls in the catch. The end.

Today, in my lab, we are finding that it is possible to access the harpooning event, not by taking the reagents and bringing them together, but by forming a loose complex which is in the configuration of the transition state, that is to say, by starting in the middle of the reaction. That is what we are currently doing. And that is indeed a lineal descendent of my father's interests.

I am, however, only one of many who have seen the extraordinary possibilities offered by harpooning reactions. For example, Dudley Herschbach began his life as a dynamicist by studying that type of reaction. One should also add that my father himself was part of a continuous progression. What drew him to sodium reactions was that Fritz Haber had been studying an unexplained chemiluminescence from them. This was in Berlin and my father was in Haber's Institute as a young researcher. The history, as is usual in science, constitutes an unbroken chain.

Was he the determining influence in the direction you took in science?

JCP: He personally wasn't. But where I trained for six years was. If the question is whether he was the determining influence in my going into science, then, yes, but I should qualify that answer. At the time when I learned most from my father, in my late teenage years, his interests were even livelier in non-scientific fields than in scientific ones. He had another son, George, who went into the humanities, equally under his influence. I could just as easily have gone into economics or philosophy or theology and have ascribed it to my father's stimulus. He was, of course, delighted to see me go into science, just as he would have been delighted to see me go in many other directions.

Perhaps I am being disingenuous. I can only say that if he steered me towards science, I didn't notice.

How did he make the transition from physical chemistry to philosophy? Were you a witness to this?

JCP: We seem destined to discuss transition states. Yes, I witnessed this one directly. I got back to England right at the beginning of my fifteenth year, and until I was well into my twenties I saw a good deal of my father. That was the time, beginning in 1944, when he was making the transition. The fact that he made that transition isn't so surprising. There are a lot of scientists who have started to ruminate about how discoveries are made, how people learn anything, and the role of logic in this as compared with faith. And all this was of interest to him too.

What is striking, in my view, is the originality and impact that he had in his new field of epistemology, the theory of learning. He would have said confidently that what he did in that area was much more important than what he did in science.

I have a sense of wonder at all he did in science, and yet I believe he may easily have been right that his contribution to epistemology will turn out to be more lasting. The sales of his books and the interest in his ideas continue to be great. Eventually his name will, of course, be forgotten, but his philosophical ideas will live on as a significant contribution to the development of philosophical thought.

What is remarkable, then, is the quality of the contribution he made in his decades as a philosopher. Actually, his first book on a nonscientific theme was being conceived in the 1930s when he attacked the Russian economic system and at the same time confronted the leading British social scientists of his day, Sydney and Beatrice Webb, who'd published a learned volume explaining how the Soviet five-year-plan constituted a superb innovation and was bringing prosperity to the USSR. My father took this thesis apart in a series of essays, which became a book in 1940, that went far beyond economics and inquired why it was that British liberals, the so-called Fabians, were so careless of the freedoms that they enjoyed; the book was called *The Contempt of Freedom*. It was an influential book and a prescient one. It is forgotten today. His best known book is, instead, *Personal Knowledge*.

As with new scientific theories, my father's thinking was initially rejected by the professionals. He was not embraced by the philosophers of his day, who felt that he was an ignorant outsider. This lasted for a large part of his time in philosophy. The people who paid attention to his work were closer to theology. This was in part because the philosophy of the time was "linguistic analysis." That brand of philosophy, centered on the study of the structure of language, passed. I don't know whether my father contributed at all to its passing. It is an interesting question. Whatever the case, there followed a school of philosophy far more friendly to his ideas.

Wigner and R. A. Hodgkin penned Michael Polanyi's obituary in the *Biographical Memoirs of Fellows of the Royal Society*. It relates to the above when they noted that "The picture one gets of Michael as a parent is of a father powerfully influencing the young towards truth and towards being enterprising wherever they were, always with an emphasis on thoroughness." [14]

Researcher and Pedagogue

In 1995, I talked with Dudley R. Herschbach about Michael Polanyi, among other topics [15]. Herschbach, John C. Polanyi and Yuan T. Lee [16] jointly received the Nobel

Prize in Chemistry in 1986 "for their contributions concerning the dynamics of chemical elementary processes."

Michael Polanyi was an early influence on Dudley Herschbach. He cherished the memory of all his five meetings with Polanyi. The first time they met was in 1962 when Michael Polanyi came to Berkeley to give some lectures. Polanyi visited Herschbach's laboratory and Polanyi was telling him stories about his son John. Polanyi was surprised that John became a scientist because, he said, John in his teenage years used to bitterly criticize his father, saying that he was writing papers, all the time, that were not connected with the real world.

At the time of Michael Polanyi's visit to Berkeley, in 1962, he had already switched to philosophy. Herschbach had read some of Polanyi's books, among them *Personal Knowledge*. Herschbach thought that Polanyi's books helped making people aware of what scientists really do. Scientists get excited about their ideas and they want to see them work. Yet they have the discipline, and they must have the discipline because the scientific community as a whole insists on it, to test their ideas. These ideas do not always pass the test and the scientists have to give them up or modify their ideas. In contrast to John C. Polanyi, who came from an exceptional family of intellectual giants, Herschbach came from a family where he was the first scientist, possibly even the first university graduate. It hurt but he was not handicapped by it.

Considering John's and Dudley's backgrounds, the third co-recipient of the 1986 Nobel award, Yuan T. Lee, considered his in the middle: "Mine was somewhere in between. My father and mother were school teachers." [17] Lee met Michael Polanyi in 1968 when Lee started his career at the University of Chicago and they both were attending a conference in Toronto.

The pedagogue Michael Polanyi influenced many more outstanding scientists than those few Nobel laureates mentioned above so far. Wigner and Hodgkin's obituary quoted W. Mansfield Cooper, Vice-Chancellor of Manchester University that "There is no doubt that the good student got much from him, but the remarkable thing is that the poor ones were happily carried along." Wigner and Hodgkin attributed this "to Polanyi's systematic coverage of detail, through handouts and guided reading, which he combined with profound exposés of major problematic themes in lectures."[14, p. 424]

One of Polanyi's disciples, Erich Schmid, who later served as president of the Austrian Academy of Sciences, had this to say about Polanyi's pedagogical qualities: "Just as he was for his collaborators the paradigm of the scientist constantly seeking for fundamental explanation, so, along with his charming wife, he also taught them to bear with good humour, or even to overlook altogether, the difficulties and limitations of the time." [14, p. 420]

Ilya Prigogine (1917–2002) received the Nobel Prize in Chemistry in 1977 "for his contributions to non-equilibrium thermodynamics, particularly the theory of dissipative structures." In 1998, he remembered Michael Polanyi with the following words: "I admired him very much. He was interested in my early work in thermodynamics and invited me to Manchester when he was still Professor of Physical Chemistry. It was some time between 1945 and 1948. It was an exceptional period in Manchester. In addition to Polanyi, there was also Evans and Turing and others." [18]

George Porter (Lord Porter, 1918–2002), shared the Nobel Prize in Chemistry in 1967 jointly with Manfred Eigen and R.G.W. Norrish "for their studies of extremely fast chemical reactions, effected by disturbing the equilibrium by means of very short pulses of energy." Porter considered himself a scientific grandson of Polanyi's [19]:

> One of the early workers who advanced this concept [uncovering the mechanism of chemical reactions] originally, M. G. Evans, was one of my teachers at Leeds who greatly inspired me. He himself studied under Michael Polanyi at the University of Manchester. I met Michael Polanyi in my first year as an undergraduate, at the age of 17. I was given the daunting task, as the secretary of the student chemical society, of proposing a vote of thanks to Michael Polanyi for his lecture. I didn't really understand the lecture very well but I managed somehow to say what a marvelous lecture it was, and that even I could understand some of it. I met him many years later when his son, John took me along to dine with him at the Athenaeum Club after a Faraday Society meeting. By this time, he had become a social scientist.

The Loneliness of the Discoverer

Making a discovery implies that the discoverer, at least for some time, will be alone as he or she knows something that nobody else does. This loneliness may be a heavy burden and it may last a short or a long while [20]. Making premature discoveries certainly prolongs this loneliness. Michael Polanyi must have experienced this loneliness on more than one occasion. In his book, *Paradoxes of Progress*, the late molecular biologist Gunther Stent used the story of Polanyi's discovery in adsorption to illuminate some points about premature discoveries along other examples, such as Gregor Mendel's discoveries related to genetics and Oswald T. Avery's discovery that DNA is the substance of heredity [21]:

> Cases of delayed appreciation of a discovery exist also in the physical sciences. One example (as well as an explanation of its circumstances in terms of the concept to which I refer here as prematurity) has been provided by Michael Polanyi on the basis of his own experience. In the years 1914–1916 Polanyi published a theory of the adsorption of gases on solids which assumed that the force attracting a gas molecule to a solid surface depends only on the position of the molecule, and not on the presence of other molecules, in the force field. In spite of the fact that Polanyi was able to provide strong experimental evidence in favor of his theory, it was generally rejected. Not only was the theory rejected, it was also considered so ridiculous by the leading authorities of the time that Polanyi believes continued defense of his theory would have ended his professional career if he had not managed to publish work on more palatable ideas. The reason for the general rejection of Polanyi's adsorption theory was that

at the very time he put it forward the role of electrical forces in the architecture of matter had just been discovered. Hence there seemed to be no doubt that the adsorption of gases must also involve an electrical attraction between the gas molecules and the solid surface. That point of view, however, was irreconcilable with Polanyi's basic assumption of the mutual independence of individual gas molecules in the adsorption process. It was only in the 1930s, after a new theory of cohesive molecular forces based on quantum-mechanical resonance rather than on electrostatic attraction had been developed, that it became conceivable that gas molecules could behave in the way Polanyi's experiments indicated they were actually behaving. Meanwhile Polanyi's theory had been consigned so authoritatively to the ashcan of crackpot ideas that it was rediscovered only in the 1950s.

Pioneering in X-ray Crystallography

X-ray crystallography has been a success story in science for over a hundred years. The technique has kept renewing itself and although for many tasks more powerful approaches have emerged, X-ray crystallography has kept its position. Polanyi would welcome and enjoy the development of the past few decades whereas crystallography has greatly expanded its scope under the name of generalized crystallography [22]. Polanyi placed the discovery of X-ray crystallography into an intriguing context in his *Personal Knowledge* [5, p. 277]:

> ...The power to expand hitherto accepted beliefs far beyond the scope of hitherto explored implications is itself a pre-eminent force of change in science. It is this kind of force which sent Columbus in search of the Indies across the Atlantic. His genius lay in taking it literally and as a guide to practical action that the earth was round, which his contemporaries held vaguely and as a mere matter for speculation. The ideas which Newton elaborated in his *Principia* were also widely current in his time; his work did not shock any strong beliefs held by scientists, at any rate in his own country. But again, his genius was manifested in his power of casting these vaguely held beliefs into a concrete and binding form. One of the greatest and most surprising discoveries of our own age, that of the diffraction of X-rays by crystals (in 1912) was made by a mathematician, Max von Laue, by the sheer power of believing more concretely than anyone else in the accepted theory of crystals and X-rays. These advances were no less bold and hazardous than were the innovations of Copernicus, Planck or Einstein.

Robert Olby, the renowned chronicler of the story of the double-helix discovery, has pointed out Polanyi's merits in the X-ray diffraction investigation of fibers. Polanyi was rather ignorant about X-ray crystallography when he joined Fritz Haber's Institute for Physical Chemistry and Electrochemistry in Berlin, but soon enough he was already working on and solving fundamental problems in this field [23].

Incidentally, the Kaiser Wilhelm Society early on realized the importance of fiber science and established a research institute for fiber chemistry (Faserstoffchemie) and Polanyi continued his research there for a while. He had ideal conditions for his work. In his words, his studies were assisted "...with every facility for experimental work, most precious of which were funds for employing assistants and financing research students. In this I was incredibly lucky. I was joined by Herman Mark, Erich Schmid, Karl Weissenberg, all three from Vienna, by Erwin von Gomperz and some others..." [24]

Herbert Morawetz in Herman F. Mark's obituary referred to Polanyi's achievements [25]:

> Polanyi found that the X-ray diffraction from cellulose fibers indicated the presence of crystallites oriented in the direction of the fiber axis and that an analogous crystal orientation existed in metal wires. A full structure analysis of cellulose seemed beyond the experimental possibilities of the time, but Mark and Polanyi noted that the increase in the modulus of cellulose fibers on stretching seemed similar to the reinforcement of metal wires during cold-drawing. They embarked, therefore, on a detailed analysis of the changes accompanying the cold-drawing of a zinc wire.

Polanyi's discoveries gain special importance in the light of the state of the related chemistry at the time. In the 1920s, it was still debated whether biological macromolecules existed or the relevant systems consisted of colloidal components. Many held the view that macromolecules did not exist and that molecules could not be larger than the elementary cell in the crystals. Polanyi was willing to brave the hostile reactions to his views that came as conclusion from his X-ray crystallography studies. It was a case in point what happened when he gave an account at institute director Fritz Haber's seminar. This is how Polanyi communicated the event with enviable self-irony [23, p. 30] and [24, p. 631]:

> The assertion that the elementary cell of cellulose contained only four hexoses appeared scandalous, the more so, since I said that it was compatible both with an infinitely large molecular weight or an absurdly small one. I was gleefully witnessing the chemists at cross-purposes with a conceptual reform when I should have been better occupied in definitely establishing the chain structure as the only one compatible with the known chemical and physical properties of cellulose. I failed to see the importance of the problem.

To Conclude

Michael Polanyi (Figs. 16, 17 and 18) did not continue his studies in crystallography after a while and from his perspective at the time, they may have not seemed sufficiently promising. In the 1920s, crystallography was immersing itself further deep in the study of fully crystalline systems. The study of less ordered structures appeared esoteric and when the British J. Desmond Bernal and William Astbury divided the field between themselves, Bernal thought that by choosing the crystalline ones he had the best of it. The development of science proved him wrong. Bernal confined the investigation of nucleic acids to their crystalline

Fig. 16 Michael Polanyi addressing a meeting on cultural freedom in 1956. On the right, the French philosopher Raymond Aron (1905–1983). Courtesy of John C. Polanyi.

Fig. 17 Memorial plaque honoring the Polanyi Family and especially Karl Polanyi and Michael Polanyi at 2 Andrássy Avenue in Budapest in the 1980s with the Egyptian-American chemistry Nobel laureate (1999) Ahmed Zewail. Courtesy of Ahmed Zewail.

components, the nucleosides. His Norwegian associate, Sven Furberg determined the structure of cytidine, which was important but served only as one of several components from which Crick and Watson constructed the double-helix structure of DNA. Bernal later wrote, "A strategic mistake may be as bad as a factual error," [26] referring to his gentleman's agreement with Astbury. Had Bernal not honored this agreement, the story of the discovery of the double helix might have turned out differently.

Bernal's words, "A strategic mistake may be as bad as a factual error," reverberate in my ears when I think about Polanyi's exceptional achievements in science. I cannot help wondering whether Bernal's self-critical observation might not have been applicable to some of Polanyi's decisions in taking turns and choosing directions when his road in science appeared bifurcating, or multi-furcating, in front of him. He, who was so good in giving advice to others, might have found himself short of good advice himself.

In some ways, although Michael Polanyi never received a Nobel Prize, he appeared in full force—alas, only symbolically—in Stockholm twice over the years. In 1963, Wigner remembered him as his mentor and quoted him in his precious two-minute speech about what science really is (see

above). In 1986, Michael Polanyi appeared there through his son, John C. Polanyi, and through the science of the three awardees in chemistry that could be considered a continuation of his own work. John C. Polanyi's evaluation of his father's works is engagingly realistic, yet gentle.

According to John, Michael Polanyi learned medicine and became a professional in it, but did not care for it. He stayed an "amateur" in everything else, where he became successful, such as chemistry, physics, economics, philosophy, and even a few other areas. He never had a mentor in any of these fields and he was the sole author of his first 15 papers, with only one exception. As Polanyi's career was at the very beginning, the mathematician George Pólya remarked: "Michael walks alone; he will need a strong voice to make himself heard." [27]

Further, according to John, Michael Polanyi stayed an outsider and chose the topics of his inquiry with great freedom. He started doing research in thermodynamics and in adsorption, and when his premature discoveries did not gain acceptance, he moved on. He was successful in crystallography as far as he went. Finally, he arrived at the ultimate

Fig. 18 The latest plaque (2012 photo by the author). Anti-Semitic vandals repeatedly destroyed the Polanyi memorial plaque.

puzzle in chemistry of what makes molecules stable and what makes and how do chemical reactions happen? He succeeded in providing an insight that did not merely prove correct, but turned out also excitingly attractive. As an irony of Polanyi's fate, his most fruitful period of scientific creativity was the years of his forced transition from Nazi Germany to democratic Britain: 32 papers appeared indicating both Berlin *and* Manchester as the venues of his work.

Role Model?

We cannot recommend to anyone to follow Michael Polanyi's footsteps, because one would need too large shoes to fit for doing so. But he has served and will be serving as inspiration in doing research in science; in maintaining interest in more than one culture; and in watching out for our fellow human beings. We have no doubt that Polanyi's thoughts expressed in his *Personal Knowledge* and elsewhere will be remembered "long after his contributions to science will have joined the melting pot of anonymity." [28]

Acknowledgments I am grateful to the late Harold P. Hanson (1921–2016), then chair of the Department of Physics of the University of Texas at Austin, for arranging a position of visiting research associate at the physics department for me and for bringing me together with Michael Polanyi in 1969.

I thank the late former Commissioner of the US Atomic Energy Commission Clarence Larson and his wife, Jane Larson, for donating to us their unique collection of video recordings with famous scientists and technologists. I thank Balazs Hargittai for years of cooperation in transcribing and editing the Larson tapes.

I appreciate the kindness of John C. Polanyi on many occasions to help me understand his science and the science and philosophy of his father's and for the images he has lent me for this review and for my various other publications.

I am grateful to Andrew Szanton for giving me permission to quote extensively from the book *The Recollections of Eugene P. Wigner as Told to Andrew Szanton* and for his comments on the manuscript.

I thank Éva Gábor and László Füstöss for their dedicated assistance and for the images they have lent me.

I have benefited from Bob Weintraub and Irwin Weintraub's suggestions in improving my manuscript.

I first summarized some of the information in this account for a talk at a symposium organized by the now defunct Michael Polanyi Society of Liberal Philosophy in 2003 in Budapest [Istvan Hargittai, "Polányi Mihálytól tanultak …" (in Hungarian, "They learned from Michael Polanyi …") *Polanyiana* 2003, Issue 1–2, pp. 21–39].

I thank Bretislav Friedrich for his kind invitation extended to me to participate in the Polanyi symposium on October 5, 2016, in Berlin.

References

1. Pais A (2000) *The Genius of Science: A Portrait Gallery*. Oxford University Press, Oxford, UK, p 334
2. Mehra J, Ed (2001) *The Collected Works of Eugene P. Wigner, Volume Vii: historical and Biographical Reflections and Syntheses*. Springer-Verlag Berlin, Heidelberg, New York, p 313
3. Polanyi M (1929) *Pesti Futár*, pp 37–38 (English translation from the Hungarian by IH). Ferenc Tangl (1866–1917) was a professor of medicine at Budapest University. Ignác Pfeifer (1867–1941) was a chemical engineer and as professor was charge of the chemistry institute of the Budapest Technical University. Pfeifer was dismissed under Horthy and moved to private industry, and in the late 1930s, he lost his employment due to anti-Jewish legislation. István Széchenyi (1791–1860) was a Hungarian aristocrat who had innovative ideas to move the feudalistic Hungary, under the Habsburgs' rule, onto a path of progress. He and a group of fellow aristocrats initiated the Hungarian Academy of Sciences and Széchenyi offered one year income of his estate in support of this initiative. Ferenc Kazinczy (1759–1831) was a writer and poet who is most remembered by his activities in the reform of the Hungarian language. The Hungarian language replaced Latin as the official language of Hungary in 1844.
4. The last paragraph demonstrates Polanyi's optimism that Hungary will one day change and that the democratic way of life in Germany will one day become the way of life in Hungary as well. We now know that Polanyi's anticipation was tragically wrong and instead of what he had hoped for, in a few years' time, Hitler and the Nazis took over Germany.
5. Polanyi M (1958) *Personal Knowledge Towards a Post-Critical Philosophy*. The University of Chicago Press, Chicago
6. Hargittai I (2004) *Our Lives: Encounters of a Scientist*. Akadémiai Kiadó, Budapest, pp 134–137
7. Hargittai I and Hargittai M (2006) *Candid Science VI: More Conversations with Famous Scientists*. Imperial College Press, London, "Nicholas Kurti," pp 554–565; actual quote, pp 557–558
8. Szanton A (1992) *The Recollections of Eugene P. Wigner as Told to Andrew Szanton*. Basic Books, Cambridge, MA, p 81.
9. Wigner EP (1963) *Symmetries and Reflections: Scientific Essays*. Indiana University Press, Bloomington, IN, "City Hall Speech—Stockholm, 1963," pp 262–263

10. Calvin M (1999), "The path to carbon in photosynthesis." In *Nobel Lectures Chemistry 1942–1962*. World Scientific, Singapore, pp 618⁻644; actual quote, p 618

11. Hargittai B and Hargittai I (2005) *Candid Science V: Conversations with Famous Scientists*. Imperial College Press, London, "Melvin Calvin," pp 378⁻389; actual quote, pp 382⁻384. These are from the transcripts of a 1984 conversation of Melvin Calvin with Clarence Larson

12. Hargittai M and Hargittai I (2004) *Candid Science IV: Conversations with Famous Physicists*. Imperial College Press, London, "John A. Wheeler," pp 424⁻439; actual quote; p 435

13. Hargittai I (2003) *Candid Science III: More Conversations with Famous Chemists*. Magdolna Hargittai, Ed, Imperial College Press, London, "John C. Polanyi," pp 378–391; actual quotes, pp 385–389

14. Wigner EP and Hodgkin RA (1977) "Michael Polanyi 12 March 1891 – 22 February 1976." *Biographical Memoirs of Fellows of the Royal Society*, pp 413–448; actual quote, p 423

15. Hargittai I (2003) *Candid Science III: More Conversations with Famous Chemists*. Magdolna Hargittai, Ed, Imperial College Press, London, "Dudley R. Herschbach," pp 392–399

16. Hargittai B and Hargittai I (2016) "Lee, Yuan Tseh." In *Berkshire Dictionary of Chinese Biography*, Volume 4. Berkshire Publishing Group, pp 268–273

17. Hargittai I and Hargittai M (2006) *Candid Science VI: More Conversations with Famous Scientists*. Imperial College Press, London, "Yuan T. Lee," pp 438–457; actual quote, p 440

18. Hargittai I (2003) *Candid Science III: More Conversations with Famous Chemists*. Magdolna Hargittai, Ed, Imperial College Press, London, "Ilya Prigogine," pp 422–431; actual quote, p 431

19. Hargittai I (2000) *Candid Science: Conversations with Famous Chemists*. Magdolna Hargittai, Ed, Imperial College Press, London, "George Porter" pp 476–487; actual quote, p 486

20. Hargittai I (2007), "The loneliness of the scientific discoverer." Struct Chem 18:1–3

21. Stent GS (1978) *Paradoxes of Progress*. W. H. Freeman and Co, San Francisco, pp 99–100

22. See, eg, Hargittai I (2017) "Generalizing crystallography: a tribute to Alan L. Mackay at 90." Struct Chem 28: in press, DOI https://doi.org/10.1007/s11224-016-0766-1

23. Olby R (1994) *The Path to the Double Helix: The Discovery of DNA*. Dover Publications, New York (original publication, Seattle: University of Washington Press, 1974), pp 28–30

24. Polanyi M (1962) "My Time with X-rays and Crystals." In Ewald P, Ed, *Fifty Years of X-ray Crystallography*. International Union of Crystallography, Utrecht, pp 629–636

25. Morawetz H (1994) "Herman Francis Mark." *Biographical Memoirs*, Vol. 68. National Academy of Sciences, Washington, DC

26. See, eg, Hargittai I (2007) *The DNA Doctor: Candid Conversations with James D. Watson*. World Scientific, Singapore, Appendix 4, Bernal JD, "The Material Theory of Life," pp 206–210; actual quotation, p 208; published originally as a review of J. D. Watson's *The Double Helix* in a little known periodical, *Labour Monthly* July 1968, pp 323–326

27. Polanyi JC (2003) "Michael Polanyi, the Scientist." *Polanyiana* Issue 1–2, pp 117–121

28. Paraphrased of what Polanyi wrote to Leo Szilard referring to Szilard's booklet *The Voice of the Dolphins*, "Maybe ... you will be remembered by these light-hearted fancies long after your contributions to science will have joined the melting pot of anonymity." See, Bernstein BJ (1987), "Introduction." In Hawkins HS, Gerb GA, and Weiss Szilard G, Eds, *Toward a Livable World: Leo Szilard and the Crusade for Nuclear Arms Control*. MIT Press, Cambridge, MA, p lvii

Glenn T. Seaborg at the time of the spring 1995 meeting of the American Chemical Society in Anaheim (photograph by Istvan Hargittai). Note that his tie displays the periodic table of the elements. The snapshot in the following article also shows him at the same meeting donning a different tie, one demonstrating atomic orbitals of electrons.

Glenn T. Seaborg; Discoveries; and the Capital of Knowledge[a]

Balazs Hargittai and István Hargittai

"... knowledge capital—a product of basic research—... might also allow us to compensate somewhat for declining physical capital and higher cost resources."

Glenn T. Seaborg [1]

Abstract

Ten years from Glenn T. Seaborg's death we remember his achievements; his teaching about the importance of basic research is as timely as ever.

Glenn T. Seaborg (1912–1999) with ion-exchanger column of actinide elements in 1950 (courtesy of Lawrence Berkeley National Laboratory).

[a]Originally published in *Structural Chemistry* 2009, 20:355–359.

B. Hargittai (✉)
Saint Francis University, Loretto, PA, USA
e-mail: bhargittai@francis.edu

I. Hargittai
Department of Inorganic and Analytical Chemistry and Materials Structure and Modeling Research Group of the Hungarian Academy of Sciences, Budapest University of Technology and Economics, Budapest, Hungary
e-mail: istvan.hargittai@gmail.com

A giant of chemistry departed a decade ago the importance of whose oeuvre extends much beyond anniversaries; yet 10 years from his passing away provides a nice opportunity to make a special remembrance of him. He was born in a little mining town Ishpeming in Northern Michigan 1912, where his father was a machinist, which Seaborg thought was as close to science as somebody could be in that environment. His entire family was Swedish and Swedish was the first language Seaborg learned to speak. In 1951, he started his Nobel address in his mother tongue. He shared the chemistry award with Edwin M. McMillan "for their discoveries in the chemistry of the transuranium elements." It was a long way from Ishpeming to Stockholm.

When Seaborg was 10 years old, the family moved to California, where he graduated from high school in Los Angeles in 1929. He became a student of the University of California at Los Angeles and received his bachelor's degree in Chemistry in 1934. For graduate studies, he moved to Berkeley and took his Ph.D. degree in Chemistry in 1937. He wrote his thesis about the inelastic scattering of neutrons. Following the receipt of his doctorate, Seaborg served as Gilbert N. Lewis's personal assistant at Berkeley for 2 years. Seaborg wrote warmly about this unique experience [2]. When he was asked to identify the greatest scientists he met during his long career, he named Lewis and Enrico Fermi.

Seaborg worked with an unusually large number of people on his many discoveries. He contributed to the discovery of ten new elements and over a hundred new isotopes of elements. Much of his career was at the University of California at Berkeley where he became instructor of chemistry in 1939, assistant professor in 1941, and professor in 1945.

Edwin McMillan led a group, which discovered element 93 by making uranium capture a neutron and, following beta-emission (the ejection of an electron from the nucleus) the element of atomic number 93 was formed. They called it neptunium, Np, after the planet Neptune orbiting next, outwards, after Uranus. After McMillan's departure for other defense-related research, Seaborg and his colleagues took over the project. They detected the next transuranium element, formed by another beta-emission; it had atomic number 94. It was given the name plutonium, Pu, after Pluto, orbiting next outside Neptune, which at that time was considered to be a planet though today it no longer is. The nuclear reactions are depicted here in short-hand notation:

$$^{92}U\text{-}238 \ + \ n \rightarrow \ ^{92}U\text{-}239$$
$$^{92}U\text{-}239 \rightarrow \ ^{93}Np\text{-}239 \ + \ \beta$$
$$^{93}Np\text{-}239 \rightarrow \ ^{94}Pu\text{-}239 \ + \ \beta$$

In 1941, Seaborg, together with Emilio Segrè and Joseph W. Kennedy, showed that plutonium was fissionable and it became the fuel of the second atomic bomb exploded over Nagasaki in 1945. In 1942, Seaborg joined the Manhattan Project and became a group leader at the Metallurgical Laboratory at the University of Chicago. Here it was that the non-fissionable uranium-238 isotope was converted into plutonium-239. The procedure was further developed at the Clinton Engineer Work in Oak Ridge, Tennessee, and served as the basis for the breeder reactors at the Hanford Engineer Works in Washington.

During World War II, there were frenetic activities in the research of the properties of newly discovered trans-uranium elements. Manuscripts describing the results were duly compiled and submitted to journals, but were voluntarily withheld from publication until the end of the war. Thus, for example the pivotal paper "Properties of 94(239)" was received by The Physical Review on May 29, 1941, but appeared only in the combined numbers 7 and 8, Volume 70, in October 1946.

After World War II, Seaborg returned to Berkeley, but remained also part of national politics through his much appreciated advising from President Truman to President Reagan. He was a member of the General Advisory Committee (GAC) at the time of the great debate about the issue whether the United States should embark on an accelerated program of developing the hydrogen bomb.

The GAC was an advisory body consisting of important scientists, which augmented the Atomic Energy Commission created after the war for directing American policy in matters of nuclear energy. The GAC held long sessions at the end of October 1949 and the outcome of the GAC meeting concerning the development of the hydrogen bomb could not have been easily predicted. On the one hand, there was the Soviet menace whereas on the other hand, the hydrogen

bomb, utilizing thermonuclear reaction of fusion of light nuclei was promised to be a thousand times more powerful than the atomic bombs. Gradually, however, the scale during the sessions was increasingly shifting toward opposing a crash program to develop the thermonuclear bomb. The only dissenting voice was Glenn T. Seaborg's, who was the only member absent from the meeting, but who had sent a letter to the chairman of the GAC.

There were two crucial sentences in Seaborg's letter that showed unambiguously his stand in the matter of the discussion. Both sentences were formulated with utmost care and one can almost sense the tormenting hesitation of their author: "Although I deplore the prospects of our country putting a tremendous effort into this [the thermonuclear bomb], I must confess that I have been unable to come to the conclusion that we should not." Then, a little later in the letter, "My present feeling could perhaps be best summarized by saying that I would have to hear some good arguments before I could take on sufficient courage to recommend not going toward such a program."

Concerning his dissent from the rest of the GAC members in his letter to Oppenheimer, Seaborg could have raised his objections upon his return, during November and December, but he did not. At the time, Seaborg was a junior member of the GAC, who, eventually, would develop into a seasoned diplomat in addition to being a world-renowned scientist. Apparently he preferred to keep quiet for the duration of this debate. As is well known President Truman decided to have the hydrogen bomb developed. For a long time it was not known, but we know it today, that at the time of the American debate, the Soviet Union had already been deeply involved in developing its thermonuclear weapons.

Seaborg served as chairman of the U.S. Atomic Energy Commission for longer than anybody, between 1961 and 1971. During this decade he spent a lot of time in Washington, DC, whereas at other times he continued his research and educational activities at Berkeley. Considering Seaborg's principal role in the discovery of plutonium and in the determination of its properties, an embarrassing scene played out at a Senate hearing in 1970. It demonstrated the ignorance of the chairing senator when he asked Seaborg derisively, "What do you know about plutonium?" [3]. However, such episodes were rare and Seaborg enjoyed being involved in high politics for decades. Seaborg served ten American presidents. He started keeping a journal at the age of 14, which was at the time of the Coolidge administration and published his documents and lessons from his encounters in 1998 [3].

Seaborg received many awards and distinctions, but none gave him as much joy as having an element named after him. In 1995, he was greatly disappointed when it seemed that this would not happen on account of his being alive and the appropriate organizations did not want to name an element after a living person. This followed a long story of sorting out the priorities in the discovery of element 106, because the discoverers have the right to propose a name for a new element.

The original discovery happened in 1974, and in 1993, the eight discoverers—members of the Lawrence Berkeley, including Seaborg, and the Lawrence Livermore Laboratories—were asked to suggest a name for the element. The votes diverged greatly; suggestions included Luis Alvarez, Frédéric Joliot, Isaac Newton, Thomas Edison, Leonardo da Vinci, Christopher Columbus, Ferdinand Magellan, Ulysses, George Washington, Peter Kapitza, Andrei Sakharov, and the country Finland. The group (without Seaborg) soon came together in a unified suggestion to name the element seaborgium after Seaborg. The final decision was made in Geneva on August 30, 1997 and seaborgium was adopted for element 106. Alas, Seaborg could not enjoy this new fame for long; 1 year later he suffered a stroke and died in half a year.

Of course, Seaborg's name is commemorated not only by element 106, but also by his many other discoveries. He gave the periodic table of the elements its final form in that he designated the actinides their proper place. The actinides with atomic numbers 89, 90, 91, etc., are characterized by their 5f electron shell being gradually filled. The series starts with actinium just as the lanthanides start with lanthanum. The actinides have similar chemical properties; absorption spectra in aqueous solution and crystals; crystallographic characteristics; magnetic susceptibilities; and spectroscopic data [4]. When Seaborg came to his new theory of the actinide series, he shared it with some colleagues before he published it. People warned him that publishing his theory might ruin his reputation. This sort of caveat is common when discoverers come to revolutionary ideas. Seaborg, however, felt very sure of the correctness of his theory; besides, he did not think he had yet gained much reputation yet to ruin. He published his theory and gave a new appearance to the periodic table of the elements [5].

Seaborg's fascination with the new elements was shared by others. An article in the magazine Discover in 1998 compiled an address of an imaginary letter to Seaborg using only element names in the following way:

seaborgium (addressee: Seaborg)
lawrencium (Lawrence Laboratory)
berkelium (Berkeley)
californium (California)
americum (United States)

Seaborg himself gave the answer to the question of why it was so important for him to have an element named after him. He said that even one thousand years from now it will still be seaborgium whereas by then what he did would probably have long before disappeared in oblivion. He would have gladly traded away his Nobel Prize for having the element

106 named after him, had such an exchange been possible. In the end all turned out to be all right. Incidentally, when Seaborg noted that sooner or later his works will disappear from collective memory, he did his best to slow down such a process. He himself edited the publication of his selected papers and furnished the collection with his commentaries. Characteristically, the volume was titled Modern Alchemy referring to his feat of turning even ordinary elements into gold, alas, the economy of the process was not viable [6].

Glenn T. Seaborg and István Hargittai at the Springer-Verlag booth of the American Chemical Society spring 1995 meeting in Anaheim, California (by an unknown photographer).

Part of Seaborg's legacy is his teaching and this is why, in conclusion, we quote Seaborg's thoughts for the future, which he sent one of us shortly before he passed away [1]:

Some Thoughts for the Future An important factor in the future, transcending the science of chemistry, will be the new public attitudes toward basic science and science in general—that is, the growing attitude toward ethical and human value considerations. The focus of this concern often is not on the question of whether the work is worth doing but instead on whether its potential harmful impact may outweigh any good it could do —that is, whether the research or project should be initiated at all. This attitude is affecting work on energy resources and technologies, biological research, aircraft development, and advances in the social sciences and education. This is going to have an increasing effect on the support and conduct of science, and I think most scientists are recognizing this.

As in the other cases of new influence, it is going to have its good and bad effects. Essentially, it is vital that science does serve the highest interest of society and contribute to the fulfillment of human values. And I believe that the science community for the most part is acting very responsibly and responsively in this direction. In many areas of research, such as genetic experimentation, atmospheric work, and the effects of chemicals on human health and the environment, it has taken the lead in placing human concerns above all.

But it should be realized that while there are certain values and ethical codes of a universal nature, there are also values that are more closely associated with the tastes, likes and dislikes, habits, and culturally induced beliefs of various individuals and groups attuned to certain so-called lifestyles. In a democratic society—and particularly one of growing advocacy and activism—there are bound to be many conflicts over these. And science and technology, with their increasing influence on life in general, certainly will be caught up in many of these. If this is the case, it may be essential that we find a way to establish some broad codes of conduct and values by which we can use science and technology to maximize human benefits within a framework of some type of consensus value scale. It seems to me that we must do this in order to avoid being paralyzed by a kind of case-by-case value judgment of all that we do. This does not mean that technology assessments and risk/benefit studies of individual concepts should not be conducted. Nor does it mean that science should not maintain a most profound sense of responsibility toward safeguarding society from possible errors on its part or misapplications of its work. It does mean, however, that we must find a way to avoid having a "tyranny of parochial interests" when it comes to the possibility of advancing the general good through scientific progress.

Perhaps I can summarize by suggesting that future directions of chemistry, and science and technology in general, may be influenced by two broad goals: more fully establishing the boundaries—physical, environmental and social—in which we can operate; and providing the knowledge capital that will allow us to operate within them. That knowledge capital—a product of basic research—upon which we have drawn so heavily in the recent past and which we must replenish with new ideas might also allow us to compensate somewhat for declining physical capital and higher cost resources.

Finally, a few general thoughts. Our success in chemistry, and science in general, over the past century, and especially the last few decades, has brought us to a high level of material affluence, but this success also has fostered many new problems for the world. It also has given many people the notion that science should move us toward a utopian, problemless, riskless society. But this is a false notion. We live and always will live in a dynamic situation, amid problems whose solutions will breed other kinds of problems, and in a society where the leaps of progress will be proportionate to the risks taken. Even within the bounds of a "steady-state society," a "no-growth society," or any other scheme of population-resource-energy equilibrium we might achieve, there always will be change and creative growth that will challenge the human intellect. There always will be dangers, risks, and increasing responsibilities that will drive us toward a new level of excellence in all we do or try to achieve. This is the process of human evolution at work, a process that started with man's ascendancy and will continue for some time.

Acknowledgment Our research is being supported in part by the Hungarian Scientific Research Foundation (OTKA No. T046183).

References

1. Hargittai I (2003) Candid science III: more conversations with famous chemists "Glenn T. Seaborg". Imperial College Press, London, pp 2–17
2. Seaborg GT (1995) Chem Intelligencer 1(3):27–37
3. Seaborg GT (1998) A chemist in the White House from the Manhattan Project to the end of the Cold War. American Chemical Society, Washington, DC
4. Seaborg GT (1949) Nucleonics, November, 149–169
5. Clark DL, Hobart DE (2000) Los Alamos Science 26:56–61
6. Seaborg GT (ed) (1994) Modern alchemy: selected papers of Glenn T. Seaborg. World Scientific, Singapore

Paul Lauterbur and Peter Mansfield shared the Nobel Prize in Physiology and Medicine for 2003 "for their discoveries concerning magnetic resonance imaging." Istvan visited both shortly following the award. This image shows Istvan and Peter Mansfield in January 2005 at the Sir Peter Mansfield Magnetic Resonance Centre, University of Nottingham, England (courtesy of Magdolna Hargittai).

Paul C. Lauterbur (1929–2007)[a]

István Hargittai

Paul C. Lauterbaur in his home in Urbana, Illinois, 2004 (photograph by I. Hargittai).

Paul C. Lauterbur, [1] who is best known as one of the principal developers of magnetic resonance imaging, had an unusual career in academia. He earned a bachelor's degree in chemistry at the Case Institute of Technology in Cleveland in 1951, but did not continue his studies right away, because he was so much interested in research that he preferred starting it rather than attending courses. Only later when lacking a PhD degree appeared to be an impediment to his advance, did he decide to go for his doctorate, which he earned from the University of Pittsburgh in 1962.

Lauterbur came across NMR spectroscopy in the early 1950s when he was working at the Mellon Institute and Varian Company gave a demonstration of the possibilities of their NMR machine; he also attended a talk by H. Gutowsky about his NMR studies of substituted methanes.

At that point Lauterbur suggested a joint project to Gutowsky in which Lauterbur would make silane derivatives and their NMR studies might yield information about the relationship between carbon and silicon. It was a good example of Lauterbur's drive and interest and Gutowsky agreed to do it, but the project did not happen because Lauterbur was drafted into the Army. There he was first involved with chemical warfare, but also had the possibility to work with NMR, and published four papers during his stint in the military. After returning to civilian life, he continued to work at the Mellon Institute and continued his NMR studies. Having read a paper by Raymond V. Damadian about the different responses of healthy and cancerous tissues, Lauterbur's attention turned to the study of biological objects by NMR. One of his motivations was his concern for sacrificing animals in operating on them for research purposes. This was typical of Lauterbur's approach to life; rather than protest animal use, he decided to do something to eliminate the need for it. Soon he came to some ideas about how the problem could be attacked by imaging and he wrote them down, and had them witnessed for potential patent purposes. This is interesting because when he worked out the technique and he had truly something patentable, eventhough he did not succeed in doing so. His private attempts failed and when he turned to his university employer, it ruled that it would cost more to apply for a patent than they could ever conceivably be making from it. They proved to be wrong by several orders of magnitude, but Lauterbur "made virtue out of necessity" and opened his laboratory to anyone who wanted to learn about his discoveries. Lauterbur realized later that his decision probably contributed to academic research in the field, but having a patent might have accelerated industrial development. Some of the companies have no interest in developing products that are not being covered by an enforceable patent that provides the advantage of keeping competition away.

[a]Originally published in *Structural Chemistry* 2007, 18:529–530.

I. Hargittai (✉)
Department of Inorganic and Analytical Chemistry, University of Technology and Economics, Budapest, Hungary
e-mail: istvan.hargittai@gmail.com

Lauterbur ideas and experiments were pioneering and this made them difficult to be accepted by others. When he submitted his seminal paper to *Nature*, the magazine rejected it. Lauterbur did not give up and argued in a long letter to the periodical and offered to expand on his ideas. The decisive comment by a referee tipped the scale in Lauterbur's favor, according to which his suggestions seemed crazy, but he had a good reputation having never done anything crazy before; so the revised manuscript was printed [2]. A similar story happened with one of his research proposals to the National Institutes of Health. The first reviews were negative because they found his proposal crazy; when they took a second look, though they still found it crazy, they could not find anything wrong with it, so it got funded.

Lauterbur received his long deserved Nobel Prize jointly with Peter Mansfield who also discovered MRI and who did patent his inventions [3]. There was some controversy about the award because Damadian took out full page advertisements in the most famous American newspapers protesting the Nobel decision, which did not include him. There is a long and interesting story behind this [4]. Although the Nobel Prize came rather late, there was plenty of other recognition and he was a star scientist for many years, but not in his behavior; he remained unassuming, modest, and somewhat withdrawn. After the Nobel Prize, when there might have been so many temptations to lure him away from research, he continued, focusing mainly on his pet project that had interested him from his youth, the origin of life. Paul C. Lauterbur's legacy in addition to his family and scientific oeuvre is in the millions of lives he helped preserve and make easier with his MRI.

References

1. Hargittai B, Hargittai I (2006) Paul C. Lauterbur. In: Candid science V: conversations with famous scientists. Imperial College Press, London, pp 454–479
2. Lauterbur PC (1973) Nature 242:190
3. Hargittai I, Hargittai M (2006) Peter Mansfield. In: Candid science VI: more conversations with famous scientists. Imperial College Press, London, pp 216–237
4. Hollis DP (1987) Abusing cancer science: the truth about NMR and cancer. The Strawberry Fields Press, Chehalis, Washington; see in particular, pp 174–175

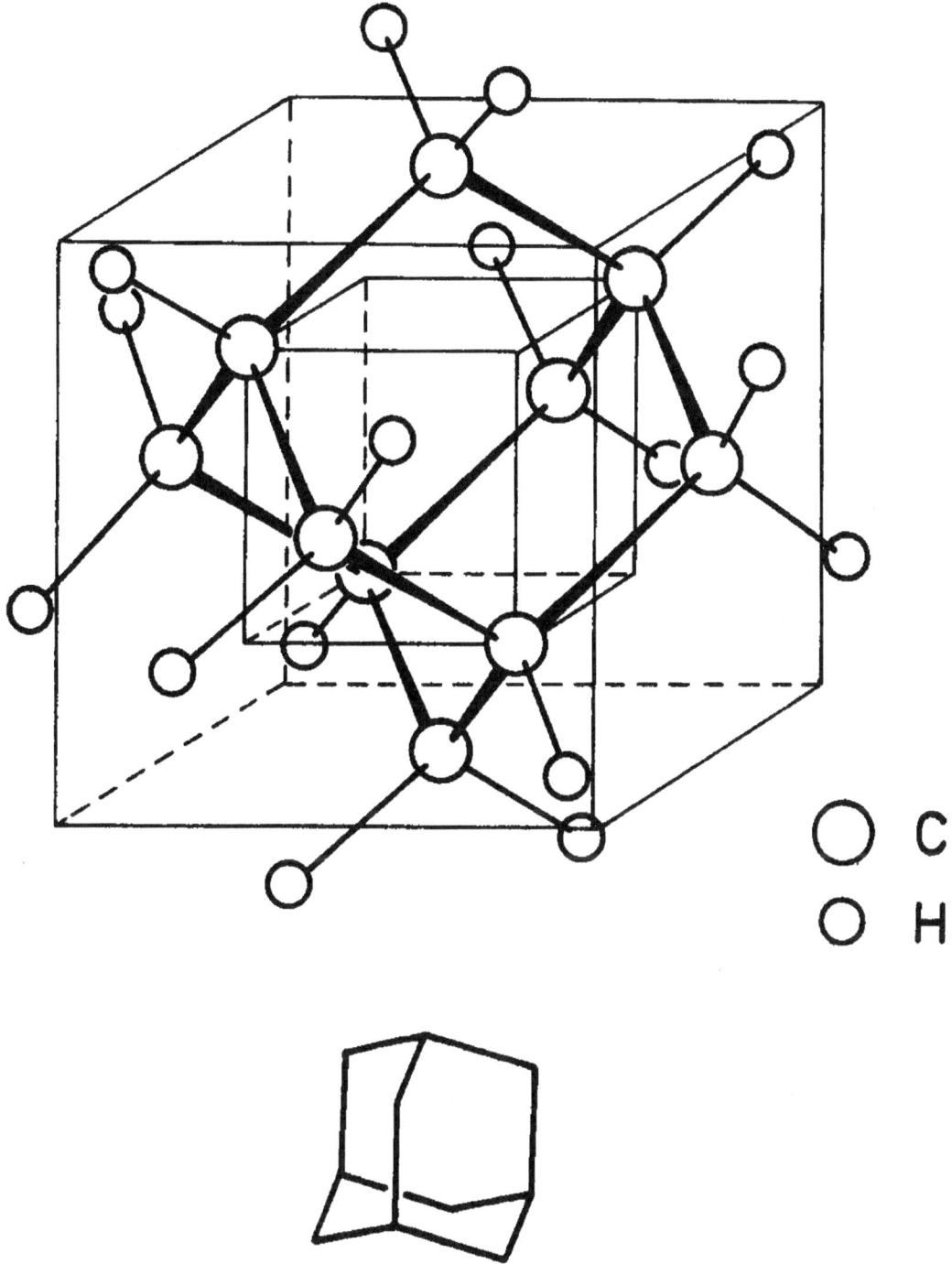

| O | C |
| O | H |

The adamantane molecule, $C_{10}H_{16}$ in two representations. The structure of this highly symmetrical molecule can be described by four imaginary cubes packed one inside the other. Only two of these imaginary cubes are indicated in the top drawing. I have hesitated to include this illustration here because the general idea about this book is to be non-technical. However, for Istvan, this molecule has been very important. It also provides a link between Istvan and Paul Schleyer. Schleyer was not the first who synthesized adamantane, but he was the first who produced it in large quantities and made it available for research. Apart from a less accurate study long before, Istvan determined its structure accurately in 1969. As he was working on this beautiful symmetrical molecule, it came up in his conversations with Eugene Wigner in 1969 in Austin, Texas. Wigner was interested in solid state structures and the adamantane geometry could be looked at as if having taken out of the diamond lattice.

Paul von Ragué Schleyer (1930–2014)[a]

István Hargittai

Abstract

This is how Paul von Ragué Schleyer, a physical organic chemist and computational chemist, summarized his most important scientific achievement: "[I] have increased chemists' awareness, by the examples of my research, of the power of computational methods and of their potential for widespread applications in chemistry."

There was something symbolic in that Paul Schleyer in one of his last papers [1] returned to the problem in whose solution he had already distinguished himself half a century before. It concerned the question about the structure of the 2-norbornyl cation, whether it had a "classical" or "nonclassical" structure. The paper published in 2013 in *Science* provided unambiguous X-ray crystallographic proof, in concert with high-level quantum chemical computations, of the bridged, nonclassical geometry. At one time, there used to be fierce debates over the classical or nonclassical structure of the 2-norbornyl cation whose main protagonists were Saul Winstein and Herbert C. Brown [2].

Paul von Ragué Schleyer in 1995 in Vicksburg, Mississippi (photo by Istvan Hargittai).

Schleyer started as an experimentalist and he was already a renowned organic chemist when he first got a taste of computational chemistry and was hooked. He maintained for the rest of his life that he was still an experimentalist and even when doing computations, his approach to research using computational methods remained that of an experimentalist.

My present brief sketch relies on our published conversation from a recording in 1995, augmented in 1997 [3], and it

[a]Originally published in *Structural Chemistry* 2015, 26:1–4.

I. Hargittai (✉)
Department of Inorganic and Analytical Chemistry, Budapest University of Technology and Economics, Budapest, Hungary
e-mail: istvan.hargittai@gmail.com

incorporates the comments of Paul's Nobel laureate friend and colleague, George A. Olah [4].

Paul was born in Cleveland, Ohio. His father's ancestors were half German and half British and his mother's, German. The last of Paul's ancestors who was not born in the United States immigrated to America in the 1880s. Paul grew up as Paul Schleyer and von Ragué was his middle name to honor his grandmother. Only when Paul moved to Germany, he somehow became "von Schleyer," not only in Germany, but also in the United States, and he did not protest.

Paul received a chemistry set from his mother in 1935 when he was five years old, and started experimenting with sulfur and other substances. When he was twelve years old, he bought himself a larger set, and continued with more involved experiments in his growing basement laboratory. He needed an adult's approval for purchasing chemicals and his grandmother went with him to buy the chemicals that he could then obtain with no consideration for safety. This is something unimaginable today.

During his undergraduate studies at Princeton, Paul had an inspiring teacher in analytical chemistry, Clark Bricker, who introduced him to physical organic chemistry, kinetics, UV spectroscopy, and polarography. For graduate school, Paul went to Harvard and Paul Bartlett was his PhD mentor. There were stellar members among his professors, Fieser, Woodward, E. Bright Wilson, Stork, Moffitt, Kistiakowsky, Rochow, and Wilkinson. R. B. Woodward was the dominating influence, not so much for his natural products chemistry, but for his lectures, seminars, demonstrations of experiments, and the use of physical techniques. Paul left Harvard in 1954 and returned to Princeton, but it took him three additional years before he wrote up and defended his PhD dissertation in 1957. He spent 21 years in professorial appointments in Princeton rising to a named professorship.

Soon upon his arrival in Princeton, he had great success in producing adamantane in a one-step rearrangement. Adamantane, $C_{10}H_{16}$, a beautiful, highly symmetrical molecule was—according to Schleyer—"the C_{60} of its day, the first cage hydrocarbon." Discovered in Czech petroleum, Vladimir Prelog had synthesized it in milligram amounts. Paul's method made this substance available for research, triggered interest in cage hydrocarbons, and propelled his career.

The cage rearrangement reactions induced Paul's interest in computations. This is when he met John A. Pople and fell under the influence of Pople's computational chemistry. He augmented Pople's chemical interests by extending the computations to boron, beryllium, and lithium. Schleyer demonstrated the applicability of Pople's techniques to larger and more complex systems than anybody had before him. He showed, for example that CLi_6, in which six lithium atoms surrounded carbon octahedrally, should be feasible to produce. This work necessitated a growing amount of computing time and in this Princeton proved not equal to the task.

As Paul had spent two sabbaticals in Germany, he learned and liked the German system, and was ready to move when Erlangen offered him virtually unlimited computational possibilities and a number of assistants' positions without having to apply for outside funding. In 1976, he left Princeton for the University of Erlangen.

Paul's first computational papers were readily accepted for publication, but when he branched off in the direction of seemingly impossible molecules, he faced difficulties. In 1978, he reported to a meeting about rule-breaking structures of lithium compounds, such as with planar tetracoordinate carbon, perpendicular ethylenes, and bridged acetylenes. The participants of the stereochemistry conference shunned him, but eventually he turned out to be right.

His consistent work and the development of computational techniques eventually brought him recognition, and he built up a large research group. At the time of our first conversation, in 1995, he had 25 co-workers, including seven or eight postdoctoral fellows. Gradually, however, he downsized his group as he was approaching the German mandatory retirement age in 1998. At that point, he moved back to the United States and accepted a position at the University of Georgia in Athens where he had been Visiting Professor since 1980. His group found a new home in the recently established Center of Computational Chemistry along with Fritz Schaefer, Lou Allinger, and Phil Bowen and their groups. When a few years later I visited the Center, he seemed comfortably established there.

There could be no more authentic chemist today to assess the significance of Paul Schleyer's oeuvre than George A. Olah who authorized me to communicate his comments:

Paul Schleyer's passing is a great loss to chemistry, particularly computational mechanistic-structural chemistry. He more than anybody helped to make it an integral part of the study of chemistry. Paul started his career as an organic chemist making from the beginning major contributions such as the first practical syntheses of adamantane, which started the significant field of cage hydrocarbons. He increasingly concentrated his independent work first at Princeton and subsequently at Erlangen, followed upon his retirement at Athens, Georgia, increasingly applying computational methods to fundamental structure-mechanistic studies. He not only contributed to fundamental chemistry, but established schools of highly talented young investigators building up what is now an integral part of chemists around the world. Paul was a friend for years and a highly respected colleague with whom we collaborated on many problems, including at the time highly controversial non-classical norbornyl ion structure, a key to the field of delocalized new 2e–3c bonding in hypercarbon chemistry. His main collaboration was, however, with John Pople, an outstanding mathematician with whom Paul, one of the great chemists of our time, formed a mutually rewarding formidable team. Paul himself never received the full recognition, which he so much deserved, but future generations will remember him as one of the great pioneering chemists of our time. We will all much miss him, but his contributions and heritage to chemistry will last [4].

In 1995, I asked Paul to assess the standing of computational chemistry in academia and here is his response. From what he said we can form a picture of the situation in 1995 and, even more interesting, can judge the extent of changes, if any, since.

I feel very strongly that computational chemistry is still not receiving adequate recognition in terms of its scientific and educational capabilities, to say nothing of its incredible future. Chemistry is a very conservative science, with a long and almost exclusively experimental tradition. The ability to gain valid chemical information computationally is recent and is not still accepted by most chemists. The strong prejudice against computational chemistry remains. Such skepticism no longer is healthy. Older chemists in influential positions have not kept up with rapid developments in computers and programs and are not well informed about the status and potential of computational chemistry, but they make prejudicial pronouncements and influence negative decisions. Most university faculties have no professors of computational chemistry. When new faculty members are hired, this area is overlooked. Consequently, the conservatism is "programmed" to continue. The importance of computational/theoretical chemistry—particularly its didactic advantages—is not reflected in the curriculum. Several undergraduate as well as graduate courses in this area should be among the requirements in every chemist's education.

The user of ab initio programs quickly realizes the enormous sophistication of molecules. Chemistry is revealed in more detail, and problems are conceived more basically. One appreciates how atoms are joined in a very intimate way; structures can be visualized directly, as can the complex dynamics of transformations from one conformation to the next, or from one molecule to another. How do bonds break and make? What is the nature of chemical bonding? A new chemical education awaits the user of electronic structure programs, which are far more than a black box to be manipulated.

Few colleagues appreciate this. I attempt to point out that the future of chemistry will be more and more computational, but, ..., this is interpreted as "Schleyer says that nobody should do experiments any more, but should only compute." The transformation is obvious. Chemistry is becoming a computational science but not rapidly enough [Ref. 3, pp., 93–94].

It was still in his Erlangen period when we invited Paul for a visit to Budapest. He gave an excellent presentation. He was very persuasive about the importance of computational chemistry and he urged us to throw out all experimental techniques and devote all our resources to computational work. Fortunately, the people who supposed to allocate funding for our experimental projects did not pay attention to Paul's recommendation. By then, we had been combining our experimental research with computations and the trend kept strengthening; whether or not due to Paul's warning, it is hard to assess. There is no doubt though that his activities made a great positive impact, in general, on moving chemical research in the computational direction.

Here is how he talked about his very personal relationship to computational chemistry:

I like to create what I study. An idea flashes into my mind. I imagine a new molecule with a weird structure and unusual bonding. Selected combinations of component elements give the proper size relationships, electron occupancy, and symmetry, but will the idea survive the test of a good-level quantum-chemical computation? The jobs take only a few minutes to set up and submit, and the first part of the run (the "initial guess") quickly reveals if the prediction is on the right track. If the first geometry optimization cycles still look favorable, I admire details of the structure—my own new invention—plotted on the monitor. How exciting—to study a completely new composition of matter no one else has thought of before. Refinement must follow and can lead to disappointment. But what a magnificent opportunity, to be able to test one's own ideas personally in great detail, so quickly, easily, and reliably! [Ref. 3, 94, p.]

Our own journal, *Structural Chemistry*, has greatly benefited from Paul's expertise in and devotion to his subject. He was a conscientious and reliable reviewer and we relied on his unselfish readiness to help. We will miss him and so will the chemistry community for his science, his wit, and his human qualities. The memory of his enthusiasm, in addition to his achievements, will serve us as inspiration for many years to come.

References

1. Scholz F, Himmel D, Heinemann FW, Schleyer PVR, Meyer K, Krossing I (2013) Crystal structure determination of the nonclassical 2-norbornyl cation. Science 341:62–64
2. Hargittai B (2012) H. C. Brown Centennial (1912–2005). Struct Chem 23:939–941
3. Hargittai I (2003) Paul von Ragué Schleyer. In: Hargittai M (ed) Candid Science III: more conversations with famous chemists (Chap. 6). Imperial College Press, London, pp 80–95
4. Olah GA (December 3, 2014) E-mail communication to I. Hargittai

Istvan
Hargittai

JUDGING
EDWARD TELLER

A CLOSER LOOK AT ONE OF THE MOST INFLUENTIAL
SCIENTISTS OF THE TWENTIETH CENTURY

Cover of István's balanced Teller biography. István learned about Teller so much that he said, it is almost improper to know so much about someone.

Edward Teller—Guardian of Freedom or Dr Strangelove?[a]

István Hargittai

. . .[Teller] fought obstinately for what he believed in.
I may have disagreed with his tactics but never with his goals.
John A. Wheeler (1911–2008) (M. Hargittai, I. Hargittai, Candid Science IV, Conversations with Famous Physicists. Imperial College Press, London, 2004, pp. 424–439, p. 429 (John A. Wheeler), the most versatile physicist of the 20th century (L. Motz, J. H. Weaver, The Story of Physics. Avon Books, 1992)

After the political changes of 1989, people will have to reevaluate Teller's role because of his impact in bringing down the Soviet Union.
Manfred Eigen (1927–), Nobel laureate (I. Hargittai, Candid III, More Conversations with Famous Chemists. Imperial College Press, London, pp. 368–377, p. 377 (Manfred Eigen))

Edward Teller (1908–2003) was the youngest and most ambitious of the Hungarian Martians—a group of Jewish-Hungarian scientists that also included Theodore von Kármán (1881–1963), Leo Szilard (1898–1964), Eugene P. Wigner (1902–1995) and John von Neumann (1903–1957). They were born in Budapest, where they got their high school education and started their university studies. They benefited from the sizzling intellectual life of fin-de-siècle Budapest that by Teller's time had mostly disappeared. Anti-Semitism and the lack of prospects drove them to Germany where they completed their education and became members of the scientific elite in their own fields. National Socialism then forced them out of Europe and they all ended up in the United States. What made them a group was that they all had a deep concern for the Free World and were relentless in their efforts to defend the United States, even at risk of their careers in science. The label "Martians" has long figured in anecdotes about these extraordinary and talented Hungarians, who seemed to be extraterrestrial visitors in disguise.

The contributions the Martians made to twentieth-century science had fundamental value as well as military significance. These contributions range from fluid mechanics to quantum mechanics, from the stored-program computer to molecular biology, and from nuclear chain reactions to game theory. Von Kármán had a key role in developing the U.S. Air Force; Szilard initiated the American program for the atomic bomb, called the Manhattan Project; Wigner was instrumental in building the first nuclear reactors; von Neumann had multifaceted contributions to defence-related projects; Teller was the father of the American hydrogen bomb and the main proponent of the Strategic Defence Initiative, popularly known also as Star Wars. Their activities extended well into the Cold War era, but had mostly come to an end by the mid-1960s, due to death or diminishing interest and capacity. Teller was the only one who carried on and stayed active and influential through the 1980s. The present account, after a brief introduction, focuses on some highlights of his political-military oeuvre.

Teller went to the legendary Minta (Model) Gimnázium in Budapest. The school was part of the system of secular high schools developed by Maurice Kármán, Theodore's father, during the last third of the nineteenth century. The eight long years Teller spent there were painful for him. His teachers did not appreciate his talent and penetrating interest. He started his university studies at the age of 17 at the Budapest Technical University, and moved to Germany when he reached 18. First he went to Karlsruhe where he majored in chemical engineering, then transferred to Munich. After 2 years he became Werner Heisenberg's doctoral student at the

[a]Originally published in *Hungarian Quarterly* 2008, 48 (189):108–122.

I. Hargittai (✉)
Department of Inorganic and Analytical Chemistry, Budapest University of Technology and Economics, Budapest, Hungary
e-mail: istvan.hargittai@gmail.com

University of Leipzig. He obtained his doctorate in physics at the age of 22 and continued for 3 years as instructor at the University of Göttingen. With Hitler's accession to power in Germany, he left for Copenhagen, then London, and finally, in 1935, became Professor of Physics at George Washington University in Washington, D.C. He participated in the Manhattan Project in Los Alamos, though not in a leading role; his attention was more on the next generation of nuclear weapons than the first atomic bombs. Following World War II he continued part time in Los Alamos, but his principal job became a professorship at the University of Chicago.

He had a leading role in developing the American hydrogen bomb, and he initiated the second weapons laboratory of the United States in Livermore, California. In 1954, he testified against Robert Oppenheimer, the legendary scientific director of Los Alamos, and thereby lost the friendship of most of his fellow physicists. He was appointed to a professorship at the University of California and later became a Senior Research Fellow at the Hoover Institution at Stanford University. His influence as an adviser to military leaders and conservative politicians of the United States was ever increasing. He maintained his influence through the Reagan-Bush presidencies and was awarded the highest distinctions. He died on September 9, 2003.

Even Teller's dry biographical data sound exciting. He lived in "interesting times," as the Chinese proverb says, and took up, with zest, the challenges presented to him. He was significant as a scientist, but his oeuvre has had a longer-lasting impact in the political and defence arena than his discoveries in science. His principal achievements came in nuclear physics and physical chemistry. However, apart from the greatest discoverers, scientists have fleeting fame. Scientific achievements are different from artistic creations. What one researcher does not discover, another will, sooner or later. *Guernica* could not have been painted by anybody other than Picasso. No other composer could have written Beethoven's symphonies. The famous B.E.T. equation describing the adsorption of gases was established by Brunauer, Emmett and Teller. Had Teller and his colleagues not arrived at it, others certainly would have. In preserving his name Teller was even "lucky" because many effects carry his name—always together with others to be sure—because he created best in the company of colleagues. Incidentally, the significance of some of the effects bearing his name has recently even gained in importance.

Teller's political awakening was gradual. It started on the eve of World War II. Szilard made two visits to Albert Einstein on Long Island in the summer of 1939, during the preparation of Einstein's letter to President Roosevelt warning him about the possibility of the atomic bomb. Eugene Wigner accompanied Szilard on the first visit, providing him company but also acting as Szilard's chauffeur. Teller played this role on Szilard's second visit. Thus Teller was involved in the American atomic bomb effort from the beginning, but his role was more accidental than initiating. In May 1940, Teller attended a large meeting of scientists where President Roosevelt spoke about human rights, the blessings of democracy and the progress made by science. The president called on the scientists "to protect and defend by every means at our command, our science, our culture, our American freedom and our civilization."[1] Teller felt as if Roosevelt was speaking directly to him and he was mindful that, among the huge audience, he might have been the only one thinking of the atomic bomb. From this point on, he believed his path had been charted, along with those of the other Martians, because—in his words—"We five were survivors of a shipwreck and found a lifeboat. Of course, we were eager to protect it against all dangers."[2]

Teller joined the atomic bomb project from its outset and when preparation of the bomb moved to Los Alamos, he was there. He worked in the theoretical group, but only part of his time and efforts were devoted to the atomic bomb. His real interest lay in working out the possibility of a yet more devastating weapon, the thermonuclear bomb, also referred to as the Super. Some time in 1942, a fateful discussion took place between Enrico Fermi and Teller about whether an atomic explosion might produce a thermonuclear reaction. Fermi was one of the giants of twentieth-century physics and one of the leading physicists on the Manhattan project. In the atomic bomb, fission takes place, that is, an element is split into two smaller ones. Fermi raised the question whether an atomic explosion might produce sufficient energy to join two elements together and produce a larger element, a process referred to as fusion. Fusion would liberate much larger amounts of energy than fission. However, it would take enormous energies to overcome the repulsion of nuclei, and the necessary energy could only be provided by the fission bomb.

It was not feasible to build a fusion bomb before the end of World War II; even the simpler fission bomb was barely ready in time to be deployed against Japan. The efficacy of the atomic bombs in bringing the war to quick conclusion has been debated ever since. My own view is that there is overwhelming evidence that it served this purpose eminently, and

[1] *The Public Papers and Addresses of Franklin D. Roosevelt,* with Special introduction and Explanatory Notes by President Roosevelt (Samuel I. Rosenman, comp.), 1940 vol.: *War—And Aid to Democracies.* MacMillan, New York, 1941, pp. 184–187, as quoted in Teller, *Memoirs,* pp. 149–150.

[2] Attributed: Edward Teller to the co-author of his *Memoirs,* Judith Shoolery. Private communication in conversation with Judith Shoolery in Half Moon Bay, California, February, 2004.

saved hundreds of thousands of American lives and many more Japanese.[3] There was yet another goal behind their use, namely to issue a warning to the Soviet Union.

When the war was over, the question arose whether to continue research on the thermonuclear bomb, or shut down Los Alamos. Teller would have stayed at Los Alamos had it been promised to him that vigorous efforts would be made to develop the fusion bomb or at least to further perfect the atomic bombs. He received no such assurance and he left for Chicago. He was not alone in his views, however, because there were others who thought that the nuclear advantage of the United States should be maintained and guarded by all possible means. It was decided, for example, that there would be continuous monitoring of radioactivity in the atmosphere.

Teller's situation after the war was markedly different from that before it. From an almost playful physics professor, he was gradually taking up the role of trusted advisor, and was even invited to testify before the U.S. Congress on questions of national security. His main arena of activity, however, was still in academia. He was a respected professor of physics at the University of Chicago and scientists of later renown studied under him. His involvement in politics focused on committee work concerned with nuclear reactor safety. At the same time, he never abandoned his interest in the development of thermonuclear weapons, known as 'hydrogen bombs' because the fusion reaction involved the joining of two hydrogen nuclei. During the period 1945–1948, there was much hesitation on the part of the American administration as to how to continue with the development of nuclear weaponry. On the one hand, the apparent atomic monopoly as well as scientific and material superiority of the United States gave the impression that an intensified program was not needed. On the other hand, there were some who advocated increased efforts to enhance the defence capabilities of the country. The Hungarian scientists recognized the necessity of serious preparations. Von Neumann went so far as to advocate pre-emptive strikes against the Soviet Union. Others, like Szilard were merely concerned with maintaining full preparedness. To clarify Szilard's position is especially important. He opposed the deployment of the atomic bombs against Japan and later made efforts to convince the superpowers to ban the testing of nuclear weapons. This was combined with his advocacy of negotiations with the Soviet Union. All this made later authors writing about the Cold War period assume that Szilard must have opposed the development of the hydrogen bomb. We will see below that this was not the case.

The years 1949–1950 were a time of awakenings, tragically combined with the rise of Joseph McCarthy's star in the U.S. Senate. McCarthy misused his position and authority to spread fear and to level baseless accusations of Communist conspiracies even within such venerable institutions as the State Department and the U.S. Army. From 1947 on, the U.S. monitored the atmosphere for unusual radioactivity; this vigilance yielded a spectacular dividend in 1949, when a Soviet nuclear explosion was detected. On January 27, 1950, the Americans learned from the British that one of their physicists at Los Alamos during the war, Klaus Fuchs, had passed nuclear secrets to the Soviets. On June 25, North Korea invaded South Korea. Senator McCarthy's first charges about Communist infiltration of the State Department were made on February 9, 1950. When Teller heard about the Soviet nuclear explosion, he called the former scientific director of Los Alamos, Robert Oppenheimer to ask, "What do we do now?"[4] Although Teller knew that Oppenheimer had opposed the development of thermonuclear weapons, the former director of Los Alamos still represented authority for him. Oppenheimer brushed him off by saying, "Keep your shirt on." At this point, Oppenheimer ceased to be a yardstick for Teller because he was convinced that this was a new situation calling for immediate action.

Edward Teller had a historic role in facilitating the United States embarking on a crash program to develop the hydrogen bomb. A brief summary here will help in understanding the enormity of difficulties Teller had to overcome in fulfilling his self-imposed task. The decision about such a program ultimately rested with the president of the United States. He relied in part upon the recommendations of the Atomic Energy Commission. The AEC was not a scientific body, but was assisted by a nine-member General Advisory Committee, headed by Oppenheimer and consisting of first-rate physicists, chemists and engineers. Teller was not a member of this committee. He had some allies, however, outside of the committee. Ernest O. Lawrence, the Nobel laureate discoverer/inventor of the cyclotron and his subordinate, the future Nobel laureate Luis Alvarez sided with Teller in his crusade to convince politicians and military leaders that the United States had to have the hydrogen bomb. In the fall of 1949, the GAC came to its decision concerning the development of the hydrogen bomb. It was not unanimous, but both the majority and the minority opinion rejected the program of building a hydrogen bomb. The majority opinion strongly recommended against an all-out effort to develop the hydrogen bomb. It condemned it as a tool of genocide, which, of

---

[3] See, e.g., discussions in Chapter IV in I. Hargittai, *The Martians of Science: Five Physicists Who Changed the Twentieth Century.* Oxford University Press, 2006.

[4] E. Teller (with Judith L. Shoolery), *Memoirs: A Twentieth-Century Journey in Science and Politics.* Perseus Publishing, Cambridge, Massachusetts, 2001, p. 289.

course, it is; it also stated that "In determining not to proceed to develop the super bomb, we see a unique opportunity of providing *by example* some limitations on the totality of war and thus limiting the fear and arousing the hope of mankind."[5] (my italics) Wonderful thoughts, but they reflect naiveté about and ignorance of Soviet intentions and determinations. The minority report, signed by two Nobel laureate leaders in physics, Fermi and Isidor I. Rabi, stated, "...we think it wrong on fundamental ethical principles to initiate a program of development of such a weapon. At the same time it would be appropriate to invite the nations of the world to join us in a solemn pledge not to proceed in the development or construction of weapons of this category."[6] (my italics) Again, the naiveté and trust are staggering. By then an accelerated Soviet program was underway.[7]

The Free World did not know about the weapons development in the Soviet Union, and there was a tendency to underestimate the Soviets in general and the Soviet physicists in particular. In the comfort of Western democracies, it was also hard to imagine that there was no discussion in the Soviet Union about the question of the development of the hydrogen bomb and that thousands of slave workers would be used to complete the program under inhuman and unsafe conditions. If not then, today it is ridiculous to maintain that the United States could have stopped the development of the hydrogen bomb by its own example of refraining from it. Relying on the recommendation of star scientists from GAC, the Atomic Energy Commission (AEC) voted to advise the President of the United States not to decide on a crash program to develop the hydrogen bomb. President Truman then asked for advice from his National Security Council (NSC) consisting of three aides, the Secretary of Defence, the Secretary of State and the Chairman of the AEC. Teller and his colleagues had succeeded in their self-imposed task of alerting the politicians and the generals about the looming danger of non-action. As a result, the trend reversed in the NSC as it voted two to one in favour of the development. It was a great victory for Teller when President Truman announced his decision on January 31, 1950, giving directions to continue research on all atomic weapons, including the hydrogen bomb. He augmented this decision in March 1950 with a secret directive to intensify work on the hydrogen bomb. It is important to stress that after Truman's decision, all physicists that were needed for the work,

including some who had fiercely opposed it, quickly and without duress converged on Los Alamos and resumed their research. Teller as a physicist went on to play a decisive role in solving the scientific and technological problems connected with the development of the hydrogen bomb. The mathematician Stanislaw Ulam made comparable contributions, especially in moving beyond the theoretical dead-end that at a certain point seemed to make the project hopeless. In this, the relative weight of Teller and Ulam's contributions cannot yet be assessed because the relevant documents are still classified.

It must have been lonely for Teller, even when a few others had joined him, to go against the tide of the majority of his distinguished colleagues. It was not his first experience of loneliness, and not the last either. There seems to be a discrepancy between the perception of Teller the public figure and the figure that emerges from a closer scrutiny of his life. The public perception of Teller is of someone fiercely arrogant and headstrong, sure of himself, winning all his debates, someone of strict principles that emanate from tremendous inner strength. However, it could be argued just as convincingly that he craved acceptance by his peers, wanted to please his superiors and was torn by self-doubt. I do not wish to imply that he was misunderstood: he himself cultivated the public image he became identified with. Donald Glaser tells about a flight he shared with Teller, on which they had an amicable and meaningful conversation. However, after deplaning, in the presence of an audience, Teller immediately put on a show of loud behaviour.[8]

Teller was teased and bullied by his classmates at the beginning of his *gimnázium* years and strove to gain the friendship of his fellows. During the same period, his virtually boundless respect for authority was inculcated in him by his revered maternal grandfather, who taught him that laws must be obeyed without exception. He never accepted that personal responsibility might override the law even if it worked against one's conscience and never subscribed to the American tradition of preferring to break a law rather than doing something against one's conscience.[9] This is why it is uncomfortable to imagine how Teller might have conducted himself in a Soviet or a Japanese environment had he chosen to emigrate eastward rather than westward after the Nazi takeover in Germany.

Teller enjoyed his years in Germany, where he flourished in the community of physicists. He was never part of German society, but he was of the society of German and other physicists, and this sufficed. Similarly, he felt very comfortable during the second half of the 1930s at George

[5] See in full in G. T. Seaborg, *A Chemist in the White House: From the Manhattan Project to the End of the Cold War.* American Chemical Society, Washington, DC, 1998, pp. 42–43.

[6] Ibid.

[7] I. Hargittai, M. Hargittai, *Candid Science VI: More Conversations with Famous Scientists.* Imperial College Press, London, 2006, pp. 808–837 (Vitaly L. Ginzburg).

[8] I. Hargittai, *The Martians of Science: Five Physicists Who Changed the Twentieth Century.* Oxford University Press, New York, 2006, p. 224.

[9] Ibid., p. 11.

Washington University in Washington, D.C. There, George Gamow was his colleague, and many others joined them for the annual meetings they organized in theoretical physics. When Gamow characterized him as "helpful, willing, and able to work on other people's ideas without insisting on everything having to be his own," he was referring to this period. What a contrast to the Teller of later years!

Los Alamos at the time of the Manhattan Project was not to Teller's liking. There were many more important physicists around him and he was relegated to a sub- ordinate role under his friend, Hans Bethe. He declined this role, and it was mainly due to Oppenheimer's tactful approach to Teller's insistence on working only on the fusion bomb that there was no major clash between Teller and other physicists. Teller was never comfortable with Oppenheimer, but heeded his advice and declined to sign the petition Szilard sent him that protested against deployment of the atomic bombs in the summer of 1945. In Los Alamos, Teller also found solace in von Neumann's friendship during the latter's periodic visits to the weapons laboratory. After the war, Teller was dissatisfied with the performance of Los Alamos.

However, upon becoming Professor of Physics at the University of Chicago, in spite of the prestige and comfort of the position, he could not find the contentment in peacetime he had had at George Washington University before the war. He was gradually becoming involved in politics. In addition to the development of the hydrogen bomb, he fought for the establishment of a second weapons laboratory (today the Lawrence Livermore National Laboratory). In this quest for the second weapons laboratory, he teamed up, again, with Lawrence, who was more right-wing than Teller, but was a better politician in that he tried to avoid antagonizing his fellow physicists. Both Enrico Fermi and von Neumann warned Teller against getting too close to Lawrence, but to no avail. Teller went to California, and at Livermore, he was surrounded by many younger colleagues whose jobs were largely due to him (i.e., to Teller).

Teller's ultimate loneliness came as a consequence of the role he played in Robert Oppenheimer's security hearing. He could have declined to testify, or, if he felt that impossible, he could have acted similarly to von Neumann, who was not a great friend to Oppenheimer, but was diplomatic in his testimony. Instead, Teller chose to be unambiguous in expressing his view that Oppenheimer was a security risk. When I.I. Rabi sarcastically congratulated Teller on the "brilliance" of his testimony and the "extremely clever way" in which he had expressed his opinion that Oppenheimer was a security risk, it signified his third and final exile. Rabi was a doyen of American physics, and he ignored Teller's proffered hand (just as Mikhail Gorbachev did to Teller decades later).

Teller's first exile was from Hungary, from where he was driven out by anti- Semitism and the lack of prospects for building a meaningful life. Hitler and Nazism forced him out of Germany and Europe. Each of these first two exiles were involuntary and both led to an improved situation and greater opportunities. The third exile was different. It was a consequence of his own actions and it isolated him from what was most important to him, the community of physicists. This would have been difficult for any scientist, but it was especially hard on Teller as he thrived on working in collaboration with others. Teller's great loss was compounded by the deaths of Lawrence and von Neumann, as well as Wigner's gradual marginalization. Szilard's death in 1964 was another blow. Szilard had remained a friend to the end, in spite of them being political adversaries.

Teller was haunted throughout the rest of his life by his performance at the Oppenheimer hearing. In his *Memoirs* he stated that he "never wanted Oppenheimer's opinion on the hydrogen bomb to count in the decision on his security clearance". He further declares that revoking Oppenheimer's clearance was not justified. This is a puzzling statement in view of his testimony. One wonders whether his judgment by then was so clouded that he expected his readers to believe this, while the same book includes excerpts from his testimony, providing evidence to the contrary. To the question whether he believed that Oppenheimer was a security risk, he concluded his response by saying that "...I feel that I would like to see the vital interests of this country in hands which I understand better, and therefore trust more. ...I would feel personally more secure if public matters would rest in other hands." At a later stage of the testimony, to a similar question, he said: "If it is a question of wisdom and judgment, as demonstrated by actions since 1945, then I would say one would be wiser not to grant clearance."[10]

In his *Memoirs,* which appeared decades after the Oppenheimer hearing, Teller claims that his misjudgements were due to not thinking his testimony through. But there is evidence from Teller's prior testimony to the FBI that what he said at the Oppenheimer hearing was not something uttered on the spur of the moment but a reflection of long-held views. More convincing is the letter he wrote to his friend and former pupil, Maria Goeppert-Mayer, shortly after the hearing in 1954. In it, he writes about backbone: "I seemed to get along fine without one. Now there seems to be some growing pains. I also wonder whether it is growing in the right direction."[11] One must read this tormenting revelation several times to believe one's eye, it sounds so incredible. Teller was close to Goeppert-Mayer, and it is difficult to imagine him opening up to this extent to anybody else at that time. In the *Memoirs,* he shares this thought with the world, but its effects are dampened by his assertions about his testimony.

[10] Teller, *Memoirs*, pp. 208 and 209.
[11] Ibid., pp. 399; 400.

Edward Teller and the author in the Tellers' home in Stanford, California, in 1996 (photograph by and courtesy of Magdolna Hargittai)

Returning to Teller's loneliness in the initial debates about the hydrogen bomb: if anyone could understand a situation that seemed hopeless, but in which the importance of action seemed overwhelming, it was Leo Szilard. He had had a similar experience in 1939, when he took it upon himself to alert the American leadership about the feasibility of the atomic bomb and the danger of Germany acquiring it. Szilard reflected on the debate of 1949 in a little-known speech in 1954. He posed the question, "Why did America come so close to missing out on the hydrogen bomb?"[12] According to him, the United States "would have missed out on the hydrogen bomb altogether had it not been for the accident that there was still one man left who—for a variety of reasons—still liked to think about the problems of the bomb." He meant Teller, and he called it accident because if there had been only one man left, there could just as easily have been no-one left. Szilard mentions that the enemies of the United States could have easily managed to discredit Teller through false accusations rampant during the McCarthy years. Thus the situation was precarious indeed. Szilard also noted the physicists' readiness to join the program once there was a presidential go-ahead. Szilard left no doubt that in his eyes "there can be only one justification for our development of the hydrogen bomb and that is to prevent its use." The aptly named doctrine of "Mutually Assured Destruction" was just what Szilard had hoped for.

Parallel with Teller's de facto exclusion from American physics, he was becoming more and more part of the Establishment of the American military and defence industry. From being a part of a circle of irreverent scientists, he shifted to people for whom subordination prevailed and where hierarchy mattered most, and where his views and ideas were neither scrutinized nor criticized. He was eagerly welcomed into this company, and he was fast becoming a much sought-after advisor of politicians and generals. Perhaps it could not be recognized in the process, but with hindsight, this had tragic consequences for Teller, beyond the missing collaborators. Within the collegial community of physicists, Teller was always willing to discuss his ideas. In his third exile, he was no longer surrounded by colleagues, but by military and political leaders. This made Teller's isolation from criticism total: even if he had wanted to discuss his ideas with them, they would have been unable to contradict him. The consequences became all too apparent during the debates about the Strategic Defense Initiative (SDI), popularly known as "Star Wars."

Teller has been accused of having brought SDI to the White House through the back door, that is, without having it subjected to the criticism of his peers.[13] However, he had had a negative experience with the judgments of most of his fellow scientists in the hydrogen bomb debate. In addition, by the time of SDI, he had got out of the habit of exposing himself to the scrutiny and criticism so common in science but so little practiced in defence circles. There is plenty of evidence for Teller's influence on the White House, but very little for his personal involvement with President Reagan during the early years of his presidency. His influence can be assessed by the fact that President Reagan's science advisor between 1981 and 1986, George A. (Jay) Keyworth, was appointed at Teller's recommendation.

President Reagan gave his famous speech announcing the SDI programme on March 23, 1983; it was televised from the White House and was delivered in front of a select audience that included Edward Teller and, among others, Charles Townes, the Nobel laureate co-discoverer of the laser. Powerful X-ray laser beams were to be the main tool for knocking out enemy ballistic missiles. Later in the 1980s, the idea of the X-ray laser had to be abandoned and other concepts were adopted. There was never a shortage of intriguing buzz words, like "Brilliant Pebbles," which were small, kinetic-energy arms to be deployed in swarms. The political aim remained the same in that rather than subscribing to the policy of "mutually assured destruction", it was meant to be a policy of "assured survival". Mutually assured destruction—as awful as it sounds—maintained peace between the

[12] L. Szilard, "The Sensitive Minority among Men of Science." In *Leo Szilard Centenary Volume* (G. Marx, Ed.). Eötvös Physical Society, Budapest, 1998, pp. 176, 182, 184.

[13] See, e.g., Recorded conversations with Peter D. Lax, New York City, June 2007.

two superpowers for decades. As early as 1945, Albert Einstein stated that "…atomic energy … may intimidate the human race into bringing order into its international affairs, which, without the pressure of fear, it would not do."[14] The Soviet Union felt that SDI broke the precariously maintained balance. The paradox that a new means of defence was an act of offence must have appeared valid to the Soviet leadership, and all the more so because the Soviet Union by the mid-1980s was no longer capable of matching the United States in the technological advances that SDI necessitated. The Soviet Union was lagging behind in what SDI relied upon most: electronics, computerization and miniaturization. This was not for want of excellent scientists; on the contrary, selected branches of physics in the Soviet Union thrived. What was lacking were the means and mechanism of translating scientific advances into technologies. In the past, the Soviet Union had been able to catch up with the West by making sacrifices in other segments of their economy. By now there was not very much more to sacrifice and even if there had been, no lowering of Soviet living standards could have been translated into high technologies. This is why Soviet President Mikhail Gorbachev was so apprehensive of SDI. It was under these circumstances at the Reykjavík summit in October 1986 that the Soviet president offered his American counterpart an incredible compromise: he suggested that the two superpowers eliminate all offensive strategic arms and limit SDI to the laboratory. This signalled a tremendous success for SDI regardless of whether it was scientifically, technologically or militarily feasible. To rid the world of all the offensive strategic arms would have been the achievement of the century and would have justified the expenses of SDI many times over. However, President Reagan declined the offer. We might say that he gambled with the fate of the Planet and, in hindsight, he won, but he took a tremendous risk. The Soviet Union collapsed and a new world order has taken shape. To the extent that SDI contributed to the demise of the Soviet Union, the initiative seems to have, bizarrely, paid for itself. Again, to the extent this is valid, Edward Teller contributed considerably to the collapse of the Soviet Union. Thus it seems justified to distinguish between the validity of claims and criticism of SDI and Teller's zealous promotion of it, and its eventual dividends, whether as directly consequential or by default.

It will take some time for historians to make an objective judgment over Edward Teller and his oeuvre. His *Memoirs* have to be read with circumspection. Teller wrote them late in life and the book appeared in 2001, when he was 93 years old. The book has been judged to be rather self-serving. The

Memoirs suffered from further handicaps. One was that a stroke prevented Teller to be as active in writing the second half of the book as he was for the first half.[15] The second half became much more the product of his co-author, Judith Shoolery than the first half.[16] Sadly, I failed to recognize Teller's distinctive style of narration in the *Memoirs*. Expressions in the spirit of "I claim" and "I don't know, but I'll tell you" are missing. Apparently, it was not only that the written text is different from spoken narrative. Again, according to Ms. Shoolery, an over-conscientious editor at the publishing house weeded everything that was not deemed the most proper usage of English. When the edited text arrived, Teller faced the dilemma of whether to request a revision, but he felt too tired to make the book suffer yet another delay. A further handicap was that the text did not undergo editorial scrutiny for contents and the spelling of Hungarian words. The latter is sad in view of Teller's excellent Hungarian, which he maintained through his old age. As to factual content, I mention here just one innocent example, the episode in the *Memoirs* in which Teller drives Szilard to Einstein's home on Long Island, and they are having a difficult time finding it. In reality, this happened when Wigner was driving Szilard on his first trip to Einstein. When Teller and Szilard went together, Szilard already knew the way. However, the story of the Wigner-Szilard trip has become part of the lore and must have replaced what Teller might have remembered about his own trip.

Soon after the appearance of Teller's *Memoirs*, Peter Goodchild published a book on Teller entitled *The Real Dr Strangelove*.[17] Goodchild tends to rely on Teller's *Memoirs* and other publications, rather than new materials. The reference in the title of the book is to Stanley Kubrick's 1964 movie, *Dr Strangelove or: How I Learned to Stop Worrying and Love the Bomb*. It is about the start of a nuclear holocaust and involves an insane ex-Nazi warmonger high in the echelons of the United States high command. I do not think it fair to assign this label to Edward Teller for the following reasons. As I have argued, the creation of the American hydrogen bomb helped to prevent rather than to initiate a nuclear holocaust. Teller was never a Nazi and thus could not become an ex-Nazi. Dr Strangelove's irrational behavior was not like Teller's at the time of the debate over whether to develop the hydrogen bomb. It would be more difficult to avert such portrayal for Teller in the debates about SDI or

---

[14] A. Einstein, "Atomic War or Peace." *Atlantic Monthly,* November 1945, quoted in the *Expanded Quotable Einstein*. Collected and edited by A. Calaprice. Princeton University Press, 2000, p. 176.

[15] According to Teller's co-author, Judy Shoolery. Private communication in conversation with Judith Shoolery in Half Moon Bay, California, February, 2004.

[16] It is inexcusable that Judith Shoolery's name was omitted in the Hungarian translation of the book.

[17] P. Goodchild, *Edward Teller: The Real Dr Strangelove*. Weidenfeld and Nicolson, London, 2004.

"Star Wars," but that came in the 1980s, long after the movie was made. There have been other suggestions for a model for Dr Strangelove, but it seems more probable that the negative traits of several individuals were combined in Strangelove, who has become a symbol of reckless warmongers.

Edward Teller is certainly someone who figures in questions of "What might have happened if?" In the August 14, 2007, issue of *Japan Times,*[18] there was an article about the Japanese atomic bomb project during World War II. The project was confined to the laboratory, but the fact that it existed is of interest. The article was prompted by the recent publication of letters and documents of Yoshio Nishina the project director. The article mentions a letter to Nishina by a German physicist, dated April 21, 1933, saying that "Edward Teller was hoping to stay in Japan after fleeing Nazi Germany." (There is no mention of the name of the German physicist and there is no mention of such intentions in Teller's *Memoirs*.)

It would be only too easy to dismiss the suggestion of the letter. However, it came years before Japan joined the Axis alliance. At this point it is worth mentioning the fact that at the same time Caltech was persuading Theodore von Kármán to move there in the second half of the 1920s, von Kármán received invitations from Japan through the Japanese Embassy in Berlin. He was offered generous terms, which he declined by requesting yet even better terms—and the Japanese accepted them. So he went and later he claimed (with false modesty), "I do not wish to take too much credit—or perhaps in this case, blame—but I believe I was also the man who introduced Japan to metal airplane propellers."[19] Sugita poses the question, "If Teller . . . had joined Nishina's group, would Japan have been the first to produce the bomb?" I think it is safe to say that Japan would not have been the first. First, even if Teller and/or others had joined the Japanese scientists at the time of or soon after Hitler came to power in Germany, that is, around 1933, the bomb program could have not started until after 1938, that is, after the discovery of nuclear fission. From 1939 till 1945, Japan's resources would not have sufficed to enrich uranium and create the bomb. Japan could not have competed successfully with the United States, even if a few physicists of Teller's caliber had joined in.

But there is an alternative scenario. Nuclear fission could have been discovered in 1934. In fact, Enrico Fermi in Rome fissioned uranium, but misinterpreted the results of his experiments. Also, Leo Szilard dreamed of the nuclear chain reaction and, had he made a systematic search for the fissionable element, he could have hit upon uranium as early as 1934. This would have given the Japanese more time to enter a race for the atomic bomb and it is unlikely that the Western democracies would have made as much of an effort and devoted as many resources to a similar program during the second half of the 1930s as they did during the war. However, had either Fermi or Szilard discovered nuclear fission in the mid-1930s, it can be supposed that Nazi Germany would have seized the opportunity to build an atomic bomb and, again, it would not have been Japan to succeed in it first. But would it have mattered whether Germany or Japan built the bomb first? Either outcome would have led to unimaginable calamity. This is why Leo Szilard suggested many years later that Fermi and he deserved the Nobel Peace Prize, not for something they had discovered, but for something they had missed.

In the final analysis, Edward Teller's "fathering" of the American hydrogen bomb contributed to the lasting peace of the four decades of the Cold War. His advocacy of SDI may have been misguided, but its consequences contributed to the collapse of the Soviet Union and the liberation of Central and Eastern Europe, including his native Hungary. Perhaps Teller's self-evaluation could be condensed in what he himself expressed as his goal: "Protect the free world with whatever means".[20]

---

[18] Hiroki Sugita of Kyodo News in *Japan Times,* August 14, 2007.

[19] T. von Kármán (with Lee Edson), *The Wind and Beyond: Theodore von Kármán, Pioneer in Aviation and Pathfinder in Space.* Little, Brown, Boston, 1967, p. 133.

---

[20] Attributed.

Elizabeth L. and James D. Watson in 2002 in the City Park, Budapest (photograph by Istvan Hargittai).

James D. Watson 88: The Discovery of the Double Helix was an Iconic Event in Structural Chemistry[a]

István Hargittai

It is structure that we look for whenever we try to understand anything.
Linus Pauling (1950)

Abstract

The ingenuity of James D. Watson and Francis Crick, the convergence of the advances in X-ray crystallography, the accumulated knowledge of structural chemistry, and the breakthroughs in chemical methods of analysis led to the discovery of the double helix structure of DNA. The discovery catapulted Watson to a career that helped DNA and the applications of the knowledge about its structure triumph in biomedical sciences. Watson's eighty-eighth birthday is an occasion to have a look at his path to success, his personality, and assess his legacy.

Introduction

The discovery of the double-helix structure of DNA (Photo 1) in 1953 was a seminal event in the history of science and a great achievement for structural chemistry [1]. The discoverers, Francis Crick (1916–2004) and James D. Watson (1928–), *suggested* a structure; they did not say they had determined it. It took another two decades of painstaking research when Crick and Watson's proposal received hard experimental evidence.

It happens often, when a scientist makes an important discovery in his or her youth, a less remarkable career

[a]Originally published in *Structural Chemistry* 2016, 27:419–428.

Dedication: This Editorial is dedicated to the great scientific partnership of Francis Crick and James D. Watson on the occasion of Francis Crick's birth centennial and Watson's 88th birthday.

I. Hargittai (✉)
Department of Inorganic and Analytical Chemistry, Budapest University of Technology, Budapest, Hungary
e-mail: istvan.hargittai@gmail.com

follows. In contrast, Crick and Watson remained at the top of science for the next half century. This alone would warrant a closer examination of their activities. In this Editorial, I am going to have a closer look at the lessons Watson's personality and career might offer.

I have been interested in twentieth-century scientists and their discoveries and this has included a fascination with James D. Watson. We met for the first time when my wife Magdi (short for Magdolna) and I visited him in 2000 in his office at the Cold Spring Harbor Laboratory (CSHL). I was recording our conversation and I had an uneasy feeling that everything appeared superficial in our exchange when we had already passed half an hour of the planned one-hour taping. Then, suddenly, things changed and the exchange became meaningful and exciting. We could not stretch much the planned one-hour meeting because and we had to start for the airport—it was the last day of our visit in the United States. Watson took us to the train station and made us promise that we would return for a more substantial visit. I had learned enough about him to know that he would not say such things out of politeness.

Later in the same year Watson and his wife Elizabeth—Liz—visited us in Budapest. Their brief stay included sightseeing, lunch in a Hungarian restaurant, sampling of ice-cream, dinner in our home followed by a meeting, still in our home, with leading Hungarian intellectuals—just as Watson had requested.

We next met in 2002 when Magdi and I spent three months at CSHL as the Watsons' guests. The purpose of the stay was to facilitate my work on my semi-autobiographical book, *Our Lives* [2]. During the subsequent years, we had brief meetings, such as in 2003 during the fiftieth-anniversary celebration of the discovery of the double helix in Cambridge, UK; in 2004 during our visit with

Photo 1 Double Helix—sculpture by Bror Marklund in front of the Biomedical Center of Uppsala University (© 1997 Istvan Hargittai).

Matthew Meselson in Woods Hole, MA; and other occasions. When in the spring of 2007, I was working on my small book, *The DNA Doctor* [3], based on previous conversations with Watson, I experienced some hesitancy in our interactions. When I asked him to give me permission to quote from among his statements in other publications, he declined. Moreover, he did not do this himself but asked one of his associates to call me and tell me about this. This associate was embarrassed conveying Watson's message. Watson's decision, however, was consistent in that he preferred using his material in his own books as he had told me.

Watson's *Avoid Boring People* appeared later in 2007 [4]. As he was preparing for launching the book, he gave an interview to a journalist, who had worked before at CSHL. They spent several hours together. During the interview, Watson made disparaging comments about Africans. When these statements appeared in print, the reactions were devastating for Watson. The CSHL reacted by attempting to dissociate itself from him. When Watson later told me about this experience, he repeatedly used the word "sordid" in characterizing the reaction from CSHL. As I was reading about Watson's humiliation, I wrote him a letter expressing my friendship.

The next time we met was in the spring of 2008 during another of our visits in the United States. This was the first time Magdi and I had been in their Manhattan home. It was then that I fully understood that his ordeal was heavier than I had suspected and it was not over yet. A few days later, Watson asked me to be present at an interview arranged for him by the publicist who had been hired for him. This turned out to be a depressing experience. I knew that CSHL had retired him from his position and the circumstances of the interview with an apparently ignorant journalist were such as if even his independent thinking had also been taken away from Watson.

In contrast, our next meetings in the spring of 2010 and in the fall of 2014, both in their Manhattan home, were uplifting. My impression was that Watson was recovering from his ordeal.

It is possible to view Watson's life in a consistent way, which I attempt below by breaking it into eight periods (Photo 2).

Photo 2 James D. Watson with a double helix model in his left hand in June 1953 at Cold Spring Harbor Laboratory (photo by and courtesy of Karl Maramorosch).

Preparation, 1928–1951

Watson was born April 6, 1928, into a non-practicing Christian family with mostly Irish and Scottish roots. He left his mother's Catholic faith by the age of twelve. The family lived in a not very well-to-do neighborhood of the south side of Chicago. The parents were determined to get a good education for their two children—Watson had a sister, Elizabeth. Watson in his succinct style referred to this as growing up in a quasi-Jewish atmosphere where books were more important than material goods.

Watson went to schools that were not especially remarkable and he breezed through them at an accelerated pace. Although no child prodigy, he was successful in quiz programs on television. He graduated from high school at the age of fifteen and enrolled at the University of Chicago under its maverick president Robert Hutchins who placed the Great Books in the focus of instruction. This broad-based education proved beneficial to Watson. He was more ambitious than most of his peers. When he found a subject that interested him, he was keener to learn about it than anybody else. He did not mind seeing others that were more talented than he was; on the contrary, he sought out their company. He learned from others if there was something to learn, and imitated others when he found people worthy of imitation.

Photo 3 James D. Watson in 2000 in the Hargittais' home in Budapest (photo by I. Hargittai).

He read Erwin Schrödinger's *What Is Life?* and this book more than anything contributed to Watson's transformation from a bird-watcher zoologist into a geneticist. He completed his undergraduate college education by the age of nineteen and began looking for a graduate school. The big-name schools were not kind to him, perhaps because they could not see anything remarkable about him—eagerness can hardly come through in written applications. He ended up at Indiana University in Bloomington in 1947, but Indiana at that time was probably the best place for his further development. It could offer him top graduate education in modern biology. It had the recent Nobel laureate Hermann J. Muller and two future Nobel laureates—three if including Watson—in the same department. This department provided Watson a diverse international environment with a strong European flavor. Watson had a compressed youth because his and Crick's seminal discovery catapulted him early into the big league of science and world fame. His maturity followed more slowly (Photo 3).

Double Helix: The Discovery, 1951–1954

Upon having earned his doctorate, Watson left for Denmark for postdoctoral studies. He was not lucky with his first assignment so he moved to another laboratory, but the project there did not go well either. In the spring of 1951, he attended a meeting in Naples where he listened to Maurice Wilkins talking about the X-ray work on DNA at King's College in London. Watson glimpsed at Wilkins's photograph of an X-ray diffraction pattern, and decided to work on the structure of DNA in Britain.

This was not a decision taken lightly. The funding agency for Watson's postdoctoral fellowship opposed his move, yet Watson was undeterred even when he lost the support that was supposed to sustain him. At this point, he hardly knew anything about X-ray crystallography, let alone its application to biological macromolecules. This was the time when some giants of science were struggling with solving the structure of proteins at the edge of feasibility.

In hindsight, Watson's decision was a sign of genius, but his ignorance must have contributed to making it. Of course, he was not ignorant in many aspects of his subsequent research and it could not be ascribed to ignorance either that he recognized the importance of uncovering the structure of DNA beyond the importance of DNA, the substance. However, he was not clear about the possibilities and limitations of structural chemistry at the time and in particular about those of X-ray crystallography. A certain amount of ignorance is useful when a scientist embarks on an ambitious project. Rita Levi-Montalcini might have had Jim Watson in mind when she stressed in her autobiography the importance of underestimating the "difficulties, which cause one to tackle problems that other, more critical and acute persons instead opt to avoid" [5].

Once Watson had decided on his project, he had to choose the venue for it, and he ended up in the best place for his purpose, in the Cavendish Laboratory in Cambridge. The change from the periphery of science in Denmark (periphery, that is, in molecular biology and not in Niels Bohr's physics) to a world center was to Watson's liking. No sooner had he arrived than he teamed up with Francis Crick, who had a background in physics, was full of ideas, and had been engaged in an unexciting project. They formed one of the most remarkable partnerships in the history of science.

In April of 1952 in Oxford, Watson—as a proxy—presented the results from the experiments of Hershey and Chase of the CSHL. These results reinforced Avery et al.'s findings that DNA was the substance of heredity. Also in 1952, the biochemist Erwin Chargaff visited the Cavendish Laboratory, and told Watson and Crick about his seminal experiments. The essence of Chargaff's discoveries had direct relevance to them: DNA was organism-specific, but the DNA bases adenine (a purine) and thymine (a pyrimidine) occurred in roughly equal amounts as did the bases guanine (a purine) and cytosine (a pyrimidine) in all DNAs, regardless from which organisms they had been extracted.

Scientists congregated in Cambridge, and were anxious to share their latest findings with the researchers there, as if seeking their approval. It was another fortunate circumstance that Linus Pauling had sent his son, Peter, there, and he became friendly with Watson and Crick. The young Pauling was happy to carry the news from his father about progress at Caltech to his new friends. Then, Watson and Crick received a roommate at the Cavendish in the person of the American chemist, Jerry Donohue, who put them on the right track about the preferred chemical forms of the bases in DNA. Watson hardly knew any chemistry at the time of the double helix discovery, but he was always ready to learn what he needed to know.

Watson and Crick did not do experiments, but had access to Rosalind Franklin's diffraction pattern. When Wilkins shared Franklin's observations with Watson, he did so as part of his angry revenge against her rather than in an altruistic move for the sake of advancing science. Wilkins considered Franklin an intruder into his research turf and resented her style. Then, through Max Perutz, Watson and Crick had access to the laboratory report with Franklin's discussion of her experiments. There has been much effort to demonstrate that there was nothing wrong with having communicated Franklin's data to Watson and Crick, but it has been questionable at least whether it was "legal" or not [6]. Nobody has ever suggested that the way Watson went about it was "moral."

In addition to Franklin and her student Raymond G. Gosling's X-ray patterns, Watson and Crick utilized Pauling's approach of relying on all available structural chemistry in their quest for the DNA structure. This was, of course, perfectly legitimate and constituted a brilliant example of how the next discovery builds on previous discoveries about which it utilizes published data and techniques. What Watson and Crick needed to do was "only" to put together all the relevant information after they had done the most crucial act by having posed the right question.

Watson and Crick's paper in April 1953 [1] was barely longer than one page in *Nature* and it stressed that its authors merely *suggested* a structure. However, it had important novel features. One was that it consisted of two helical chains, each coiling around the same axis, but having opposite direction, and thus complementing each other. The other novel feature was the manner in which the two helices were held together through hydrogen bonds between the purine and pyrimidine bases. The bases were joined in pairs, a single base from one helix paired with a single base from the other helix. The two bases in a pair lay side by side, and the complementary pair of a purine base was always a pyrimidine and vice versa. A majestically simple sketch illustrated the report. The structure was consistent with all the information available by then: X-ray crystallography, model building, and chemical analysis of DNA.

Watson and Crick's approach to research was very efficient, but unusual at that time. It was using other people's measurements, techniques, experimental results, and conclusions. Science works this way. Isaac Newton explained that he saw farther than his predecessors, because he stood on the shoulders of others. This is what Watson and Crick did, except that Franklin was their contemporary and they failed to inform her that they had stepped onto her shoulder. In any case, Watson and Crick did not want to let themselves get bogged down with details.

Watson and Crick's working style appeared unorthodox to many. They seemed sloppy, did not seem hard working, and appeared as if they had plenty of free time for entertainment. At times they behaved as if they were underemployed—not the usual image of the mad scientist who lives for his work day and night. Furthermore, they seemed too interested in scientific gossip and not enough in learning from the scientific literature. However, there is no definition of what constitutes the most efficient approach to research, and the unconventional features of Watson's and Crick's approach turned out to be an excellent way to attack the problem they were working on.

Max Delbrück formulated his idea about the usefulness of limited sloppiness, according to which if one is very sloppy that is bad, but thriving for too much rigor might hinder advances. Crick formulated his idea about the advantages of listening to gossip because the grapevine might bring in crucial information that had not yet reached the degree of perfection that would fit publishing it. Finally, hard work and

Photo 4 James D. Watson's portrait of 2003 in Cambridge by Magdolna Hargittai on a book cover [3].

hard thinking do not necessarily appear the same on the surface while the latter may not be less needed in research than carrying out yet another experiment or computation. Not all environments in the world would have so easily tolerated Watson and Crick's way of doing science as the Cavendish Laboratory (Photo 4).

Transition, 1954–1962

During the period from 1954 to 1962, Watson was seeking his role for the rest of his life. It was a transition between the great discovery and Watson's becoming an impresario of science. He first tried to emulate his and Crick's big success in research, but it did not work. He was a good researcher, but unremarkable if compared with his early achievement. He distinguished himself as a professor at Harvard University, but just being a Harvard professor did not satisfy him (while for most it would be a dream position). He did not seem comfortable in a situation, in which however distinguished he

could be, there were others around him similarly distinguished. He built up an excellent laboratory at Harvard and attracted to it first-rate scientists, among them Walter Gilbert, a theoretical physicist and future Nobel laureate for his biological discoveries.

Watson was increasingly recognized for the 1953 discovery by such road posts as the Lasker Award and membership of the US National Academy of Sciences. In 1962, Watson, along with Francis Crick and Maurice Wilkins received the Nobel Prize for the double helix. By then Franklin had died. Had she lived, it is not at all certain that she might have been included in the award (a three-person limit in any category of the Nobel Prize is rigorously observed). In the early 1960s, her contribution to the double-helix discovery was not yet recognized to the extent that it has since.

At this time, Watson embarked on textbook writing that would result in his exceptional *Molecular Biology of the Gene* [7]. It was a first both for its subject and for its unusual, creative style.

The Double Helix: The Book, 1962–1968

The book *The Double Helix* [8] was long in the making, and the story of its publication is symptomatic of Watson and of the environment in which he operated. It appeared in 1968, following clashes with fellow discoverers and with the Harvard authorities for his unconstrained and subjective style. The book became a success and a defining contribution to twentieth-century literature on science. His negative portrayal of the late Rosalind Franklin sparked a re-evaluation of her contribution to the double-helix story and led to its enhanced recognition. The end of this period brought Watson a long-awaited marriage and his initial appointment to the Cold Spring Harbor Laboratory (CSHL).

Cold Spring Harbor Laboratory, 1968–

Initially, Watson was CSHL's part-time director, but in 1976, he left Harvard and became full-time director of CSHL. He transformed CSHL from a dilapidated and impoverished laboratory to an institution of world leadership in biological and cancer research.

Ever since Watson's dedication to it, the CSHL has enjoyed the fruits of his exceptional fund-raising abilities.

Watson reshaped not only CSHL's scientific profile but also its physical appearance to universal satisfaction. In this, his architectural historian wife, Elizabeth L. Watson, proved to be a creative and dedicated partner. Simultaneously with his taking command of CSHL, Watson was one of the leaders

Photo 5 James D Watson lecturing on June 15, 2010, at Moscow State University. Courtesy of MASTER-MULTIMEDIA Ltd. © 2010 Felix O. Kasparinsky.

in molecular biology whose importance had been reinforced by the fast emerging biotechnology. Watson contributed to the movement of scientists that publicly faced the potential hazards of genetic engineering. This movement led to the memorable Asilomar meeting in 1975 that discussed the scientific safety and ethical ramifications of biotechnology. Subsequently, he was instrumental in calming the runaway hostile sentiments by some segments of the public toward genetic engineering. In 1988, Watson stepped onto the national scene in a major way for his next undertaking (Photo 5).

Human Genome Project (HGP), 1988–1992

The HGP became central to Watson's thinking and efforts from the mid-1980s. It is an oversimplification to ascribe the roots of the Human Genome Project to the discovery of the double helix, but it is easy to do so because the structure has such an easily perceived and beautiful appearance. Other factors, most notably the cracking of the genetic code by Marshall Nirenberg and others as well as Frederick Sanger's (and to a smaller extent, Walter Gilbert's) works in creating the techniques for sequencing complex biological macromolecules, played decisive roles in this.

From the mid-1980s, increasingly loud voices called for deciphering the human genome, pointing to the potential

benefits in biomedicine. When the project became a national program in the United States, Watson assumed its administrative leadership in 1988, which proved crucial for the success of the HGP. It was characteristic for Watson's anticipatory thinking and innovative approach that from the start, he had a percentage of the budget of the HGP assigned to the study of societal and ethical issues related to the project. Although Watson was forced out of the HGP leadership in 1992, he has remained a staunch supporter.

Elder Statesman, 1992–2007

For the next decade and a half, Watson continued in a somewhat reduced role at both the CSHL and nationally. In 1993, he resigned from his directorship of the CSHL and became its president, thus removing himself from the day-to-day running of the Laboratory. There was no doubt, however, that he could get involved in micromanaging at any point at the CSHL, and he often did. His dominating presence prevented other strong personalities to consider a leading position at the CSHL. But the Laboratory has thrived. At some point, Watson even felt the presidency superfluous for him and he became chancellor, continuing fundraising and being a major presence but without administrative duties. He had book writing projects of recording everything in minute detail about his own life. This was to change along with everything else in his life in October 2007.

Exit and Twilight, 2007–

In October 2007, there was the scandal that I have already referred to in the Introduction and that had been in the making for many years if considering Watson's recklessness in making politically-not-correct statements. This time, however, he overstepped an important boundary and appeared as if he were a racist, which he definitely was not. Watson underwent the most critical period of his life. He appeared to be no longer master of his fate, and not even of his thoughts. This state continued for several long months. Lately, the situation has slowly consolidated, but Watson's fierce independence seems to be gone for good. In time, Watson has resumed his fundraising activities for CSHL.

Assessment and Legacy

Any student of Watson's life may seek to answer a plethora of questions. Here is a sampler, but no attempt raising all possible questions, let alone answering them all. It will be the task of a future biography.

What does it mean that Watson is a genius (something few would doubt)?

How could someone, not obviously a great scientist, rise to the top in science?

How could Watson stay at the top in science for half a century?

What explains his tremendous authority in spite of his lack of oratorical abilities and in spite of his lack of many positive human qualities?

What is the explanation for the tremendous popularity of the double helix?

Did Watson "make DNA" or did DNA make Watson? How did it happen that Watson has become identified not just with the double helix, but also with DNA itself?

What kind of role model does Watson represent?

What will his legacy be and how far will his influence extend into the future?

The closing sentence of Watson and Crick's seminal paper about the double helix has become a celebrated quotation in the scientific literature: "It has not escaped our notice that the specific pairing [of the bases] we have postulated immediately suggests a possible copying mechanism for the genetic material" [1]. Today, this is commonplace whereas in 1953, it was revolutionary. The double helix structure of DNA came within a decade after the discovery that DNA was the genetic material. When Oswald Avery and his two associates first pronounced it in 1944, few people noticed it and it impacted yet fewer. When, in 1952, Alfred Hershey and Martha Chase showed the same, its acceptance was enthusiastic and broad.

The discovery of the double helix structure of DNA opened a new era in science with a direct route to the Human Genome Project four decades later, and its beneficial consequences in human medicine we cannot yet fully fathom. For years, Watson had doubts about the structure. Only in the early 1970s did reliable crystal structure determinations of DNA, finally, confirm Watson and Crick's original suggestion. It was only then that Watson, finally, had his first good night's sleep about the double helix.

The 1953 discovery catapulted the twenty-five year old Watson to the pinnacle of twentieth-century science. He was an ambitious young man who himself wondered in retrospect about how could it happen to him to "go beyond [his] ability and come out on top" [9]. He had doubts about whether he was bright enough, whether he would at all be able to solve a problem, and whether he would ever have original ideas. He was much sooner a genius than a great scientist, and what happened to him was the fortunate confluence of many factors of being at the right place at the right time, and above all, of being the right person for his self-ordained task.

It certainly was not sheer luck, because it was his decision about what to do and where to continue his career when he faced branching points. Circumstances, too, favored what he decided doing. Watson was very lucky, but he worked hard at finding his luck. He always had the right mentors; supporters; partners; ultimately, the right wife; the right venues for remaking a research place into his own image; and most of all, the right shoulders to stand on in order to look farther. Peter Medawar, the great immunologist, remarked, "Lucky or not, Watson was a highly privileged young man" [10]. It was less his background at home than the environments he eventually sought out for himself that made him privileged.

Watson and Crick never explicitly acknowledged that Watson had had access to Franklin's data, not even in the April 1953 *Nature* paper, and this omission was as much a breach of ethics as snatching the information itself. Watson ignored—whether knowingly or just because he did not care—many minor and not so minor societal conventions. Some of this was on purpose. Legend has it that he was so absent-minded that he often forgot to tie his shoelaces, but it has been observed, when arriving at a party, just before entering the house, Watson untied his shoelaces.

His idiosyncrasies might have made Watson unwanted company, but the opposite happened; they enhanced his popularity. So did many of his manners that went against accepted norms. He mumbled in his lectures, often speaking to the blackboard rather than to the audience, and in a voice hardly audible, yet his audiences eagerly awaited and attended his talks. He was a poor dresser, but was sought out to attend gatherings. He was clumsy and awkward with girls, but the Cambridge ladies threw themselves into helping him find dates and girlfriends.

For six decades, Watson basked in success and it was not a casual relationship, because he thought a great deal about how to succeed in science. He wanted success and he thought about the Nobel Prize already early in his career. Fame was a driving force for him; he set up rules that assured success, and he practiced them. Watson summarized his prescriptions in over a hundred rules in his book *Avoid Boring People*, and that title was one of his favorite rules [4]. The near-obsession has remained with him and on a recent, June 2010, visit to Russia, he enumerated his rules to his eager Russian audience. The students of Moscow State University took his advice very seriously.

When I used to lecture about Watson in my course on the great discoveries in the twentieth century, I told my students that if Watson opened the door to our auditorium and looked for a place to sit down, he would single out the person in the audience whom he would find most interesting. At this moment, usually there was a little commotion; my students looked around as if assessing themselves and their peers, and sometimes one of them shifted in his seat as if making room for Watson (it was invariably a he rather than a she) (Photo 6).

It is a Watson maxim that if you are the smartest person in the room, you are not in the right room. Watson and Crick were roommates at the Cavendish Laboratory in Cambridge and Watson felt comfortable about it. They fortunately complemented each other. Their contributions blended to such a degree that when Crick had to decide about the topic of his long overdue dissertation, he better thought of choosing something form protein structural work rather

than the discovery of the double helix where it proved impossible to disentangle their contributions. Watson and Crick were very different not only as human beings but even more so as researchers. For example whereas both were curious and ambitious, Crick's curiosity was stronger than Crick's ambitions whereas Watson's ambitions were stronger than Watson's curiosity. Crick was a great scientist willing to attack even risky problems if he was sufficiently curious about them. Watson was a great scientist whose ignorance contributed to his decision to study the structure of DNA—he was not fully cognizant of the then possibilities of X-ray crystallography and even of the state of analyzing biologically important macromolecular structures. He was, however, fully aware of the importance of elucidating the structure of the substance of heredity. His going for it against all odds was a stroke of genius.

Watson's keys to success are comprised of a broad domain of traits. They included the ability to distinguish between the important and the unimportant, and he always found time for relaxation. He economized with his time, but when he was doing something that he judged truly needed doing, he spent his time on it liberally. He was very patient when he was cutting out his paper models of the bases for his model as he was on the verge of the discovery of base-pairing in DNA. He paid meticulous attention to the minutest details in writing his textbooks. He devoted a lot of time to the back-and-forth exchanges with his colleagues and friends as he was preparing the publication of his book *The Double Helix*. He paid the most careful attention to all aspects of the planning of new constructions and renovating old buildings at Cold Spring Harbor Laboratory.

It is equally noteworthy what he did *not* do. There are scientists who once they find a fertile area of research, exploit it to the fullest; once they establish a new methodology, apply it to whatever it may be applicable. Others may feel in retrospect that they had moved away too quickly after they had made a discovery. For Watson, it was never a problem to determine when his work would become repetitious without, however, under-utilizing the potentials of an area. After the discovery of the double helix, but only after having made sure that everybody saw its biological implications, he moved on. His negative experience with the study of the structure of RNA and with the quest for the messenger RNA strengthened his determination that instead of trying to top his previous feat in research, he should be seeking his success elsewhere.

He became immensely successful in his new avocations, directing science and authoring books. His next success, his textbooks, covered new grounds and were innovative not only for their contents but for their style as well. His account of the double helix discovery showed the process of scientific research in a way that nobody before him had been capable of or dared. Cold Spring Harbor Laboratory did not merely become singularly successful, including its Watson Graduate

Photo 6 James D. Watson with Istvan Hargittai in 2010 in the Watsons' home in Manhattan, New York (photo by and courtesy of Magdolna Hargittai).

School; it has also become Watson's shrine. However, only time will show whether it will become a lasting success after Watson is gone. He had generated hostility at CSHL due to his methods of enticing success through competition between members of the same group, between groups of the same laboratory, and so on. On occasion, it seemed to his associates that nothing was too sacred to him for the sake of success.

Watson was seldom a player in politics at the national level, but there were exceptions. When President Nixon declared his "War against Cancer," Watson pointed out the futility of the project. He showed that they could spend the money more wisely if they first reached a basic understanding as to the causes of the different cases and the mechanism of actions. He acquired a prominent role in the Human Genome Project between 1988 and 1992, a brief period, but crucial as it was the start of the project. Otherwise, he was seldom involved in politics. His public appearances made headlines for some shocking, but inconsequential statements like the one that fat women have better sex lives than slim women do. Mostly, he was restrained as one who knew what he could say publicly and where to draw the line to keep his views private, with due consideration for his fundraising role for the Cold Spring Harbor Laboratory. This restraint was absent in his 2007 debacle. Due to his age, the scandal could have signified the closing of his career and would have made a sad ending.

Watson, however, was not done yet: he persevered. He managed a comeback. The former whiz-kid, now an octogenarian, has lately been active again, traveling, giving talks, and raising funds—for CSHL. James D. Watson is still going strong. He continues shaping his legacy, which he sees primarily in his books and in CSHL. His image building has long focused on making him identified with DNA. He knows that the fame of an individual based on scientific discoveries is fragile. His haunting experience in 2007 reinforced the necessity of a stronger basis for his legacy than an institution.

Sydney Brenner, one of the architects of modern molecular biology, stated: "Worrier or Warrior, Jim has been the guardian of DNA for the past 50 years" [11]. Watson's legacy may be dependent on his success in having become identified with DNA, not just its structure, but also the substance. Nobody could ever destroy DNA—it is eternal.

Acknowledgments Magdi and I are grateful to Jim and Liz Watson for their friendship and hospitality extended to us over the years. I thank Robert Weintraub and Irwin Weintraub of Beersheva for critical reading of the manuscript and for helpful suggestions.

References

1. Watson JD and Crick FHC (1953) A Structure for Deoxyribonucleic Acid. *Nature* April 25, pp 737–738
2. Hargittai I (2004) *Our Lives: Encounters of a Scientist* (Budapest: Akadémiai Kiadó)
3. Hargittai I (2007) *The DNA Doctor: Candid Conversations with James D. Watson* (Singapore: World Scientific)
4. Watson JD (2007) *Avoid Boring People (Lessons from a Life in Science)* (New York: Alfred A. Knopf), pp 343–347
5. Levi-Montalcini R (1988), *In Praise of Imperfection: My Life and Work* New York: Basic Books, p 5
6. See, e.g., the notes by Max Perutz and others (1969) in *Science* June 27, pp 1537–1538, following the publication of Watson's book *The Double Helix*.
7. Watson JD (1965) *Molecular Biology of the Gene* (New York: WA Benjamin), First Edition
8. Watson JD (1968) *The Double Helix: A Personal Account of the Discovery of the Structure of DNA* (New York: Atheneum)
9. "Our Future Scientists (Panel Discussion)." *The New York Academy of Sciences Magazine* 2009 Spring, pp 22–24; actual quote, p 22
10. Medawar P (1982) "Lucky Jim." In *Pluto's Republic* (Oxford: Oxford University Press), pp 270–278; actual quote, p 275
11. Brenner S, "Jim and Syd." In Inglis JR, Sambrook J, and Witkowski JA, Eds (2003) *Inspiring Science: Jim Watson and the Age of DNA* (Cold Spring Harbor, NY: Cold Spring Harbor Laboratory Press), pp 67–69; actual quote, p 69

Istvan Hargittai and Marshall W. Nirenberg in 2001 in Stockholm during the centennial festivities of the Nobel Prize (photograph by Magdolna Hargittai).

Deciphering the Genetic Code: Marshall Nirenberg[a]

István Hargittai

Marshall Nirenberg in 1999 in his office at NIH (photograph by Istvan Hargittai).

Marshall Warren Nirenberg (b. 1927)[1] is Chief of the Laboratory of Biochemical Genetics, National Heart, Lung and Blood Institute, National Institutes of Health (NIH) in Bethesda, Maryland. He received his B.S. degree in zoology and chemistry in 1948 and his M.S. degree in zoology in 1952, both from the University of Florida, and his Ph.D. degree from the University of Michigan in 1957. He has been at NIH since 1957 and has held his present position since 1966.

Dr. Nirenberg received the Nobel Prize in physiology or medicine in 1968, sharing it with Robert W. Holley (1922–1993) of Cornell University and Har Gobind Khorana (b. 1922), then of the University of Wisconsin, "for their interpretation of the genetic code and its function in protein synthesis." He has been a member of the National Academy of Sciences, the National Academy of Medicine, the Pontifical Academy of Sciences, and others. He has received the Molecular Biology Award of the National Academy of Sciences (1962), the National Medal of Science (1965), the Prix Charles Leopold Meyer (French Academy of Sciences; 1967), the Lasker Award (1968), and many other distinctions.

Dr. Nirenberg first made the announcement of the "first word to be identified in the genetic code: 'One or more uridylic acid residues [poly (U)] appear to be the code for phenylalanine'" to the Fifth International Congress of Biochemistry in Moscow in 1961. That announcement catapulted him to fame and recognition.

Marshall Nirenberg's autobiography is one of the shortest on the Nobel web site. I wonder if Marshall Nirenberg opens up often. I think not. So I felt privileged to be able to spend the afternoon with him on April 1, 1999. As we met in the hall outside his laboratory, I recognized him at once from a 35-year-old photograph. He had not changed much, except that his dark hair had turned white. As we started talking in his cluttered and windowless office, he made me feel comfortable, as if I had known him for a long time.

Nirenberg puts down his ideas every night in his notebooks. So far he has collected thousands of ideas in 40 volumes, which he is going to donate to the National Library of Medicine. Nirenberg compares his ideas to darts thrown at a target on the wall. You not only cover the target

[a]Originally published in *The Chemical Intelligencer* 2000, 6(4):48–53.

[1] Marshall Nirenberg died in 2010.

I. Hargittai (✉)
Department of Inorganic and Analytical Chemistry, Budapest
University of Technology, Budapest, Hungary
e-mail: istvan.hargittai@gmail.com

but you cover the entire wall with darts. A few of them, though, will hit the bull's-eye. A good part of Nirenberg's life must be in those 40 volumes, so future scholars will probably have easier access to him than his contemporaries.

Nirenberg was born in New York in 1927. His father went to medical school for several years, but when he got married, he dropped out and joined the family business. Nirenberg's paternal grandfather came to the United States from Odessa, Russia, when he was 12 years old. He got a job in a shirt factory, where he slept on the floor; eventually, he owned the shirt factory and built it into a nationwide business. He had five sons and a daughter, and all the sons went into the shirt business.

Nirenberg's maternal grandparents came from somewhere in Poland. His mother was a university graduate. His father always regretted that he hadn't finished medical school, and he encouraged his son to study. Both of Nirenberg's parents died before he had visible success, but he'd told them about his work.

When Nirenberg was a child, he went to the synagogue on the High Holidays but he is not religious. After the genetic code was deciphered, he wondered whether the code was the same in *E. coli*, in amphibians, and in mammals. They prepared tRNA, they did all the experiments, and they found that the code was the same, that it was essentially a universal code. He was familiar with evolution and Darwin and understood them, but this was on a different level. When looking outside the window, seeing the trees and seeing the squirrels, and knowing that the genetic code of these organisms was the same, or essentially the same, as the genetic code in him, he found that a very powerful philosophical concept. It had a profound effect on him, and he found an almost religious significance in this unity of Nature.

Nirenberg had had rheumatic fever from age 8 to 13. At that time, nobody knew what caused it, it was a big killer among children, and bed rest was all they could prescribe for him. At one time, he spent a whole year in bed. They also thought that a warm climate would be better for him, so his father decided to give up the shirt business and the family moved to Orlando, Florida. His father started a new business there.

Nirenberg has been interested in biology from early childhood. When he went to camp as a 6- or 8-year-old, he always won the nature award. Florida was still unspoiled when his family moved there. They lived on the outskirts of Orlando, which was a small town at that time, way before the arrival of Disney World. There was a tremendous swamp nearby that went on for 20 miles without a house in sight. Every afternoon, after school, he roamed the swamp. He collected spiders for the American Museum of Natural History.

Nirenberg was in high school during World War II. There was a big air base at Orlando, and professional biologists from all over the United States were brought in to teach a jungle survival course to pilots who were going to the South Pacific. One of them was an ornithologist, and Nirenberg was very much into identifying birds at the time. With two other boys, he went to talk with this ornithologist, Lieutenant Frank McAmey, who invited them on birding expeditions. On one such trip, they went to Merit's Island, later renamed Cape Canaveral, to see a rare bird called the dusky seaside sparrow, which is now extinct.

After receiving his B.S. degree in zoology and chemistry from the University of Florida, Nirenberg joined his father in a candy manufacturing business, but he didn't like it and decided to go back to the University of Florida for another two years to earn his M.S. degree in zoology. Nirenberg had taught comparative anatomy as an undergraduate, and while he was in graduate school, he had a part-time job as a technician in the nutrition laboratory, working with radioactivity. This was his first exposure to biochemistry and he got hooked. Looking for a graduate school for doctoral studies in biochemistry, he wrote to three places, Wisconsin, Michigan, and Duke. He was accepted by all three, but Michigan offered him a fellowship to work in the lab and teach medical students. He enjoyed the years in Ann Arbor, where Jim Hogg was his mentor.

Nirenberg started working at the NIH right after his doctorate, on a postdoctoral fellowship. Upon its completion, he wrote to François Jacob because he wanted to spend some time in Paris at the Pasteur Institute. Some of the most exciting work that came out of molecular biology in the 1950s was that carried out by Monod and Jacob at the Pasteur Institute, using a model system to study how genes were regulated. Their studies involved genetic experiments on *E. coli* bacteria. But nobody knew from a molecular point of view how genes were regulated. Nirenberg and his colleagues called these elegant and beautiful experiments the latest Parisian fashion. Alas, Jacob, apparently flooded by applications, did not take Nirenberg. Thus, Nirenberg never spent any considerable length of time away from the NIH. In 1961, Nirenberg married Perola Zaltzman. She is a biochemist, and although they never worked together, they discussed their work all the time.

To Nirenberg, in the late 1950s, protein synthesis was the hottest field in biochemistry. The best biochemists in the world were working on the biosynthesis of proteins. They had just discovered transfer RNA and the amino-acid-activating enzymes that catalyzed the activation of transfer RNA to link an amino acid to a particular species of transfer RNA. They also knew that proteins were synthesized on ribosome particles in the cells. But nobody knew anything about the messenger.

This was the first problem that Nirenberg worked on as an independent investigator at the NIH, where he stayed on after his postdoctoral fellowship. He asked himself, "What chance do I have as a single person against the best people with big groups in the best laboratories of the world who are working on protein synthesis?" He thought that within two years protein synthesis would be solved and then he would set up a protein synthesizing system and would study how to regulate gene expression by proteins in a cell-free system.

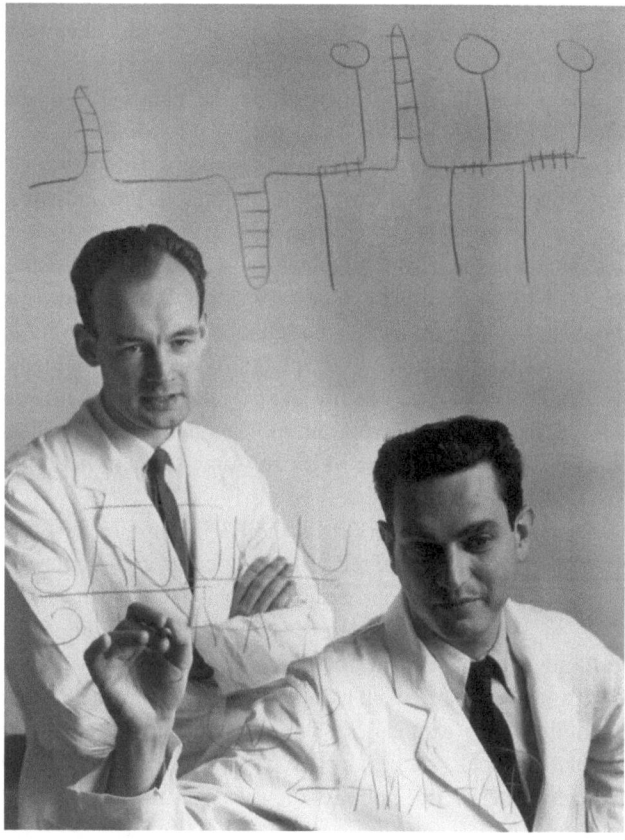

Heinrich Matthaei (left) and Marshall Nirenberg (right) about the time of their discovery (from a display at the National Institutes of Health).

As he started his research, Nirenberg hypothesized that probably RNA was transcribed from DNA and that probably RNA was the template for protein synthesis. But he thought that protein synthesis could also come directly from DNA. So he tried both, and his first experiment worked. He realized that RNA was stimulating protein synthesis. From there on, he just had to improve the system. Heinrich Matthaei,[2] a plant pathologist from Germany, was Nirenberg's first postdoctoral fellow. In 1960, Matthaei came to the NIH to work on protein synthesis, and at that time Nirenberg was the only person there who was working on protein synthesis.

Nirenberg speaks about him with admiration and regrets that Matthaei decided not to stay at the NIH when his postdoctoral fellowship ended. Rather, he returned to Germany, got a large laboratory, and did a lot of work at the Max Planck Institute in Göttingen until his retirement. In their joint project, they ran concentration curves of everything in the reaction to get the optimum conditions. First, they used ribosomal RNA and then viral RNA, which enhanced their efficiency by several orders of magnitude, and then they tried synthetic polynucleotides, synthetic RNAs.

Nirenberg had found that viral RNA was a tremendous stimulator of amino acid incorporation into a protein. Heinz Fraenkel-Conrat at Berkeley, a great authority on tobacco mosaic virus, had some mutants that Nirenberg thought were directing the synthesis of viral coat proteins. He went to Berkeley to work with Fraenkel-Conrat for a month, using viral RNA to direct protein synthesis. He packed a bag of enzyme-containing extracts in a picnic basket and brought it on the plane. This was at the beginning of 1961. Before he left, he wrote a whole series of protocols for Matthaei, using synthetic polynucleotides in a cell-free system, and Matthaei was actually the person who did the first poly(U) experiment. When this happened, he immediately called Nirenberg, who returned to Bethesda at once.

Nirenberg goes out of his way to stress that deciphering the code was a group project. Many people contributed to it at the NIH, like Maxim Singer and Leon Heppel and the postdoctoral fellows who came to work with him. When the book *The Eighth Day of Creation* comes up in conversation, Nirenberg calls it an awful book. He finds the description of how the code was deciphered terrible and thinks that the author, Horace Judson, didn't capture the atmosphere in the lab at that time in the book, although they had talked several times. Since Nirenberg does not elaborate on his criticism, I can only guess that he might have found the description of Matthaei's contribution exaggerated in the book. He stresses over and over that deciphering the code was teamwork. It took about six years, between 1961 and 1966. They worked as hard as they could, and he had about 20 postdocs on the project, at most nine at any one time. The atmosphere in the lab was exhilarating, everything they touched worked, and discoveries were coming fast and furious. He compares it to walking into a toy store and seeing all the toys on the shelves and being able to pick whichever one they wanted to play with.

There was great excitement when they discovered that a sequence of U's in poly(U) is the RNA code word for phenylalanine. They then started to make randomly ordered polynucleotides with different bases, mixtures of bases, and so forth. They did not know the sequence, but they knew the base composition. Nirenberg was sure that it had to be a triplet code because there are four bases and a doublet code could only provide 16 permutations, not enough for the

---

[2] Heinrich Matthaei (1929–) lives in Germany.

20 amino acids. It took about three years to determine the base compositions of codons for the 20 amino acids. Then they had to determine the base sequences. For this, they tried at least ten different approaches. The approach that really worked well was a very simple one. He posed the following question: "What's the smallest message that we could detect?" They could indeed determine the sequence in each codon by simple experiments, allowing them to decipher the code between the nucleic acids and the proteins. Most of the trinucleotides had never been prepared before, and the major problem was to make the 64 triplets for testing.

As we neared the 1961 Moscow announcement of the discovery in our conversation, it was like approaching the crescendo in a musical movement. He knew that he really had a terrific thing to report at the meeting. However, being unknown in the field, in fact, in any field, he was assigned 10 min in a tiny room with a giant-size projector, and there was only a handful of people. He had introduced himself to James Watson before the talk, somewhere in a hallway, and told him what they had done. Watson sent a colleague, Arnold Tissières, to Nirenberg's talk, and he then corroborated what Nirenberg had told Watson. Word soon reached Francis Crick, who chaired a large session on nucleic acids at the Moscow meeting, and Crick invited Nirenberg to give his talk again, at the same meeting, but this time to a much bigger audience. When Nirenberg gave the talk the second time, he was overwhelmed by the response.

This was indeed the first time that anybody had shown definitively in an in vitro system that RNA directs protein synthesis. It was obvious that they had identified the first codon. They had shown that polyuridylic acid, poly(U), directs the synthesis of polyphenylalanine, a protein. A series of U's in RNA corresponded to the amino acid phenylalanine. This was the beginning of the deciphering of the genetic code, the translation between the structure of nucleic acids and the structure of proteins. A synthetic RNA, containing only one kind of base out of four, when added to their cell-free protein synthesizing system, directed the synthesis of a protein consisting of only one kind of amino acid, out of the 20 amino acids. This is what they proved and this is what Nirenberg presented at the meeting.

As soon as Matthaei first reported to Nirenberg the results of his experiment, Nirenberg understood that this was a staggering discovery and that they had to be very convincing. He knew he had to "prove it upside, downside, and every which way." He wanted an unambiguous physical characterization of the polyphenylalanine product. They isolated it, demonstrated its stoichiometry, that all of it was incorporated into a product that precipitated in trichloroacetic acid like most proteins do.

Marshall Nirenberg in 1967 (courtesy of Marshall Nirenberg).

If the discovery was serendipitous, proving the identity of the protein wasn't lacking serendipity either. Nirenberg wanted to find out something about the physical properties of polyphenylalanine. On his way to the library, he went by Christian Anfinsen's lab, which was just one flight below his. Anfinsen was an expert on protein structure and he was to win the Nobel Prize for his work on the structure of ribonuclease in 1972. Nirenberg didn't find him in the lab; instead, he found a young visitor, Michael Sela, who was spending his sabbatical with Anfinsen. He later became President of the Weizmann Institute. Sela was very knowledgeable about synthetic peptides and he used them in immunology. He told Nirenberg that polyphenylalanine was very insoluble, that it wasn't soluble in normal solvents, but it did dissolve in 15% hydrobromic acid dissolved in glacial acetic acid. Sela just happened to have this solvent at hand and offered a sample to Nirenberg.

Only about 20 year later did Nirenberg find out that Sela was the only person in the world who knew about that solvent. It so happened that Sela had once made a mistake when he was using polyphenylalanine as a reagent to study carboxy-terminal amino acids of proteins. Instead of doing his assay in the right way, he poured a suspension of polyphenylalanine in the wrong solution, which was this particular combination, and polyphenylalanine dissolved to Sela's surprise. As it turned out, on his way to the library, Nirenberg asked the only person in the world who could answer his question.

The discovery that was the first step in deciphering the genetic code raises an interesting question with broader implications about scientific discovery. Here was a tremendously important question with enormous competition among the best laboratories in the world in looking for the answer, and a beginning scientist not only hits on such a seminal problem, on his own, but also finds the solution. However, Nirenberg hastens to point out that there was yet at least one other person dealing with the same question independently. It was Severo Ochoa, one of the world's leading biochemists, who had won the 1959 Nobel Prize in physiology or medicine for his discovery, with Marianne Grunberg-Manago, of polynucleotide phosphorylase. Apparently, Ochoa was on the right track but did not have the serendipity that helped Nirenberg and Matthaei so much.

For Nirenberg and Matthaei, the experiment with poly (U) proved to be lucky indeed. The sample was a gift from a colleague, Dan Bradley, at the NIH. Bradley was a physical chemist who studied the properties of nucleic acids. He synthesized poly(U), which would not have been available commercially at that time. Nirenberg remembers with gratitude that Leon Heppel, one of the few great experts in the world on nucleic acid chemistry, was also at the NIH and he was a tremendous help, and so were others. Heppel gave Nirenberg his first triplets, and he also gave him the idea of an enzymatic method to synthesize triplets, all 64 of them.

Another fortunate feature of Nirenberg's situation was that the NIH supported his research. He could do whatever he wanted to do, and he didn't have to write grant applications. The NIH didn't mind either that he went on without a paper for 2 years. Had he applied for a grant, chances are that he would have been turned down because of his complete lack of any experience related to what he would have been proposing to do. When he applied to attend a Cold Spring Harbor symposium, prior to the Moscow meeting, he was turned down because he was unknown in the field.

Immediately before Nirenberg left for Moscow though, he had written two papers for the *Proceedings of the National Academy of Sciences (PNAS) of the U.S.A.* Since he was not a member of the Academy, Joseph Smadel, the Vice-director of the NIH, submitted them. However, when he later asked Smadel to sponsor yet another of his papers in *PNAS*, Smadel said, "Nirenberg, I've done enough for you."

Nirenberg also tells about an experience with Leo Szilard in this connection. Szilard was a member of the National Academy of Sciences and he was in Washington, so Nirenberg had asked him to sponsor his first papers in *PNAS*. When Nirenberg called him, Szilard invited him down to the Dupont Hotel, where Szilard lived at that time, on Dupont Circle. His office was the lobby of the Dupont Hotel. He was deeply involved in questions of defense, he knew all the people from the Pentagon and from official Washington. They would come to the hotel, and he would confer with them. So Nirenberg came, and since Szilard was a physicist, he asked Nirenberg to explain the work to him. Nirenberg spent the day talking to Szilard in the lobby of the Dupont Hotel, explaining to him what they'd done and what

the implications were. People passing through kept interrupting them all the time. At the end, Szilard said, "It's too much out of my field. I'm sorry, I can't sponsor it." This was an unexpected ending to this charming story because Szilard was famous for his foresight and, at the time of their meeting, he had already been much involved in biology. This makes this encounter even more interesting. Nirenberg and Szilard later developed a cordial relationship and saw each other on numerous occasions when Szilard went to the NIH to learn about molecular biology.

It was upon completion of the deciphering of the genetic code but before the Nobel Prize that Nirenberg decided to change fields and go into neurobiology. He was attracted by the challenge and he was always interested in the nervous system. When about a year later he won the Nobel Prize, it did not change his life much because for him the continuity of work was the major determinant. He continues doing basic research to this day. He does not patent, never has.

His interest is in the development of the nervous system, and he speaks about his present work in neurobiology with enthusiasm. He finds the way the nervous system develops in *Drosophila* incredible. The homeobox genes have provided the opportunity to investigate the early events in the development of the nervous system. They encode proteins that bind to DNA and regulate gene expression. They turn genes on and they turn them off. Thomas Morgan found two of the homeobox genes in *Drosophila* 90 years ago. One is *Antennapedia*, which converts the antennae to a pair of legs, and the other is *Ultrabithorax*, which converts the fly with two wings to a fly with four wings. These single mutations have become the favorite topics for genetic exploration.

Nirenberg and his associates discovered four homeobox genes in *Drosophila*. One of them turns on neural development in part of the central nervous system. It is called NK2. They were the first to clone it, but it was discovered genetically earlier and was called VND, the ventral nervous system defective. During evolution, the NK2 gene has been duplicated in mammals. In the mouse, there are seven copies of this gene that have different roles. Mutations in two of them have been studied. Knocking out one of these genes results in mouse embryos that lack the ventral part of the hypothalamus in the brain, their lungs are missing, and the thyroid is missing. When the other gene is knocked out, the ventricle of the heart doesn't develop. So at least one of the genes has a key role in the development of the brain and the central nervous system, and the other, in the development of the heart.

Nirenberg compares his laboratory of about 25 people to a university department. There are four independent investigators, each having a group. In his group, there are eight postdoctoral fellows and two technicians. One of them has been with him for decades.

The long afternoon is ending, but before we leave, Nirenberg patiently subjects himself to a photography session. He photocopies some hard-to-find papers for us and sees us off. After we bid each other good-bye, I see him walking back down the hallway. Soon, he will be sitting at his huge desk in his cluttered, windowless office, beginning slowly, comfortably, today's entry in his latest notebook.

Jeanne and Frank Westheimer at Owl's Head on Squam Lake, Center Harbor, New Hampshire (photograph by István Hargittai). In the following article István describes how moving it was when Westheimer talked about his interaction with James B. Conant. István himself becomes visibly moved when he narrates the story. He told me that there was one more moving moment in his encounter with the Westheimers: Jeanne Westheimer told István that she hoped István would write nicely about her husband.

Frank H. Westheimer (1912–2007)[a]

István Hargittai

Frank Westheimer was an influential and versatile chemist whose activities cut through the traditional borders of different branches of chemistry. He did his studies in the 1930s, receiving his Ph.D. at Harvard University in 1935. He started his academic career at the University of Chicago and continued it at Harvard University. During World War II, he worked at the Explosives Research Laboratory of the National Defense Research Committee (NDRC) in Bruceton, Pennsylvania. He chaired the Committee of the National Academy of Sciences of the U.S.A. in 1964–1966 that surveyed chemistry and produced the famous Westheimer Report.

Frank Westheimer at Owl's Head on Squam Lake in Center Harbor, New Hampshire, 1995 (photograph by I. Hargittai).

Westheimer's father was a stockbroker and he was expected to go into his father's business. During his college years, however, he fell in love with chemistry, and he never left it for the rest of his life. Few structural chemists are aware of the fact that Westheimer was a pioneer of what we today call "molecular mechanics." It happened after the war when he returned to the University of Chicago. He had no ongoing research and no students and he stumbled upon the idea of calculating the steric effects from first principles and classical physics, relying on known values of force constants and van der Waals constants. In his energy minimization calculations he asked for advice from Joseph Mayer and molecular mechanics got started from their interaction. There were similar attempts by others, but Westheimer provided a "reduction to practice" as he called it and therefore it had considerable impact.

[a]Originally published in *Structural Chemistry* 2008, 19:361–362.

I. Hargittai (✉)
Department of Inorganic and Analytical Chemistry, Budapest University of Technology and Economics, Budapest, Hungary
e-mail: istvan.hargittai@gmail.com

His interest in the steric effects manifested itself in biochemical studies, such as his work on direct and stereospecific transfer of hydrogen in the enzymatic oxidation of ethanol. There, alcohol dehydrogenase could distinguish between the two hydrogen atoms on the methylene group. Although Westheimer was generally counted as an organic chemist, at one time he served on the Editorial Board of the *Journal of Chemical Physics*. In fact, he started out as a physical chemist, became a physical-organic chemist, and, finally, a biochemist.

He had many achievements in various branches of chemistry and was richly decorated by important awards, including the Willard Gibbs Medal (1970), the Robert A, Welch Award (1982), the Cope Award (1982), the National Medal of Science (1986), the Priestley Medal (1988), and others. In the 1960s, the National Academy of Sciences of the U.S.A. set up a series of committees to survey the state of the various sciences. One of the purposes was to advise the government about financing pure research. Westheimer was selected as chairman of the committee for chemistry. The title of their report was Chemistry: Opportunities and Needs. He solved his task splendidly although it was a great challenge. As he said, preparing a report about physical science is not itself physical science; it is social science. Among others, they pointed out the importance of pure research as being the roots of important chemical inventions. They selected 20 of the most important industrial and 20 of the most important pharmaceutical inventions, and studied the first papers where the inventions were reported. Judging by the references in those papers they could see what the inventors themselves considered to be the important underpinnings of their own inventions. They determined that the vast majority of the references came from academic, fundamental research.

I recorded a long conversation with Frank Westheimer during my wife's and my visit with him and his wife in their summer house at Owl's Head on Squam Lake in Center Harbor, New Hampshire, on July 23, 1995. At that time his main interest was in RNA catalysis. He was fascinated by the notion that the first living organisms on earth may have used RNA rather than DNA and from this followed that evolution would progress from RNA to DNA. He was telling about his research, again, describing the fundamentals and the potential applications.

In 1995, I was at the beginning of my interviewing project, so I did not realize it then that my Westheimer interview would remain one of my most memorable interviews ever. At some point, I asked him about people who had the strongest influence on his career. He singled out James Bryant Conant. Westheimer had originally wanted to be Conant's doctoral student, but by the time he had arrived at Harvard, Conant had become Harvard's president. So he did his work with someone else. When he was finishing, he won a prestigious fellowship for postdoctoral work. From this point I let Westheimer take over the narrative:

About that time, Conant called me into his office. He said that he knew I was getting my doctorate and was interested in my career. What was I going to do? I told him I'd won this fellowship and explained with great pride the problem I'd submitted and was going to work on. Conant had the habit of putting the tips of his fingers together and rocking back and forth while he thought, and he put the tips of his fingers together and rocked back and forth, and then he said, "Well, if you are successful with that project, it will be a footnote to a footnote in the history of chemistry". As I walked out of his office, I realized what he had told me.

Really, it was two things. One was, of course, that my project wasn't very important. The other was—and it may have been pretty stupid that I had never thought this until that moment—that I was supposed to do important things. Chemistry was a lot of fun; it was great entertainment, and I was going to be paid—or at least so I hoped—for entertaining myself with it. Yet Conant had essentially told me that I was expected to do things that were scientifically important. The interview with Conant provided a vital kick in the pants for me. It changed the way I thought about my future. At Columbia, I did the project that I had proposed, and it worked out beautifully. But it was exactly what Conant had said it was, a footnote to a footnote in the history of chemistry.

Then I set my sights higher—much too high, as a matter of fact. As a physical-organic chemist I was concerned with general acid—general base catalysis and had decided that enzyme catalysis was probably caused by simultaneous general acid—general base catalysis. I was going to demonstrate this in my next piece of research. Amino acids, with their combination of acid and base in the same molecule, should prove especially active catalysts themselves. So I tried their catalysis of the mutarotation of glucose, but it turned out that there was nothing special about them. The project was obviously enormously ambitious, and although I was fundamentally correct about enzymes, demonstrating it was much too big a project for me at the time. That attempt came to nothing, but at least Conant wouldn't have been able to object that the attempt was directed at a footnote to a footnote. Eventually I settled down to things that were more important than footnotes to footnotes, but not as grandiose as the youthful project that I just mentioned. I never discussed my research with Conant again, but I did restrict myself to things that he might have approved of.

Many years later, after I was a Professor at Harvard, and after Conant had retired from his many careers, I was working in my office one Saturday when someone knocked on my door. I opened it, and there he stood. He looked at me and said, "Do you remember me?" Needless to say, I did.

When Westheimer quoted the sentence, "Do you remember me?" he was so moved that we could not continue the recording. He added the last sentence in our subsequent correspondence. His story about Conant was an example that even a "kick in the pants" can serve as a life-time mentor. Sometimes I repeat what Westheimer told me about his experience with Conant in my talks about my interviews. When I reach the point of Conant standing in the doorway to Westheimer's office I am much moved, too.

Reference

1. Hargittai I (2000) Frank H. Westheimer. In: Hargittai I Candid Science: Conversations with Famous Chemists. Imperial College Press, London, pp 38–53

Giulio Natta with his daughter, Franca Natta Pesenti (courtesy of Franca Natta Pesenti). She was very helpful to the authors in collecting the material for the following article.

Centennial: Giulio Natta[a]

Winner of the 1963 Chemistry Nobel Prize was First to Make Synthetic Stereoregular Polymers

István Hargittai, Angiolina Comotti, and Magdolna Hargittai

Giulio Natta in the workshop experimenting (courtesy of Franca Natta).

In the cemetery in Bergamo, Italy, an unusual sculpture stands at the edge of a green plot to the right of the entrance. The model of a helical molecule decorates the burial place of Giulio Natta and his family. Natta shared the 1963 Nobel Prize in Chemistry with Karl Ziegler "for their discoveries in the field of the chemistry and technology of high polymers."

Born 100 years ago this month, Natta's fame transcends awards: In everyday life, we use many products that he discovered how to produce. His co-laureate Ziegler had found new methods of polymerization using organoaluminum compounds as catalysts, leading to new synthetic materials. Natta determined that certain types of organoaluminum compounds—known as Ziegler–Natta catalysts—result in stereoregular polymers; that is, macromolecules with spatially uniform geometry. In such a molecule, called isotactic, all the side chains point in the same direction.

The model of a helical polymer as Giulio Natta's tombstone in the cemetery of Bergamo (photograph courtesy of Magdolna Hargittai).

[a]*Chemical & Engineering News* 2003, February 10, pp. 26–28. © 2003 American Chemical Society.

I. Hargittai (✉) · M. Hargittai
Budapest University of Technology and Economics, Budapest, Hungary

A. Comotti
University of Milano-Bicocca, Milan, Italy

Stereoregular macromolecules had been known in nature, but Natta was the first to produce them in the laboratory and, consequently, in industry. Arne Fredga, the Swedish organic chemist who introduced the chemistry recipients at the 1963 Nobel award ceremony, said: "Nature synthesizes many stereoregular polymers, for example, cellulose and rubber. This ability has so far been thought to be a monopoly of Nature operating with biocatalysts known as enzymes. But now Professor Natta has broke *this* monopoly."

Education and Family Natta was born in Imperia (then Porto Maurizio), Italy; near the French border, on Feb. 26, 1903. His father was a judge in Genoa, where the family spent the winters. A child prodigy, Natta started school at an early age. When he was eight years old, he had already decided to become a chemist rather than a judge, which disappointed his parents. Instead of playing, he preferred to experiment with chemicals in the basement of their home, to write poems, and to read. One of his favorite books was Dante Alighieri's "Divine Comedy."

Natta kept a diary in which he made sharp observations. At age nine, he described the snail as a modest and deserving animal whose shell has intrigued architects and whose protruding and withdrawing horns, which resemble a telescope, have intrigued astronomers. Examining the snail may have given him his first exposure to the spirals and helices that became important for him in polymeric structures.

Natta studied mathematics at the University of Genoa for two years. He moved to Milan in 1921, enrolling in the course of industrial engineering (chemistry) at Milan Polytechnic Institute. He started doing research in 1922 at Polytechnic's Institute of General Chemistry under the direction of Giuseppe Bruni and his associate Giorgio Renato Levi. Jacobus H. van 't Hoff—the first chemistry Nobel laureate—had been one of Bruni's teachers. In addition to his work in the university laboratory, the young Natta set up a well-equipped laboratory at home. He graduated at the age of 21 and continued as Bruni's assistant.

From the beginning, Natta could see in Bruni's laboratory the concerted interest between purely scientific problems and industrial applications. Natta accepted as his own Bruni's maxim: "The only difference between theoretical and industrial problems is that the latter are more difficult to resolve because you have to take into account a lot of factors that can be neglected in the former." He also subscribed to the maxim attributed to Confucius: "The essence of knowledge is, once you've got it, apply it." For Natta, chemistry's scope ranged from studying molecules in the laboratory to building industrial plants to producing substances of interest.

Natta married Rosita Beati, a graduate of the arts, in 1935. Their honeymoon included a visit to Herman Mark, a polymer chemist and X-ray crystallographer in Vienna. Rosita

was devoted to Giulio, doing everything she could to facilitate his work. She ran an efficient "public relations" office at home that took care of sending Christmas cards (about 800 annually), subtle lobbying for the Nobel Prize, and answering the congratulations afterward.

Devoted: Natta with his wife, Rosita, around 1963 (courtesy of Franca Natta).

It was Rosita to whom Natta turned for advice in giving generic names to the stereoregular polymers. Natta knew Greek names were preferable, but he had not learned Greek. His wife was well versed in classical languages and helped devise the names atactic, isotactic, and syndiotactic. An atactic polymer has randomly oriented side chains, an isotactic one's side chains all point in the same direction, and a syndiotactic polymer's side chains are distributed in a regular left-right sequence.

According to his son, Giuseppe, Natta lived for his science alone and was more a father to his students than to his children. According to his daughter, Franca, however, Natta was always ready to give her advice. In the summer, they used to go for long walks in the woods and collect mushrooms. She learned a great deal from Natta during these walks. But he was absentminded; on a walk with his son, he lost him and came home saying, "I remember I went out with something in my hand."

It was a tremendous blow for Natta when, in 1968, he lost his wife. From the late 1950s, Natta had suffered from a debilitating illness but continued working. (Fredga referred to this in the Nobel Prize presentation speech, expressing "the admiration of the academy for the intensity with which you are continuing your work in the face of difficulties.") Natta was already very ill when his wife died. He re tired in 1973 and spent his last years in the loving care of his daughter in Bergamo, where he died in 1979.

Career Moves Natta started his research career in the field of preparative organic chemistry; studying the formation of mercaptans and alkyl sulfides and the analogous selenium and tellurium compounds. After graduation, he synthesized various homologs of mustard gas during his military service at the laboratories of the Military Chemical Center in Rome. Showing his passion for research, he tested these compounds on himself to prove their blistering properties; the scars could be seen on his wrist for years. This work showed his superiors that they could entrust him with independent work, and they transferred him to Milan for the rest of his service. Subsequently, he worked on the synthesis of methanol and higher alcohols and studied the structure of various catalysts used in this work. He made important observations on the physical, surface, and volume properties of these catalysts.

Natta established new processes for the production and separation of butadiene that contributed to the production of synthetic rubber in Italy during World War II. His research was also judged to be very useful in industrial plants producing fertilizers, fuel, and explosives.

During the first years of his research career, Natta was involved in X-ray and electron diffraction studies of solid materials. Supported by a grant from the Volta Foundation, he went to Freiburg im Breisgau in Germany to carry out electron diffraction studies. An unplanned benefit of his stay in Freiburg was that he met Hermann Staudinger and his group, who infected him with their enthusiasm for macromolecular substances. At that time, many scientists still opposed Staudinger's theories about biological macromolecules and favored the notion of colloidal systems. Staudinger did not receive the Nobel Prize until 1953 for his pioneering work carried out decades before. Upon his return to Italy, Natta continued his electron diffraction studies on some macromolecular samples from Staudinger.

Natta had a meteoric career in Italian academia, with professorships in Pavia, Rome, and Turin. In 1939, he became the chair of industrial chemistry at Milan Polytechnic Institute, where he remained for the rest of his academic career.

Natta's appointment to this professorship was due to Italy's anti-Jewish laws. The Ministry of National Education drew up a list of Jewish professors who were to be removed from their positions. This list included Mario Giacomo Levi, whom Natta replaced. Levi had directed the Institute of Industrial Chemistry; where he was an expert on fuels, and had worked on synthetic fuels to help make Italy less dependent on imports. During the last critical years of World War II, when the Italians had deposed Mussolini and Italy had turned against Hitler, the Nazis still ruled in northern Italy, and Levi escaped to Switzerland. He turned up in Lausanne, where he taught industrial chemistry to Italian students in an engineering school.

When Levi returned after the war, a cordial relationship with Natta was reestablished. The Natta family album holds a photograph of Levi with an inscription for Natta, expressing friendship and admiration on the occasion of Natta's 50th birthday. According to his former pupils and his family, Natta was probably never a fascist Although the family album contains a picture of Mussolini's visit to the university and Natta is seen wearing a black shirt, this was almost a uniform for students at that time.

The first two decades following his appointment to Milan Polytechnic Institute were the golden era for Natta's research. His seminal discoveries included those in the 1950s that brought him the Nobel Prize. After the war, he established contacts with the chemical industry in the U.S. On his first trip to the U.S., he was impressed by the number of scientists who worked at research laboratories and by the way their research was organized. This visit inspired him to enter an agreement with Montecatini Co. (later Montedison) to train large numbers of graduates in chemistry and industrial chemistry to work in the Montecatini laboratories. The support received from the industry in return helped to equip the university and Natta's laboratory with the most sophisticated instrumentation.

Natta thrived on industrial connections, and he differed greatly from his co-laureate in this respect. Ziegler had been reluctant to move from a university job to the Institute for Coal Research in Mülheim, Germany, because he preferred pure research to product-oriented work. One of the other extensive industry-related projects taken on by Natta was the industrial applications of the chemistry of carbon monoxide.

Research Accomplishments A singular determination characterized Natta's scientific activities, which he directed to the new field of the stereochemistry of polymers, in which he has been recognized as founder. His pupils stress the importance of his contributions to the asymmetric synthesis of optically active polymers and copolymers and the first asymmetric synthesis in the solid state. His arsenal of research tools encompassed state-of-the-art physical techniques, such as electron diffraction and X-ray diffraction; synthetic organic chemistry; and the study of reaction mechanisms.

Natta's close association with Montecatini provided funding for his work, as well as coworkers and apparatuses for spectroscopic, analytical, and various other physicochemical characterizations of polymers, plastic materials, fibers, elastomers, and films. It also helped Natta in establishing his interactions with Ziegler. When Ziegler gave a lecture in Frankfurt in 1952 at the meeting of the German Chemical Society, Natta and one of his coworkers attended. The audience did not seem too impressed by the lecture, but Natta and his colleague were.

Natta may have recognized even before Ziegler did that the German scientist had found a new principle of synthesizing polymer chains, although Ziegler went as far as coining a new name for his reaction—*aufbaureaktion*. Polymer synthesis was not achieved by monomer addition to a free radical or to a carbonium ion; rather, it occurred by monomer insertion between a metal atom and the growing chain bound to it. Natta convinced Montecatini to invite Ziegler to Milan. The visit resulted in Montecatini's purchasing the rights for industrial development in Italy of Ziegler's

discoveries. Under the agreement, Natta sent three of his coworkers to Mülheim to work with Ziegler.

Also under the agreement between Montecatini and Ziegler, Natta received the text of Ziegler's patent on the catalytic preparation of ethylene high polymers with catalysts formed by causing titanium tetrachloride and alkyl aluminum compounds to react. Natta's group already was familiar with alkyl aluminum chemistry, so it could immediately start further research. Although Natta continued the ethylene project, he sensed that Ziegler could make more rapid progress because of his previous experience. Also, Natta was interested in high-molecular-weight polypropylenes. Therefore, he decided to focus the efforts of his group on the polymerization of propylene. As early as in the spring of 1954, they had produced some yellowbrown gummy product, which convinced everyone that the project was worth pursuing. Within months, they had the first fibers of polypropylene.

Industrious: Natta (center) visited the first Italian plant for methanol production in Coghinas in 1930 (courtesy of Franca Natta).

Natta's previous experience in structure research helped him to initiate an extensive program to probe the structural characteristics of the new polymer and to deter mine the conditions for producing stereoregular polymers. The sequence in which the structural elucidations and the production steps followed each other is well documented. The records witness an exemplary model of concerted utilization of concepts and experiments, chemical and physical approaches, and perhaps most important of all, Natta's creative skills.

After having seen Natta's first report on polypropylene in manuscript, Paul J. Flory, who himself would win the 1974 Nobel Prize in Chemistry for his work on macromolecules, wrote in response: "The results disclosed in your manuscript are of extraordinary interest; perhaps one should call them revolutionary in significance. The possibilities opened up by such asymmetric polymerization are of the utmost importance I am sure." Incidentally, the manuscript Natta submitted to the *Journal of the .American Chemical Society* on Dec. 10, 1954, was originally rejected by a referee who criticized the lack of details about the catalyst Natta used. The referee's opinion was overruled by Flory, who was an editor of the journal. The paper does lack detail about the catalysts, which was due to Natta's close association with Montecatini.

Herbert Morawetz of Brooklyn Polytechnic Institute remembers that when Natta lectured in Brooklyn in 1958, many of his slides were illegible; some of the audience suspected that this was intentional. Natta's relationship with Montecatini may also explain why he did not tell Ziegler about his discovery when he went to Mülheim just before the appearance of his paper on the stereoregular synthesis. This omission was strongly resented by Ziegler after all the help Ziegler had given to Natta's team.

Natta had a tremendous insight into polymer chemistry. Soon after the discovery of stereospecific polymerization, he told some of his associates that if they could disturb the tendency of polyethylene to crystallize, they could produce a good elastomer. He suggested putting some propylene units into the polyethylene chain in a random fashion; the result would be copolymerization of ethylene with propylene. Be cause the scientists knew how to homo-polymerize each of the two monomers with the same type of catalyst, the copolymer was easy to prepare. In a few days, the first ethylene-propylene copolymer was produced, and it showed elastomeric properties, as predicted by Natta. This copolymer is still prepared on an industrial scale and is used for high-impact materials.

In less than 10 years following the discovery of polypropylene, Natta and his associates synthesized about 130 new types of stereoregular polymers; determined their crystal structures; investigated their physical, chemical, and mechanical properties; identified the most suitable catalysts for the production; and studied the reaction mechanisms. Natta published 610 articles and filed 316 patents. He had numerous pupils who have faithfully continued his legacy in chemistry and in polymer chemistry, in particular. He lived for his science. He concluded his last paper in 1972 with the following sentence: "Should I begin anew, I would devote my life to research." He may have sometimes been oblivious to the happenings around him outside his laboratory, but, paraphrasing Alfred Nobel's words, he certainly conferred a great benefit on mankind.

The authors are grateful to Franca Natta Pesenti and Giuseppe Natta for their hospitality and conversations in Bergamo and Milan in October 2002, and to Giulio Natta's former pupils.

The cover of the sixth and concluding volume of the *Candid Science* series.

Candid Chemistry[a]

István Hargittai

Interviews with eminent scientists often provide for excellent oral history, a more engaging way to learn about science. As science has become increasingly impersonal, interviews allow readers to become personally acquainted with great scientists. Also, they provide a forum for these scientists to profess their opinions on various issues. Current chemical literature is often so terse and journal space is at such a premium that it is often impossible for authors to describe their unsuccessful attempts on the way to their discoveries. However, in an unhurried conversation, scientists are willing to reminisce about their failures as well as their successes. With this article, I want to share some of the experiences that I have gathered while conducting interviews.

From left to right: István Hargittai, Arthur Kornberg (Nobel Prize in Physiology or Medicine, 1959), and James D. Watson (Nobel Prize in Physiology or Medicine, 1962). Photograph by Magdolna Hargittai.

There is a plethora of questions that one may ask of great scientists. In addition to asking them about their family background, education, and their most important achievements in science, many other questions, such as the following come up:

What turned them to science? What was the determining factor in their success? Did they recognize the importance of their discovery right away? Was it easy to publish their groundbreaking findings? And so on. I have found most scientists communicative with a few exceptions. In most cases, once we immerse ourselves into the conversation, I have to ask few questions to get a lot of information. In some other cases, a little prodding is needed. These are in-depth interviews, sometimes going on for hours.

My technique of interviewing is the following. I contact the scientist informing him (or her) that I will be in his neighborhood because of a conference or some other reason and would like to record a conversation with him. If we can arrange a meeting, we record a conversation. I live in Budapest, so this is almost always connected with a trip. Before the interview, I try to do as much homework as possible in my preparation for the interview. Usually, the more I know in advance, the more new information I can get during the encounter. However, for the sake of my future readers, I must try not to boast about my knowledge. To me the best interviewer is almost invisible. I would like to stress that no two interviews and no two sets of questions are ever the same. There are exceptional cases, when there is no possibility for any preparation.

[a]Originally published in *Chemistry International* 2002, 24(5):11–14; ©1997–2002 International Union of Pure and Applied Chemistry.

I. Hargittai (✉)
Department of Inorganic and Analytical Chemistry, Budapest University of Technology and Economics, Budapest, Hungary
e-mail: istvan.hargittai@gmail.com

Glenn T. Seaborg (Nobel Prize in Chemistry, 1951). Photograph by István Hargittai.

In 1994, while I was attending an American Chemical Society meeting in Anaheim, California, I came across Glenn Seaborg, who was taking a leisurely walk alone. I introduced myself and asked him for an interview. The only available time was right then. Fortunately, my camera and miniature Dictaphone were with me, so we found a relatively quiet corner and recorded a conversation. The interview was brief but good, thanks to his gracious, cooperative nature. I later followed up with some additional questions in writing. I was lucky to have interviewed Seaborg when I did, because there was no other opportunity for me to meet with him before he passed away a few years later.

Upon my return home I prepare the transcript of the conversation, edit it slightly, and send the material to the interviewee for checking and changing. We repeat this process until the interviewee feels comfortable with the text. I consider the original recording merely a framework for the interview, which takes its final shape at a more leisurely pace. It may be argued that by doing so, some of the original spontaneity is lost. Also, when I feel during the conversation that a topic seems uncomfortable for the interviewee, I rather drop it than try to force getting more information. My experience is that it would not work anyway. A scientist is willing to reveal more about his inner self to an understanding colleague than to an aggressive investigating reporter. A famous American chemist wrote me after I had sent him the edited transcript of our conversation that, having read several of my previous interviews, he had decided to be on his guard and be reserved. When he received the transcripts, he was astonished that he had told me about things that he had not thought about for a long time and never disclosed, even to his wife. After that remark I would have expected him to delete substantial parts of the interview before publication, but he hardly touched the text.

I would like to sample here some of the answers to the example questions I mentioned above. What turned today's great chemists to chemistry? The most frequent answer is either Paul de Kruif's book Microbe Hunters or a chemistry set. It is interesting that de Kruif's book is about pioneers of science, but not so much about chemists. On the other hand, a chemistry set has been the original source of interest not only for many future chemists but for many future physicists and biologists as well. De Kruif's book first appeared in 1926 and has remained in print ever since. Its popularity has faded though. I suspect that nowadays the computer, let alone television, is a great competitor for books among youngsters (we are talking about early teenagers) and, besides, Microbe Hunters may be somewhat too romantic for the modern young adult. Chemistry sets may have also lost some of their luster. Today they are less popular than they used to be and part of the reason may be safety precautions that have excluded some of the most spectacular experiments from their repertoire.

To the question about the determining factor in their success, most great scientists name one or two mentors. The determining period of their lives as scientists is their graduate studies in most cases. The venue where they happen to do their graduate work or where they start their independent research career is also a determining factor. The mentor effect may come as a result of a tight interaction with one's thesis supervisor, but it may also be just a casual remark by someone whose impact may then last for a whole research career. My favorite example is Frank Westheimer's story. Westheimer went to pursue his graduate studies at Harvard University because he wanted to work with James B. Conant, who, in the meantime, had become the president of Harvard. Thus, Westheimer had another supervisor, a rather indifferent one. When he finished and was about to leave, Conant called him into his office and asked him about his plans. Westheimer told Conant what he was planning to do as his research project. Conant's reaction was devastating for Westheimer: "If you are successful with that project, it will be a footnote to a footnote in the history of chemistry." At that point Westheimer realized that he was supposed to do important things. For his long and successful career, Westheimer measured everything against Conant's words. Conant, in his turn, had a great career as a public servant. They met one more time when Conant had retired and Westheimer was a professor at Harvard. He was working in his office when someone knocked on his door. It was Conant. He looked at Westheimer and asked, "Do you remember me?" At this point in the interview, Westheimer became so moved that we could not continue for some time, but it was also a good finishing point for the conversation. Whenever I re-read this interview or tell others about it I can't help being deeply moved as well.

There are many examples of the importance of the venue for the start of research careers. I found Sidney Altman's description of his postdoctoral stint in the late 1960s at the

Laboratory of Molecular Biology (LMB) in Cambridge, United Kingdom, especially interesting. Altman was a co-recipient with Thomas Cech of the chemistry Nobel Prize in 1989 "for their discovery of catalytic properties of RNA." He felt in Cambridge as he thinks young physicists must have felt in Copenhagen in the 1920s at the dawn of modern physics. Altman narrates how everyone went to tea at the LMB, according to the English custom, twice a day, and the "gods" of molecular biology were there. They were accessible in fact, eager to discuss things with everyone else in the lab. The scientists included Francis Crick, Sydney Brenner, Frederick Sanger, César Milstein, Max Perutz, Hugh Huxley, and Aaron Klug. It was a very formative atmosphere indeed.

The question about whether a discoverer recognizes the importance of his discovery right away is less trivial than it sounds. It does happen sometimes that the recognition comes much later after some other people had done additional work in the field. The story of buckminsterfullerene provides a conspicuous example. Eiji Osawa in Japan proposed the C_{60} molecule of the truncated icosahedral shape 15 years before its discovery. However, he did not recognize its importance and restricted himself to publish about it in the Japanese language only, although he published his other works almost exclusively in English. A few years later two Russian authors, Gal'pern and the late Bochvar reported this structure from their quantum chemical computations. Although their Russian-language article had been translated into English, nobody had noticed it until after the actual discovery in 1985 by Kroto, Curl, Smalley, and their students. The original idea of the truncated icosahedral shape came from a colleague of Bochvar and Gal'pern, Ivan Stankevich, but they failed to include him among the authors because they, and Stankevich too, thought that this piece of work was of no particular interest. Stankevich used to play soccer and the shape of the soccer ball gave him the idea. The interviews with Osawa and with Gal'pern and Stankevich, along with those with the Nobel laureates Curl, Kroto, and Smalley, reflect a lot of human drama.

I have gradually warmed up to asking famous scientists whether it was easy to publish their groundbreaking findings. Originally, I thought that the best journals would be eager to publish Nobel Prize-level discoveries. However, the experience is different, although seldom can one read about it. For some, even for Nobel laureates, it is an uncomfortable topic to narrate about rejections by stern editors of studies that later would merit the highest recognition. Yet it is instructive to observe that perhaps mediocre papers have the easiest way onto the printed page. Very poor papers get filtered out, of course. However, real groundbreaking papers often have their hurdles because of their pioneering character.

These are but a small sample of questions and the answers. There is great diversity among the fates and personalities of great scientists just as among the rest of us. It sounds commonplace, but the most succinct way to characterize great careers is for the right person to be there in the right place at the right time. In addition to being gifted though, many of the greatest scientists worked very hard to be there just when it was the right time and to move around until they "happened to be" in the right place. Recently, Alan MacDiarmid of the University of Pennsylvania and co-recipient of the chemistry Nobel Prize in 2000 summed up his philosophy in the guise of a Chinese proverb: "I am a very lucky person and the harder I work the luckier I seem to be." So far, two volumes of my Candid Science series have appeared, Candid Science: Conversations with Famous Chemists (2000) and Candid Science II: Conversations with Famous Biomedical Scientists (2002). The third volume is now coming out, Candid Science III: More Conversations with Famous Chemists. Each volume contains 36 interviews, and more than half are Nobel Laureates. The Candid Science series published by Imperial College Press London is open ended, and one more volume has been contracted (Candid Science IV: Conversations with Famous Physicists). Beyond that, I already have material for yet another volume. I have been doing this interviewing mostly during the past half a dozen years.

It was the interaction with Linus Pauling that initiated this project in 1993. However, my very first interview with a famous scientist was 1965 with Nikolai Semenov (1896–1986), the Russian Nobel laureate of 1956. I was asked to do this interview by the science section of Radio Budapest. Semenov came to Budapest to receive an honorary doctorate from the University of Technology. The interview was not only broadcast on radio, later it was also printed in the Radio and Television Yearbook, a volume of the best programs. Recently, I purchased a copy of the original tape from the Archives of Radio Budapest and, after more than 35 years, I still found it interesting. I was lucky that Semenov was as experienced in such matters as he was gracious. I was as inexperienced as one could be, but had the self-assurance of an ignorant beginner. The Radio supplied me with a tape recorder of the size of a trunk and a technician who operated it. One of the interesting features of the interview was that I asked Semenov to prognosticate about science from the perspective of the mid-1960s, and he did. From today's perspective, he did not say anything extraordinary, but this is what makes his prognostication so valid. I was happy to include Semenov in my first interviews volume.

For six years (1995–2000) I published most of my interviews in The Chemical Intelligencer, a now defunct magazine. I have now an interview in each issue of the magazine Chemical Heritage published by the Chemical Heritage Foundation. These interviews have been my second university education and I am happy to share all that I have learned from them with everyone.

The building of the Presidium of the Russian Academy of Sciences, 14 Leninsky Avenue in Moscow (photograph by István Hargittai).

Politics of Remembrance

Scientist Statues in Moscow[a]

Istvàn Hargittai and Magdolna Hargittai

Moscow has an extraordinary number of statues, busts, and memorial plaques honoring scientists, most from Soviet times, and a few, recent. Some, among them recent ones, represent individuals whose memory serves to preserve Soviet-era values.

Moscow has many hundreds of memorials of scientists, perhaps more than any other city in the world. Some represent world-renowned greats, such as, for example, the mathematician Nikolai I. Lobachevsky, the chemist Dmitry I. Mendeleev, the physiologist Ivan P. Pavlov, and even the nuclear tsar Igor V. Kurchatov. There are memorials of other greats and lesser scientists as well as others who might have received such commemoration for purely political reasons in Soviet times.

In view of the current political discourse in the United States about what to do with the statues of Confederate generals and the like, and having the centennial of the 1917 Russian revolution, it may be of interest to see how the Russians deal with the memorials of their past. The circumstances are so different as to create doubt whether any direct lesson can be learned from the examples we present, yet they might still be instructive. Thus, establishing sculpture parks for gathering monuments of discredited persons whose statues may help to understand the past and may represent the artistic values of the time, is worthy of consideration. Our presentation is mostly limited to examples of memorials of scientists. This is not a severe limitation on our selection, because, as we hinted at

it above, there are so many statues of scientists in Moscow.

As the Soviet Union was collapsing, the question of how to deal with the memorabilia of the Soviet era arose immediately. There was a suggestion for a sculpture garden for monuments that no longer should be exhibited in public spaces, but should not be destroyed either. Actually, we know of only one memorial that had disappeared without trace—one depicting Trofim D. Lysenko (see below) and Iosif V. Stalin—but this had happened long before the collapse of the Soviet Union.

In 1992, a sculpture park called the *Muzeon* was initiated.[1] It was established near the center of Moscow on a prominent site; the address is Krymsky Embankment 2, and it is easy to reach by walking from the subway station *Oktryabskaya*. The Muzeon has been a great success. However, its principal concept has gone through a metamorphosis over the years as not all the statues on display represent Soviet-era monuments, and new artistic creations keep being added. Conversely, some statues of discredited political leaders have been left in their original public space. Suffice it to mention Stalin's bust in front of the Kremlin Wall on the Red Square, over his grave. There are always freshly cut flowers on its plinth. Stalin's latest bust in Moscow was unveiled in September 2017 in a park of a museum, along with busts of other twentieth-century leaders of Russia.

The transformation of the Muzeon sculpture garden into a mixed-theme exhibition has its benefits. It makes possible

---

[a]Based on a somewhat shorter version in Hungarian in *Magyar Tudomány* 2017/9:1088–1097. Used with permission. All photographs in this article are by the authors and are protected by copyright, © Hargittai.

The authors are at the Budapest University of Technology and Economics. They are currently working on their new book, *Moscow Scientific*.

I. Hargittai (✉) · M. Hargittai
Department of Inorganic and Analytical Chemistry, Budapest
University of Technology and Economics, Budapest, Hungary
e-mail: istvan.hargittai@gmail.com

[1] Such a sculpture garden opened in 1993 in Budapest whose function has not changed since.

Left: Bust of Nikolai I. Lobnachevsky (by N. V. Dydykin) on the Alley of Scientists, Vorobevy Hills. *Right*: Statue of Dmitry I. Mendeleev in front of the Department of Chemistry, Lomonosov Moscow State University.

Left: Statue of Ivan P. Pavlov (by M. G. Manizer and E. A. Yanson-Manizer) in the tower of Lomonosov Moscow State University. *Right*: Monumental bust of Igor V. Kurchatov (by I. M. Rukavishnikov, 1971) on Kurchatov Square in front of the main building of the Research Center "Kurchatovsky Institute".

curious thought-juxtapositions of contrasting statues. A good example is the positioning of the statues of Feliks E. Dzerzhinsky (1877–1926) and Andrei D. Sakharov (1921–1989) in each other's vicinity. The distinguished physicist Sakharov was a dedicated and conspicuous human rights activist who was awarded the Nobel Peace Prize in 1975 and had become a beacon of human rights. The Soviet authorities tried to silence the "father of the Soviet hydrogen bomb" and one of the most decorated citizens of the Soviet Union. They exiled him to Gorky (before, and now, Nizhny Novgorod) in January 1980. He was kept there under constant harassment until the end of December 1986.

laboratory, preventing them from publishing their findings, not allowing them to attend scientific meetings (and not only abroad, but even within the country) was a heavy, but common penalty.

In Stalin's time physical annihilation was often the end. Stalin's Terror came in several waves; the victims were in the millions, including many scientists. However, only a few of the memorials in Moscow honor individual martyrs of those terrors. At the time of the sixtieth anniversary of the 1936–1938 Great Purges, a group of citizens wanted to commemorate the victims with memorial plaques on the buildings where they had lived. The mayor of Moscow turned them down, fearing that the facades of the city would be overlaid by memorial plaques. Even in those few cases when there is a memorial, an uninformed observer would not learn that the person commemorated had fallen victim of the ruthless dictatorship. Sadly, both during Stalin's reign and in the post-Stalin period, there were always ample members of the scientific community who assisted the authorities in bringing down fellow scientists either out of fear or for career advancement.

Nikolai Vavilov

Statues of Feliks Dzerzhinsky (*left*, by E. V. Vuchetich, late 1950s) and Andrei Sakharov (*right*, by G. V. Pototsky, 2008) in the Muzeon Sculpture Park.

Dzerzhinsky was the founder of the infamous Soviet security police, which had a variety of names over the years, and was best known as the KGB. Dzerzhinsky's statute used to stand in the center of Lubyanka Square, in front of the headquarters of the secret police, and it was transferred to the Muzeon after the political changes. Sakharov with Dzerzhinsky in the background reminds us of the heroic struggle the defiant Sakharov had taken up with inhuman repression. This struggle must have seemed hopeless at the time yet this fragile man defeated the seemingly invincible powers that be. Under today's Russia, it appears still undecided whether Sakharov's victory was final or merely transient.

In Sakharov's time, the Soviet regime punished its critics by prison terms, exile, incarceration in psychiatric wards, and suchlike. For scientists barring them from their

Statue of Nikolai I. Vavilov (by L. N. Matyushin, 2015) at the Timiryazev Academy.

The internationally renowned plant geneticist Nikolai I. Vavilov (1887–1943) was one of the best-known scientist victims of Stalin's Terror. Vavilov studied at what is today the Timiryazev Agricultural Academy. His statue on the campus represents him as a student and there is no indication of his tragic ending.

Vavilov had a brilliant career at home and internationally. He directed research institutes and led international expeditions to build up a valuable grain bank. His theory of the relationships among plants was compared to Mendeleev's Periodic Table of the Elements and to Linné's taxonomy. Unfortunately, Vavilov's fate became linked to the pseudo-scientific charlatan Trofim D. . Lysenko (1898–1976). Lysenko was an enthusiastic and diligent agronomist and Vavilov supported his rise on the academic ladder. Lysenko in time got carried away and introduced spurious innovations from which the Soviet leadership expected help lifting the country out of its food shortages. The peasant-scholar Lysenko became a mascot for Stalin's anti-intellectual campaign in the mid-1930s. Lysenko promoted the idea that changing external conditions would change the inherited characteristics of plants and animals—consonant with Stalin's goal of transforming the people of the Soviet Union quickly. Vavilov eventually caught up with Lysenko's lies and crimes and turned against him. By then, it was too late; scientists critical of Lysenko were arrested, sentenced, and executed, or simply, disappeared. This was to be Vavilov's fate too; he was arrested in 1940 and following years of suffering, including torture to extract a confession from him, he died of hunger in prison and was buried in an unmarked mass grave in 1943. Today there are several memorials devoted to him in Moscow and elsewhere.

Aleksandr Chayanov and Nikolai Kondratiev

There are many memorial plaques of former professors on the buildings of the Timiryazev Academy. Only two of them honor repressed martyrs, but the inscriptions on the plaques reveal nothing about their tragic fate.

Aleksandr V. Chayanov (1888–1937) graduated from the Timiryazev Academy in 1911 and completed his education in Western Europe. He was appointed professor at his Alma Mater, but in 1926, he was accused of anti-Marxist interpretation of the agricultural policy. In 1930, he was arrested for alleged anti-Soviet activities and was sentenced to imprisonment. Eventually, his sentence was changed to exile, but in 1937, he was arrested again, and executed. In 1987, the Supreme Court of the Soviet Union determined that all the sentences had been illegal, including the methods—obviously, torture—used to extract from Chayanov the admission of guilt.

Nikolai D. Kondratiev (1892–1938) created a theory of economic cycles and gave a theoretical foundation to the New Economic Policy (NEP) in the early 1920s. The communist methods of governing the economy had proved disastrous and the NEP meant their temporary relaxation. Kondratiev's and Chayanov's teachings, in which features of planned and market economies figured, were criticized at the highest level. Stalin considered their views to represent an anti-Soviet movement. The NEP lasted until about 1928. Kondratiev had already been arrested briefly in 1922 and was arrested again in 1930, was sentenced to 8 years in prison in 1932. In 1938, he was sentenced to death and executed on the same day. He was first rehabilitated in 1963, but the decision about it remained classified and so were his views. His complete rehabilitation followed in 1987.

Sergei Korolev

There were scientists who were arrested, sentenced to prison terms, labor camps, exile or death, but survived. The lucky ones ended up in specialized labor camps, so-called *sharashkas*. There the conditions were better than in ordinary labor camps and the prisoners were engaged in meaningful occupation. Some of these prisoners eventually became celebrated contributors to creating the formidable Soviet military power.

Memorial plaques to Aleksandr Chayanov (*left*) and Nikolai Kondratiev (*right*) at the Timiryazev Academy.

One of Sergei P. Korolev's memorials (S. A. and S. S. Shcherbakov, 2008) in Moscow. This one is on the Alley of Cosmonauts, next to the enormous monument "Explorers of Cosmos" at Ostankino Park.

Sergei P. Korolev (1906–1966) was in charge of the development of the Soviet rocket-cosmonautics system at the zenith of his career. He began his college education in the Kiev Polytechnic Institute, then moved to Moscow and continued at Bauman University under Andrei N. Tupolev's (1888–1972) mentorship. Tupolev later became one of the most famous airplane designers of the Soviet Union. Korolev and his colleagues built devices, but Korolev was arrested in 1938 on trumped up charges and underwent rough interrogations. According to unconfirmed descriptions, this included psychological abuse and torture during which both his jaws may have been broken. In 1940, Korolev was sentenced to eight years of incarceration.

Korolev worked in a special camp, again under Tupolev, who in the meantime had also been arrested and sentenced to a prison term. Korolev designed devices for airplanes and torpedoes that proved useful for the military. He was then moved to another special prison where he worked on rocket motors. Korolev was freed from prison in 1944, but was not rehabilitated. From this time Korolev was in top positions in the projects of the Soviet rocket technology, including the ballistic missiles, and directed the program of automated satellites and manned spaceships. There were a number of chief designers in the Soviet rocket technology and Korolev was not just one among them; he chaired the council of the chief designers.

The public celebration of the cosmonauts by the leaders of the Soviet Union was Korolev's triumph too, but he was kept in the shadows for security reasons until he died. The accolades and honors after his death were not only in stark contrast to his prison life, but also with his obscurity during his active life. His resting place is in the Kremlin Wall and his villa has been turned into Korolev's museum. His memorials nowhere mention his years of repression.

Aleksandr Prokhorov

Aleksander M. Prokhorov's memorial (by Ekaterina Kazanskaya, 2015) on Prokhorov Square—the intersection of Leninsky Avenue and Universitetsky Avenue.

Aleksandr M. Prokhorov (1916–2002) was born in Australia where his family fled from the repressive tsarist regime. In 1923, they returned to the Soviet Union. Prokhorov graduated in physics from Leningrad University in 1939. He fought in WWII, and was seriously wounded in 1944. From 1946 till 1998 he worked in the Physical Institute of the Academy of Sciences and from 1982 he was also the founding director of what is now the Prokhorov General Physics Institute of the Academy. In 1964, Prokhorov shared the Nobel Prize in Physics with his colleague Nikolai G. Basov and the American

Charles H. Townes for their work in quantum electronics that contributed to the development of lasers.

It is a stain on Prokhorov's vita that he was one of the *four* academicians (see below) who in 1983 published a letter entitled "When honor and conscience are lost," condemning Andrei Sakharov in the most damning terms. It was at the time when Sakharov was in exile under constant harassment by the KGB. Publishing letters of condemnation was a favorite technique of the Soviet authorities to discredit individuals fighting for human rights. There were usually many signatories some of whom may have volunteered to sign; others may have been cajoled to do so, yet others did not even know what was in the letter under which their signatures appeared. In this particular case the very small number of signatories indicated that the letter was genuine. Prokhorov's Nobel Prize added weight to his action. There have been recent signs of adulation of Prokhorov's memory more than some other physicists of comparable scientific stature.

Andrei N. Tikhonov

Memorial plaque of Andrei N. Tikhonov on the façade of the Faculty of Computational Mathematics and Cybernetics of Moscow University.

Andrei N. Tikhonov (1906–1993) graduated from Moscow University majoring in mathematics and stayed on at his Alma Mater. From 1933, he continued at the department of mathematics of the Faculty of Physics of the University, serving as its head between 1938 and 1970. He also held leading positions at three institutions of the Academy of Sciences. In 1970, Tikhonov initiated the Faculty of

Computational Mathematics and Cybernetics and served as its first dean. This was a long way from the condemnation of cybernetics as an imperialist conspiracy alien to Marxism-Leninism of around 1950. In 1983, Tikhonov was one of the four who published a letter of condemnation of Andrei Sakharov's activities (see above). Besides Tikhonov and Prokhorov, the computer technologist A. A. Dorodnitsyn and the veterinarian G. K. Skryabin signed the letter. Dorodnitsyn and Skryabin also have memorials in Moscow.

Ivan Vinogradov

Ivan M. Vinogradov's memorial in the Steklov Institute of Mathematics. There are other memorials of Vinogradov elsewhere in Moscow.

Ivan M. Vinogradov (1891–1983) was the director of the Steklov Institute of Mathematics of the Soviet Academy of Sciences, the country's leading institution for research in mathematics, for half a century. The Institute came to existence in about 1934 in Leningrad and almost immediately moved to Moscow. Vinogradov's directorship acquired international notoriety for anti-Semitism that manifested itself in his rigorously discriminative hiring practices. The Soviet authorities practiced ill-masked anti-Semitism, especially in hiring, but Vinogradov went even further than what was expected of him.

Lev Pontryagin

Memorial plaque of Lev S. Pontryagin on the façade of the Faculty of Computational Mathematics and Cybernetics of Moscow University. There is a similar memorial plaque of Pontryagin on the façade of the building where he lived at 7 Leninsky Avenue.

Lev S. Pontryagin (1908–1988) lost his eyesight at the age of 14 when an oil-stove exploded and his face was burned badly. His mother became his eyes and this is how he completed his education. Pontryagin graduated from Moscow University in 1929 majoring in mathematics. He stayed at the University and was appointed professor, and he was also a leading associate at the Steklov Institute. Pontryagin was very successful in the applications of mathematics for practical problems. There was a dark side of his career. He was a dedicated anti-Semite and was active in preventing even the most outstanding Jewish mathematicians from attending international gatherings and getting elected to the Academy of Sciences. In his actions he was helped by his high positions in the hierarchy of mathematicians; by other high-positioned anti-Semites, such as Ivan Vinogradov; and by the general anti-Semitic policies of the Soviet State and the Communist Party.

Nikolai Burdenko

Nikolai N. Burdenko's bust (by Sergei Merkurov, 1949) in the garden of the Burdenko Institute of Neurosurgery, 16 Fourth Tverskaya-Yamskaya Street.

Nikolai N. Burdenko (1876–1946) studied in a divinity school, but for higher education, he opted for medical school. He attended Tomsk University, but was expelled for participating in a radical student movement. Finally, he graduated as a physician in 1906 at the University of Yuryev (today, Tartu, Estonia). By 1917, he was back there as professor, but not for long as Estonia was to become independent and he and his clinic moved to Voronezh in Russia proper. Burdenko was interested in military medicine and in the preparedness of the medical personnel of the Red Army. In 1923, he moved to Moscow and he was appointed head of surgery at what is today the Burdenko Clinic of Surgery of the Sechenov Medical University. In addition, in 1929, he was appointed to be director of what is today the Burdenko Institute of Neurosurgery. He initiated and supervised the establishment of a network of neurosurgery clinics in the Soviet Union. In WWII, he treated wounded soldiers returning from the German front and suffered a stroke when meeting such a train of wounded Red Army personnel. He initiated the establishment of the Academy of Medical Sciences in 1944 and was its first president.

A dark stain on Burdenko's career is his chairmanship of the Extraordinary State Commission charged in 1943 with investigating the killings of twenty-two thousand Polish army officers, doctors, professors, lawmakers, police officers, and other members of the Polish intelligentsia in the Katyn Forest in 1940. The "Burdenko Commission" "found" overwhelming evidence that the Germans committed the crime. However, over the ensuing decades, doubts and evidences kept surfacing pointing to falsification by the Burdenko Commission of what happened, as in reality, the Soviet security organs committed the murders. Mikhail Gorbachev admitted Soviet responsibility, but would not accuse Stalin directly with complicity. Under Boris Yeltsyn's presidency the execution order by Stalin was declassified, but the policy of stressing Soviet achievements rather than uncovering the atrocities prevailed. Finally, in 2010, seventy years after the events and as a result of extended investigations in Soviet archives, the State Duma—the Russian Parliament—accepted a resolution. It established that the Katyn massacre was carried out on Stalin's direct orders and with the direct involvement of other top Soviet leaders.

Vasily Vilyams (Williams)

Statue of Vasily Vilyams (Samuil Makhtin, 1947) at 49 Timiryazevskaya Street.

Vasily R. Vilyams (1863–1939) whose father, by the name of Williams, immigrated to Russia from the United States, was one of the founders of soil science in Russia. Vilyams studied at what is today the Timiryazev Academy and upon graduation he was sent for a study trip to Western Europe and later to North America. He started teaching at the Timiryazev Academy in 1891. In 1904, he initiated a nursery garden in which he collected species of cereals and leguminous plants. He established an experimental station for studying animal fodder, which is the current Vilyams Fodder Research Institute.

Vilyams followed up Vasily Dokuchaev's work of making large plains often suffering from the worst draught into fertile areas. Dokuchaev proposed to regulate rivers, establish water-basins, and forestation for improving the rainwater economy. Vilyams declared that Soviet science can turn any soil into a high-producing one and he worked out a grass-arable rotation system. He observed that fallow land can be restored by planting certain grasses. Such an approach may prove workable, but this requires a long period of time. Vilyams wanted to shorten this period by alternately planting grasses and perennial papilionaceous plants that would loosen the soil and enrich it in nitrogen.

There was scarcity of chemical fertilizers and agricultural machinery in Russia, and Vilyams's approach could have helped in the short term though it might have impeded progress of agriculture in the long run. Furthermore, this approach was recommended, even enforced, not only for restoring fallow land but everywhere, including the East-European countries that had become parts of the Soviet sphere of influence following WWII. For this development, Vilyams was not responsible, as it continued after his death.

There was a dark side of Vilyams's career. He conducted sharp debates with other scientists and in the mid-1930s such disagreements could lead to tragic consequences. Thus, when he had a disagreement concerning certain techniques in farming with academician Nikolai M. Tulaikov (1875–1938), Tulaikov was arrested and disappeared. The internationally renowned geochemist Vladimir Vernadsky wrote in 1943 that Vilyams was leaving behind a bad and distorted school and that he used inaccurate data, which sometimes contradicted reality. According to Vernadsky, Vilyams's membership of the [communist] party did not make him an authority and Vernadsky predicted that Vilyams would be soon forgotten. Judging by the memorials honoring Vilyams, he is not forgotten.

Ivan Gubkin

Ivan M. Gubkin's statue (by Andrei Kovalchuk, 2011) in front of the main building of the Gubkin University of Oil and Gas, 65 Leninsky Avenue.

Ivan M. Gubkin (1871–1939) attended the Mining Institute in St. Petersburg and by the time of his graduation in 1910 he was almost forty years old. He participated in geological explorations and in 1917–1918, he was a member of a delegation visiting the United States to learn about the American petroleum industry. In the 1920s, Gubkin was rapidly rising in positions in academia and beyond. From 1920, he was professor and from 1922, rector of the Moscow Mining Academy and from 1930, rector and department chair at the Gubkin Institute of Oil, the institution named after him from its inception.

Gubkin was elected full member of the Soviet Academy of Sciences—academician—in 1929. He was 57 years old; not too young to become an academician, but unusually early if considering the scarcity of his research output. However, he and some other members of the Communist Party were elected with the obvious purpose of "Sovietizing" the Academy. In a parallel development a large-scale repression began against members and officials of the Academy. The main thrust of this repression happened in 1929–1930, and Gubkin was an active participant in persecutions. Scientists were sentenced to death, to exile, or to long prison terms, or they simply vanished. There were no trials; the secret police determined their fate. The repression ended only when the authorities were satisfied that the Communist Party had taken over the Academy.

From 1931, Gubkin was also in charge of the main geological authority of the Soviet Union. His 65th birthday was widely celebrated in 1936, together with the celebration of the 40th anniversary of his scientific research activities—although he had only graduated as an engineer 26 years before. In the same year, he was elected vice president of the Academy of Sciences. His fame today is based on the myth created about his greatness in the late 1940s and early 1950s, during the period of anti-science terror and the triumph of Lysenkoism.

Concluding Comments

Russian science has tremendous traditions and the Soviet era was especially rich in achievements in specific domains of science, those related to defense, including their derivative of space science. Other fields, notably, agricultural science, biology and what comes under today's umbrella term of information technology suffered heavy losses. The fate of memorabilia of controversial scientists cannot be understood without considering the general political situation in Russia. There appears to be hardly any attempt to judge the painful segments of Soviet science history including the devastation caused by Lysenko's long-lasting reign. In two conspicuous aspects the situation of science in Russia has worsened as compared to Soviet times. One is that the luster of science for gifted youth has dimmed. In the Soviet era science had a near-exclusivity of attraction for them; today, there are many other areas where their talent can flourish. The other is that the prestige of the Academy of Sciences has diminished in Russian society.

The Russian Academy of Sciences was founded in 1724 in St. Petersburg as the St. Petersburg Academy of Sciences, then, the word Imperial figured in its name from 1747 until 1917. Some leading academicians and Vladimir Lenin, the founder of the fledgling Soviet state both recognized the mutual benefits of maintaining the Academy in the new order. For a time, the Academy was under the department of mobilization of science of the ministry of education. Lenin promised political and financial support for the Academy and the academicians promised to assist in the reconstruction and development of the national economy. From 1925, the organization was renamed the Academy of Sciences of the USSR, in short, the Soviet Academy of Sciences (SAS), and it operated independently of any ministry. There used to be also the Academy of Medical Sciences and the Lenin Academy of Agricultural Sciences. The membership in the Academy of Sciences was more prestigious than the membership in the other two academies.

A new federal law in 2013 united the three academies under the Russian Academy of Sciences (RAS). This change—along with the academy properties having become state properties—is an indication of a diminishing role of science in the new Russian order. The 2013 law created a Federal Agency of Scientific Organizations (FASO). It was also determined that FASO will be in charge of all properties of the expanded RAS. In 2014, the Russian government approved the list of those 1,007 (!) institutions that were placed under the jurisdiction of FASO. All these institutions used to belong to the three previously independent academies.

The Russian Academy of Sciences is a vast organization and there is no attempt here to present its history, accomplishments, and organization. The Academy of Sciences was an integral part of the Soviet system. The academy membership was probably the most coveted recognition, and the grip of the Communist Party on its activities was tight. At the same time, there was probably no other institution in the Soviet Union that could have produced such early cracks in the Soviet system as the members of the Academy. Just think of the initiatives of the physicists in 1958 when they wanted the President of the Academy to give an account of his activities before he might be re-elected for a second term. Of all the Soviet institutions and organizations, the Science Academy was unique in preserving some features of democracy in its operations. The instructions of the Communist Party were not always fully complied with in the elections of the new members. In the post-Stalin era, the members of the Academy could not be excluded even if they had lost employment. This meant that they continued receiving their substantial remuneration as academicians. Andrei Sakharov was an example; this was his source of income during his long years of exile between 1980 and 1986.

The current political leadership of Russia has introduced drastic changes in the status and ownership conditions of the Academy. The changes happened without consultations with the interested parties and virtually overnight and there was hardly any protest or dissent strong enough that would have made the political leadership blink. The diminishing status of the Academy of Sciences has its reflection in the appearance of the physical plants of some of its leading research institutes. There is a great contrast between the general impression one gains from the freshly renovated outlook of downtown Moscow and the dilapidated state of the buildings of some of the research institutes. By the way, they take up some of the choicest sites of Moscow, for example in the wooded area of the Vorobevy Hills. The institutes have no means to renovate their buildings and a possible solution might be to let their lots be privatized and their laboratories relocated to less expensive areas. Fortunes could be made from such transactions.

The Academy of Sciences was created under the auspices of Peter the Great. Stalin used it to further his goals, reworked its membership to limit free thinking, but left its framework intact. The next Soviet leader, Nikita Khrushchev, saw in the Academy a seed of opposition and he tried to change it. He was not strong enough to do so, hence he threatened to abolish it. Khrushchev soon had to go, and the Academy stayed. The laws in 2013 and 2014 introduced changes, the kind that Stalin did not need to consider and that Khrushchev might have only dreamt of.[2]

[2] We thank Walter Gratzer and Bob and Irwin Weintraub for their helpful comments on the manuscript.

"Emanuel Tree of the Holocaust" (by Imre Varga, 1990) in the Raoul Wallenberg Memorial Park, in the garden of the Great Synagogue of Dohány Street in Budapest (photograph by István Hargittai).

Unaccounted For: Scientist Martyrs of Hungary in the Horthy Era and the Holocaust[a]

István Hargittai

The anti-Semitism of the Horthy regime and the Hungarian Holocaust caused tragic losses in Hungarian scientific life. The author—a chemistry professor and Holocaust survivor— calls for a reliable assessment of the losses preceded by thorough research. He laments the absence of proper remembrance of the martyrs who have already been identified.

Introduction

A few years ago, my wife and I collected the images of the memorials of science in Budapest—statues, busts, and memorial plaques. Augmented with related stories, we published a guidebook [1]. Our collection was inclusive as in addition to fundamental science we considered applications, technology and innovation, medicine, and education. On the backdrop of the large number of memorials, it was conspicuous that hardly any scientist victim of the Hungarian Holocaust has been remembered in such a way. During the preparation of the book, I published an account of this observation in *Magyar Tudomány*, the periodical of the Hungarian Academy of Sciences [2]. It appeared in 2013, amid the preparations for the 70th anniversary of the deportation of Hungarian Jewry in 1944. I hoped that my paper would turn attention to the losses of Hungarian science in the Holocaust and research would be initiated to take account of those losses. It did not happen.

Ours was a limited consideration as our attention was directed to existing memorials and it was limited to Budapest.

The present writing is an expanded version of my 2013 paper. I have added emphasis to the need of a comprehensive project, but my considerations still did not expand beyond Budapest.

The Hungarian scientific establishment owes such a comprehensive project to the martyrs and it owes such a project to itself as well. The Hungarian Holocaust usually refers to the events of the 1944–1945 period and I stress that a meticulous discussion should also extend the time period it covers. What happened in Hungary in 1944–1945 was the culmination of the history of the entire Horthy era lasting 25 years between 1920 and 1944.

I regret to say that the situation today (summer 2017) is worse than it was in 2013. In 2016, an account, "History of the Hungarian Academy of Sciences," appeared on the official web site of the Hungarian Academy of Sciences [3]. Rather than remedying the situation, this document aggravates it. There is no mention in it of the anti-Semitic discrimination in scientific life during the Horthy era and no mention of any losses in the Holocaust, not even that there was a Holocaust (or a Second World War, for that matter).

> Following the war and revolutions, the Academy's work restarted with a lot of difficulty as war inflation had consumed many of its assets. After the Treaty of Trianon came into effect the institution was supported with regular state subsidies administered by Count Kunó Klebelsberg, Minister for Education, who planned to provide pivotal roles for culture and science in the reconstruction of the country. The Academy did not regain its financial stability until the late 1920s when Count Ferenc Vigyázó left his entire estate to it. From this point on the yearly income of 500–600 thousand pengős from the Vigyázó legacy was spent on scientific purposes.
>
> Between the two world wars, the Academy was characterised by a peculiar duality. Its spirituality and its leaders stubbornly stuck to the conservatism of the late nineteenth century. At the same time internationally renowned natural scientists became members of the Academy, including biochemist Albert Szent-Györgyi (the first Hungarian Nobel Prize Winner), mechanical engineer Kálmán Kandó and chemist Géza Zemplén.

[a]First appeared in Hungarian in the Jewish cultural periodical, *Múlt és Jövő* (Fall 2017, pp. 102–124). Photographs, unless indicated otherwise, are by the author (© István Hargittai).

I. Hargittai (✉)
Department of Inorganic and Analytical Chemistry, Budapest University of Technology and Economics, Budapest, Hungary
e-mail: istvan.hargittai@gmail.com

I called the attention of the Department of Communication of the Hungarian Academy of Sciences to this misrepresentation. The response was that the inclusion of the items whose absence I lamented (that is, that there was anti-Jewish discrimination, that there were losses in the Holocaust, and that there was a Holocaust in the first place) would not be possible in the limited framework of the treatise [4]. The question arises how much should this section be expanded to make it possible mentioning the impact of anti-Semitism of the Horthy era, the destruction of Hungarian Jewry, and World War II (and not only on the Jewish component of Hungarian scientific life)? Albert Szent-Györgyi was highly critical of the Hungarian Academy of Sciences during the Horthy era and pointed to its considerable responsibility for the national catastrophe. This he stated and elaborated in his letter of November 30, 1945, addressed to the Secretary General of the Academy [5].

The Nobel laureate American scientist James D. Watson at (the closed gate to) the Emanuel Tree of the Holocaust memorial in 2002.

In Budapest, there are moving memorials of the victims of the Hungarian Holocaust, but there is no memorial in public space to the over four hundred thousand Hungarian Jews who perished in Auschwitz. The "Emanuel Tree of the Holocaust" (by Imre Varga, 1990) in the Raoul Wallenberg Memorial Park in the garden of the Great Synagogue of Dohány Street is not in public space and it has limited opening hours. The Emanuel Foundation of New York sponsored the establishment of the Emanuel Tree, which remembers all the Jews killed during the Holocaust. I attended its inauguration and

one of its speakers was the then Prime Minister elected recently in the first democratic elections after the fall of communism. Mr. József Antall was mourning the victims, but he was mourning them not as Hungarians that they also were. He appeared to divide the people present at the dedication into two groups, "You" and "We," and made it clear that this memorial was "Yours," and not "Ours."

The Nobel laureate Israeli scientist Aaron Ciechanover at the "Shoes on the Danube Bank" memorial in 2005.

The "Shoes on the Danube Bank" memorial (by János Can Togay and Gyula Pauer, 2005) honors the Jewish victims shot into the river by Arrow Cross men. The Arrow Cross was the Hungarian Nazi movement that took over the country on October 15, 1944. The memorial tablets in Hungarian, English, and Hebrew read: "To the memory of the victims shot into the Danube by Arrow Cross militiamen in 1944–1945." It is not mentioned that the victims were Jews; it is just assumed as obvious—I think it should be mentioned.

Memorial to the victims of fascism at Viza Street, District XIII.

There is a memorial to the victims of fascism on the bank of the Danube in Viza Street (by Makris Agamemnon, 1986). It is a half-size copy of the original, which stands at the Mauthausen Nazi concentration camp in Austria. According to the English translation of the inscription of the Budapest version, it commemorates "the resistance fighters, deserters, and the persecuted whom the Fascists murdered on the Pest bank of the Danube in the winter of 1944–1945." This dedication was typical of the ambivalence of the communist regime toward the Hungarian Holocaust. The principal victims were Jews and the murderers were Hungarian Nazis—the Arrow Cross—but János Kádár's regime did not consider spelling this out. The memorial was once again unveiled in 2010, with unchanged inscription. In 2012, unknown perpetrators painted anti-Jewish slogans of hate on the memorial. The euphemism of the text of the monument did not mislead them.

The expression "victims of fascism" is ambiguous at best, but is really, misleading. There was Mussolini's Italian fascism and Hitler's German National Socialism. In Hungary, there was Horthy's autocratic and anti-Semitic regime, which started in 1920 with the introduction of post-WWI Europe's first anti-Semitic legislation. It continued with increasingly harsh anti-Semitic legislation in the late 1930s and early 1940s, and culminated in the deportation of countryside Jewry still under Horthy's reign. Speaking about the "victims of fascism" is a camouflage masking Hungarian responsibility.

The Arrow Cross movement took over Hungary from Nicholas Horthy on October 15, 1944. By then, under Horthy, well over four hundred thousand Jews, mostly from the countryside, outside of Budapest, had been deported, primarily to Auschwitz. The discrimination against Jews,

their persecution, and annihilation had been going on in Hungary even before the German invasion of the country on March 19, 1944, even if by means of less efficient technologies than those employed by Germany. Thus, when, from 1941, the slave laborer Jews were sent to the Eastern Front, their guards knew that none of their charges were expected to survive the ordeal and they did their best to comply with this expectation.

In Budapest, the Arrow Cross murdered an estimated one hundred thousand Jews. They hunted out their victims and shot them into the Danube often after they had tortured them. There were scientists among the murdered.

I describe a few individual fates below so that we see them beyond the statistics. In the existing memorials in public space, the fact that the victims were Jews is usually masked. Camouflaging this information is misleading and for future generations it will mean a falsification of history. This falsification is though consistent with the efforts to ascribe the crimes against the Jews exclusively to the Germans and to the Arrow Cross.

Victims

Memorial plaque of Gedeon Richter (by István Buda) on the façade of 21 Katona József Street 21, District XIII. On December 30, 1944, Richter was taken from here and murdered.

The pharmacist entrepreneur Gedeon Richter (1872–1944) [6] was born into an assimilated Jewish landowner family, in an East Hungarian village, Ecséd, near the town of Gyöngyös. Before completion of his high school studies, he started working in a pharmacy in Gyöngyös. It was at the time that new legislation required university training for pharmacists, and he successfully completed the prescribed higher education. After this, he spent 2 years working in various pharmacies to acquire the qualifications to open his own pharmacy. He soon found his real vocation in the creation of an independent Hungarian pharmaceutical industry. He spent 4 years in Italy, Germany, France, and England, preparing himself for the task. He was 29 years old when he bought a pharmacy in Budapest and started his own manufacturing laboratory using the money from the sale of his family's estate. The pharmacy he ran is still there, at the corner of Üllői Avenue and Márton Street in District IX.

Richter focused on deficiency diseases. The necessary ingredients for the therapy, called organotherapia, were extracted from animal endocrine glands. Richter built up a strong research section in his laboratory and kept close contact with the medical profession. By 1902, he had already started publishing an information bulletin and was distributing it among medical doctors, free of charge. The laboratory soon proved inadequate for his goals and he founded a plant for manufacturing his preparations, which extended to plant extracts and synthetic drugs. One of his associates, Emil Wolf, soon left him to found another future giant pharmaceutical company, Chinoin.

The Richter chemical factory acquired fame for some long-lasting products, such as kalmopyrin (comparable with aspirin), insulin, and the Glandtrin injection, which contained oxytocin and proved efficient in gynecological applications. Insulin was discovered in 1921 with its first testing on humans in 1922, and Richter was already manufacturing it by 1926. In the early 1930s, Richter became one of the leading worldwide manufacturers of estrogens. Richter created subsidiaries and expanded internationally. In 1923, the Richter chemical factory became a shareholders' company with Gedeon Richter retaining the majority of shares. By the mid-1930s, Richter products were marketed in 100 countries. By the late 1930s, the Richter Company was second only to the United Incandescent Lamp Company (Tungsram) among Hungarian exporters.

From the late 1930s, increasingly harsh anti-Jewish legislation was hindering the normal operations of the company. It was placed under military control; Richter was under attack and had to resign as Chairman of the Board. In 1942, he was stripped of the rest of his functions in the company, and soon he was banned from entering the plant. He offered his services without remuneration, but his offer was rejected. For some time, he was still participating covertly in directing his company through his faithful associates. While it was still possible, he rejected suggestions that he should escape. His only concern was how to save the company.

On December 30, 1944, the Arrow Cross took him to the embankment of the Danube and murdered him. His corpse was never found. The company has survived; today, it carries Gedeon Richter's name.

Bust of Emil Wolf (by Dávid Tóth, 2010) at the corner of István Avenue and Nyár Street, District IV.

Emil Wolf (1886–1947) studied in Munich and graduated in 1910 as a chemical engineer. He and György Kereszty (1885–1937) co-founded the chemical factory from which the Chinoin Chemical and Pharmaceutical Company was created in 1913. They produced mainly synthetic drugs in close cooperation with Géza Zemplén, professor of organic chemistry, at the Technical University. In 1944, the Jewish Wolf was deported to a concentration camp in Germany. He survived and returned, continuing to direct Chinoin, but died in 1947.

Bust of Imre Bródy in the garden of the Imre Bródy Gimnázium in Ajka.

Imre Bródy (1891–1944) was born in Gyula, in southeastern Hungary. Between 1909 and 1914, he studied mathematics and physics at Budapest University. He began teaching in a high school, and conducted research in theoretical physics at the University. In 1918 he earned his doctorate and the university invited him to be an assistant professor. Following the period of WWI and the revolutions, the virulent anti-Semitism engulfing the country prevented a university career for Bródy. In 1920, he left for Germany, and joined the great physicist Max Born in Göttingen. Many of the world's best young physicists, among them Werner Heisenberg, congregated around Born. Therefore, Born's words with respect to Bródy's talent carry exceptional weight: "There was the little Hungarian Jew, E. Bródy, perhaps *the most gifted of them all*, who could solve intricate problems." [7] (my emphasis). Here E. stands for Emerich, the German equivalent of Imre. Bródy seemed destined for a great career, but after 2 years he returned to Hungary. He joined the research laboratory of the United Incandescent Lamp Company, known by its trademark, Tungsram. The sagacious director general, Lipót Aschner, developed a research laboratory under the leadership of Ignác Pfeifer.

Bust of Ignác Pfeifer (by Sándor Mikus, 1975) on the campus of the University of Technology and Economics.

Ignác Pfeifer (1868–1941) was a chemical engineer who at one time was in charge of the chemistry institute of the Technical University (today, Budapest University of Technology and Economics). His area of research was water softening. He was accused of involvement in the 1919 communist regime, and had to leave the Technical University. He was invited to head the research laboratory of Tungsram, where Pfeifer

created a modern research organization, the first such entity in Hungary. At the end of the 1930s, the anti-Jewish laws jeopardized his employment at Tungsram. While Pfeifer was with Tungsram, he attracted excellent scientists to work there, such as Bródy, Pál Selényi (see below), and others. Even expatriate scientists, such as Michael Polanyi (physical chemist in the UK, later, philosopher) and Dennis Gabor (physicist in the UK, future Nobel laureate inventor of holography), took up part-time appointments as external advisors.

Bródy demonstrated great versatility when he transformed himself from a successful theoretician into an innovative technologist. His best-known invention was the krypton lamp, which had great advantages over the previously used argon lamp. The application necessitated new ways of producing the expensive krypton gas, in which Bródy cooperated with others, such as Ferenc Kőrösy (physical chemist, later in Israel) and Polanyi. The krypton lamp became an international success, and in 1937, the company built a new manufacturing plant for the lamp near the town of Ajka.

The intensifying anti-Semitism in Hungary and the increasingly harsh anti-Jewish legislation reached Tungsram as well. The company had to replace both Pfeifer and Aschner. The new man in charge, Zoltán Bay, was both an excellent scientist and a caring human being. He achieved special status for the company as being important for defense. A number of leading researchers and engineers, including Bródy, received exemption from the slave labor service that Jewish men were subject to. This saved them for some time. However, when Bródy's wife and daughter were deported, he gave up his exempt status. The Nazis arrested him on July 3, 1944. Inmates sighted him at various camps, including Auschwitz–Birkenau. His wife and daughter probably perished upon their arrival in Auschwitz; by the end of 1944, Bródy too was dead.

Lajos Steiner medal (by Miklós Borsos) of the Hungarian Meteorological Society (from unknown source).

Lajos Steiner (1871–1944) was a meteorologist and geophysicist. He studied at the University of Budapest and for some time he was Loránd Eötvös's associate. Loránd Eötvös was the doyen of Hungarian physicists, a famous researcher of gravitation. Between 1892 and 1932, Steiner served at the Royal Hungarian Central Institute of Meteorology and Earth Magnetism; from 1927, as its director. His principal research concerned the theory of geomagnetism. His main achievement was the introduction of the weather forecasting service in Hungary. In 1917, he was elected a corresponding member of the Hungarian Academy of Sciences. On April 2, 1944, he committed suicide, thus escaping further anti-Jewish persecution.

Nándor Mauthner (1879–1944) was a chemical engineer who studied at the ETH Zurich and graduated in 1902. In 1903, he received his doctorate at the University of Geneva. For a few years, he did research in Emil Fischer's organic chemistry institute in Berlin. From 1911, he worked at the University of Budapest with an interruption between 1917 and 1918 when he served as a military chemist in Vienna. He received a promotion in May 1919, during the communist dictatorship, and this caused hurdles in his subsequent career. His main field of interest was sugar chemistry. In 1934, he was elected a corresponding member of the Academy. After the mid-1930s there is no information about Mauthner, except for the rumors that on May 21, 1944, he committed suicide as an escape from anti-Jewish persecution.

Memorial of Adolf Káldor (by Erzsébet Schaár) at Duna Street 2, District XXII.

Adolf Káldor (1882–1944) was born in Modor (now, in Slovakia). He was the first municipal physician of the town Budafok—today, part of District XXII of Budapest. He was popular among his patients. Together with his family he was deported early summer 1944 and perished in Auschwitz.

Grave stone of Pál Selényi and his wife in the Kozma Street Jewish cemetery (courtesy of József Varga). The inscription remembers their two martyr sons.

The physicist and inventor Pál Selényi (1884–1954) survived the slave labor camp, but his two sons were murdered, György Selényi (1915–1944) and Tamás Selényi (1923–1944). Pál Selényi graduated as a high school teacher of physics and mathematics and embarked on a career at the University. Following Loránd Eötvös's death he was appointed lecturer in experimental physics. In 1919, he was a member of the leadership of scientific societies and museums. After the collapse of the communist dictatorship, he could not hold a state supported position. He became an associate of Tungsram and excelled with numerous inventions. One of these is considered to be the forerunner of photocopying. The anti-Jewish legislation forced him into retirement in 1939. He returned from the slave labor camp severely ill, but continued his work. After Liberation, he was elected to the Hungarian Academy of Sciences and taught at the Technical University.

Fellner Frigyes (source: http://www.magyarzsido.hu, 12.05.2017).

Frigyes Fellner (1871–1945) was an internationally renowned economist and statistician. He trained at the law school of the University of Budapest, and in 1897, he earned the qualification of lawyer. He achieved high positions in banking, but when he embarked on a career in academia, he withdrew from direct involvement in finance. He pioneered the determination of gross national income in Hungary. He was a professor both at the University of Budapest and the Technical University. In 1917, he was elevated to nobility. He was a member of the Hungarian Academy of Sciences. In 1927, soon after the Upper House of Parliament formed, Fellner became an alternate member, and in 1938, he became a full member, when Pál Teleki resigned his membership (due to Teleki's election to the lower chamber). Fellner was never directly involved in politics.

Fellner's spectacular career ended abruptly. The timing coincided with the introduction of the anti-Jewish legislation in 1939. It may be that he had kept his Jewish origins secret or might not have been aware of them. His membership in the Upper House of Parliament was terminated although such membership was for a lifetime. His publishing activities came to a halt in 1939, at which time he had 130 publications. There is no information about his last 5 years. He was arrested soon after the German invasion of Hungary on March 19, 1944. He was taken to the Mauthausen concentration camp where in early 1945, he starved to death.

Discrimination in the Academy Elections

Three members of the Hungarian Academy of Sciences mentioned above (Steiner, Mauthner, and Fellner) perished between in 1944–1945 because of their Jewish roots. We cannot know whether there were more. It appears that relatively few Jewish members of the Academy of Sciences had fallen victim of persecution. However, this is a misleading impression. The fact is that due to its discriminative elections, hardly any Jewish scientists became members of the Academy during the Horthy era. Mauthner's election to corresponding member was a rare exception. Fellner's election is irrelevant in this respect because his possible Jewish background was not known.

The following list contains the names of academicians elected during the Horthy era whose background and religion might have *any* relevance to Jewish origin. It is based on a data bank [8], which, of course, may be incomplete. It does not contain information about Nándor Mauthner, elected corresponding member in 1934 (for whom the document of nomination stressed his religion as Roman Catholic) [9]. If only Christian religion is indicated, it means that already the parents had converted. According to this data bank, for the 25-year period between 1920 and 1944, only two scientists of Jewish religion were elected; both were world-renowned mathematicians. The elections during the Horthy era are distinguished by italics.

András Alföldi (1895–1981, Lutheran) archeologist, *corresponding member, 1933*, full member, 1945; he was excluded in 1949, and reinstated in 1989.

Dávid Angyal (1857–1943, converted, 1895) historian, corresponding member, 1902, full member, 1917, *honorary member, 1936* [10].

Lipót Fejér (1880–1959, Jewish) mathematician, corresponding member, 1908, *full member, 1930*, honorary member, 1946.

Frigyes Fellner (1871–1945, converted, 1903) economist, corresponding member, 1915, *full member, 1936*.

Béla Földes (1848–1945, converted, 1879) economist, corresponding member, 1893, full member, 1901, *honorary member, 1933*.

Alfréd Haar (1885–1933, Jewish) mathematician, *corresponding member, 1931*.

Sándor Korányi (1866–1944, Roman Catholic) physician, *honorary member, 1935*.

István Möller (1860–1934, converted, 1882) architect, *corresponding member, 1927*.

Gusztáv Rados (1862–1942, converted some time between 1875 and 1884), corresponding member, 1894, full member, 1907, *honorary member, 1937*.

Frigyes Riesz (1880–1956, converted) mathematician, corresponding member, 1916, *full member, 1936.*

Lajos Winkler (1863–1939, Calvinist) chemist, pharmacist, corresponding member, 1896, *full member, 1922.*

Anti-Jewish bias in the elections to the Hungarian Academy of Sciences is difficult to investigate. Data on the ethnicity and religion of the members of the Academy are scarce, and elections depend on so many factors that demonstration of bias can hardly be unambiguous. The conspicuously small number of Jewish scientists elected to the Academy suggests a strong anti-Jewish bias. I offer only one example of a failed election in spite of the candidate's strong credentials.

Bust of John von Neumann with Edward Teller visiting it on the campus of the University of Technology and Economics in the early 1990s (courtesy of János Philip).

In 1934, eight members of the Academy nominated John von Neumann for membership. The nominators were outstanding scientists whose scholarship enabled them to appreciate von Neumann's achievements in mathematics and theoretical physics. In addition to a descriptive recommendation, they listed von Neumann's 49 publications in support of their nomination. The signatories were Ottó Titusz Bláthy, Gusztáv Rados, Radó Kövesligethy, Károly Tangl, Lipót

Fejér, Béla Pogány, István Rybár, and Rudolf Ortvay. Alas, von Neumann was not elected. He was 31 years old at the time and looking back from today's perspective when new members of the Academy are often over 60 years of age, one might think that he may have been too young for this honor. However, this was not the case. Just consider von Neumann's nominators, who, apart from Bláthy, were all elected at a young age; Fejér at 28 (in 1908), Pogány at 31 (1918), Rados (1894) and Rybár (1918) both at 32, and Kövesligethy at 33 (1895). In 1937, von Neumann naturalized as a U.S. citizen and in the same year, he was elected a member of the National Academy of Sciences of the U.S.A.

A worthy candidate did not need to be Jewish to be declined. Thus, for example, Albert Szent-Györgyi was elected to full membership in 1938, only after he had received the Nobel Prize in 1937.

I am aware of two academicians of Jewish background who had been excluded from the Academy on the pretext of their participation in the revolutions of 1918–1919. They were Bernát Alexander (1850–1927) and Manó Beke (1862–1946). Alexander was a philosopher who was elected corresponding member in 1892 and full member in 1915. He was excluded on November 24, 1919. His membership was reinstated posthumously in 1989. Beke was a mathematician, elected corresponding member in 1914; he was excluded in 1920, and his membership was reinstated in 1945.

Numerus Clausus

On June 4, 1920, the so-called Trianon Peace Treaty was concluded—named after the Versailles palace where it was signed—with tragic consequences for Hungary. It dismembered historic Hungary by giving independence to Croatia and Slovakia and carved out large chunks of the country to add them to Romania, with some territory even given to Austria. In the "happy peacetime" between 1867 and 1914, the presence of an assimilated Jewish population strengthened the Hungarian population, which was appreciated, because the Hungarian population was in a minority in this multiethnic country. After Trianon, there were hardly any sizeable minorities within the new borders.

Many young Hungarians, who suddenly found themselves living outside of Hungary as a consequence of the Trianon Treaty moved to Budapest seeking higher education. The relatively large Jewish student body became a target and the infamous Law XXV in 1920—usually referred to as *numerus clausus*, that is, closed number—was enacted. This was the first anti-Jewish legislation in post-WWI Europe and it severely limited the number of Jewish students at the universities. This happened under Prime Minister Pál Teleki, a noted scientist in geography. His reign was short lived, but he would be back as prime minister toward the late 1930s, with more severe anti-Jewish legislation.

The exodus of young ambitious Jews who were eager to acquire a higher education was a direct consequence of the *numerus clausus* law. Even some Jewish and non-Jewish youths who might not have been directly impacted by this law, such as John von Neumann and (the future Nobel laureate theoretical physicist) Eugene P. Wigner left Hungary in the 1920s because they envisioned their future hopeless had they stayed.

This law remained in effect throughout the Horthy era in spirit, even if its stipulations loosened toward the late 1920s. Kuno Klebelsberg (1875–1932), the minister of religion and public education between 1922 and 1931, proposed replacing references to racial and national origin in the law by national loyalty and moral reliability. Teleki, Klebelsberg, and their colleagues understood that the *numerus clausus* legislation reflected negatively on Hungary in the civilized world, and Klebelsberg practised a sophisticated double talk. In the West he said what he supposed was acceptable there and at home he continued the rhetoric that served domestic politics. In 1924, he made this cynical statement in the Hungarian Parliament, as though shifting the responsibility to the West: "Give us back the old Greater Hungary and then we will be able to revoke numerus clausus." [11]

Klebelsberg had a broad vision for the dominance of Hungary in the region through Hungarian "cultural superiority." He aimed at bringing back some of the talent that had left the country. It is doubtful though that he would have welcomed the return of Jewish expatriate scientists. None of the attempts to gain professorial appointments for expatriate Jewish physicists or mathematicians turned out to be successful, although they had in the meantime become top players, internationally, in their fields.

Max Born, the doyen of the famous Göttingen School of physics tells an amusing story. It happened in about 1930 that Klebelsberg visited this famous university and its chief administrator, J. T. Valentiner, gave a luncheon party for the distinguished visitor to allow him to meet with the luminaries of the school. This is how Born remembered it when the high guest asked him what he [Born] "thought about the Hungarian mathematicians and physicists. I replied with a hymn of praise for my Hungarian colleagues, mentioning first my old friends [Alfred] Haar and [Theodore von] Kármán, then [George] Polya in Zürich and others I cannot now remember, and finally the young generation who were at present in Göttingen: John von Neumann, Eugene Wigner and Edward Teller. At this point, I got a fearful kick on the shin from [James] Franck, whereupon I stopped and let him continue the discussion. I did not understand what he meant by this violent interruption until he explained it to me after lunch. All I had mentioned were Jews, and therefore, in the eyes of a representative of that anti-Semitic government, not Hungarians at all." [12]

Bust of Eugene P. Wigner with the Hungarian-born Israeli Nobel laureate Avram Hershko on the campus of the University of Technology and Economics in January 2005.

I show another example of how well understood it was that the Horthy regime did not consider Jewish scientists and scientists of Jewish origin to be genuinely Hungarian. In 1935, there was an opening for a physics professorship at the University of Szeged. The former Szeged physics professor, now, Professor of Physics of the University of Budapest, Rudolf Ortvay, mentioned this possibility to Eugene P. Wigner. By then, Wigner was a recognized physics professor in the United States. In his letter of January 13, 1936, Wigner wrote to Ortvay, that exactly those non-scientific considerations, that is, anti-Semitism, that would prevent his [Wigner's] appointment to this position made this position undesirable for him [for Wigner] [13].

Leo Szilard (courtesy of the US Department of Energy Photography).

The law of *numerus clausus* was enacted in accord with the anti-Semitism of a considerable portion of the society. There was no numerus clausus yet when nationalistic students prevented Leo Szilard from continuing his studies at the Technical University in fall 1919. Szilard started his studies as an engineering student in the academic year 1916/17. Then he had his military service in the Austro-Hungarian Army in 1917/18 after which he wanted to resume his studies. However, nationalistic students kicked him down the stairs in front of the main entrance of the School. Szilard had converted and wanted to show his certificate about it but the nationalist students were not interested in it. Szilard was Jewish and no conversion changed this fact for them. It was at that moment that Szilard decided to leave the country. Even the limited amount of time Szilard spent as a student of the Technical University is noteworthy and a reason for pride on the part of the University. For years, in the early 2000s, I kept suggesting to the leadership of the University to erect a bust for Szilard in the sculpture garden of the campus, but it did not happen.

The beating of Jewish students by their nationalistic peers was a daily event for years in the 1920s. Eugene P. Wigner who also spent a short period of time at the Technical University maintained that he did not remember any of it, but he added, had he remembered, he would have denied that it happened. He was a proud Hungarian who felt embarrassed for the actions of those nationalist students. When pressed for remembering, he admitted an episode when three students attacked and beat him, but he hastened to add that it did not hurt too much.

In the mid-1930s the anti-Semitic incidents at the universities intensified. This was going on even before the increasingly harsh and comprehensive anti-Jewish laws were introduced starting in the late 1930s. It happened, for example, that all non-Jewish students, including the converted ones, a total of about 250, refused to sit together with the handful of Jewish students during a lecture of Tivadar Huzella at the medical school. The only exception was Count Miklós Wenckheim who kept sitting with the Jewish students and ignored the verbal abuse directed at him. Professor Huzella was a humanitarian and kept lecturing, paying no attention to the protest. Other professors though joined in the frequent demonstrations of overt anti-Semitism.

The *numerus clausus* law prevented the higher education of many. Those who could afford it attended universities in foreign countries. Some stayed away for good, but many returned to Hungary after graduation. There are no data of either category and it has not been considered polite to find out whether or not someone might have acquired a higher education abroad because of the *numerus clausus*. Considering memorials, three examples of renowned engineers follow who had to attend universities of technology abroad and became successful professionals. Their biographers duly record the fact that they studied abroad though not the reason why they had to acquire their qualifications outside of Hungary.

Busts of László Kozma, László Heller, and István Barta on the campus of the University of Technology and Economics.

László Kozma (1902–1983) was rejected because of the numerus clausus when in 1921 he wanted to enroll at the Technical University. For years, he worked as an electrician. Between 1925 and 1930, he studied at the Brno Technical University (Czechoslovakia at the time; today, Czech Republic), and graduated as an electrical engineer. He worked in Belgium, but in 1942, upon the German invasion, he lost his job. He returned to Hungary and in 1944, he was deported to the Mauthausen concentration camp. He survived, but was severely ill when he returned to Hungary. He worked in industry and taught at the Technical University until 1949, when the communist dictatorship arrested him on false charges, tried, and sentenced him. He spent 5 years in prison and was freed in 1954. In 1955, he was rehabilitated and continued teaching at the Technical University. In 1958, he built the first digital and programmable computer in Hungary. He worked out plans for modernization of the Hungarian telephone system. He reformed the training of electrical engineers, and initiated the specialization of electronic technology.

László Heller (1907–1980) studied mechanical engineering at the Swiss Federal Institute of Technology (ETH) in Zurich and graduated in 1931. He stayed in Zurich for two more years before returning to Hungary where he worked in private engineering business. We have no information about him during the last years of the Horthy era and the Holocaust. In 1948, he defended his doctoral dissertation at the ETH. Heller and his associate, László Forgó, invented a cooling technique utilizing air condensation. Power stations still use this method. He founded the department of energy studies at the Technical University. He is buried in the Kozma Street Jewish Cemetery.

István Barta (1910–1978) studied in the technical universities of Vienna, Brno, and Karlsruhe following his high school graduation in Budapest. He received his Diploma in 1933 and his doctorate (1934) as an electrical engineer from the Technical University of Karlsruhe. He returned to Hungary and worked first for the Ericsson Company, then for Tungsram. After the war he worked in industry and was a professor and chair of the department of communications engineering at the Technical University. His research was on radio and television techniques, as well as acoustics. He was instrumental in the development of the Hungarian electronics industry. We have no information about how he survived the last years of the Horthy era and the Holocaust.

In the late 1930s and early 1940s, the *numerus clausus* gradually became *numerus nullus*, and there was increasing discrimination against the remaining Jewish students. As war was approaching, paramilitary instructions became more rigorous. From 1942, Jewish students had to wear yellow armbands and the converted Jewish students white armbands. Thus, such visible anti-Jewish discrimination predated the German occupation of Hungary (on March 19, 1944) after which the Jews in Hungary had to wear the yellow six-pointed star. In the early summer of 1944, many Jewish university instructors and students were deported to concentration camps, where most of them perished.

Bálint Hóman, the leading politician of education and cultural affairs of the second half of the Horthy era advocated the extension of numerus clausus to include the high schools. His ministry of religion and public education issued a classified circular in summer 1939, which extended the numerus clausus to high schools [14].

Slave Labor and Deportation

From 1939, the Horthy regime conscripted Jewish men and from 1941, after Hungary entered the war, used them as auxiliary troops—in reality as a punitive slave labor force—on the Eastern Front and elsewhere. In addition to the hardship and dangers that accompany wartime conditions, they suffered from humiliation and sadistic treatment by their supervising Hungarian army personnel, the so-called skeleton staff. To be sure, there were exceptions. Many of the inmates of slave labor camps never returned, and there were among them representatives of Hungarian scientific life.

Memorial to the Jewish slave laborers, 1939–1945, at 2 Bethlen Gábor Square, District VII.

There is a memorial to the Jewish slave laborers at 2 Bethlen Gábor Square. Its inscription says that they stood unarmed on the minefields. This is in reference to Miklós Radnóti's poem, A la recherché . . .". . . standing unarmed on distant and freezing minefields; . . ." [15] The inscription applies, among many others, to the father of the author, Dr. Jenő Wilhelm, who was made to sweep a minefield with bare hands and a mine blew him apart in the fall of 1942; his remains rest in a mass grave in Western Russia.

We have no specific information about the scientists who perished in slave labor camps, but we offer the list below as a symbolic reminder. It enumerates the names of former slave laborers who later became members of the Hungarian Academy of Sciences.

György Ádám (1922–2013), physiologist
Pál Benedek (1921–2009), chemical engineer
Frigyes Csáki (1921–1977), mechanical engineer
Ervin G. Erdős (1922–), external member, pharmacologist
Jenő Ernst (1895–1981), biophysicist
Dávid Rafael Fokos Fuchs (1884–1977), philologist
László Fuchs (1924–), external member, mathematician
Tibor Gallai (1912–1992), mathematician
János Gergely (1925–2008), physician, immunologist
István Hahn (1913–1984), historian
Péter Hanák (1921–1997), historian
Róbert Hoch (1926–1993), economist
Miklós Julesz (1904–1972), physician
László Kalmár (1905–1976), mathematician
László Kardos (1898–1987), historian of literature
Béla Kellner (1904–1975), physician, oncologist
György (Georg) Klein (1925–2016), honorary member, physician, immunologist
Károly Lempert (1924–), chemist
József Lukács (1922–1987), philosopher
Károly Marót (1885–1963), classical philologist
Gyula Mérei (1911–2002), historian
Zsigmond Pál Pach (1919–2001), historian
Alfréd Rényi (1921–1970), mathematician
László (Ladislas) Robert (1924–), external member, biologist
Pál Selényi (1884–1954), physicist
István Simonovits (1907–1985), physician
Bence Szabolcsi (1899–1973), musicologist

Sándor Szalai (1912–1983), sociologist
Pál Turán (1910–1976), mathematician
György Vajda (1927–), mechanical engineer
Imre Vajda (1900–1969), economist
Tibor Vámos (1926–), electrical engineer
Andor Weltner (1910–1978), jurist
Ervin Wolfram (1923–1985), chemist
László Zsigmond (1907–1992), historian

The next list contains the names of former deportees who later became members of the Hungarian Academy of Sciences. Let this list be again a symbolic reminder of those deportees who perished.

Iván Berend T. (1930–), economist historian, Dachau
Samu Borbély (1907–1984), mathematician, mechanical engineer, ?
Ervin G. Erdős (1922–), external member, pharmacologist, Sachsenhausen
János Frühling (1937–2015), external member, physician, oncologist, Strasshof
István Hargittai (1941–), chemist, Strasshof
Avram Hershko (1937–), honorary member, physician, biochemist, Strasshof
Miklós Julesz (1904–1972), physician, Buchenwald
Béla Kellner (1904–1975), physician, oncologist
József Knoll (1925–), physician, pharmacologist, Auschwitz
László Kozma (1902–1983), electrical engineer, Mauthausen-Gunskirchen
Károly Lempert (1924–), chemist, Mauthausen
Géza Mansfeld (1882–1950), physician, pharmacologist, Auschwitz
Pál Pándi (Kardos) (1926–1987), historian of literature, Laxenburg (Austria)
Rezső (Ruben) Pauncz (1920–), external member, chemist, Strasshof
György Ránki (1930–1988), historian, Auschwitz
Gábor Szabó (1927–1996), biologist, Auschwitz
Andor Weltner (1910–1978), jurist, Buchenwald, Dachau

A few names appear in both lists. Some survivors of the slave labor camps were then deported to concentration camps.

Mathematicians

Memorial tablet for mathematician victims, V. Reáltanoda utca 13–15.

There is a plaque with two sets of names in the entrance hall of the Rényi Institute of Mathematics of the Hungarian Academy of Sciences. One of the sets is "Our Greats" and the other is "They Embarked on the Road of Creating." There is then a third list that is conspicuously missing, because it would be for "those young talents that prepared for their start."

> "Our Greats"
> Mihály Bauer 1874–1945
> Pál Csillag 1893–1944
> Géza Grünwald 1913–1943
> Dénes Kőnig 1884–1944
> Simon Sidon 1892–1941
> Adolf Szücs 1884–1945
>
> "They Embarked on the Road of Creating"
> Ervin Aczél –1942
> István Ádám 1925–1944
> Ervin Feldheim 1912–1944
> József Krausz –1944
> Dezső Lázár 1913–1943
> Gyula Sándor 1921–1945
> Miklós Schweitzer 1923–1945
> István Valkó 1904–1945
> László Waldapfel 1911–1942

The memorial plaque was unveiled in 1976, that is, much earlier than most of the holocaust memorials. A memorial to the Jewish victims outside of Jewish institutions was such a novelty that the euphemism of "victims of fascism" could be easily overlooked. The former slave laborer Pál Turán spoke at the ceremony. There is a Miklós Radnóti quote on the plaque, "... my words, they will yet ring out by those new walls; ..." [16]

From left to right: Dénes Kőnig (courtesy of Vera T. Sós), Géza Grünwald (courtesy of Éva Gergő), and Dezső Lázár (source: http://www.komal.hu).

Here, we say a few more words about some of the people on the lists. Mihály Bauer studied at the Budapest Technical University where Gusztáv Rados and Gyula Kőnig were among his teachers. Bauer started writing mathematics papers at the age of eighteen. In 1918, he received his professorial appointment at the Technical University. In 1922, Bauer was the first recipient of the newly established Kőnig Gyula Prize of the Eötvös Loránd Mathematical and Physical Society. In contrast, he was kept at a low-level position at the Technical University, where anti-Semitic students often disturbed his lectures. In around 1936, the university forced him into retirement. In 1944, he was taken to a concentration camp from where he returned, but he died in February 1945.

Dénes Kőnig was the son of Gyula Kőnig (1849–1913), a former rector of the Technical University. Dénes studied in Budapest and Göttingen. In 1907, he received his doctoral degree and joined the Budapest Technical University. He was a pioneer of graph theory, and in 1935 he reached the rank of full professor. In 1936, he published a monograph, *Theorie der endlichen und unendlichen Graphen*. It has remained a fundamental treatise in the field and in 1990 it was published in English translation, *Theory of finite and infinite graphs* (Birkhauser). Following the Nazi occupation of Hungary on March 19, 1944, Dénes Kőnig helped persecuted mathematicians. A few days after the Arrow Cross takeover, on October 15, 1944, Dénes Kőnig committed suicide.

Adolf Szücs studied in Budapest and Paris, taught in high school, and in the late 1920s he received an appointment at Gusztáv Rados's department at the Technical University. His research concerned variation calculations and differential equations. On February 3, 1945, a group of the Arrow Cross took him from his home. He was last seen on February 4, 1945.

Pál Csillag studied with Lipót Fejér and earned his doctorate when he was 21 years old. The Budapest Goldberger Textile Works employed him as a mathematician, but they had to retire him in 1938 because he was Jewish. He perished in 1944, but there is no exact information about his death.

Géza Grünwald (1910–1942) [17] studied in Szeged and earned his doctorate in 1935. Approximation theory was his principal research interest. He was member of a circle of young mathematicians that held weekly meetings in the City Park at the statue of Anonymous. He was murdered in one of the slave labor camps that were often indeed death camps. Today, the Bólyai János Mathematical Society has a Géza Grünwald memorial medal. This annual award is given to mathematicians under 30 years of age who have demonstrated considerable achievements in fundamental mathematical research.

Dezső Lázár began his studies in Budapest, but had to continue in Szeged because of the *numerus clausus* law. He could not find employment following graduation and worked as an apprentice to a cabinet maker. Then, he was offered a job in Kolozsvár (today, Cluj-Napoca in Romania) which had again become part of Hungary. In 1942, he was ordered for slave labor. He had one printed publication while he lived; it was in set theory, in the early 1930s. Pál Erdős found this paper important and showed it to John von Neumann who arranged for its publication in the periodical *Compositio Mathematica*. In 1947 Lázár had another publication, posthumously, arranged for him by his surviving friends.

We learned more about Dezső Lázár from the renowned late mathematician László Fejes Tóth [18]: "My dear friend, Dezső Lázár, turned my attention to the area of research that I worked in during my entire life; this was the project of arranging and covering the surface.... About Dezső Lázár I would like to say that when I moved to Kolozsvár, he was there, working as a teacher in the Jewish Gimnázium. Later, he was called upon for forced labor service. He was made to detect mines, was wounded in the thigh, and left to bleed to death. While he was in the forced labor service, we kept in close contact with his family. We often visited his wife and two small children. We learned about what happened with him from his wife. I don't remember when this exactly happened, because the years have become blurred in my memory. His wife was a refined, beautiful lady and the thought horrifies me to this day that she was dragged away in a box-car and after a lot of suffering they murdered her with her two small children in the gas chambers of Auschwitz." [19]

Dániel Arany (courtesy of the Révai Miklós Gimnázium, Győr) and the grave stone of Arany and his wife over one of the mass graves of Holocaust victims of the Heroes' Cemetery in the garden of the Dohány Street Synagogue.

We conclude with a few words about the mathematician Dániel Arany (1863–1945). While working as a high school math teacher, in 1893, he founded a mathematical magazine, *Középiskolai Matematikai Lapok*, for high-school pupils He edited this for a few years; then, he passed the editorship over to László Rátz. The magazine has fostered ever since young talent in mathematics. Between 1905 and 1919, Arany taught in a technical college, but after 1919, he was forced into retirement for alleged involvement in the communist dictatorship. He never found employment again. He continued his research in probability theory and game theory, and co-authored a monograph in actuarial mathematics. His most important contribution remained in the area of high school mathematical education. In 1944, the Jewish Arany and his wife were incarcerated in the ghetto where they both died. Before moving into the ghetto, Arany donated his valuable collection of mathematical books to the Eötvös Loránd Mathematical and Physical Society. Today, the mathematical competition of 1st and 2nd grade high school pupils carries Dániel Arany's name.

Invisible Memorial in Trefort Garden

This is one of the 198 names of martyrs on a metallic strip in between the bricks of the wall at the Trefort Garden of Eötvös Loránd University.

In fall 2014, an unusual and moving memorial was unveiled in the Trefort Garden campus of Eötvös Loránd University. Its title is "Sign in the Garden." The memorial, created by a group of architects and others, is a long metallic strip inserted in between the bricks of two buildings. The introduction in the strip says in Hungarian, English, and Braille that it honors the memory of those university members who were victim of the anti-Jewish laws, the Holocaust, and World War II. The overwhelming majority of the victims were Jewish, but the strip gives only the following information: names, dates of birth and death, and status. The latter appears in one of the following designations: deported, slave laborer, deported slave laborer, civilian victim, taken by force, died by suicide, forced laborer, officer-candidate corporal in reserve-titular lance sergeant, officer-candidate sergeant in reserve, vanished, murdered, shot, or simply, nothing. The designations demonstrate considerable insensitivity: How could a deported or a slave laborer be different from a civilian victim? Were not those deported to Auschwitz or murdered on the bank of the Danube, civilian victims? When I asked about this one of the spiritual leaders of the memorial, the response was that the designations followed the usage of *designations of the time*. This is, however, unacceptable, because succumbing to the usage of designations of the time will only facilitate conserving the *judgments of the time*.

Sadly, the memorial strip stays practically invisible to anybody who is unaware of its existence. Apart from the strip itself, there is no indication of the memorial; it is as if it should be kept in strict confidence. When I visited the memorial soon after its inauguration, I thought an information tablet or something directing the attention to the strip would be added later. In May 2017, there was still nothing.

Professor Géza Komoróczy raised the issue of the missing text in Hebrew on the strip at the inauguration. This omission symbolizes the ambiguity of the memorial. Subsequently, Komoróczy [20] interpreted the ambiguity of the memorial as a *limited* apology from the University. (my emphasis) According to him, the University restores the university membership of the victims, all the victims. However, this restoration concerns them all, as Hungarians, but not as Jews, for which they had been murdered in the first place. According to Komoróczy, and I fully agree with his assessment, this is an example how remembrance is subordinated to the "politics of remembrance."

So Many More to Remember

Memorial tablet for 41 dermatologists in the waiting room of the Dermatological Clinic of Semmelweis University, 41 Mária Street, District VIII.

A unique memorial tablet was unveiled in 2014 in the waiting room of what is called officially, the Department of Dermatology, Venereology, and Dermato-Oncology. It is for 41 Hungarian Jewish dermatologist victims of the Holocaust 1944–1945. The uniqueness of this memorial is that it commemorates not only dermatologists of this University, but the dermatologist victims from all over Hungary. This circumstance is as uplifting as is painful to think that this is the only such commemoration among all the clinics of Semmelweis University. Browsing the tablet in the crowded waiting room, the viewer involuntarily has the unreal sensation that the waiting time in this crowded venue might be reduced had these dermatologists been still around.

This thought association brought me to thinking of the many other Jewish physicians who perished in concentration camps and in punitive slave labor units. The anti-Semitic hatred caused the guards of these units and responsible army officers to prevent the incarcerated doctors from helping wounded soldiers in nearby units of the Hungarian Army, who were often left untreated.

Mária Flóra Zoltán's portrait of Dr. István Zoltán (courtesy of Mária Flóra Zoltán).

István Zoltán (1899–1945) attended the famous Lutheran Gimnázium in Budapest and then graduated as an MD from the Royal Pázmány Péter University in 1923. He became an ear, nose and throat specialist and specialized also in surgery and in pathology. At the invitation of Professor Zoltán Lénárt, he joined the Clinic of what the Department of Otorhinolaryngology, Head and Neck Surgery is today. Dr. Zoltán was a clinician, a researcher, and in charge of the histology laboratory. His publications and activities in congresses drew international attention to his achievements. He was expected to succeed Professor Lénárt upon his retirement scheduled for 1941. Alas, this was not to happen because Dr. Zoltán was Jewish. International invitations offered an escape for him from the ever-worsening situation, but he declined to leave his aging parents and his patients. The Arrow Cross took him in 1944 and he perished in the Mauthausen concentration camp only days before its liberation. His artist daughter painted his portrait and donated it to her father's former clinic. When after years it was still not displayed, the daughter took back the painting. There is thus no remembrance there of the martyr physician István Zoltán.

There is a memorial brick honoring István Zoltán in front of the apartment house in Andrássy Avenue where the Zoltán family used to live. The family would like to erect a memorial plaque on the façade of the building and they have the consent of the district authority. However, the principal proprietor of the building has so far prevented this commemoration.

An internationally renowned Hungarian-American scientist told me about his horrible experiences in Budapest in 1944–1945. However, he asked me for discretion to avoid his Hungarian friends learning about his Jewish origins. He said he would not like to lose his friends in Hungary. I wondered how good friends those people can be. It is also possible that he supposes their anti-Semitism without foundation.

The misuse of the label "Martians" comes to mind. The label originated at the Manhattan Project in a conversation between Enrico Fermi and Leo Szilard. They joked that the conspicuously large number of Hungarian participants were in reality Martians and they spoke Hungarian for camouflage. The emigration of Theodore von Kármán, Leo Szilard, Eugene P. Wigner, John von Neumann, and Edward Teller—the five truly Martians—is often ascribed to their curiosity and thirst for adventure. Thus the label becomes a euphemism to mask the real reason for their emigration. This misleading approach was compounded by Edward Teller when he turned to his audience at a meeting with the following words: "Véreim, Magyarok!" meaning "My blood brother Hungarians!" This could have been the playing out of a childhood fantasy. In high school he suffered from anti-Semitism and he could have not used this expression at the time. In the early 1990s, so soon after the political changes, this expression, however dated it sounded, was consistent with the gradual return to the Hungary of the 1920s and 1930s.

To this day, mentioning someone's being Jewish or being of Jewish origin is considered impolite in Hungary. Teller noted that "In Hungary, even if you were born there, being a Jew you do not feel at home." [21] In this, Teller referred to his high school years. The question is, has the situation changed? There was an interesting case in connection with Teller's election to the Hungarian Academy of Sciences in 1990, right after the political changes. Here I quote the story on the basis of published reference and archival material in the Teller Folder at the Archives of the Hungarian Academy of Sciences [22].

A memorandum was prepared in February 1990 for Teller's election to honorary membership of the Hungarian Academy of Sciences. It was at the time of the great political changes in Hungary when the one-party system was giving way to a multi-party democratic system. In this memorandum, there are four references to Teller's being Jewish. One mentions that one of Teller's high school teachers used to address the class as "Gentlemen and the Jews!" Another is that Teller's father did not see the conditions in Hungary encouraging for his Jewish son's scientific career. The third mentions that Teller's family had to wear the yellow star under Nicholas Horthy, was incarcerated in a ghetto under the Hungarian Nazi leader Ferenc Szálasi, and was exiled from Budapest to the countryside in the early 1950s under the communist leader Mátyás Rákosi. The fourth reference is that Teller always stressed his Hungarian and Jewish origins.

The final, "official" nomination does not contain any of these four references to Teller's being Jewish. Teller was elected honorary member in 1990.

Saviors and the Saved

In paying tribute to the saviors—symbolically, to all saviors—we single out two individuals among the heroes of the Hungarian Holocaust. They were the Lutheran minister Gábor Sztehlo (1909–1974) and the Swedish trade representative Raoul Wallenberg (1912–1947?).

Left: One of Raoul Wallenberg's memorials in Budapest, Szent István Park, District XIII. It is a copy by Sándor Győrfi (1999) of Pál Pátzay's original work (1949). Right: Raoul Wallenberg, November 26, 1944 (photograph by Tamás Veres, courtesy of the late László Ernster).

Left: Gábor Sztehlo's memorial plaque on the façade of the Lutheran Church, Bécsi kapu Square, District I. Right: Sztehlo's memorial (by Tamás Vígh and Barnabás Vinkler, 2009), Deák tér, District V.

Among the saved, there were many young people who then became renowned scientists. For example, Gábor Sztehlo saved the Hungarian-born American Nobel laureate chemist George A. Olah (1927–2017) and Wallenberg saved the internationally famous Hungarian-born Swedish biochemist Lars Ernster (1920–1998).

George A. Olah as a high school student (courtesy of the late George A. Olah) and Olah among other Nobel laureates in Stockholm in 2001 (photograph by Hans Mehlin, © and courtesy of the Nobel Foundation). The conspicuously tall Olah is in the middle of the picture.

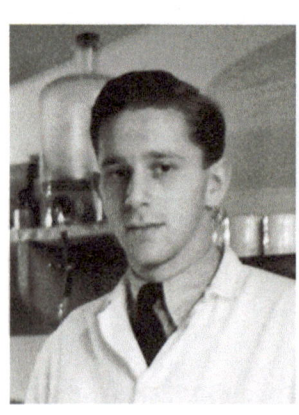  

Left: László Ernster in the Jewish Hospital as Dr. Alfréd Lajta's assistant. Middle: Dr. Lajta, who was murdered by the Arrow Cross with all the other doctors and the patients of the Jewish Hospital (Ernster was elsewhere at the time). Right: The Queen of Sweden at the Nobel Prize award ceremony in 1978. Ernster is her right (all three photographs courtesy of the late Lars Ernster).

Sad Conclusion

It appears that hardly any scientist martyr of the Hungarian Holocaust is remembered as victim of the Holocaust by a personal memorial. In the second half of the 2010s this is the more gruesome observation, because the number of memorials honoring anti-Semitic authors, politicians, and other public figures of the Horthy era steadily increases. There is a strengthening impression that it is not only that official Hungary avoids facing the Past, but that beyond the falsification of history, it carries on the political legacy of the era between the two world wars that led to a national catastrophe.

Acknowledgments I thank Magdolna Hargittai, Diána Hay, Géza Komoróczy, János Kőbányai, and András Simonovits for their assistance in the preparation of the original Hungarian version of the manuscript (see Footnote 2). I thank Bob Weintraub and Irwin Weintraub as well as Tamás Magyarics for their thoughtful suggestions for the English version of this paper.

Recently, the paragraph on the Horthy era in the "History of the Hungarian Academy of Sciences" (see the first page of this chapter) was augmented by mentioning the anti-Semitism of the Horthy era and the consequences of *numerus clausus*. These changes were made following the publication of Istvan Hargittai's article whose title in English translation is "Inconvenient Truth. About the History of the Hungarian Academy of Sciences" [*Magyar Tudomány* 179, 435–442 (2018].

Two in the lists of those future academy members who had been slave laborers or deportees have died recently, László Róbert and József Knoll, both in 2018.

References

1. Istvan Hargittai and Magdolna Hargittai, *Budapest Scientific: A Guidebook* (Oxford, UK: Oxford University Press, 2015). Since then, we have published a similar volume for New York: Istvan Hargittai and Magdolna Hargittai, *New York Scientific: A Culture of Inquiry, Knowledge, and Learning* (Oxford, UK: Oxford University Press, 2017). We are currently completing Istvan Hargittai and Magdolna Hargittai, *Moscow Scientific: Memorabilia of a Research Empire*.
2. Hargittai István, "Hetvenéves fehér folt." *Magyar Tudomány* 2013, 1035–1045.
3. http://mta.hu/data/MTA_Tortenete_ENG.pdf (last downloaded, December 30, 2017).
4. Communication by e-mail from Soma Rédey, deputy-head of the Department of Communication, June 13, 2017.
5. Letter of November 30, 1945, from Albert Szent-Györgyi to the Secretary General of the Hungarian Academy of Sciences. The letter is stored under registry number 48/1946 in the Manuscript Archive of the Library of the Hungarian Academy of Sciences.
6. Pillich Lajos, "Richter Gedeon (1872–1944)." In: Novák Takács Krisztina – Hermecz István (szerk.): *Esti beszélgetés: Magyar gyógyszerészkutatók portréi.* Magyar Gyógyszerésztudományi Társaság, 2005, Budapest, 69–82.
7. Max Born, *My Life: Recollections of a Nobel Laureate.* Charles Scribner's Sons, New York, 1978, p. 214.
8. *Diszkrimináció, Emancipáció – Asszimiláció, Diszkrimináció: Magyarországi egyetemi tanárok életrajzi adattára 1848–1944.* I. Zsidó és zsidó származású egyetemi tanárok. Szerkesztette Kovács I. Gábor. Összeállították Kovács I. Gábor – Kiss Zsuzsanna – Takács Árpád. ELTE Eötvös Kiadó, Budapest, 2012.
9. Tagajánlások 1934-ben, Magyar Tudományos Akadémia, Budapest, 1934 (kézirat gyanánt), p. 62.
10. Prior to 1949, honorary membership was given to officers of the Academy, prominent members of public life, politics, the churches, and the financial world. Also, elder academicians were recognized for their achievements by this distinction. See: *A Magyar Tudományos Akadémia tagjai 1825 – 2002*, MTA Társadalomkutató Központ – Tudománytár, first volume, Budapest, 2003, p. 14.
11. *Nemzetgyűlési Napló*, 1922–1926. XXIV. kötet, 295. ülés, 320. old. (1924. június 4), in Kovács M. Mária, *Törvénytől sújtva: A numerus clausus Magyarországon, 1920–1945.* Napvilág, Budapest, 2012, p. 50.
12. Max Born, *My Life: Recollections of a Nobel Laureate.* Charles Scribner's Sons, New York, 1978, p. 236.
13. Letter labeled K785/139 in the Archives of the Hungarian Academy of Sciences; quoted in István Hargittai, "Wigner Jenőről." *Élet és Irodalom* 2014. October 10, p. 12.
14. K. Farkas Claudia, *Jogok nélkül. A zsidó lét Magyarországon, 1920–1944.* Napvilág, Budapest, 2010, in Kovács M. Mária, *Törvénytől sújtva: A numerus clausus Magyarországon, 1920–1945.* Napvilág, Budapest, 2012, p. 218.
15. Thomas Orszag-Land (translator), *The Witness: Selected Poems* by Miklós Radnóti. London: Tern Press, 1977, p. 42.] Other translations exist.
16. Quote from "The Protector" by Miklós Radnóti, Thomas Orszag-Land (translator), *The Witness: Selected Poems by Miklós Radnóti.* London: Tern Press, 1977, p. 35. Other translations exist.
17. Éva Gergő, Géza Grünwald's daughter, informed me about the correct years of birth and death.
18. Hargittai István, "Fejes Tóth László." *Magyar Tudomány* 2005, 166 (március), 318–324.
19. In 1942, my father was killed the same way as Dezső Lázár. Then, our story continued as that of his family's, but only up to a point. In June 1944, my mother, my brother, and I were put into a box-car and the train started for Auschwitz. Along the way, somewhere, however, the train stopped, moved back for a while, and continued in a different direction, toward Vienna. What happened was that a train that had been destined for Austria had already left for Auschwitz, by mistake. Our train was the replacement. My brother was ten years old and I was not yet three. I. Hargittai, *Our Lives: Encounters of a Scientist*, Budapest: Akadémiai Kiadó, 2004; pp. 52–54.
20. Komoróczy Géza, "Öt héber szó." *Élet és Irodalom*, 2015, December 4.
21. Istvan Hargittai, *Judging Edward Teller: A Closer Look at One of the Most Influential Scientists of the Twentieth Century* (Amherst, NY: Prometheus Books, 2010), p. 48.
22. Balazs Hargittai and Istvan Hargittai, *Wisdom of the Martians of Science: In Their Own Words with Commentaries* (Singapore: World Scientific, 2016), p.191.

Sculpture of a chair with an open book beneath it by David Saunders (1987), in the Owen F. Dolen Park in the Bronx (photograph by I. Hargittai).

Introduction

In this Section, I collected book reviews. Father has published a large number of these reviews. In addition to other journals, he was often asked by the international journal *Structural Chemistry* for which Mother has been the Reviews Editor. When Father edited *The Chemical Intelligencer*, he often assigned to review books to himself. Father has had a special relationship with books. His father was killed when Father was barely one year old. All what remained after his father was a few photos and a book. It was a monograph on the legal aspects of unfair competition, which my grandfather co-authored. Father considered this book as his father's legacy, which he never read, but frequently talked about. Reading books for Father meant several things. It was entertainment, source of information, and examples of writing that he might want to emulate.

Father started reading very early, well before school, and he was always a veracious reader. He read very few children's books and his reading started more with what would qualify as young adults' literature. Even that stage did not last long and he fast moved to the classics. Books like Heinrich Mann's *The Loyal Subject* and Somerset Maugham's *The Razor's Edge* came early on and innumerable books followed. He also liked to reread books that he had liked and enjoyed rereading because there was always plenty to discover anew. Books were possibly the most frequent topics of our dinner conversations.

Father started writing scientific monographs in his thirties. Already his second book was Mother and Father's joint project. Father co-authored a book with someone else only once. It was Ronald J. Gillespie, a British Canadian chemistry professor, the originator of a most popular model of molecular geometry. Father had expanded the applicability of this model and they decided to co-author a monograph about the model. Gillespie was a great pedagogue and Father learned from him a didactic approach in writing. The experience of the joint authorship, however, turned out to be frustrating and Father vowed never to co-author a book with anybody again, with the exception of Mother. I have turned out to be another exception. Father continued though co-editing books with others.

Writing a book, even a technical one is very much an individual activity. Father has his own approach, probably not very economical, but it has worked for him. He prepares the first draft relatively fast, because he wants to see the project in its entirety even if nothing of the initial draft remains in the final version. There is tension and round-the-clock work, to the extent his other obligations allow it, until this initial draft appears. Then the painstaking improvement and refinement and addition and subtraction and organization follow. At the end, Father has often expressed his wish that he had had the concept of the final draft around at the start because in that case he could have limited the entire effort to a single version. Once, at the book launching of one of his most successful books, *The Martians of Science*, someone asked Father whether he experienced frustration and tension in the process of writing. Father responded without hesitation that he did not. It was true that by then he had genuinely forgotten all the initial difficulties that had lasted a mere 10–12 months. Mother, however, remembered clearly and took some effort on her part not to correct Father then and there.

Father likes to write reviews about books that he likes and usually gives up producing a review if he does not like the book. This has caused some problems with journal editors who might consider such a no show as a sign of unreliability even if the reason is explained. Father does not like negative reviews and there must be a special reason for him to publish one. What he likes even less is when people substitute hagiography for a review. A good review wets the appetite to read the book, gives an idea what the book is about, singles out some merits, points to some shortcomings, if there are any that justify mention, and delineates the likely circle of readership. He welcomes editorial assistance in improving style and wording but resists when an editor tries to communicate his or her opinion about the book in the disguise of an independent reviewer's comments. In any case book reviews is another important genre for Father and the selection is compiled to illustrate the diversity of his approach to it.

Matisse's *La Danse* (1909) and its admirers, Museum of Modern Art, New York (photograph by Istvan Hargittai). In his review, Istvan is taking issue with Arnheim's criticism of this painting. For a long time Istvan has used it for demonstrating a hard-to-grasp chemical concept (pseudorotation).

Rudolf Arnheim, The Split and the Structure: Twenty-Eight Essays[a]

University of California Press, Berkeley, CA, 1996, 184 pp

István Hargittai

This is the fourth volume of Rudolf Arnheim's essays.[1] More than two-thirds of the 28 pieces were originally published in the 1990s, the latest in 1996, and there are also earlier pieces, the earliest from 1966. Three of the essays are not dated and may have been prepared for this volume. Only a few of the essays will be commented upon in this review.

Arnheim writes in the essay "A Maverick in Art History" that during his 50 years of having been involved in the field he has gradually changed his observation point. In the early years he was exploring the arts for convenient illustrations, only to move eventually into the territory of the arts themselves.

What is truly remarkable, though, is that he has done so while keeping up with the sciences and integrating scientific concepts and discoveries into his work.

Arnheim is himself an example of an exception to what C.P. Snow labeled as the split of the "Two Cultures." There is no more appropriate place than *Leonardo* to take notice of this achievement.

In the foreword and in the introductory essay, "The Split and the Structure," Arnheim explains how artists grope for structure in order to shape powerful and enlightening images, and how the scientist's search for truth is a search for structure. Nobel laureate physicist Eugene Wigner's description of the most important method of science comes to mind:

> [S]cience begins when a body of phenomena is available which shows some coherence and regularities; that science consists in assimilating these regularities and in creating concepts which permit expressing these regularities in a natural way. . . . It is

this method of science rather than the concepts themselves (such as energy) which should be applied to other fields of learning.

This quotation is from a brief speech Wigner gave during the Nobel ceremonies in 1963 in Stockholm. With this statement Wigner was also paying tribute to his teacher Michael Polanyi, the physical chemist-turned-philosopher from whom Wigner had originally learned this idea.

As Arnheim enumerates the most important properties of structure, such as range/space, interacting forces, growth and dynamics, each attribute is treated in such a way that is more general than might be expected from an art historian. As he speaks of the range of structure as determined by the amount of space a structure needs and can accommodate for its best functioning, it is obvious that space is much more to him than mere abstraction. With his observation that the structure—occupying the available space—is held by interacting forces, we are reminded of Buckminster Fuller's physical geometry. Fuller was far from being a (natural) scientist but his design (science) was all about structure. There is also excellent resonance between Arnheim's comments about space and Francesco Borromini's articulation of space.

Structure and split are discussed at various levels by Arnheim. He gives much more emphasis to the structure than to the split. Yet the two do not appear disparate, and we get a glimpse of how the split, in this case in a non-abstract manner, may be an integral part of structure. This is represented by Arnheim's quoting the story of the Chinese cook who never has to sharpen his knife because when he has to carve the meat of an ox, he puts his hand on it, presses with his shoulder, his foot and his knee, and right away the skin splits and the knife slides smoothly between the natural sections of the body.

In "The Way of Crafts" Arnheim relates his experience with Japanese structures. He notes the affinity of Japanese design to nature's own way—the best possible space utilization. He also remarks on the close relationship between functionality and beauty in the best designs. Arnheim notes that he was fortunate that his year of teaching in Japan in

[a]Originally published in *Leonardo* 1997, 30:236–237. © 1997 by the International Society for the Arts, Sciences and Technology (ISAST), published by the MIT Press.

[1] Rudolf Arnheim (1904–2007).

I. Hargittai (✉)
Department of Inorganic and Analytical Chemistry, University of Technology and Economics, Budapest, Hungary
e-mail: istvan.hargittai@gmail.com

1959 was early enough to allow him to witness much of the tradition of the arts shaping both the style of daily living and the objects of practical use instead of being confined to museums.

More than three decades later I found that this is still the case in Japan, probably less so than in 1959 but noticeably more so than in Western cultures. I find this important to stress because a visit to Japan still has rewards in such aspects. Arnheim mentions the Japanese tendency of giving exposure to unpretentious shape and his example is drawn from the Katsura palace in Kyoto. In fact the Katsura Imperial Villa is an almost inaccessible place for Japanese (especially) and foreign visitors alike. On special occasions there are organized tours for small groups of foreigners, and this is how I got to see it, as I happened to be there on the right day and to have the right connections. But for any practical purpose, it can be considered closed to the general public.

However, examples of using natural forms and at least seemingly untreated materials abound in Japan. Invoking the Katsura Imperial Villa for illustrating something that could have been done much more easily is an example of what I find at places a little awkward in this essay volume. Some examples in the discussion convey elitism, others are simply unfortunate. A case in point is in the otherwise very interesting discussion of order and disorder in "From Chaos to Wholeness." Following some criticism for the lack of comprehensible order in some artworks, Arnheim proceeds to speak about locally arising troubles when elements are not clearly identified as belonging either apart or together. He quotes one example, Henri Matisse's *La Danse*, of which two versions are known. The first is from early 1909 and is in the Museum of Modern Art, New York. The second was done a few months later and is at the Hermitage in St. Petersburg. The essay reproduces the earlier version, but Arnheim's comments are directed at both. The pictures show a round dance of five figures holding hands. Arnheim complains about a chaotic disorientation resulting from the hands of the frontal figures interfering in their reaching for each other by the knee of the figure behind them. I do not think that there is any confusion about which of the dancers are in front and which are behind, though some how the gap between the two hands conveys the feeling of the dance slowing down unnecessarily at that point.

This may have bothered Matisse as well: in the more elaborate second version of the painting, the two hands touch or almost touch; behind these hands the other dancer is in a slightly different position and is more sharply drawn than in the first version (in fact, so is the whole picture) . To me the first version appears as if it were a study for the second. Although the first is not ambiguous as far as which dancer is in front and which is behind, whatever ambiguity there may be is absent in the second version. I am surprised that Arnheim did not make the distinction between the two versions of *La Danse* in this respect.

All in all, these 28 essays are as many gems to read and muse about for experts and mere art-lovers alike.

A selection of snowflake images from W.A. Bentley collection (W.A. Bentley, W.J. Humphreys, Snow Crystals, New York and London: McGraw-Hill, 1931). One of the contributors to the book edited by Gombrich finds snowflakes boring. Istvan in his review takes issue with this comment as snowflakes come in as many variations as many snowflakes there are.

Richard Woodfield, Ed. Gombrich on Art and Psychology

Manchester University Press, Manchester, UK, 1996, 271 pp[a]

István Hargittai

I am always amused when there is a company and X says something and Y tries to explain that what X really meant was this and this, and a heated discussion ensues about what X may have meant, while the somewhat bewildered X is looking over, being ignored completely. This discussion of Gombrich on art and psychology is not like that.[1] All contributions were submitted to Gombrich and on two occasions Gombrich felt compelled to make comments, which are duly printed following the relevant chapters. Referring to some detailed analysis of the extent of his belonging to the "Vienna School" of art history, Gombrich remarks, "I don't think it matters to what extent I belong to the Vienna school or any other school, of course [we] all absorbed views and problems from our teachers, but my present interests lie on a very different plane, as any reader of my writings (and possibly of my future writings, if I live so long) will be able to judge."

It is truly fortunate that Gombrich could review the contributions. Before the seemingly careful plot of the story of Gombrich's patricide of his "spiritual grandfather," Alois Riegel, might take off (see the contribution "The Vienna School's Hundred and Sixty-Eighth Graduate: The Vienna School's Ideas Revised by E.H. Gombrich," by Jan Bakos), Gombrich sets the record straight. He states that the fact that he knows the principal works of Riegel better than many of those who write about him does not mean that he is obsessed with Riegel.

Another tour de force of the book is that one of the 13 contributions is actually by Gombrich himself, titled "Four Theories of Artistic Expression."

To me Gombrich's teachings on visual perception are of special interest in this volume (see, for example, the chapter "Form and Its Symbolic Meaning" by Chang Hong Liu and John M. Kennedy). This is also where modern science and technology may provide further raw material if not the answers to the questions in the study of the psychology of art. In the discussion of form and its symbolic meaning, the consensus shown by subjects playing Gombrich's game with circles and squares is examined. Current research employing magnetic resonance imagery (MRI) provides mapping of the responses of the human brain to the visual experience of circles and squares. Significant differences in brain laterality are observed, for example, in the responses by females and males. It appears also of interest to investigate the dependence of responses on the actual conditions of the mind at the time of the recordings. It is expensive research, but an unexpected bonus may be a by-product that helps the analysis of visual perception and art appreciation. The first step should, of course, be to see whether there is any discernible correlation between the measurements and the psychologists' and art historians' findings. Whether or not there is a correlation, the matter seems worth pursuing.

In another contribution, "Orders with Sense: Sense of Order and Classical Architecture" (by Joaquin Lorda, University of Navarra), the infinite symmetric patterns created in the kaleidoscope are characterized, in agreement with Gombrich, as at first fascinating and then soon becoming boring. This is contrasted with architectural design, which is capable of arousing (presumably permanent or, at least, long-lasting) interest and pleasure. The discussion then moves on to snow crystals, likening them to the patterns in the kaleidoscope rather than to architectural marvels. Lorda even illustrates snow crystals with a drawing that, very much to the point, looks like, and is, an ornament. "A game of this

[a]Originally published in *Leonardo* 1997, 30:160–161. © 1997 by the International Society for the Arts, Sciences and Technology (ISAST), published by the MIT Press.

[1] Ernst Gombrich (1909–2001).

I. Hargittai (✉)
Department of Inorganic and Analytical Chemistry, University of Technology and Economics, Budapest, Hungary
e-mail: istvan.hargittai@gmail.com

kind," Lorda laments, "does not have human rules; our sense of order can read little in this redundant glacial geometry; the sense of meaning becomes paralyzed: there is nothing here to understand."

Before I comment on this description of snowflakes, I would like to introduce two quotations related to snowflakes and architecture. Thomas Mann gives a most eloquent description of the beauty and symmetry of snowflakes in *The Magic Mountain:* "the exquisite precision of form displayed by these little jewels, insignia, orders, agraffes— no jeweler, however skilled, could do finer, more minute work . . . Yet each in it-self-this was the uncanny, the anti- organic, the life-denying character of them all-each of them was absolutely symmetrical; icily regular in form. They were too regular, as substance adapted to life never was to this degree—the living principle shuddered at this perfect preci- sion, found it deathly, the very marrow of death—Hans Castorp felt he understood now the reason why the builders of antiquity purposely and secretly introduced minute varia- tion from absolute symmetry in their columnar structures." The last sentence is very telling about the potentials of beautiful and perfect architecture becoming boring and lifeless.

With it, a much earlier Japanese statement resonates exceedingly well. The following quotation is from *Essays in Idleness,* a translation of *Tsurezuregusa* by Kenko Yoshida (1924–1931) (quoted here after D. Keene, 1981, C.E. Tuttle, Tokyo) , "In everything . . . uniformity is undesirable. Leav- ing something incomplete makes it interesting, and gives one the feeling that there is room for growth. . . . Even when building the imperial palace, they always leave one place unfinished."

I have two comments. One is that architecture may acquire all the good and negative features of the kaleidoscope if we are not watching out and avoiding the traps of perfect and virtually endless repetition and other features of geometrical symmetry. The other is that, at closer inspection, snowflakes have not only tremendous variety in their general shapes— this has been noted by many—but, at close enough inspec- tion, there are minute variations in the six directions of even the seemingly most perfect snowflake. These tiny variations diminish nothing in their beauty and symmetry but may suffice for the interested eye to enhance their intriguing intricacy of the execution of design. Then, depending on the background of the observer, the association of the exter- nal shape with its origin in the internal arrangement of the water molecules, all interconnected by the fragile yet rigor- ously distributed hydrogen bonds, may be the source of further marvel and contemplation—anything but boring.

In taking issue with Mann and Lorda, or, for that matter, with Gombrich, I am suggesting that the patterns of the kaleidoscope and patterns in architecture have potentially more in common with each other than snowflake designs have with either, unless the snowflake design comes from a master drawing or the computer rather than from nature. In the symmetries involved, I am suggesting to delineate geo- metrical symmetry, which is rigorous, and "material" sym- metry (using the term suggested by the Russian crystallographer and symmetrologist, A.V. Shubnikov), which allows imperfections. The strict symmetry concept in the geometrical sense restricts its utilization to giving "yes" and "no" answers only, whereas there is a wide range for symmetries if we follow Hermann Weyl, and, indeed, the ancient Greeks, in relaxing the meaning of this term and extending it to include harmony and proportion.

Returning to what Joaquin Lorda has to say about snowflakes, he concludes with an important caveat: "were the forms [of snowflakes] sketched in greater detail they might become more interesting." This is an important warn- ing about the significance of scale and resolution that plays a decisive role in our perception of various forms. This is but an example of how thought-provoking Joaquin Lorda's chap- ter, this whole volume and, ultimately, the teachings of Ernst Gombrich are.

Roald Hoffmann in 1994 in Budapest (photograph by Istvan Hargittai). Hoffmann and Robert B. Woodward extended the application of the symmetry principles to the description of chemical reactions.

Symmetry by Numbers[a]

Mario Livio, The Equation that Couldn't be Solved: How Mathematical Genius Discovered the Language of Symmetry, Simon & Schuster 2005, 368 pp

István Hargittai

The equation in the title is the quintic equation, the mathematical genius is Évariste Galois (1811–32), and the language of symmetry he discovered is group theory. Symmetry combines both beauty and science, and can easily be seen in the world around us. But before he could use it in science, Galois had to create the necessary mathematical tools. The world was slow to listen, and it took almost a hundred years for the practical value of group theory to be truly appreciated. Galois, meanwhile, was killed in a duel at an early age. In *The Equation that Couldn't be Solved*, Mario Livio follows his brief existence like a sleuth.

Born into a scholarly family in a Paris suburb in Napoleonic France, Galois was educated at home before being sent to a boarding school in Paris that rivaled the English schools of the time for austerity and rigid discipline. He was not a great success at school, but soon found satisfaction in mathematics, which became his sole occupation by the time he was 16. Having failed to gain entrance to a more prestigious college, he continued his studies in a high school.

Galois was still only 17 when he continued work started by the Norwegian mathematician Niels Henrik Abel, showing in general terms whether an equation is solvable by a formula or not. For this he introduced the seminal concept of a group, and created a new branch of algebra now known as Galois theory. His first publication appeared in 1829, but a combination of neglect and egotism prevented senior mathematicians of the day from giving him the exposure he deserved. When the work was finally introduced to the French Academy of Sciences, it was hardly appreciated.

Nonetheless, Galois continued his creative work, against all the odds. He failed another entrance examination as people greatly inferior to him could not appreciate his work, and lost his adored father, a Republican, who was driven to suicide by his royalist political opponents. Young Galois also had a passion for Republican revolution and served a prison term for his political activities. He fell in love with an undeserving girl and was killed in a duel that was related to this unfortunate entanglement. During the night before the tragedy, Galois hurriedly wrote a profound description of his group theory, remarking in the margin: "I have no time."

The Equation that Couldn't be Solved covers a remarkable number of different topics, including biographies of scientists and mathematicians. It also covers the Rubik cube and other puzzles; string theory; supersymmetry; the origin of creativity; the relationship between the external symmetry of the human face and body, and mate selection and sex life; and much more. Livio examines the contributions of others that led up to Galois' discovery, and gives a panoramic view of the direct, as well as quite remote, applications of group theory.

Very little escapes Livio's attention, especially in twentieth-century physics. But one omission is the contribution to the story of Eugene Wigner. He applied group theory to quantum mechanics in the 1920s, when most of his contemporaries were yet to value it: Wolfgang Pauli called it "die Gruppenpest"—roughly translated as "that pesky group business". Wigner was awarded a Nobel Prize in 1963 for this work.

Another omission from the book is that, in discussing crystallography, Livio stops at the classical notions of symmetry and defines crystallography as "the science studying the structures and properties of assemblies made of very large numbers of identical units". This idea supposes regularity and periodicity, and was largely a result of the tremendous success of X-ray diffraction in the twentieth century. Recently, however, the field has embraced other structures, such as the newly discovered quasi- crystals of regular but non-periodic

[a]Originally published in *Nature* 2005, 437:34.

I. Hargittai (✉)
Department of Inorganic and Analytical Chemistry, University of Technology and Economics, Budapest, Hungary

patterns. It was an early suggestion by British crystallo-grapher Alan Mackay that the rules describing 'crystal' structures be relaxed—and they have been. They now include structures that fall beyond the 230 space groups, and the new rules do not necessarily form groups.

Overextending the inferences from symmetry can be restrictive. As the historian of mathematics E. T. Bell said:

"The cowboys have a way of trussing up a steer or a pugnacious bronco, which fixes the brute so that it can neither move nor think. This is the hogtie and it is what Euclid did to geometry."

The book seems a little biased in places when it emphasizes the omnipresence of symmetry, but it nevertheless makes a lively and fascinating read for a broad audience.

One of John Emsley's exhibits is the penicillin molecule whose discovery is commemorated by a maroon plaque on Praed Street, W2, in London (photograph by Istvan Hargittai).

Chemistry Gallery

John Emsley, Molecules at an Exhibition: Portraits of Intriguing Materials in Everyday Life, Oxford University Press, 1998, 250 pp[a]

István Hargittai

Marcel Berthelot once pointed out that chemistry resembles the arts. It is unique among the natural sciences in that it creates most of its objects by synthesis. The exhibits in this book include both natural and man-made substances, selected for being important, either useful or harmful, in our everyday life. Some, such as nitric oxide, have made a remarkable transition in our judgment. Eight galleries group the substances, from foodstuffs to vital components of our body, from illicit drugs to raw materials and energy sources, from agents that destroy us and our environment to those that give us pleasure.

Chemistry books, even those written for the general public, are often burdened with complex formulae. Fortunately, here John Emsley provides only a few familiar formulae, such as H_2O and CH_4. Knowing the formulae of the more complex molecules he describes would not make it easier to understand their function.

A case in point is his description of the way we excrete unwanted nitrogenous material from the body by molybdenum-containing xanthine oxidase, a mammalian enzyme, producing uric acid. If uric acid is overproduced it accumulates in the form of sharp crystals in the joints, resulting in the painful illness of gout. The mechanism is understandable as described, and including the relevant formulae would not make it any easier to grasp the idea.

Most of us are largely unaware of the roles of many of the exhibited molecules in our everyday life. For example, phenyl ethylamine makes us feel good when we eat chocolate, sodium azide explodes on the impact of a car crash to save our lives, and thallium will poison us if the food of cows whose milk we drink had been exposed to thallium sulphate, which is used to kill rats.

Engaging stories are sprinkled through- out the galleries. Emsley tells us about the British discovery of penicillin and why British citizens had to pay US companies to use it. He explains that every time a young man thinks of sex—four times an hour, on average—his thoughts generate nitric oxide to help him fulfill his desires. And he provides a balanced assessment of the uses and dangers of DDT, one of the most worshipped and most feared molecules.

The descriptions are accurate without being pedantic and the captivating short stories didactic without appearing patronizing. But this is a chemistry book nevertheless, albeit an unusual one in that it provides a lot of natural and cultural history along with its chemistry. A broad audience, regardless of whether it has a background in chemistry, will enjoy browsing and reading it. The composer Modest Mussorgsky's "Pictures at an Exhibition" is a classic; Emsley's own 'exhibition' will also be receiving many visitors for a long time to come.

[a]Originally published in *Nature* 1998, 393:641

I. Hargittai (✉)
Department of Inorganic and Analytical Chemistry, University of Technology and Economics, Budapest, Hungary
e-mail: istvan.hargittai@gmail.com

Swedish postage stamp commemorating the centennial of the Nobel Prize.

Erling Norrby, Nobel Prizes and Life Sciences

World Scientific, New Jersey, etc., 2010, xv + 317 pp[a]

István Hargittai

After a hundred and ten years, the Nobel Prize continues stimulating scientists and fascinating the general public. In many aspects the award has been considered to be somewhat arbitrary, sometimes even a matter of luck, but much less so in the sciences than in other fields. There are a number of books that present factual material about the Nobel Prizes, interesting statistics, including data about nominations and nominators, but very few providing analysis of the awards and the merits of awardees and those who did not make it to the Nobel roster. The book under review presents such analyses for selected cases in the category of physiology or medicine and joins other treatises covering the science prizes, viz., the present reviewer's *Road to Stockholm* [1] and Robert Marc Friedman's *Politics of Excellence* [2].

Erling Norrby (1937–) is eminently qualified to produce *Nobel Prizes and Life Sciences*. He was a long-time professor and department head of virology at the Karolinska Institute and for two decades he was involved with awarding the Nobel Prize in Physiology or Medicine. Then, for another considerable period he served as permanent secretary of the Royal Swedish Academy of Sciences responsible for the Nobel Prizes in physics and chemistry and for the Sveriges Riksbank's prize in economical sciences in memory of Alfred Nobel. Beside having been involved in the workings of the Nobel Prize institution, he was the pupil or colleague of many others who had also been involved in Nobel Prize dealings, and he had the privileged situation of narrating about their activities from a close personal perspective. For this project, he collected materials in the Nobel archives both at the Karolinska Institute and the Center for the History of Science

in Stockholm. He has lectured on the subject matter of this book all over the world.

The book focuses on selected topics rather than providing inventories and this makes it an especially enjoyable read whereas while this approach does not deduct anything from it being a rich source of factual information. Its eight chapters are compiled in such a manner that each could be read and appreciated on its own and, indeed, most had resulted from separate and independent studies sometimes in cooperation with others. Chapter 1 provides the historical background of the establishment of the Nobel Prizes, Chapter 2 describes examples of serendipity in scientific discoveries, Chapter 3 focuses on the awards related to the virus concept and the next chapter may be considered its follow-up as it discusses the only Nobel Prize awarded for a virus vaccine. Chapter 5 tells us about the awards related to polio. Chapter 6 collects some cases of unusual Nobel Prizes. Chapter 7 describes the Nobel Prizes related to discoveries in nucleic acid research, and the last chapter is about prion research with additional comments on the personalities of successful scientists. From this enumeration it is obvious that a considerable fraction of the book is based on the author's direct experiences not only with respect to the Nobel Prize, but also to the scientific areas covered, and these are the best parts of the book with a lot of interesting information and forward-looking conclusions. Only a few of the aspects discussed in the book will be highlighted below.

Alfred Nobel (1833–1896) came from a family that had its ups and downs for its economic situation. He had very little formal education, but a great deal of private education and various international experiences, and he spoke several languages. He could be considered a chemist or chemical engineer. His most famous invention was dynamite, and he acquired a tremendous fortune by the time he died. He was born in Sweden, lived in Russia, France, and Italy; was a successful inventor, but a lonely and reclusive man who never married, and died without descendents. He hardly

[a]Originally published in *Structural Chemistry* 2011, 22:483–487.

I. Hargittai (✉)
Department of Inorganic and Analytical Chemistry, Budapest University of Technology and Economics, Budapest, Hungary
e-mail: istvan.hargittai@gmail.com

published anything but his engineering and innovative work was recognized by Swedish academia; he was elected a member of the Royal Swedish Academy of Sciences, and was otherwise also distinguished.

Nobel established the award (named only later the Nobel Prize) in five categories, physics, chemistry, physiology or medicine, literature, and peace. The prize awarding institutions are the Royal Swedish Academy of Sciences (physics and chemistry), the Nobel Assembly at the Karolinska Institute, the Swedish Academy (which has 18 members in literature), and a committee established by the Norwegian Parliament, respectively. There is a selection process, which is well-defined and broad-based, but which is also guarded by secrecy. There is a wide distribution of Nobel laureates among the nations, which the United States, Great Britain, and Germany dominate. There was a distinct change in the distribution at the top, the United States replacing Germany in the second half of the twentieth century. U.S. preeminence has been facilitated by the democratic atmosphere in and substantial support for American science. The input by Jewish refugee scientists from Nazi Germany and elsewhere in Europe was also considerable. The tremendous success and unprecedented world-wide prestige of the Nobel Prize institution has been the subject of surprise, contemplation, and also of emulation.

Serendipity has been a component in many discoveries, probably more than one would suppose without looking into their actual histories. On the other hand, Louis Pasteur (1822–1895) noted that chance favors the prepared mind. Of the numerous interesting examples in the book, I single out here the one about the discovery of background radiation. In 1964, Arno Penzias (1933–) and Robert Wilson (1936–) of Bell Labs communicated their serendipitous, but stunning observation of the cosmic microwave radiation amounting to three kelvins. Their finding was viewed as a confirmation of the Big Bang model. However, Penzias and Wilson were not concerned with models about the origin of the universe; they only wanted to understand the mechanism by which the galaxy radiates and to illuminate the electrodynamics of the Milky Way. They started with calibrating their equipment by pointing it to the sky, and observed about four kelvins, which is not much, but for them the sky appeared much "hotter" than it should have been. They spent then a whole year to make sure that what they were observing was not something of an artifact [3]. Finally, they convinced themselves of the reality of their measurement. In a remarkable coincidence, other scientists at Princeton and elsewhere at about the same time turned one more time to the possible models of the origin of the universe. Penzias and Wilson learned about the cosmological interpretation of their observation work by the Princeton group, and the two teams published their papers back to back in *Physical Review* in 1965. In 1978, Penzias and Wilson shared half of the physics Nobel Prize for their discovery of cosmic microwave background radiation.

If there is a central scientific topic in this book, it is about viruses whose modern concept has evolved over many decades in the twentieth century. It may refer to vastly different things in different fields. In life sciences, it refers to an ultrafiltrable agent whereas in computations, it refers to something disturbing operations. From ancient times, the name has referred to unpleasant and dangerous phenomena. The first virus identified as infecting humans was the yellow fever agent. Two chemistry prizes were awarded for research on Tobacco Mosaic Virus, six in Physiology or Medicine for studies of animal viruses, including prions, four in the same category for genetic discoveries using viruses attacking bacteria (bacteriophages, or simply phages), and another five for basic biological discoveries using animal or plant viruses. Norrby singles out a set of basic principles that emerged from the virus studies. They are presented here in a somewhat simplified manner:

- Molecules involved in life processes are simply organic substances (contrary to the old concept of vitalism).
- Viruses differ from bacteria and are distinguished by their own mode of replication involving both extracellular and intracellular processes.
- There are many different kinds of viruses differing in their targets of infection, but can still be grouped together.
- The genes of viruses are nucleic acids, either DNA or RNA.
- The genome of some viruses can become part of the cellular genetic material of their host organisms.

An interesting feature about virus research is the enormous number of virus species (their numbers may reach into the billions) and that their significance may by far not be limited to playing the role of pathogens! This is an area of research that might lead well into the future and might lead to future Nobel Prizes, eventually.

Remaining in the past, the only Nobel Prize that was awarded for a virus vaccine was related to yellow fever and it was given in 1951 to Max Theiler (1899–1972). The viral nature of yellow fever was identified in the first decades of the twentieth century and then Theiler developed the strain of an attenuated virus from which he produced the vaccine. Norrby follows the history of yellow fever epidemics and exposes the historic events with which they became intertwined. In this light, Theiler's discovery is seen as even yet more significant. The story of his research and the events leading to Theiler's award is also of interest and reads almost like a novel.

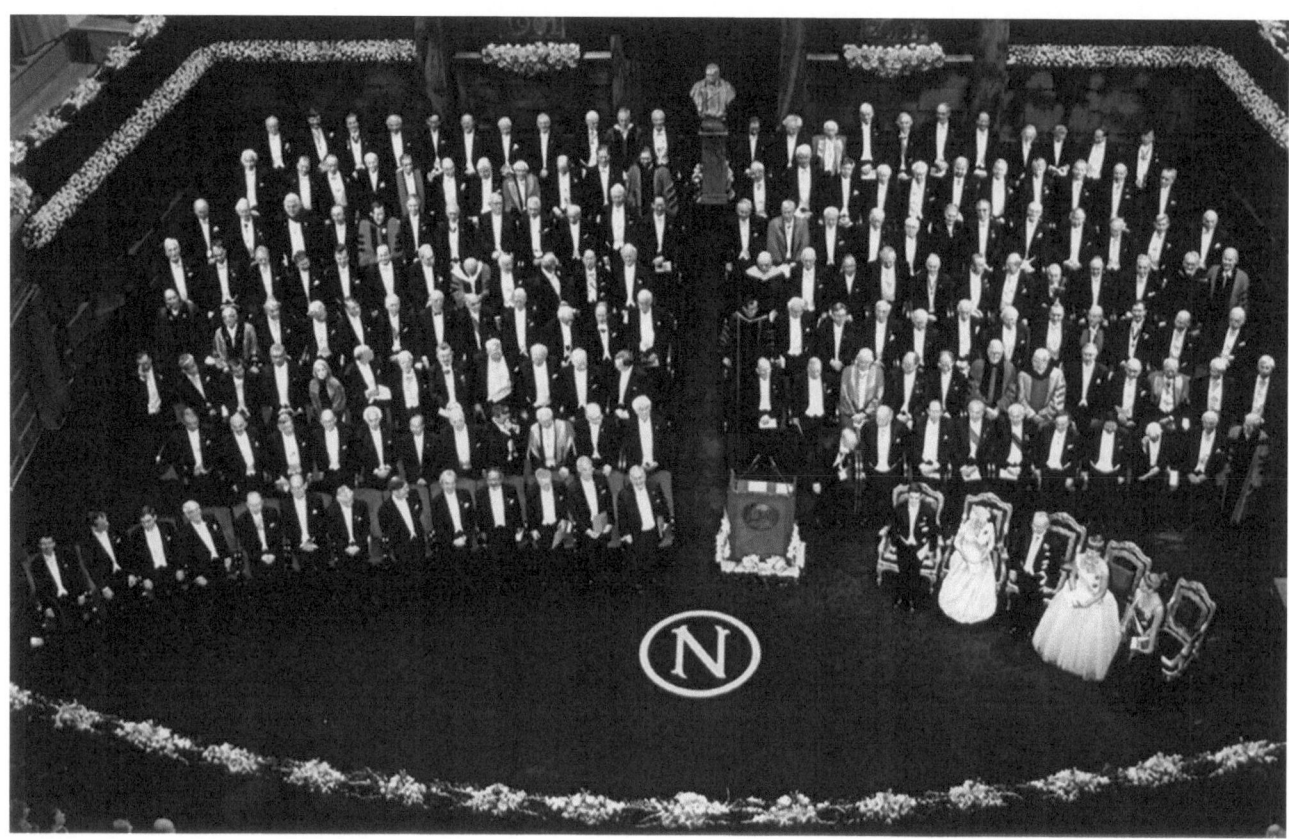

This image captures the concluding moment of the prize-awarding ceremony on December 10, 2001, in the Concert Hall in Stockholm, during the Nobel Prize Centennial. On the left, in the first row, are the Nobel laureates of 2001, from the left, the three physics laureates, the three chemistry laureates, the three laureates in physiology or medicine, the laureate in literature, and the three laureates of the "Sveriges Riksbank's prize in economical sciences in memory of Alfred Nobel" (for the Nobel Peace Prize, there is a separate prize-awarding ceremony in Oslo). On the right, in the first row, are the members of the Royal Family. Behind the first row are the Nobel laureates of previous years who came to Stockholm to participate in the centennial celebrations. Photo by Pål Sommelius, © The Nobel Foundation, Stockholm, used with permission.

Polio was another disease that was eradicated in the twentieth century and it involved the work of many scientists and many aspects of discoveries. However, only one Nobel Prize was given out in its connection. It was awarded in 1954 to John Enders (1897–1985), Thomas Weller (1915–2008), and Frederick Robbins (1916–2003) "for their discovery of the ability of poliomyelitis viruses to grow in cultures of various types of tissue." Growing the virus in tissue culture opened the way toward creating the vaccine and the prize awarders considered the former step more important than the latter. It is interesting that under poor conditions of hygiene, children used to acquire immunity early on, which protected them when exposed to infection in later life. However, as hygiene improved, the development of such immunity was delayed in children, and epidemics resulted. They first occurred in developed regions in Scandinavia, and the disease was named after two physicians, Heine and Medin. Later it became also known as infantile paralysis or poliomyelitis, or polio in short. The group under the leadership of John Enders at Harvard Medical School was not even primarily interested in growing the polio virus in tissue culture and this became merely a side project for them. It is to the credit of Enders that he used every opportunity to recognize the contributions of his two young associates and this was also acknowledged by the Karolinska Institute when their award was announced. The discovery paved the way for vaccination in which a number of scientists and physicians participated, among them the most famous Albert Sabin (1906–1993) and Jonas Salk (1914–1995). The absence of their Nobel recognition has puzzled many and Norrby's narrative answers at least some of the questions in this connection.

The story of the so-called prion, coined from "proteinaceous and infectious" by Stanley Prusiner (1942–) may also be considered to belong to the stories on viruses if the virus concept is taken very broadly. Prion signifies infectious proteins and this short and easy-to-remember name was a real hit. It identified both the field and Prusiner. But the story goes back to another scientist, D. Carleton Gajdusek (1923–2008), one of the most colorful scientists of all time who studied infectious diseases and understood that the disease called kuru in Papua New Guinea spreads by cannibalism. When cannibalism was eradicated, the disease also disappeared. Kuru was the first prion disease although this name was not yet around then. Gajdusek was awarded half of the Nobel Prize in Physiology or Medicine in 1976 along with Baruch S. Blumberg (1925–) who received the other half, and the motivation read: "for their discoveries concerning new mechanisms for the origin and dissemination of infectious diseases." Baruch's research led to the identification of the virus of hepatitis B and is discussed in Norrby's book among the serendipitous discoveries.

The nature of the prion "virus" has been a matter of fierce debates. All data point to the fact that it is a protein and no nucleic acid is involved in the infection. For structural chemistry it is an especially exciting development that the prion protein, which is mostly organism-specific, differs from its healthy counterpart merely in spatial geometry and not at all in composition. Thus, it could be considered to be caused by structural isomerism rather than in any omission or addition of one or more amino acids or a change in their sequence. In the respective healthy protein the structure is dominated by alpha-helices and in the prion protein much of the helical structure is turned into beta sheets. These beta sheets then induce other protein molecules to undergo structural changes leading to a rapid deterioration of the brain where they figure as building constituents. This mechanism is not considered as yet a settled interpretation of the infection. Prusiner was awarded the Nobel Prize in Physiology or Medicine in 1997 "for his discovery of prions—a new biological principle of infection."

Under the designation "unusual Nobel Prizes," there are both interesting data as well as stories. During the first fifty years of the Nobel Prize, it happened in eight instances that the award was made right in the year when the recipient was nominated for the first time. The interval that elapsed between the first nomination and the actual award ranged from zero to thirty years! In six cases, the award was made to scientists for whom no external nomination was received and they were nominated only by members of the Nobel Committee. Norrby describes fascinating stories that include the first Nobel Prize in Physiology or Medicine to Emil von Behring (1854–1917); the story of Alexis Carrel's (1873–1944) award in 1912 "in recognition of his work on vascular suture and the transplantation of blood vessels and organs," with tidbits of his post-award life and controversial activities; the missing prize to Gustaf Embden (1874–1933); and the famous case of the Nobel Prize "for the discovery of insulin," with special attention whether or not Charles Best (1899–1978) should have been one of the recipients of this award.

The discovery and applications of penicillin and DDT were both recognized by a Nobel Prize each, in 1945 and 1948, respectively, but their stories diverge as far as the afterlives of these discoveries are concerned. Penicillin and its derivatives are still widely used antibiotics whereas the use of DDT has shrunk to very limited conditions. Norrby's book throws light not only to Nobel-Prize-worthy discoveries and the discoverers, but also to their evaluations. Thus, for example, he attributes, at least in part, the speedy recognition of Joshua Lederberg's (1925–2008) discoveries concerning genetic recombination and bacterial genetics to their "brilliant review" by the novice member of the Nobel Committee, George Klein (1925–). There is revealing discussion of the development of the activities of the committee preparing recommendations for the Nobel Prize in Physiology or Medicine during the 1950s and 1960s.

The most conspicuous success story in twentieth century biology was perhaps the discoveries related to nucleic acids and a whole string of Nobel Prizes were awarded for them. Nonetheless, the most important omission has also occurred in that field by having left out Oswald T. Avery (1877–1955) from the Nobel roster. Avery and his two associates published the results of their seminal experiments showing unequivocally that deoxyribonucleic acid (DNA) was the substance of heredity [4]. There were outstanding representatives in Swedish science engaged in research of nucleic acids and, ironically, this might have blinded them in recognizing the importance of Avery's discovery. Similar results of a later, less thorough investigation were accepted more readily as proof of DNA being the hereditary substance, but the elucidation of the double helix model of the DNA structure was the decisive step in recognizing the role of DNA in heredity. In proving this structure and its role, a crucial experiment proved the semi-conservative manner of DNA replication. The discovery of the double helix was recognized by a Nobel Prize, but the demonstration of its semi-conservative replication was not.

Even before the chemical nature of the hereditary material was discovered, recognition of progress in genetics by Nobel Prizes had begun. They involved the role played by chromosomes; mutations induced by X-rays; the regulating roles of genes in chemical events; genetic recombination and the organization of bacterial genetic material; and the existence of mobile genetic elements. Although the discovery of the double helix showed spectacular harmony between structure and function, it took time before its importance gained Nobel recognition. No nomination had there been before 1959, thus the 1962 award for the double helix could not be considered as much delayed for the discovery made in 1953. But, from 1959, an avalanche of Nobel Prizes was given out for discoveries in the field of nucleic acids both in Physiology or Medicine and in Chemistry. They amounted to seven awards in the former and six in the latter, the last being as recently as 2009. To use the term "avalanche" is only justified here on the background of scarcity of Nobel Prizes in any given field.

In an interesting development, whereas the nucleic acids had to take over the proteins in claiming preeminence in deciphering life, with solving the problem of the genetic code, a considerable attention from them was redirected toward the proteins. The reason is that they are the embodiment of traits that are inherited through the nucleic acids from generation to generation. Frederick Sanger's career symbolizes the unity of these two classes of biological macromolecules. He has been the only scientist with two Nobel Prizes in Chemistry, the first received for producing the methodology for sequencing proteins, the second for sequencing nucleic acids. Another example providing food for thoughts along these lines was the discovery that nucleic acids could also act as catalysts—a role that before was ascribed exclusively to proteins. This was another discovery that had a hard time getting accepted, but after it had been, the Nobel Prize for it was virtually inevitable. This happened in 1989, in chemistry, and the recipients were Sidney Altman (1939–) and Thomas R. Cech (1947–).

Norrby takes a broad view in discussing the Nobel Prizes and deals with issues such as what constitutes a discovery, what are the stories of those scientists who could have or should have received a Nobel Prize, but never did, and his narrative is sprinkled by personal tidbits that make it especially appealing. Much of the book helps demystifying the institution of the Nobel Prize. Norrby is very open about his own interactions with Nobel laureates before and after they were distinguished with this recognition. At one point he even wonders in connection with the latest research achievements of a Nobel laureate "if he should not be invited to visit Stockholm a second time" (p. 240). It must be a delicate consideration for someone so intimately involved in the prize-awarding exercises, but he seems to have handled it without any hint of criticism. And we must be grateful to him that he so generously shared his knowledge and experience with us through this book, which I wholeheartedly recommend to all interested in how the Nobel recognition of discoveries and discoverers works.

References

1. Hargittai I (2002, 2003) The Road to Stockholm: Nobel Prizes, Science, and Scientists. Oxford University Press, Oxford
2. Friedman RM (2001) The Politics of Excellence: Behind the Nobel Prize in Science. Henry Holt & Co, New York
3. Hargittai M, Hargittai I (2004) Candid Science IV: Conversations with Famous Physicists. Imperial College Press, London, pp. 272–285 ("Arno Penzias") and pp. 286–297 ("Robert W. Wilson")
4. Avery OT, MacLeod CM, McCarty M (1944) J Exp Med 79:137–158

Istvan Hargittai and Craig Venter, one of the principal architects of the Human Genome Project, in 2007 in Brussels following a major address by Venter.

The Human Genome Project—A triumph (also) of Structural Chemistry: On Victor McElheny's New Book, *Drawing the Map of Life*[a]

István Hargittai

Abstract

Structural chemistry greatly contributed to the feasibility of the Human Genome Project (HGP) by the discovery of the double helix structure of DNA. Victor McElheny's new book *Drawing the Map of Life* paints a panoramic picture of the story and the expected benefits of the HGP.

The significance of the Human Genome Project (HGP) is difficult to overestimate and could be compared only to that of very few other grand projects such as harnessing nuclear energy, space exploration, interstate highway systems, transcontinental railroads, flood-control dike systems, and a few others. Its costs are lower but its long-range impact is greater than those of some of the others. For structural chemistry, the HGP is unique among these extraordinary projects because our domain of science has been part of the foundation of molecular biology through the discovery of the double helix structure of DNA.

Science journalist and biographer Victor McElheny has now published a book, *Drawing the Map of Life: Inside the Human Genome Project* (New York: Basic Books—A Merloyd Lawrence Book, 2010), which is worthy of close attention. The Sydney Brenner quote introducing it is surprising at first glance as it says that "progress in science depends on new techniques, new discoveries and new ideas, probably in that order." Intuitively one might assign preference to new ideas rather than to new techniques. However, closer scrutiny of various developments justifies Brenner's words. Thus, for example, one of the most crucial developments on the road to understanding the human genome—Frederick Sanger's discoveries of sequencing first proteins, then nucleic acids—clearly depended on new techniques in chromatography and elsewhere. Without them Sanger might have not even embarked on these tasks, but while working on his projects, Sanger himself became a great toolmaker.

Thus, at the start, McElheny justifiably focuses on the tools that eventually led to the HGP. These tools included enzymes, instruments, chemicals, and mathematical approaches, among them statistics. As molecular biology is in fact the conglomerate of all techniques used in modern biological research related to finding out about the molecular basis of life with genetics as a main focus, the bits and pieces communicated here come together as a rough history of this branch of science. And it is a highly personalized history: the discoveries and innovations are introduced along with some basic information about their principal protagonists.

At some point, there was a dilemma whether to wage a comprehensive attack in deciphering the human genome or continue to concentrate on various diseases in a piecemeal manner, one after the other and often by randomly looking for the genetic markers associated with their manifestations. It was soon recognized that everything could be done faster and more economically if the human DNA were deciphered in its totality at once. Crucial changes were taking place in the 1980s. This was not only in scientific techniques, but also in public awareness and, accordingly, in the political climate for considering the importance of what has come to be called "genomics." The involvement of politics was justified not only due to the expected benefits from the new science but also because new dangers—real or perceived—had to be handled as well.

When recombinant DNA initially became a possibility in the mid-1970s, the scientists themselves initiated policing (see, the famous Asilomar meeting), accompanied sometimes by hysterical reactions in the general public, but things eventually returned to normal. With the accumulating information from mapping the genes and from sequencing DNA, it was increasingly apparent that a larger-scale effort would be needed and that the ultimate goal should be the complete

[a]Originally published in *Structural Chemistry* 2010, 21:667–671

I. Hargittai (✉)
Department of Inorganic and Analytical Chemistry, Budapest University of Technology and Economics, Budapest, Hungary
e-mail: istvan.hargittai@gmail.com

sequencing of the human genome. Some of the opponents of such a project hastened to point out that it would not be the final goal because knowledge of the complete sequence could not yet mean medical applications. However, it became clear that this was an unavoidable step.

The work had to take place on several fronts simultaneously. Information was to be collected about the sequence, but technology improvement was to be continued if sequencing was not to be considered a single project since it was realized that for true applications, i.e., discovering the genetic basis of diseases, numerous individual DNAs would have to be sequenced.

Questions such as the relationship between government-funded work and private enterprise came into the forefront just as it did between university research and commerce. The HGP was to be enormous with a price tag of an estimated 3 billion dollars. Other questions arose as well, such as whether DNA base sequences were patentable or not. And reality was moving rapidly without waiting for the outcome of lengthy deliberations.

The first explicit calls for a project to sequence the human genome emerged as early as the mid-1980s. It was to become the task of a new generation, because the scientist most noted for his discoveries in sequencing, Frederick Sanger, retired in 1983, and the first big meeting about the HGP took place in 1985. When Sanger was asked at the time whether such sequencing should be done, he unequivocally supported such a project.

Everything seemed to be moving toward the HGP. Small organisms but of ever increasing complexity became subjected to complete sequencing and the best scientists were vocal in advocating the need for such a project and in actually participating themselves. This included people like Sydney Brenner, who had been one of the pioneers in establishing the genetic code, and Renato Dulbecco.

There were also opponents, to be sure. They were apprehensive about big science entering the traditionally small science arena of biology, but it was recognized that the scale of a possible HGP would truly have to be big science. On the other hand, there was no real danger of concentrating all the HGP in one huge center; rather, it was anticipated that big science in this case would mean a well-coordinated assembly of little science projects. Initially, there was a gap between the older scientists who supported the big project and the young ones who saw it not only as an infringement on their toiling ground but as a sponge absorbing the support that could have gone to more diverse projects. There was the promise that rather than depleting ongoing projects in biology, the HGP would be subsidized with additional funds.

One of the moving forces was another promise implicit in all the plans for the HGP and that was its potential in fighting cancer. Another was the benefits expected from the technological innovations for all biological science. Still another driving force—more influential than anybody else—was James D. Watson. Initially, in the mid-1980s, he was lukewarm toward such a project, however, he quickly warmed toward it, and he had excellent resources to back up his efforts. He had the Cold Spring Harbor Laboratory behind him and he had his ways to influence the general public and, in particular, the media as well as his fellow scientists. In addition, the presumably genetically based illness of one of his sons became publicized and added a personal touch to his involvement, and generated additional trust and sympathy toward what he advocated.

At one point, it was no longer a question of whether the HGP should be set up but rather at which institutions, with what framework, and under whose leadership it should operate. The Howard Hughes Foundation—the largest private organization of its kind—quickly bowed out. The U.S. Department of Energy (DOE) with its vast experience in big projects, such as those at Los Alamos and Livermore, was willing and interested. The DOE was not without prior involvement in biological research either. Nonetheless, the most logical choice seemed to be NIH due to its enormous funding of biomedical research and its responsibility for advances in human medicine where the HGP was expected to bring most of its benefits. Ruth Kirschstein, in charge of the National Institute of General Medical Sciences, was made responsible for studying the possibilities of the HGP on NIH's part. It was realized that the HGP would be much less expensive than the space program and probably more directly benefiting human life.

One of the preconditions for a meaningful HGP was the recognition of the universality of the genetic code, which by this time had been established. This is something we take for granted but initially it could not have been. The universality of the genetic code itself was something that had to be established. It impressed me when Marshall Nirenberg, who had accomplished the first step in cracking the genetic code, told me that it was a profound moment when he realized this universality. It had an almost religious significance on the non-religious Nirenberg. He considered it an expression of the unity of Nature.

The preparations for the HGP involved U.S. legislation and Watson along with a few others got involved in informing members of Congress. Watson was such an important component that he appeared even to have a say in which of the two principal contending agencies should be involved,

DOE or NIH, in administering the project. It was then also almost inevitable that he was chosen to initially direct the efforts with an appropriate title in the NIH administration. McElheny thoughtfully enumerates Watson's traits, both favorable and not so favorable, as a leader of the HGP. Although it could be doubted from the start whether Watson would have the stamina to carry on this function for long, it had great significance that he was at its helm at its very inception.

Watson's assuming the leader's role of HGP was advantageous because his fame added visibility to the project and generated additional trust. Now there was someone—someone well known, that is—who appeared to take responsibility for this great excursion into the unknown. Watson himself considered it important to have someone at the helm of the project who could take the blame in case of failure. An additional benefit was his authority among scientists who could be induced to join the project answering his call. He knew that success or failure depended to a large extent on the quality of cadres he would be able to attract. It was characteristic that a leading scientist suggested that offering sufficient cash would be attractive enough to recruit the right people. In contrast, Watson warned that such an incentive would attract the wrong people. He wanted to have people who were too busy to join, who had a lot of things going for them, and for whom merely a lot of money would not be decisive in making a career move. Watson's dedication to the HGP could not be demonstrated better than by pointing out that he continued his lobbying for it even after he had been "fired." One of his most telling personal imprints on the project was the allocation of a percentage of the HGP budget to studying its ethical, social, and moral issues. There was a lot of relevance to these questions and the emphasis on their studies from early on enhanced public trust in the project.

As the question of the scaling up of the genome project arose, various further considerations had to be addressed. They included multifaceted supports of the project, the dilemma between technological innovations and continuous use of more traditional methodology, and how the HGP functioned as a peculiar big science being constituted of numerous small-scale projects. Even though Watson's directorship did not last long, it appeared that it was crucial that he had had his imprint on it from the start. His scientific authority could not be questioned and his dedication was unconditional. Though he might have been forced out for reckless statements, it was his interest in some biotech companies that was used as a pretext to cause him embarrassment. Still, one has the impression that he might have been grateful for it is difficult to imagine him lasting too much longer in a bureaucrat's role. Amid bickering among various governmental branches, science kept going on, producing complete mappings and even sequences of organisms of ever increasing complexity.

Patenting appeared to be a crucial question, that is, whether the human genome and its portions could be patented at all. Curiously, there was not a clear-cut divide between those who supported patenting and those who did not. One might have expected the big pharmaceutical companies to prefer patenting, but they recognized its dangers and opted instead for the public domain approach. In contrast, I remember how dedicated Walter Gilbert appeared to me, during a personal encounter at a meeting "Frontiers in Biomedical Research" in Indian Wells, California, on February 2, 1998, to seeing his company earn money from genetic tests based on such patenting. I remember it because I had been on a low-salt diet from my youth due to my tendency toward elevated blood pressure. The test would determine whether people inclined to have such a condition would or would not benefit from a low-salt diet. When Gilbert told me about it the test had been on the market for just a few weeks.

An obvious player in scaling up was Craig Venter who had started at NIH, but moved out, and became the most conspicuous player in the private enterprise sector of the human genome race. At some point, Francis Collins succeeded Watson at the helm of the HGP. Even though Watson no longer occupied any formal position in the project, he continued his role behind the scenes and utilized his enormous authority for gathering support in Congress and elsewhere for the project. As Venter aggressively pursued his goals, he mobilized tremendous funds from the private sector for the project. It meant not only more monies but also more competition. The situation resembled the stimulating effects of excellent private universities on the state universities.

There were reasons for alarm as well. Venter's private sequencing company would release sequence data at 3-month intervals; thus, the private company would have considerable advantage in developing diagnostic tests and eventually drugs, and they would even patent genes and important functions. The British were especially vigilant in not letting patenting become a barrier to public access to the benefits of the HGP. Their efforts were financed by the Wellcome Trust, a private foundation.

A few of the principal players on the road to the Human Genome (all photos by and © of István Hargittai). From the top row, left, and in each subsequent row, from the left, Werner Arber, 2005; David Baltimore, 2004; Seymour Benzer, 2004; Paul Berg, 1999; Elizabeth Blackburn, 2003; Sydney Brenner, 2003; Erwin Chargaff, 1994; Francis Crick, 2004; Walter Gilbert, 1998; François Jacob, 2000; Ruth Kirschstein, 2000; Arthur Kornberg, 2001; Joshua Lederberg, 1999; Maclyn McCarty, 1997; Matthew Meselson, 2004; Kary Mullis, 1997; Daniel Nathans, 1999; Marshall Nirenberg, 1999; Richard Roberts, 2003; Frederick Sanger, 2001; Phillip Sharp, 2001; Maxine Singer, 2000; Hamilton Smith, 2001; Gunther Stent, 2003; John Sulston, 2003; Harold Varmus, 2002; Craig Venter, 2007; Robert Waterston, 2003; James Watson, 2000; Charles Yanofsky, 2006.

It was realized from the start that deciphering the human genome would only be the beginning in revolutionizing the medical sciences. Various aspects were coming into the fore-front that would show the way to utilization of the information from the HGP in human medicine. One of them was the determination of disease-causing variation—the change of a single nucleotide for another, called also the single-nucleotide polymorphism or SNP. A whole new area of biomedicine, "pharmacogenetics," was emerging. In the meantime, private companies flourished on NASDAQ due to the promise of personalized medicine with the yet more attractive goal of preventive medicine appearing close to reality in the not so distant future. On March 14, 2000, President Bill Clinton and Prime Minister Tony Blair took a strong stand against patenting, and, as a consequence, the shares of private companies dropped and the NASDAQ index slashed considerably, after they had, previously, skyrocketed.

Various analogies have been introduced to stress the importance of deciphering the human genome. A parallel was drawn between the periodic table of the 100 chemical elements as the guiding principle in twentieth century chemistry and the knowledge of the tens of thousands of genes of the human body in the biomedical sciences of the twenty-first century. There seemed to be a competition of superlatives in characterizing the importance of the human genome in which Watson appeared to be among the most restrained when he declared "...it's the script of life. It's the information for the play of life" (p. 161). Watson's preeminence in the project was acknowledged by President Clinton when he turned to Watson saying, "Thank you, sir" (p. 165). The occasion was a joint, electronically linked White House–10 Downing Street event on June 26, 2000, declaring the next triumph of the HGP.

The flood of information from the HGP gave hope for attacking numerous diseases, but the data had broader implications as well. It was established that well over 99.9% of the genome is the same in all humans and in this light the concept of "race" was fast losing importance. At the same time, the 0.1% still represented the possibility of 3 million differences among the 3 billion nucleotides. The possibilities of utilization were enormous, ranging from diagnostic tools to drug development, to genetic screening, forensic applications, and coming to decisive information in paternity disputes. One of the goals was the mapping of genes associated with every inheritable disease. Parallel to the virtually limitless potentials of the benefits of the HGP, possible dangers were also emerging. The information might make genetic discrimination possible in employment, insurance, and elsewhere. It might be that skin color and gender would not be the most decisive factors in discrimination, but

it could become the variations in a person's DNA. So far, seven diseases have been identified as linkable to DNA mutations that included manic depression (bipolar disorder); coronary artery disease; irritable bowel syndrome (Crohn's disease); hypertension; rheumatic arthritis; type I diabetes; and type II diabetes. This is in addition to the susceptibility genes already known for breast/ovarian cancer, colon cancer, and Alzheimer's disease that were found using the targeted disease-based approach.

It was an event announced with big fanfare when in 2007 Watson and Venter posted, simultaneously, their own full DNA sequences on the Internet. Their purpose was to combat the fear on peoples' minds when they thought about the genome project. This reality was also demonstrated by Watson. He said that he was glad he had sons rather than daughters because his genome showed familial tendency for breast cancer. Even more to the point, he held back one section of his DNA that might have revealed information to him about his chances for Alzheimer's. He declined to learn about his chances for developing this awful condition. Venter wrote a whole book in connection of the publication of his DNA, *A Life Decoded*. He called it his genomic autobiography.

Of course, for developing personalized medicine the DNAs of thousands of others will have to be sequenced. However, this appears increasingly realistic. Sequencing the first human genome cost 3 billion dollars, Watson's price tag was a mere 1 million—a drastic decrease within only a decade and a half. The most immediate goals of reaping the benefits of the HGP would be pinning down the causes of mental illness and autism. The next would be establishing diagnostic tools for various cancers and finding their treatments. Tall orders to be sure, but considering the pace of progress in recent biomedicine there is justified optimism about reaching these goals.

McElheny's book is a great service for a broad audience in disseminating reliable knowledge in an accessible way. His background eminently qualified him for producing such a book. He was a science journalist during the decades that led to the HGP and during its initial periods. He worked for years at the Cold Spring Harbor Laboratory. He wrote an excellent biography of James D. Watson, *Watson and DNA*. It appears, the importance of the topic and the preparedness of the author made a perfect match and the result is an informative and readable account of the most important scientific project of our time (the photo collage shows some of the principal players on the road to the Human Genome).

Acknowledgment I thank Dr. Doris T. Zallen (Virginia Tech, Blacksburg) for helpful consultation.

J. Michael Bishop in 2004 at the University of California at San Francisco (photograph by Istvan Hargittai).

Eyes on the Prize[a]

J. Michael Bishop, How to Win the Nobel Prize: An Unexpected Life in Science, Harvard University Press 2003, 320 pp

István Hargittai

If offered reincarnation, the Nobel laureate J. Michael Bishop would choose to come back as a musician (with exceptional talent, to be sure), because he thinks that one life-time as a scientist is enough. The son of a Lutheran minister, he grew up in rural Pennsylvania, and became enchanted with research during his last years at Harvard Medical School. He has been at the University of California, San Francisco since 1968, and worked for 15 years with his former postdoctoral associate, and ultimately fellow Nobel laureate, Harold Varmus.

Their work had its roots in the discovery of a cancer-causing virus in chickens by Peyton Rous in 1911, who was awarded the Nobel Prize fully 55 years later. Five individuals then went on to win Nobel Prizes for related work. David Baltimore, Renato Dulbecco and Howard Temin won in 1975 "for their discoveries concerning the interaction between tumour viruses and the genetic material of the cell", and Varmus and Bishop became Nobel laureates in 1989 "for their discovery of the cellular origin of retroviral oncogenes". Baltimore and Temin found the viral enzyme reverse transcriptase, which allows RNA to be copied into DNA, a reversal of the normal flow of genetic information. This discovery could have been Bishop's had he been more daring. However, he learned his lesson and was fortunate enough to get another chance.

The discovery of oncogenes (cancer genes) raised the question of whether such genes might be present in the genetic composition of normal as well as cancerous cells. Locating them carried the promise of understanding human cancer at the genetic level. At first it was thought that oncogenes were viral genes, but Bishop and Varmus discovered that they were cellular genes that had been kidnapped by the virus. It took their team four years to identify them.

Bishop quotes a beautiful description of a moment of discovery by one of their post-docs, Dominique Stehelin: "The intensity of the emotion I experienced and the intellectual clarity induced by the situation at that moment were very special." Furthermore, "I suspect that few have the privilege of enjoying such a moment when one is intensely and profoundly aware that a major step forward in Science has been made, and that one has contributed to it." Alas, the quote was from an open letter to the Nobel Committee by Stehelin, who was not among the Nobel awardees. For every Nobel laureate there are others who might have also been included but were not, and every story about how to win the Nobel Prize may have its counterparts.

Bishop does not give a recipe for winning the prize, as any attempt to emulate a particular research career would be doomed to fail. However, throughout the book, he makes important points that budding scientists may find useful. For example, it is more useful to learn from one's peers than from one's teachers. Start a research career in a place where you feel genuinely needed, rather that choosing somewhere for its prestige. Being a pioneer in research is fun, although it may bring more fame to be part of a team completing a discovery. Give a name to your discovery as soon as it is made. And finally, Bishop points out, good scientists should also market their ideas well.

Nobel laureates often seem to be standard-bearers for good causes, usually by signing petitions or making statements about issues with which they may not even be too familiar. Bishop's involvement in public causes has been different. He actively organized the participation of scientists in a non-partisan movement to increase legislative attention for science. Their high-level lobbying helped to achieve record support for research from taxpayer's money in the United States.

[a]Originally published in *Nature* 2003, 423:921.

I. Hargittai (✉)
Department of Inorganic and Analytical Chemistry, University of Technology and Economics, Budapest, Hungary
e-mail: istvan.hargittai@gmail.com

Bishop compiled his experience and ideas in this book for the general public. He also provides a crash course on the microbial world that is a gem of instruction without being condescending. And his copious use of art, including poetry, is a statement about the unity of the two cultures.

J. Desmond Bernal's work place (left) and home (right, photographs by Istvan Hargittai). The work place is Birkbeck College on Malet Street and the home with a blue plaque is at 44 Albert Street, NW1, both in London.

Andrew Brown: J. D. Bernal, The Sage of Science

Oxford University Press, 2005[a]

István Hargittai

J. Desmond Bernal (1901–1971) was one of the most original and influential scientists in the twentieth century and what he wrote in his *Science and History* in 1954 is fully applicable to him: "The greater the man the more he is soaked in the atmosphere of his time." Andrew Brown wrote a biography, which sets Bernal's life in the frame of the time he lived in.

Bernal was born in Ireland and went to school in England during the Great War (known as World War I elsewhere). He was a precocious child who also engaged in sports and outings. He became acquainted with crystallography and symmetry in high school and they remained among his interests for the rest of his life although many other interests would also be added. During his undergraduate studies he derived the 230 three-dimensional space groups, not a small feat both for crystallography and mathematics. For this he shared a college prize with a future Nobel laureate. When he was a freshman in Cambridge, he virtually overnight changed from a catholic worshipper to Marxist socialist. Perhaps this is why this highly intellectual scientist behaved politically as if it were his faith-based religion rather than conviction.

He married early and never divorced but in practice he had many wives in succession and often in parallel. Often they were not merely lovers of diversion, rather, they were both intellectual and physical partners, and he fathered children with some for whom he acted as father though mostly in absentia. Women tolerated his unorthodox behavior because they preferred having some of Bernal to none of him. In Cambridge as later at the Royal Institution in London, he had excellent peers, such as William Astbury and Kathleen Yardley—later Lonsdale—and teachers, such as William

Henry Bragg—the elder of the two Braggs. Bernal determined the complete structure of graphite and explained the difference in the properties of diamond and graphite on the basis of their structures. From the Royal Institution he returned to Cambridge where in the 1920s it must have been a unique experience to be in Ernest Rutherford's vicinity. In his last move, he returned to London, to Birkbeck College, which he developed into a world-class center of crystallography and molecular biology. Bernal's crystallographic studies encompassed both inorganic and organic substances as well as initiating organizational framework for this emerging science.

In 1934, Bernal made meaningful X-ray photographs of a protein and wandered sleeplessly all night in the streets of Cambridge. He realized that the way was open for solving life's puzzle, alas, it would not be his work bringing this premonition to fruition, but his disciples and their associates. Back in 1930, Bernal and Astbury made an arrangement to divide the work in the crystallography of biologically important substances. Bernal got the single crystals, that is, the more ordered materials, whereas the less ordered systems became Astbury's share. Today it would be less realistic to arrange for such divisions in science, but it was still possible to do so in the 1930s. The two took the division seriously and at one point Astbury warned Bernal, "... I insist that you stick to the constituents [amino acids], and don't 'snaffle' any proteins" (Brown, p. 89). Decades later, after Pauling had discovered the alpha-helix and Watson and Crick the double helix, Bernal referred to his arrangement with Astbury in his review of Watson's book, *The Double Helix*: "A strategic mistake may be as bad as a factual error. So it turned out to be with me. Faithful to my gentleman's agreement with Astbury, I turned from the study of the amorphous nucleic acids to their crystalline components, the nucleosides." (J. D. Bernal, "The Material Theory of Life." *Labour Monthly* **1968**, 323–326).

[a]Originally published in *Structural Chemistry* 1996, 17:451–452.

I. Hargittai (✉)
Department of Inorganic and Analytical Chemistry, Budapest University of Technology and Economics, Budapest, Hungary
e-mail: istvan.hargittai@gmail.com

Because of the arrangement with Astbury, Bernal's name did not figure in what he called the greatest biological discovery of the twentieth century, the double helix story, and in the 1968 review some sadness transpires in this connection. He had come close to being a participant. In the same review he mentions Sven Furberg's work in his group, who studied cytidine and according to Bernal, "the structure he found proved to have wide implications. In fact, had we realized it, it contained the key to the whole double helix story. . . ." By the time of the review Bernal had suffered from debilitating health problems. Besides, he was far from the Establishment and his review appeared in a non-mainstream periodical, the *Labour Monthly*. Characteristically, his thoughtful and valuable review did not make it to Gunther Stent's review of reviews of *The Double Helix*.

Bernal's prowess in scientific research culminated in the 1930s. He had insightful and forward-pointing ideas whether it was the suggestion that the secret of life was contained in linear arrangements, that is, the sequence, of biopolymers or the structure of liquid water. It was also at this time that he launched the scientific careers of Dorothy Hodgkin (née Crowfoot) and Max Perutz, both future Nobel laureates. Later, Aaron Klug and Rosalind Franklin extended the list of the most extraordinary discoverers whom Bernal started on their respective careers. John Kendrew, Perutz's co-laureate for protein structures met Bernal during World War II and followed his career at his suggestion in crystallography after the war.

During World War II, Bernal suspended his activities in basic research and more or less even in communist politics, and devoted his multifunctional talents to assisting the war efforts. He did useful work in solving specific problems, for example, in the preparations of the invasion of France. Moreover, he set examples to showing the utility of the scientific approach in military operations. To sense the magnitude of his achievements, suffice it to mention that upon the cessation of hostilities in 1945, he was awarded the Medal of Freedom, the highest civilian award in the United States to recognize exceptional meritorious service in the war. This is even more important if considering that he did not receive much recognition in Great Britain for his services, which were though very well known.

That Bernal was immersed in public affairs can be understood by the French physicist Langevin's credo, which he shared. Langevin stated that "The scientific work which I can do, can be done, and will be done, by others, possibly soon, possibly not for some years; but unless the political work is done there will be no science at all" (Brown, p. 125). In the 1930s, it was not unusual for Cambridge graduates to be on the left. With the Great Depression and rising Nazism and

Fascism, socialism and the Soviet experiment seemed to be a possible alternative to capitalism. With Stalin's tyranny and purges in the late 1930s, the show trials then and after the war, the famine, and other tragedies, a man of Bernal's intelligence and scientific skepticism should have developed a critical approach toward the Soviet Union. He did not.

This is perhaps the greatest puzzle of his life. It is a puzzle that remains unsolved. The only suggestion is that Bernal made the transition from a card-carrying Catholic to a card-carrying Communist too suddenly, and apparently needed faith in which he would believe unquestioningly. Sometimes people wonder how great scientists can reconcile their deep religious faith with their scientific achievements in which they demonstrate a rigorous and skeptical approach. This seems possible as long as the two domains remain strictly separate. Alas, for Bernal, the two got mixed up in some crucial instances. He did not have the will or the strength to see or admit the charlatanism of T. D. Lysenko, the unscientific dictator of Soviet biology and agriculture with literally lethal consequences for some extraordinary scientists and unknown number of lesser known researchers. This is an inexcusable stain on Bernal's ouvre. Even when the polywater story came, the ostensible discovery of anomalous water with peculiar properties, he was too quick to declare it to be "the most important physico-chemical discovery of this century" (Brown, p. 447). It is doubtful whether had it not been a claim by Soviet scientists, Bernal would have jumped to such a conclusion without further studies.

Andrew Brown provided an extraordinary service to the scientific and even broader community to collect and present the details of Bernal's life and activities. He objectively apprises his deeds, although one cannot but be somewhat biased toward the object of one's labors of five years. Yet Brown lets the reader make his or her own judgement about Bernal. What he offers is a narrative that is as captivating as a mystery novel can be, meticulous without boring, and provides ample background about other players as well as about the scientific development and the political scene that allow the reader to see the broader implications of Bernal's story. Brown does not shy away from discussing scientific issues without which it would be impossible to appreciate the significance Bernal's achievements. He does this accurately yet in an accessible way whether it is about symmetry groups or crystallographic innovations or the structural characteristics of biopolymers or other fields of scientific knowledge and research in keeping with Bernal's unusually broad interests and areas of activities. We can all be thankful to Andrew Brown for an informative and entertaining account of a most extraordinary individual of twentieth-century science.

Max Born in his thirties (courtesy of his son, Gustav Born, London).

Nancy Thorndike Greenspan: The End of the Certain World. The Life and Science of Max Born: The Nobel Physicist Who Ignited the Quantum Revolution

Basic Books, 2005[a]

István Hargittai

When Max Born (1882–1970) was presented the Planck Medal in 1948 in Göttingen, Germany, it was in recognition of his seminal work in several areas of physics and chemistry: relativity theory; the dynamics of crystal lattices; quantum mechanics; and the theory of the electron. It was not yet the end of his scientific activities, nor was it the last of his awards, but it represented a major milestone in his life. The previous decades saw him through two world wars, ever increasing anti-Semitism in Germany, forced emigration, and the barbaric killing of over 40 of Born's relatives by the Nazis. They also saw—in view of his tremendous oeuvre—far less recognition of his achievements than he should have received. At that point Born returned to his then home in Edinburgh, feeling that he would not like to live among Germans again.

Born grew up in a cultured and well-to-do Jewish family in Breslau, Germany (now Wroclaw in Poland). He was a 4-year-old child when he lost his mother. He was an average and lonely student in gymnasium, but he was exposed to a scientific environment from early on. He was also aware of the limitations that his family's being Jewish placed onto their lives in the choice of professions and if engaged in state service—such as academia—in promotions. Max Born went to the University of Breslau and made his first friends in mathematics. From 1904, he continued his studies in Göttingen where he learned from such greats as Hermann Minkowski, Felix Klein, and David Hilbert. He also became involved in practical applications of higher mathematics under Ludwig Prandtl.

Military service and a stint at the Cavendish Laboratory in Cambridge followed, where, among others, he attended J. J. Thomson's lectures. Born earned his doctorate by doing research in thermodynamics. Then he continued his work for his habilitation to earn the right to teach at university, and published papers in relativity and electrodynamics. He moved around in Breslau and Göttingen and during this time he finally decided to devote his life to physics. His prospects though for a career in German academia did not look good. Before World War I, less than two percent of full professors were Jewish in Germany.

Born made friends that included Albert Einstein, Paul Ehrenfest, Theodore von Kármán, and others. With von Kármán they further developed Einstein's theory of specific heat in 1911. It was a major advance in the understanding of the structure and dynamics of solids prior to the discovery of their structures by X-ray diffraction, which was initiated only one year later by Max Laue. Von Kármán was a Hungarian mathematician and engineer and he had long-lasting impact on Born's career. Born later credited von Kármán with teaching him "the essentials of mathematical physics," which was a determining factor on Born's way in becoming a physicist.

Nonetheless, Born had a long way to become a professor and even his conversion to become a Lutheran in 1914 did not accelerate his promotion. The same year World War I broke out, and the nationalistic fervor polarized the international world of scientists with the exception of a few. Born was too junior to be drawn into controversies, which nonetheless tormented him until Einsten won him over to oppose the war. Born had the moral strength to decline Fritz Haber's urging to join work on poisonous gas warfare; he did though war research for the artillery. His clash with Haber was eventually forgotten and the two cooperated on what became famous as the Born–Haber cycle describing the energetics of the formation of ionic crystal lattices. Gradually, Born's career in academia took off and his positions included

[a]Originally published in *Structural Chemistry* 2006, 17:157–159.

I. Hargittai (✉)
Department of Inorganic and Analytical Chemistry, Budapest University of Technology and Economics, Budapest, Hungary
e-mail: istvan.hargittai@gmail.com

professorships at Berlin and Frankfurt before receiving his most coveted appointment in Göttingen.

There he was surrounded with a most gifted and successful circle of assistants and coworkers, including Wolfgang Pauli, Otto Stern, Walter Gerlach, Friedrich Hund, Pascual Jordan, Werner Heisenberg, and others. Heisenberg served as a link to Niels Bohr and his Copenhagen School of physics. Their differences also came to surface in Bohr's reliance on intuition and Born's insistence on rigorous mathematics. This was in the mid-1920s when the revolution of quantum mechanics was taking place, with Born's coining the term quantum mechanics itself.

This was the time when Eugene P. Wigner commented on a paper by Born, Jordan, and Heisenberg that it gave hope that man would be able to grasp quantum mechanics after all. Also at this time, Paul Dirac published his studies, which pointed in the same direction. Born was the doyen of the physicists creating the new field; he selflessly helped his younger colleagues; eventually, he did not always receive the recognition he deserved. In addition to those mentioned above, he had other disciples, including Maria Göppert and Robert Oppenheimer. With the latter he formulated the famous Born–Oppenheimer approximation, which has been a cornerstone approach in studying and understanding molecular structures.

Max Born spectacularly succeeded in bringing Göttingen physics to the highest level in world science. James Franck was another professor in whose appointment he played a pivotal role and who soon became a Nobel laureate. Among Born's students, assistants, and temporary coworkers, eight received eventually Nobel Prizes, and he had other future luminaries of physics under his wings. The more shocking was then the turn of events that forced him, along with many others, to flee from Göttingen and from Germany after Hitler's accession to power in 1933.

The role and behavior of Werner Heisenberg, one of Born's principal disciples, during World War II has been the subject of much discussion and it is rife with controversy. There is no doubt, however, that his behavior toward Born gave evidence of his low moral standard, which only too politely could be called insensitivity, as is sometimes done. When Heisenberg visited England after Born's emigration there, he offered his former teacher a position in Germany. He was acting in possession of the necessary permission from the Nazi government. Born was confused and homesick and showed interest in the offer. It turned out, however, that Heisenberg's offer would not include Born's wife and children.

Later, Born wrote to his son about Heisenberg, "His philosophy of life is definitely somewhat infected by Nazi ideas. He has a kind of 'biological' creed, 'survival of the fittest' applied to human relations, and seems to regret more that the Germans have not turned out to be the fittest, than what we regard to be the sad and regrettable things."

Heisenberg was awarded the Nobel Prize in 1933 for 1932, alone, "for the creation of quantum mechanics, ..." Michael Polanyi, who subsequently examined Heisenberg's writings, condemned Heisenberg for not having clarified the seminal contributions by Max Born, which he had had ample opportunity in the period between 1926 and 1933. Much later, Born wrote to Bohr that "... not all achievements usually connected with Heisenberg's name are really his. During the Nazi times I could not expect him to put this right, but when he did nothing after the end of the war, I felt a great disappointment ...." Born was aware of the general situation according to which "the Germans have become accustomed not to acknowledge the merits of the refugees."

Born would receive his share of the Nobel Prize almost 20 years later, in 1954, "for his fundamental research in quantum mechanics, especially for his statistical interpretation of the wave function." This was important for Born, for this contribution did not have to be delineated from those of others; this was entirely his alone.

Born's philosophy of physics concerning the interplay of cause and chance that was the subject of so much of his discussions with Einstein could be summarized in his answer to the question, "Can we be content with accepting chance, not cause, as the supreme law of the physical world?"

> To this last question I answer that not causality, properly understood, is eliminated, but only a traditional interpretation of it consisting in its identification with determinism. I have taken pains to show that these two concepts are not identical. Causality in my definition is the postulate that one physical situation depends on the other, and causal research means the discovery of such dependence. This is still true in quantum physics, though the objects of observation for which a dependence is claimed are different: they are probabilities of elementary events, not those single events themselves. ... We have the paradoxical situation that observable events obey laws of chance, but that the probability for these events itself spreads according to laws which are in all essential features casual laws.

Born and his wife, though none of their children, returned to live in Germany in the 1950s. He remained a pacifist to the end, but, paradoxically, many of his superb former disciples participated in the creation of the deadliest weapons the world has ever seen.

Greenspan's book grew out of her friendship with Irene Newton-John, one of Max Born's daughters. It is a fascinating read about a life that should inspire and instruct and Greenspan succeeds in conveying both. Recently I gave a talk on the Nobel Prize at the Highgate Literary and Scientific Society in London. After the talk, Professor Gustav Born, FRS, introduced himself to me and told me about his memories in connection with people I mentioned in my talk. It must have been a unique childhood and youth to grow up in Max Born's family to whom his son was especially close. Gustav Born published a booklet *The Born Family in Göttingen and Beyond* (Göttingen Institut für Wissenschaftsgeschichte, 2002). I also mention Max Born's two volumes of autobiographical character, *My Life & My Views* (Charles Scribner's Sons, New York, 1968) and *My Life: Recollections of a Nobel Laureate* (Charles Scribner's Sons, New York, 1978).

I wholeheartedly recommend Nancy Thorndike Greenspan's book to all scientists and to all those who are interested in the cultural history of the twentieth century.

Erwin Chargaff in 1947 (courtesy of the late Erwin Chargaff).

Erwin Chargaff, Serious Questions. An ABC of Skeptical Reflections[a]

Birkhauser, Basel, 1986, 261 pp

István Hargittai

I found a curious set of cards in a popular Manhattan card store on one of my recent visits to New York. These were birthday cards, a separate card for each day of the year, listing a couple of dozen names of important people who were born on that particular day of the year. Of those listed on the card of my birthday, I could recognize only one name, Erwin Chargaff's.[1] I was impressed by the cardmaker's appreciation of science. Chargaff is one of this century's greatest biochemists, most distinguished, perhaps, for his discovery of base complementarity in DNA, which provided the key to the double-helix model. It is now generally called base pairing in reference to the DNA structure. If appearing on such a card is not a sign of popularity, then what is, at least for a scientist whose work rarely gets recognition beyond a rather confined circle of people? At about the same time, I heard that Chargaff's writings lately have appeared increasingly in German rather than in English as he finds a shrinking audience in America for what he has to say. I was surprised to hear this in the light of my experience with the birthday cards. Having read his *Serious Questions,* I am no longer so surprised.

What Chargaff has to say, for example, about American democracy is not exactly what the audience likes to read. Not that he doesn't find it (along with Winston Churchill) the best possible system in spite of it being very far from the ideal. He is disturbed, however, by the lack of healthy skepticism and finds the people being too easily manipulated by advertising, not only in the market place but also in politics. He distinguishes between civil liberties and freedom, and with plenty of the former around, he finds want for the latter. His fear is that America may become what he calls a totalitarian democracy. His concerns about America are the insider's genuine concerns, even though he calls himself an outsider on the inside. He is critical of America because America has such a great potential and may not live up to our expectations. Here I am saying 'our expectations', and I have only been a visitor in America. However, regardless of legal status, we all have a lot at stake in America.

I don't think, however, that the main reason for Chargaff's shrinking American audience is in his criticism of America. It is shrinking because he is right when he evaluates the American value system, interests and aspirations. I happened to work on this review on a library copy that I checked out of a huge library of one of America's largest schools, with 40,000 students and faculty. There was only one copy of this book, and in summer 1989 I was the first ever to check it out since the book appeared in 1986.

Chargaff deals with literally very serious questions, questions that he mostly does not and probably cannot answer. However, raising these questions is important enough. There are 33 entries in alphabetical order in this small-format, 261-page book, and that means 33 weighty topics. I am a relatively quick reader, but this book does not lend itself to quick reading. This is not because Chargaff's style is not smooth or easy flowing; indeed, it is. Rather, the topics make the reader pause, reflect, set the book aside for a while before embarking on the next topic. My attempts may be very inadequate to give at least a flavor of the topics discussed.

"Frozen Delight" tells us about the first baby born of a frozen embryo. This brought great happiness to a particular family. It is the extrapolation from this event that gives us a frightening picture. "Classics" explains why all countries, not only Russia, need a *samizdat.* It also predicts that of our time Franz Kafka has the best chance of becoming a literary

---

[a] Originally published in *Leonardo* 1991, 24:362, © 1991 by the International Society for the Arts, Sciences and Technology (ISAST), published by the MIT Press.

[1] The date is August 11.

I. Hargittai (✉)
Department of Inorganic and Analytical Chemistry, Budapest University of Technology and Economics, Budapest, Hungary
e-mail: istvan.hargittai@gmail.com

classic. "Amateurs" sings the praise of the nonexperts. We can only expect the world to be saved by them. "Holocaust" is about the inadequacy of euphemisms to describe unspeakable miseries.

Chargaff knows that hospitals may eventually build up frozen embryo banks, that Americans will keep jogging and listening to their earphones rather than reading classics, and that it is not only with our heart but also with our research and documentation that the unspeakable tragedies of the past must be remembered. What he can accomplish is to make us think before and during every action, see the broader context, the implications, and the consequences, to make the human element count more.

Chargaff is one of those amateurs (he says that he lacks any degree in thinking) in the original sense of this word who can and will save the world. His final optimism shines through all his subjects, gloomy or otherwise.

My hope and strong wish is that many people read this book.

Carl Djerassi in 2013 at Baruch B. Spinoza's memorial plaque (by Antal Czinder, photograph by Istvan Hargittai), 15 Dob Street, District VII, Budapest. The approximate translation of the Spinoza quote is Weapons can't conquer the spirit; only love and generosity can.

Carl Djerassi, Structural Chemist Turned Author/ Playwright, Has Published a New Autobiography, *In Retrospect: From the Pill to the Pen*[a]

Imperial College Press, London, 2014, 388 pp, 94 Pictures

István Hargittai

Abstract

The stereochemist and steroid chemist Carl Djerassi turned poet, novelist, and playwright, published another autobiography taking stock of his prolific production of science-in-fiction during the last decades. He has been successful in bridging the gap between science and art.

Carl Djerassi was born in 1923 in Vienna to physician parents; his mother was Austrian and his father Bulgarian.[1] Due to the annexation of Austria by Nazi Germany, the *Anschluss*, in 1938, Carl at the age of 15 became a Jewish refugee, first in Bulgaria and then in the United States. He received his education at Kenyon College in Gambier, Ohio (BA summa cum laude 1942) and at the University of Wisconsin (PhD 1945). He spent four years at CIBA Pharmaceutical in New Jersey before he joined Syntex Company in Mexico City. In 1996, he stated, "The two years I spent in Mexico were the most productive years in my scientific life." [Ref. 1, p. 78] Between 1952 and 1959, he was Professor of Chemistry at Wayne State University and from 1959, at Stanford University. He became Professor of Chemistry Emeritus in 2002. Between 1957 and 1988, he also held company positions. He did not mix the two activities; his academic research produced research papers, and his industrial work produced patents in different areas.

Carl Djerassi at 90, in 2013, in the Hargittais' home during a visit in Budapest (photo by Istvan Hargittai).

He has the rare distinction of having received both the US National Medal of Science (in 1973, from President Richard M. Nixon) and the US National Medal of Technology (in 1991, from President George H. W. Bush). His awards include among many others, the first Wolf Prize in Chemistry (Israel 1978), the Priestley Medal (American Chemical Society 1992), and 34 honorary doctorates. In 2004, the Austrian Post Office issued a Djerassi postage stamp in his honor. He has been a member or foreign member of the most prestigious scientific societies, such as the US National Academy of Sciences, the Academia Europaea, the Royal Society (London), the Leopoldina (Germany), the Royal Swedish Academy of Sciences, and many more. Today, he is a celebrated author and playwright, and theaters play his dramas (translated in over 20 languages) in different corners of the world. First, he had become one of the foremost research chemists internationally.

---

[a]Originally published in *Structural Chemistry* 2014, 25:1597–1600.

[1] Carl Djerassi died in 2015.

I. Hargittai (✉)
Department of Inorganic and Analytical Chemistry, Budapest University of Technology and Economics, Budapest, Hungary
e-mail: istvan.hargittai@gmail.com

Academically, his interest was in structure elucidation of organic substances. Uncovering the structures of complex organic compounds used to be an exceptional achievement, even though today it falls into the domain of applied spectroscopy. However, a great deal of chemical knowledge and experience went into such studies at the time of Djerassi's embarking on his career. He was always in the forefront of his science and greatly contributed to the application of physical techniques to structure elucidation. He started with ultraviolet spectroscopy, then infrared spectroscopy, followed by NMR spectroscopy and mass spectrometry as well as more specialized techniques, such as chiroptical methods, electron spin resonance spectroscopy, and others. He authored a scientific monograph about optical rotatory dispersion and its application to organic chemistry, edited a monograph on steroid chemistry, and co-authored four scientific monographs about the interpretation of mass spectra of organic compounds and about the structure elucidation of natural products by mass spectrometry.

At one point, in the 1970s, when the use of physical techniques in structure elucidation had become routine, Djerassi with his Stanford colleagues Joshua Lederberg and Edward Feigenbaum extended his arsenal of approaches to computer-aided structure elucidation. He did not just follow the lead of others; rather, as a pioneer, he was curious about the possibilities as well as the limitations of this new approach. He involved his students in these tests. He decided to confront the "computer" with the sum of information they collected about specific structures, using the physical techniques (save X-ray crystallography whose information was usually unambiguous), and see what the "computer" came up with. This was a novel way of looking for structures, widely applied today, but Djerassi's innovation was pivotal in making it into a useful tool.

Djerassi published over one thousand research papers and his main involvement was in steroid chemistry. He was one of the founders of the field of marine natural products chemistry together with Paul Scheuer. Beside his principal contribution to fundamental science, his best-known achievement from a societal point of view was the first synthesis in 1951 of an oral contraceptive, the "Pill," back in his Mexican period.

Djerassi produced several volumes of autobiography and this latest has the title *In Retrospect: From the Pill to the Pen* [2]. His transformation from a chemistry professor to writer and playwright did not happen overnight; rather, it was a gradual process. In the mid-1980s, he started writing and publishing poetry, novels, and, ultimately, dramas for the theater. When in 1996, I asked Djerassi whether he envisioned a mission for himself in bridging the gulf between the two cultures, he responded: "This is *the* mission. This gulf is one of the most important social problems today, the gulf between the scientifically literate constituency, which is a very small portion of the population, and the intelligent literate community, which is scientifically totally illiterate. This is also part of the reason for chemophobia in contemporary society. The important factor, of course, is the readership, and this is why I decided to use fiction. I call it 'science-in-fiction' because I'd like to smuggle concepts of the scientific culture or behavior into the conscience of people who are not interested in science." [Ref. 1, p. 90]

Djerassi first wrote about his career focusing on his chemistry in a volume published in 1990 by the American Chemical Society [3]. Then followed a more complete autobiography in 1992 that did no longer focus only on his chemistry [4]. The fiftieth anniversary of the synthesis of the oral contraceptive served as occasion for another look back on his life [5]. The next volume in 2008—in his opinion the best he ever wrote—was also in part of autobiographic character even though it did not even deal with scientists [6]. Djerassi is a unique author, but it is not unique that all his books of fiction also have autobiographical relevance.

The latest autobiography covers more than the two decades since the 1992 *Pill, Pygmy Chimps, and Degas' Horse*. It provides truly a retrospect of his last 25 years. Djerassi is taking a second look at his life and oeuvre, but outside of chemistry. He refers to Flaubert's saying, "An autobiography? ... Wait 20 years to write about a painful experience," but Djerassi extends his retrospective to the present day [Ref. 2, p. 3].

The transformation from scientist to writer and playwright was a long process. In 1996, he told me that if I had asked him about it five years before, "I would have said that I was a chemistry professor who was also writing fiction. Today [1996] I'm a novelist who is still a professor of chemistry." [Ref. 1, p. 90] In the autobiography under review, though, he points to a pivotal point in this transformation. In 1985, he was diagnosed with colon cancer; he took stock of his aspirations, and came to a conclusion. He decided that "... for the remaining years ... I would attempt a new intellectual life as a writer, very different from what I had done for the preceding forty-three years as a scientist: to explore another creative world, beyond science, beyond research and its applications, with which I wanted to deal directly and to do so in the seemingly most unscientific manner of them all, namely in fiction." [Ref. 2, p. 243]

Djerassi has been incredibly prolific during the past decades in his newly found profession. He has written poetry, novels, and plays and it is intriguing to skim his production through his eyes in this new volume. He is a great storyteller, which comes through the pages of this book. In 2013, he came to Budapest for a brief visit, and we had a gathering in our home in his honor. In preparation for his visit, I asked him whether he had any special wish and he responded that in addition to the Jewish Museum, he would be happy to meet with "Budapest intellectuals" (just as a decade before, James D. Watson expressed a similar desire for the program of his visit). Djerassi then kept spellbound a group of "Budapest intellectuals" for hours and as I am reading this new autobiography, I am reliving the experience in these pages.

He wrote a poem for his sixtieth birthday, "The clock runs backward," from which there is an excerpt in this new volume (the full poem is reproduced in Ref. 5, p. 7). Referring back to his forties, he says in the poem,

> But wasn't that the time
> His loneliness had first begun?
> Or was it earlier?
> Why else would one collect,
> Except to fill a void?

Now, in 2014, he augments this thought "... my current solution is producing rather than collecting; working rather than moping; moving rather than relaxing. My travel schedule ... [is] one of my idiosyncratic antidotes to loneliness." [Ref. 2, p. 325]

Keeping with the retrospective character of this latest Djerassi book, I counted that we have 15 of his volumes on our shelves at home. He discusses many of his books and plays *In Retrospect* and illustrates his description with excerpts from the plays that truly convey their flavor. Here we offer a few words about two of them.

Carl Djerassi and Roald Hoffmann jointly wrote *Oxygen*, which is about priorities of discoveries and about the Nobel Prize, using a clever plot. For the celebration of the centennial of the first Nobel Prize in 2001, the Nobel Committee decides to establish a retro-Nobel Prize for discoveries preceding those that have been considered for the existing Nobel awards. It happens that the retro-Nobel Committee has a no-easier job of selecting the winner than the judgment about current Nobel Prizes. The selection of the discovery is not difficult; they decide to choose the discovery of oxygen. However, making the choice among Antoine Lavoisier, Joseph Priestley, and Carl Wilhelm Scheele is hard. There are a number of issues to consider and high on their list is priority. Djerassi remarks in his autobiography, "One of the main themes in *Oxygen* is the preoccupation by many scientists with priority—one of the most common but also ugliest behavioral features of the scientific community." [Ref. 2, p. 277] My wife and I saw *Oxygen* in 2001 in the Riverside Studio Theater in London. It was the premiere and both authors were present. There was a sympathetic audience and there was success. The critic of *The Scientist* noted, "The only thing that *is* certain at play's end is this: Science has changed the world during the last 200 years, but the scientists, the human beings behind the discoveries, have not." [7]

The play *Phallacy* [8] is about the reattribution of a major piece of art from one period to another; thus, it is not about forgery, but scientific pride and conviction. The play involves art historians and analytical chemists. Djerassi is a chemist and an art collector and is at home in both areas. It is based on a true story in the circumstances he describes. It is also realistic if merely chemical research or scientific research is concerned, again, excluding direct fraud, and considering only the common occurrence when a researcher expects something in the course of research and does find it even though it is not there. I mention here a simple, well-documented case.

When the compound $OClF_3$ was produced for the first time, by analogy, the well-known trigonal symmetric geometry of OPF_3 was suggested for its structure. The vibrational spectra of $OClF_3$ were duly interpreted in terms of trigonal symmetric geometry with the $O=Cl$ bond and its continuation being the three-fold axis. However, Ronald Gillespie, relying on his newly established valence-shell-electron-pair repulsion (VSEPR) model, proposed instead a trigonal bipyramidal arrangement of the electron pairs of the chlorine valence shell and a lower C_S symmetry model for the geometry. Thus, Gillespie sent back the spectroscopists to extend their experimental range and reinterpret their findings. Indeed, a reanalysis of the extended spectra unambiguously confirmed the predictions of the VSEPR model [9]. This was a simple case and I offered it here to show that Djerassi's story might reflect common occurrences even in rudimentary chemical research.

Incidentally, Djerassi's choice for spelling Phallacy rather than fallacy, was, as he told me (by e-mail on August 30, 2014), "because in the play, the phallus of the Roman sculpture, played such an important role." My wife and I saw the piece in 2007 in an off-Broadway theater in Manhattan, the Cherry Lane Theater. The theater was full, Djerassi was present, and it was a success. *Phallacy* received a host of appreciative reviews in *TheNew Yorker* and other prestigious publications [10].

An incredibly exciting life offers itself for writing about it, yet a book from it is not necessarily on a par with the extraordinary character of the life. On the other hand, an excellent writer may be able to write a book, which spell bounds its readers, even if the life he/she writes about is rather common. What then, when an extraordinary author takes account of an extraordinary life? It is highly probable that a masterpiece emerges, and this is what Djerassi's *In Retrospect* is. I recommend Djerassi's probably last autobiography for the broadest circle of chemists, scientists, laypersons alike. They will feel enriched, informed, and entertained.

References

1. Hargittai I (2000) Carl Djerassi. In: Hargittai M (ed) Candid Science I: Conversations with Famous Chemists (Chap. 6). Imperial College Press, London, pp 72–91
2. Djerassi C (2014) In Retrospect: From the Pill to the Pen. Imperial College Press, London

3. Djerassi C (1990) Steroids Made it Possible. American Chemical Society Books, Washington, DC
4. Djerassi C (1992) The Pill, Pygmy Chimps, and Degas' Horse: An Autobiography. Basic Books, New York
5. Djerassi C (2001) This Man's Pill—Reflections on the 50th Birthday of the Pill. Oxford University Press, Oxford and New York
6. Djerassi C (2008) Four Jews on Parnassus—A Conversation: Benjamin, Adorno, Scholem, Schönberg. Columbia University Press, New York
7. Rayl AJS (2001) Oxygen: Putting a Human Face in Science. Renowned chemists advance science through the arts. The Scientist, October 15, New York
8. Djerassi C (2012) Chemistry in Theater: Insufficiency, Phallacy or both. Imperial College Press, London
9. Gillespie RJ, Hargittai I (2012) The VSEPR Model of Molecular Geometry. Dover, Mineola, NY
10. http://www.djerassi.com/phallacy/reviewquotes.html (accessed on 1 Sept 2014)

Martin Gardner and Istvan Hargittai (photograph by Magdolna Hargittai). My parents visited Martin Gardner in 1996 in Hendersonville, North Carolina, where he lived in retirement.

Martin Gardner, Gardner's Whys & Wherefores[a]

The University of Chicago Press: Chicago and London. 1989. ix + 261 pp

István Hargittai

Martin Gardner is best known for his column "Mathematical Games" that he wrote for 25 years in *Scientific American* and for his popular books about science and mathematics. He has also contributed numerous essays to a large number of periodicals, such as *Sports Illustrated, Psychology Today, Discover,* and even to dailies, such as *The Washington Post.* In 1983 he published a selection of essays and book reviews written over a period of forty years *(Order and Surprise,* Prometheus Books). This is a similar collection covering the time period from 1983 to 1988, with a few earlier pieces also included.

There are 16 essays in Part One on a broad variety of subjects that is so characteristic of Gardner. Everything is highly educational but as for any truly good teacher, the twinkle in his eyes is often felt. He himself says that he has a reputation as a hoaxer, in a book review published under a pseudo name about one of his own books. One of the side-benefits of his provocative popularity is the rich correspondence he seems to receive and to enjoy. The fascination with numbers, for example, may lead to interesting discoveries, genuine and otherwise. A pen pal of Gardner's called his attention at one time to two oddities in the then President Reagan's name, Ronald Wilson Reaga n, prior to the 1984 elections. One was that each of these names has six letters, yielding the Biblical number of the Beast, 666. The other was that by adding the values of the letters (using 100 for A, 101 for B, etc.), the sum is 1984. In 1983 Gardner took this to be a certain prediction that Reagan either would or would not be reelected President in 1984.

Part Two reproduces 20 book reviews originally published in *Science Digest, The Sciences, Nature, The New York Review of Books,* the *New York Times Book Review,* and other places. I feel a special fascination for Part Two. Holding the volume feels like being surrounded by an excellent mini library. The reviews are both entertaining and very informative. The first one is on Polywater (Felix Franks, *Polywater,* MIT Press, 1981). It gives the history of this bizarre non-discovery in a nutshell, and conveys the atmosphere as well as some facts. I was beginning my own research work at the time when polywater was becoming an extremely hot topic, and I knew people who were eager to jump into the field. I also remember having read a report about a scientific meeting from a time when skepticism was surfacing increasingly. Deryagin was trying to avoid answering the repeated question about the amount of polywater produced in his laboratory. Finally he gave up and said something like this, ". . . an amount enough for 15 dissertations." I don't really remember the number, but this was the most quantitative answer the skeptics could get that day.

Of course, I shouldn't stray away too much from my own task of reviewing Gardner's own reviews. This should probably be done in conjunction with the books themselves which he reviewed. I may add this: he writes about the books under review so invitingly that one certainly gets the urge to read or reread the original books.

When Gardner argues and criticizes, he does so in a gentle way. He also corrects factual mistakes, for example, that von Neumann was not a German but was by birth Hungarian, in a review on a mathematical book. He is right, of course, but the mistake is rather common and not without reason. John von Neumann is a strange mixture for a name. The original was Neumann Janos, as surnames come first in Hungarian, quite logically so. The Neumanns were among those prominent Hungarian Jewish families who were elevated to nobility in the enlightened atmosphere of the Austro-Hungarian Monarchy around the turn of the century. Neumann, as so many

[a]Originally published in *Symmetry* 1990, 1:128–129. Copyright Wiley-VCH Verlag GmbH & Co. KGaA. Reproduced with permission.

I. Hargittai (✉)
Department of Inorganic and Analytical Chemistry, Budapest University of Technology and Economics, Budapest, Hungary
e-mail: istvan.hargittai@gmail.com

other great Hungarian scientists, emigrated to the U.S. between the two World Wars, via Germany. His first name first became Johann from Janos, and *von* was added in front of his surname to indicate nobility. Johann then simply translated to John in America, but the von was retained.

Gardner gets involved in his reviews. In a review called "Physics: End of the Road?" (Heinz Pagels, *Perfect Symmetry,* Simon and Schuster, 1985 and Paul Davies, *Superforce,* Simon and Schuster, 1984). He examines the question whether theoretical physicists are really about to reach the point of providing the tools for explaining everything. Then he goes on and considers the important question, what will happen to science if this is the case indeed? The review was written in 1985 and Gardner augments it at length with a "Postscript on Superstrings" in this volume.

Gardner's book reviews are the best essays I imagine on the topics involved, giving also the impression of an opening window to much, much more. I would certainly welcome having all his book reviews collected together in a separate volume.

I have thought about Gardner's ability to write about science for the non-specialist. I can describe my own feelings best by paraphrasing him: his writings can be read with as much delight by non-specialists as non-poets can relish the poems in a good anthology. His reviews provide a unique service to us all.

Blue plaque for a former residence of Dorothy L. Sayers (1893–1957) on 24 Gt. James Street, London WC1 (photograph by Istvan Hargittai). Sayers wrote mystery stories and her book, mentioned in the following review in passing, is an outstanding example of communicating science in fiction.

Georgina Ferry, Dorothy Hodgkin: A Life[a]

Granta Books: London, 1998, viii + 423 pp

István Hargittai

The first time I met Dorothy Hodgkin was in 1971 in Manchester at a European crystallography meeting. At one of the sessions, the lecturer had finished her talk much ahead of time and the chairman was pressing for questions, but there were none, and the silence was becoming embarrassing for everybody. At some point, Dorothy Hodgkin asked inconspicuously something like whether the presenter had tried to use her results in teaching. The effect was as if a floodgate had opened, and the chairman eventually had to apologize for cutting short the discussion to announce the next paper. I vividly remember the unassuming Dorothy sitting there with an apologetic smile. The last time I saw her was in August 1993 at the airport in Beijing following the world congress of crystallographers as she was embarking her plane in a wheelchair.

To me, Dorothy Hodgkin appears alive on the pages of this book by Georgina Ferry, a nonscientist author who had never met Dorothy Hodgkin and had not written a book before.

Dorothy Crowfoot (1910–1994) was brought up to be sensitive to both her immediate surroundings and the wider environment. Science and crystallography in particular, suited her inquiring mind and provided an enriching opportunity for self fulfillment. The 1930s was an exciting time in Cambridge, where molecular biology was being born and its midwives were socially sensitive intellectual giants, such as J. Desmond Bernal. The revolutionary times of the 1920s in the Soviet Union and the rising National Socialism in Germany in the 1930s provided strong motivation for people like Bernal to become communists. The fact that he stayed

with it through the 1950s and 1960s should be an interesting topic for further study. This environment shaped Dorothy. Although she never became a party member, she was active in various front organizations and her beloved husband, Thomas Hodgkin, was a communist. Bernal was Dorothy's mentor in crystallography although the period in which they actually worked together lasted only two years. And they were lovers for some time, which was not uncommon for him, but was a determining experience for her.

Dorothy, as she was called by everybody, was in the right place at the right time to become a pioneer in crystallography. The field was lucky too to have such practitioners as Dorothy, Bernal, Max Perutz, Francis Crick, Rosalind Franklin, Aaron Klug, and many others who developed it into a very special branch of twentieth-century science. Dorothy's discoveries concerned the X-ray crystallography of proteins. Her background was in chemistry and she utilized her knowledge about the properties and behavior of substances. Her chemical intuition and experience came in handy also in her later years when she remarkably resisted methodological innovations in structure analysis. When it was announced in 1964 that she had received the Nobel Prize in Chemistry "for her determination of the structure of important biochemical substances," the news was universally welcomed.

Dorothy was a great scientist and she was also a daughter, sister, wife, mother, grandmother, great-grandmother, friend, and colleague. She was not a crusader, and neither is her biographer. There is restraint in her descriptions, and there is no hidden agenda of feminism or any other politics. As the title suggests, this book is more about Dorothy's life than her science, but there is science also, since that was a major, though not exclusive, component of Dorothy's life. It is remarkable how well Ferry explains complex scientific ideas and abstract concepts. As in everything else, here too she exercises moderation although, being a nonscientist, she could easily have fallen into the trap of over-explaining things. I found only the explanation of organic chemistry

[a]Originally published in *The Chemical Intelligencer* 1999, 5(2):59–60.

I. Hargittai (✉)
Department of Inorganic and Analytical Chemistry, Budapest University of Technology and Economics, Budapest, Hungary
e-mail: istvan.hargittai@gmail.com

superfluous. Georgina Ferry had access to excellent sources and she did her homework.

As a note in passing, I disagree with Ferry when she implies that Dorothy L. Sayers's detective stories may not have been suitable for serious discussion (p. 54). Sayers's book, co-authored with R. Eustace, *The Documents in the Case* (my copy is a Perennial Library edition, Harper & Row, New York, 1988) is an excellent example of science in fiction, making use of chirality considerations in a most forward-looking way.

Georgina Ferry's book is excellent reading. It will enhance Dorothy Hodgkin's image and popularity as a role model for young women and men and should also encourage some young enthusiasts to go into science.

J. Robert Oppenheimer's statue (detail) in Los Alamos (photograph by Istvan Hargittai).

History: Dreaming of the Bomb[a]

Ray Monk, Inside the Centre: The Life of J. Robert Oppenheimer,
Jonathan Cape, 2012, 832 pp

István Hargittai

A towering yet enigmatic figure among theoretical physicists, J. Robert Oppenheimer directed the US laboratory in Los Alamos, New Mexico, that, between 1943 and 1945, built the first atomic bombs. He earned the label 'father of the atomic bomb' and worldwide fame, and features in numerous books. In the latest, Inside the Centre, Ray Monk—biographer of Bertrand Russell and Ludwig Wittgenstein—brings a philosopher's nuanced perception to Oppenheimer's life and work.

Oppenheimer grew up in a privileged upper-west-side Manhattan family, but felt burdened by being Jewish and "tried to pretend that he wasn't", in the words of his friend, the Nobel-prizewinning physicist Isidor Rabi. A lonely childhood was followed by a troubled youth; he even showed signs of destructive tendencies. Oppenheimer was trying, as he would all his life, to discover an identity and an avocation.

Oppenheimer followed the customary path of budding US scientists of the time, completing his education in Europe. In 1925, he joined Ernest Rutherford's Cavendish Laboratory in Cambridge, UK, where he was mentored by future Nobel prizewinner Patrick Blackett. Rumours persist of a bizarre incident in which Oppenheimer left an apple laced with a chemical—believed to be cyanide—on Blackett's desk. In any case, Oppenheimer was unhappy: he had little aptitude for experimental physics. Moving to Max Born's lab in Göttingen, Germany, a hotspot of theoretical physics, he became a top player.

In 1929, Oppenheimer returned to the United States for good. He worked at the California Institute of Technology in Pasadena and the University of California, Berkeley, building up an American school of theoretical physics. Soon, an influx of brilliant scientists fleeing the Nazi takeover in Europe arrived to bolster his efforts. Among the glowing successes were contributions to what later became known as the black-hole concept and astrophysics. By the time the field could contribute to the war effort, he and his colleagues were ready.

For a long time, the well-to-do Oppenheimer was oblivious to the economic difficulties around him and had little interest in world affairs. His political awakening in the mid-1930s occurred as a consequence of the hardship he observed during the Great Depression and the intensifying persecution of Jews in Germany. He was drawn to the Communist Party, although he always denied having been a card-carrying member.

When nuclear fission was discovered in Germany in 1938, the Manhattan Project was initiated to develop an atomic weapon. Its final phase was bomb production—for which the Los Alamos Laboratory was created in 1943. This powerhouse drew in other Manhattan Project resources: brainpower from the Metallurgical Laboratory in Chicago; uranium-235 from Oak Ridge, Tennessee; and plutonium from Hanford, Washington. Oppenheimer, however, seemed an odd choice as leader, having never directed anything. What no one foresaw was his remarkable ability to inspire associates.

Oppenheimer never regretted his role in making the bombs. He saw their deployment against Japan as helping to end the Second World War quickly, saving millions of lives, despite having killed some 150,000 Japanese in Hiroshima and Nagasaki. In 1947, he declared that "physicists have known sin". Later, he clarified that he meant the sin of taking pride in their achievements rather than the sin of having caused destruction.

Once involved with the Manhattan Project, Oppenheimer gradually dissociated himself from communism. However, even while directing Los Alamos, he was constantly being investigated by US security organs over his communist

[a]Originally published in Nature 2012, 491:670.

I. Hargittai (✉)
Department of Inorganic and Analytical Chemistry, Budapest
University of Technology and Economics, Budapest, Hungary
e-mail: istvan.hargittai@gmail.com

activities and connections. In his eagerness to demonstrate loyalty to his country, Monk reveals, Oppenheimer lied despicably about friends and former pupils. For example, he unjustly accused his gifted former student, Bernard Peters, who had participated in anti-Nazi street-fights in Germany, of being a dangerous Red.

After the war, Oppenheimer was in great demand, and seen as a hero scientist. He chaired several committees, including the General Advisory Committee of the Atomic Energy Commission (AEC), which sometimes caused conflict of interest. For example, the Pentagon gave up the idea of the hydrogen bomb after Oppenheimer told them it was technically unfeasible. He then told the AEC that the Pentagon wasn't interested in developing the bomb. Spreading himself too thin also impaired his judgment: he humiliated others, made powerful enemies and hurt his chances of maintaining a leading role in government affairs, which he craved.

During the McCarthy era between 1950 and 1954, Oppenheimer's leftist past caught up with him. His concocted stories surfaced, and his only explanation was: "I was an idiot." Monk's presentation of the well-known story of the 'Oppenheimer hearing' before an AEC security panel is a highlight of the book.

Oppenheimer had the highest level of security clearance because of his sensitive position. By the time his clearance was about to expire, his loyalty and trustworthiness had been questioned by a number of people. The AEC set up a personal security board to decide on an extension and, in 1954, many scientists testified before it. The damaging testimony of nuclear physicist Edward Teller is often held responsible for Oppenheimer's downfall. The most relentless advocate for a US hydrogen bomb, Teller viewed Oppenheimer as an obstacle to his efforts. But the 'prosecution' had already destroyed Oppenheimer's veracity by the time Teller stepped into the witness stand. Teller's testimony ultimately harmed him more than it did Oppenheimer.

Oppenheimer was both a brilliant physicist and a poor politician; a sophisticated speaker and an inconsistent debater; an inspirational colleague and a disloyal friend. In this highly readable book, Monk makes great strides towards fully understanding the phenomenon that was J. Robert Oppenheimer.

Linus Pauling (courtesy of the late Linus Pauling).

A Forceful Life[a]

Thomas Hager, Force of Nature: The Life of Linus Pauling, Simon & Schuster 1995

István Hargittai

Thomas Hager first met Linus Pauling in 1984 at a presentation on vitamin C by the then 83-year-old scientist. Both men arrived early and found themselves alone in the seminar room. Pauling introduced himself and, wasting no time, proceeded to deliver an enthusiastic "minilecture on the chemical binding properties of tin." Hager's description of this encounter reminded me of my own meeting with Pauling, only a couple of years before Hager's, at the University of Oslo. Pauling lectured a packed auditorium about structural chemistry. He was deriving complicated expressions without using so much as a scrap of paper, marching back and forth in front of the long blackboard, which he covered with formulas. He kept his enthusiastic Norwegian audience in awe and only gradually did it dawn on me that the sophisticated derivations were superfluous to an understanding of the subject matter. During the luncheon after the talk, he stayed fresher and more alert than any of us.

Hager faced a daunting task in trying to document the energy and diversity that marked Pauling's long scientific career (spanning almost 70 years until his death in 1994). To the public, Pauling is probably best known for his championing of the health benefits of vitamin C. In scientific circles, however, he is most renowned as the principal architect of structural chemistry, the fundamental science of the spatial arrangements of atoms in molecules and crystals and the interactions that bond substances.

This work, crowned by his book *The Nature of the Chemical Bond* (first published in 1939), earned him the 1954 Nobel Prize for Chemistry. Although Pauling possessed

I. Hargittai (✉)
Department of Inorganic and Analytical Chemistry, Budapest University of Technology and Economics, Budapest, Hungary
e-mail: istvan.hargittai@gmail.com

only a tiny fraction of what we know today about structural chemistry, his observations have withstood the test of time. His achievements have also demonstrated that a method of collecting information and arriving at a discovery may have as lasting an impact on the development of science as a discovery itself.

Hager, a science journalist, describes Pauling's science well—not a simple task, considering its breadth. He explains complicated concepts easily yet correctly and fixes ideas in the reader's mind with succinct descriptions. For example, the ability of hydrogen to bond simultaneously to two atoms instead of the usual one-a structural feature of vital importance to chemistry-becomes "hydrogen bigamy." Just as Hager uses human terms in explaining science, he also speaks about the "chemical bond" between Pauling and a fellow scientist.

The author has exercised much restraint in condensing Pauling's exceptionally productive and inspiring life into a manageable book. Those aspects of the scientist's work that are left out of *Force of Nature* might suffice as life achievements for lesser researchers. For example, Pauling and his graduate student assistant, Lawrence O. Brockway (whose name is misspelled in the book) used gas-phase electron diffraction to determine the structure of volatile molecules during the early 1930s; they introduced a technique called Fourier transformation with which the distances between atoms can be determined directly. This approach is now applied daily in electron diffraction laboratories.

Hager also omits Pauling's role in the development of Corey-Pauling-Koltun (CPK) space-filling models, Tinkertoy-like objects whose relative sizes and connection points are based on those of actual atoms and molecules. They facilitate hands-on testing of proposed molecular structures and are still in widespread use. They were instrumental in launching the theory of host-guest chemistry, which earned Donald J. Cram a Nobel Prize in 1987, and have aided in many a chemist's education.

Force of Nature presents Pauling not only as a great scientist but also as an exceptional human being. Hager reaches back to Pauling's ancestors from Germany and Ireland, tracing his childhood in Oregon and his youthful travels to his happy and productive decades at the California Institute of Technology. (Hager also tells the love story of Pauling and Ava Helen Miller. We learn about their dogged pursuit of their goals; a pursuit they sometimes engaged in at the expense of their four children.)

Pauling played a major role in making Caltech a world center of scientific research. Yet the institute attempted to ease him out in the 1950s, when his leftist political activism began to embarrass its mostly conservative administration. His resistance to the actions of the House Un-American Activities Committee, the Federal Bureau of Investigation and the Passport Office of the State Department showed him to be a true champion of the spirit of American independent thinking. Even so, Pauling eventually felt compelled to choose between conspicuous political resistance and research opportunities, so he scaled down his political activism. There are lessons in this chapter of Pauling's story that remain important today.

Hager's book gives great emphasis to the other arena of Pauling's political work: his fight against nuclear weapons testing, for which he received the 1962 Nobel Peace Prize. In these activities he appears to have been somewhat one-sided, trying to pressure the U.S. (and Great Britain) more than the Soviet Union. Pauling explained that it was more natural for him to criticize his own country's government than that of the U.S.S.R. He apparently was fooled by Soviet propaganda and did not see the Soviet Union for what it was.

It is ironic, then, that just as Pauling was facing political problems at home, he was declared a public enemy by the Soviet chemistry establishment. Some mediocre but influential professors considered his resonance theory to be ideological heresy and managed to terrorize the entire Soviet chemistry community into reviling it. Pauling thought Soviet chemists merely needed more time to appreciate his theory. In fact, generations of talented young Russians considered theoretical chemistry hazardous and continued to shy away from it long after the resonance theory had become a nonissue.

At times, Hager appears to succumb to the temptation to make his subject larger than life. Something of this bias may be seen when Hager details Pauling's unsuccessful attempts to help the son of a German crystallographer escape Nazi Germany. The relatively large weight given to this one episode—and the lack of similar ones—suggests that Pauling's aid to victims of German National Socialism was limited. This early stance is in pointed contrast to his work on behalf of Japanese–Americans interned in the U.S. during World War II and his later dedication to other causes involving the persecuted and oppressed.

Hager does not flinch, however, from recounting some of Pauling's personal and professional relationships that became very close, only to break apart, sometimes ending in lawsuits. Pauling's attitude toward the mathematician Dorothy Wrinch and her original (albeit probably erroneous) protein model, for instance, appears anything but magnanimous.

Another such story, which Hager does not mention, involves Pauling's unbending hostility toward quasicrystals after their discovery in 1984 by Dan Shechtman. Quasicrystals are regular but nonperiodic structures that scientists once considered to be a physical impossibility; the evidence for their existence necessitated a change in the very definition of what a crystal is. Pauling never did believe in quasicrystals, and his immense influence may have hindered the broadening of crystallographic concepts. Despite the 627 pages of text and more than 50 pages of notes, it is inevitable that *Force of Nature* omits chunks of Pauling's life. Nevertheless, Pauling comes alive on the page—forceful, creative and unyielding. Hager has produced a book worthy of its subject.

Max Perutz in 2000 at the Laboratory of Molecular Biology, Cambridge, UK (photograph by István Hargittai).

Max Perutz, Science is Not a Quiet Life: Unravelling the Atomic Mechanism of Haemoglobin[a]

Imperial College Press, London, 1997, xxi + 636 pp

Max F. Perutz, I Wish I'd Made You Angry Earlier: Essays on Science, Scientists, and Humanity[b]

Oxford University Press, Oxford, England, 1999, xv + 354 pp

István Hargittai

Review of book "a"

When I recently asked Max Perutz about when he stopped doing original research, he snapped back that he has not stopped yet. He was 83 then, still working on hemoglobin, after six decades and after 36 years after he received his share of the 1962 Nobel Prize in chemistry [with his former doctoral student John Kendrew (1917–1997), "for their studies of the structures of globular proteins"]. This volume is an annotated compilation of Perutz's most important publications, selected and commented by Perutz himself. Except for the first and last chapters, the remaining ten deal with hemoglobin studies. The first chapter introduces X-ray analysis of crystal structures, as reprinted from one of Perutz's books, and the last is a set of three papers on his glacier studies.

The glacier studies are an important diversion, the only one that took Perutz away, however briefly, from his beloved hemoglobin. In 1938 he joined a small expedition to the Swiss Alps and secured a travel grant for a vacation. Out came some serious studies in which he made use of his recently acquired knowledge in crystallography and which provide another example of his meticulous and diligent working style.

The single-minded, stubborn quest for the structure of hemoglobin and the elucidation of its structure-function-mechanism relationships has been Perutz's life motif. He was born in Vienna in 1914 and studied chemistry at the University of Vienna. Then he went to Cambridge to do his doctoral work there. Within a few weeks after his arrival, he realized that Cambridge was the place where he wanted to spend his life. It also became his refuge and his family's too when they were forced out of Austria following the *Anschluss*. Although originally Perutz came to work for J.D. Bernal, when Bernal left for Birkbeck College in London, Perutz stayed at the Cavendish and continued his doctoral work under Lawrence Bragg. Bragg was delighted to have someone of Perutz's qualities making an attempt to extend the application of X-ray crystallography to the then most important molecules, the proteins. Bragg used the expressions like "tremendous worker" and "extraordinarily diligent" to characterize Perutz [Judson, H.F. The Eighth Day of Creation; Simon and Schuster: New York, 1979; p 537]. It was also Perutz who developed what was originally a two-person unit into the Laboratory of Molecular Biology of the Medical Research Council in Cambridge as we know it today.

Perutz's contributions to the study of hemoglobin are documented by the reprints of papers from 1938 through 1995, from the early studies to the triumphs of the solution of the phase problem, mechanism studies, and research on the molecular pathology of hemoglobin, the role of hemoglobin as a drug receptor, species adaptation, and folding and unfolding problems, to the recent work on hemoglobin as an oxygen sensor. Related work on polar zippers and neuro-degenerative disease are also represented. In the introduction to the chapter entitled "The Hemoglobin Battles," Perutz acknowledges that his stereochemistry of the cooperative mechanism originally convinced only a few. By the time its status changed from "controversial" to accepted, interest in it

---

[a]Originally published in *The Chemical Intelligencer* 1999, 5(3):55.

[b]Originally published in *The Chemical Intelligencer* 1999, 5(4):55–56.

I. Hargittai (✉)
Department of Inorganic and Analytical Chemistry, Budapest University of Technology and Economics, Budapest, Hungary
e-mail: istvan.hargittai@gmail.com

had long faded, yet it cost him a tremendous amount of work to satisfy his critics.

The 51 reprints present a story of one of the epoch-making research areas in what has become molecular biology, and the economically as well as entertainingly written commentaries greatly facilitate our grasp of the importance of the papers. It is a great advantage that the author himself made the selection although it may have introduced some bias, of course.

Perutz is an engaging writer, and this holds not only for his commentaries but even for his original research papers. There is much to learn from him, and this volume is an excellent teaching aid, for both the newcomer and the accomplished scientist.

*****

Review of book "b"

Max Perutz (b. 1914)[1] is famous for his painstaking work, extending over decades, on the structural elucidation of hemoglobin. The preliminary result of his work brought him the Nobel Prize in chemistry in 1962. This high distinction did not stop him from continuing his research with the same intensity as before. The involvement in the most demanding research apparently did not prevent Perutz from reading and writing about a broad range of subjects and thus exemplifying the unity of the "two cultures." That he reads with gusto and at the same time with meticulous attention to detail is evidenced by his book reviews, which constitute the bulk of the present volume. About two thirds of these reviews appeared in the 1990s. It seems to me that almost the entire book can be read and enjoyed by anybody, not only people with a background in science. This is remarkable since this book is primarily about scientists, scientific discoveries, and science. A few of the essays toward the end of the book are aimed at more science-oriented people. The book reviews and essays are augmented with a photo gallery, a collection of quotations, notes, and references, and a subject index.

The title of the volume derives from one of Perutz's own stories. Perutz was in the midst of his hemoglobin studies in 1950 when, on a Saturday morning while reading the recent literature in the library, he discovered a series of papers by Linus Pauling and found that Pauling and his associate Robert Corey had suggested a model for the structure of α-keratin and called it the α-helix. Perutz did not need much time to see that the α-helix model was correct. This realization must have been a painful one since it excluded his own numerous models for the same substance that he together with his professor William Lawrence Bragg and his student John Kendrew, had communicated just a short while before. Perutz immediately set up a control experiment which provided, that very Saturday, additional evidence of the correctness of the Pauling-Corey model, thus

strengthening the case against his own previous models. On Monday morning, Perutz rushed into Bragg's office, showed him his results, and told him that the experiment providing the additional confirmation of the α-helix originated from his furor over having missed a great discovery in the first place. To which Bragg replied: "I wish I'd made you angry earlier."

The story was familiar to me since about 2 years ago I heard it from Max Perutz himself. I could easily visualize the young and excited Perutz rushing into Bragg's office with his bittersweet news, telling Bragg on Monday morning about his Saturday experiment. Yet it puzzled me especially since he stressed so much the word "rushing," why he would have waited with this until Monday morning, so I asked him: "You did your experiment on Saturday and you rushed into Bragg's office on Monday. How could you wait so long? Why didn't you let him know during the weekend?" To which he replied: "Relations were a little more formal then than now. I wouldn't have disturbed him at home."

W.L. Bragg figures also in another essay, "How W.L. Bragg Invented X-ray Analysis." Many other great scientists come to life in these pages, mostly in book reviews, such as Fritz Haber, Lise Meitner, Leo Szilard, Andrei Sakharov, François Jacob, Peter Medawar, Albert Szent-Györgyi, Linus Pauling, Max Delbrück, Oswald Avery, Dorothy Hodgkin, Carl Djerassi, Hans Krebs, Rita Levi-Montalcini, and Alan Hodgkin. These are vivid and instructive, often inspiring accounts but Perutz does not shy away from criticism when he feels it warranted, whether in a book review or even in an obituary. Fritz Haber is, of course, and easy target because of his involvement in chemical warfare. Selman Waksman, who received the Nobel Prize for the discovery of streptomycin, is criticized for not giving sufficient credit to his junior co-worker, Albert Schatz. Linus Pauling's vanity is also mentioned, and so are Albert Szent-Györgyi's old-age delusions.

Perutz's book reviews are often spiced with personal observations. In his review of Ruth Lewin Sime's *Lise Meitner: A Life in Physics*, he comments on the cultural life of Vienna into which Meitner (and later Perutz himself) was born, and on Meitner's last years in Cambridge, which he witnessed. He shares his impressions of Werner Heisenberg, whom he met in Vienna as a student and later in Cambridge. He tells about having acquired Linus Pauling's influential book *The Nature of the Chemical Bond*, right after the publication of its first edition and about asking Pauling, decades later, about the origin of his seminal discoveries.

Perutz is a magnanimous reviewer. If he likes the book, and he happens to review mostly books he appears to like a great deal, he tells its story in such a way that the reader of the review truly gets a taste of the book and the essence of its story; he wastes no time and space on nit-picking. When he feels that strong criticism is warranted, he provides it with precision and meticulous attention to detail. He deconstructs Gerald L. Geison's deconstruction of Louis Pasteur (*The Private Science of Louis Pasteur*) and restores Pasteur's

---

[1] Max Perutz died in 2002.

image. He also refutes Marilynne Robinson's accusations of alleged carelessness in the operation of a nuclear power plant in Britain (*Mother Country: Britain, the Welfare States and Nuclear Proliferation*).

The individual chapters of this book can be read separately, yet one is eager to read them one after the other. They come together cohesively, and the glue is Max Perutz's experience and wide range of interests and his obvious willingness to share it all with his readers.

9 Tverskaya Street in downtown Moscow (photograph by István Hargittai). The noted Italian physicist Bruno Pontecorvo (1913–1993) lived here, following his flight in 1950 to the Soviet Union. This house was built in 1949; the granite blocks in its foundation were the ones from which Adolf Hitler planned to construct a victory memorial of Nazi Germany over the Soviet Union. A number of distinguished people lived in this house, and over twenty of them are remembered by memorial plaques on its façade; Pontecorvo is not among them.

A Cold War Puzzle Persists[a]

Simone Turchetti, The Pontecorvo Affair: A Cold War Defection and Nuclear Physics, University of Chicago Press 2012, 292 pp

István Hargittai

I was a teenager in Hungary when I first heard that the nuclear physicist Bruno Pontecorvo had defected from the West to the Soviet Union. The communist press praised his defection as a testament to the superiority of Soviet science and Soviet life, but to us it was a great puzzle, and it has remained one for more than 60 years. His action was unique—no other well-known scientist ever defected from the West to the East—defections in the opposite direction were less extraordinary.

The latest attempt at fathoming his actions is *The Pontecorvo Affair*. Written by the University of Manchester historian Simone Turchetti, the book provides an informative account of Pontecorvo's life up to his defection. Although it does not offer an unambiguous explanation for the event itself, it does go some way towards satisfying the historian's curiosity about Pontecorvo's motivations. Curiosity about the second half of the physicist's life, however, is left entirely unsatisfied, as the book more or less avoids discussing how he adapted to life behind the Iron Curtain.

Pontecorvo's early years contained little indication of the turmoil that would befall him later in life. He was born on 22 August 1913, near Pisa in Italy. His was a large and well-to-do Jewish family, composed of entrepreneurs and intellectuals. Young Bruno was good at tennis and science, and he became a member of Enrico Fermi's exceptional team in the Physics Department of the University of Rome while still a teenager. He would remain in the group for five years, gaining experience in looking for applications of the fundamental discoveries being made there.

Perhaps the most remarkable event during his tenure in the Fermi group was the 1934 discovery of slow neutrons, which would have far-reaching consequences for world history and for Pontecorvo personally. The discovery yielded both a patent and a research paper by a stellar group of authors, including two future Nobel laureates, Fermi in 1939 and Emilio Segrè in 1959 (1935 E Fermi, E Amaldi, O D'Agostino, B Pontecorvo, F Rasetti, E Segrè, "Artificial Radioactivity Produced by Neutron Bombardment, Part II", *Proc. Royal. Soc. Lon. Series A* **149** 522–558), and it is unfortunate that Turchetti does not cite the paper in his book.

In the early 1930s, Italian Jews like Pontecorvo experienced relatively few problems from the country's fascist government. During the second half of the decade, however, Mussolini began to adopt Germany's anti-Semitic policies, which had previously been alien to Italian society. In 1936, Pontecorvo responded to the increased tensions by moving to Paris. There he worked with Frédéric Joliot-Curie, and he also became politically aware for the first time, in concert with several of his relatives who were already card-carrying members of the communist movement.

In 1940 Pontecorvo and his family emigrated again, this time finding refuge in the US from the advancing Nazis. He got a job in Tulsa, Oklahoma, using his expertise in nuclear physics to develop novel technologies for oil exploration. Eventually, his acumen proved equally useful in prospecting uranium—the crucial raw material for producing atomic bombs. His next move came in 1943, when he became a member of the British–Canadian efforts to build a nuclear reactor at Chalk River, Ontario. The reactor reached criticality in 1947, and in 1948 Pontecorvo moved for a fourth time, this time to Harwell, England, where he began working for the UK Atomic Energy Research Establishment.

By the time he arrived in Harwell, two developments were causing Pontecorvo increasing worries. One was an intensifying investigation by the US and UK security organs into his associations with friends and family members who

[a]Originally published in *Physics World* 2012, August: 44–45.

I. Hargittai (✉)
Department of Inorganic and Analytical Chemistry, Budapest University of Technology and Economics, Budapest, Hungary
e-mail: istvan.hargittai@gmail.com

were involved in communist politics. The other was an unsettled compensation claim that the holders of the slow neutron patent had lodged against the US government. As Turchetti describes, the complex legal proceedings of the patent dispute put Pontecorvo and his colleagues in the spotlight that made Pontecorvo increasingly uncomfortable.

His troubles culminated in the summer of 1950. It was in many ways a peculiar year, one that witnessed US President Harry Truman's decision to go ahead with the development of the hydrogen bomb; the unmasking of Klaus Fuchs as a Soviet atom spy in the UK; the start of the Korean War; and the development of McCarthyism in the US. All of these events conspired to make Pontecorvo's communist connections appear a considerably heavier burden than they had been just a few years before. Under pressure from these developments—and maybe something else that we are still not aware of—Pontecorvo cracked, and he fled, together with his family, to the Soviet Union.

Turchetti gives a meticulous account of Pontecorvo's movements, his excellence in nuclear science and its applications, and the fate of the patents filed by Fermi and colleagues in the US. He also offers some useful insights into what may have been Pontecorvo's value to the Soviet Union as a scientist. In addition, he demonstrates how both British and Americans authorities attempted to make Pontecorvo's flight appear to represent a next-to-negligible breach in national security.

Ultimately, however, we are still left with an uncertain picture of the motivations that led to Pontecorvo's decision to flee. There is also very little about Pontecorvo's life in the Soviet Union; it is not promised, to be sure, yet the absence of any real analysis of this period inevitably leaves the reader with a void. There are some hints that Pontecorvo was much appreciated by the Soviets, though Turchetti mistakenly states that Pontecorvo had an honorary membership in the Soviet Academy of Science (p. 180). The "honorary" designation would have implied being a foreigner, whereas Pontecorvo became a Soviet citizen, and in 1958 he was elected corresponding member of the Science Academy and in 1964, full member—the pinnacle in Soviet scientific life. He enjoyed the perks and privileges of the highest echelon of Soviet society to the end of his life. He died in 1993. His name does not figure prominently among the movers of the Soviet nuclear projects—the impression is that to the end he was to some extent kept in the shadow.

There are some trivial inaccuracies in the book that are disturbing. Here is a sampler: Brien McMahon was not a member of the US Atomic Energy Commission (p. 109); rather, he was a US senator much involved in legislation of nuclear matters. William Borden was not the prosecutor in the Oppenheimer case (p. 130), but the author of an accusatory letter against Oppenheimer. The US decision in 1950 to develop the hydrogen bomb did not impel the Soviets to follow suit (p. 185); they had already embarked on this path. The book *The Vavilov Affair* did not have two authors, Mark Popovski and Mark Aleksandrovich (p. 273); the author was Mark Popovsky and his patronymic was Aleksandrovich.

Readers of *The Pontecorvo Affair* will find that the book boosts their appreciation of the importance of Fermi's group and of Pontecorvo's work in applied nuclear physics. Turchetti offers a good account of Pontecorvo's later discoveries and contributions, including his work in prospecting, and vividly conveys the difficulties that he and other inventors encountered in their efforts to be compensated for patents that were amply utilized for defense purposes. His description of how Western security organizations attempted to belittle the significance of Pontecorvo's flight, hints that the Soviets were not the only experts in the art of propaganda. Turchetti shows meticulously Pontecorvo's movements leading to his flight to the Soviet Union, but much less his motivations. The result is that we are still not clear on the complete picture of Pontecorvo's defection, though, thanks to this book, our ignorance has now reached a higher level of sophistication than before.

Edgar J. Applewhite (1920–2005) in 1996 in Washington, DC (photograph by István Hargittai). Applewhite was a close associate of R. Buckminster Fuller and worked with him on Fuller's opus magnum, *Synergetics: Explorations in the Geometry of Thinking*. Fuller assigned little credit to Applewhite, but this hardly bothered Applewhite. He told my father, "If you crave public recognition, you don't belong to intelligence." Applewhite had been a CIA operative for many years before he had joined Fuller (I. Hargittai and M. Hargittai, *In Our Own Image: Personal Symmetry in Discovery* (New York: Kluwer/Plenum, 2000), p. 54).

Amy C. Edmondson, A Fuller Explanation: The Synergetic Geometry of R. Buckminster Fuller. Design Science Collection[a]

Birkhauser, Boston, MA, 1987. 302 pp

Lloyd Steven Sieden, Buckminster Fuller's Universe: An Appreciation

Plenum Press, New York, NY, 1989. 511 pp

István Hargittai

Buckminster Fuller (1895–1983) was an unconventional thinker and practitioner whose best-known creation is his stable lightweight geodesic dome. He was, however, by all accounts, not a great communicator. Even many of those who attended his long lectures and were captivated by his style remained unsure whether they really grasped what he had to say. My first exposure to Fuller was when, a few years ago, I acquired his *Synergetics,* and somewhat later, *Synergetics 2.* It is an understatement that they are not easy reading. Buckminster Fuller took a new look at geometry, one that did not follow the school approach and curriculum but was consistent with nature, and applied it, in his words, to elucidate dynamic events of the physical universe, hence his *Synergetics.* Fuller had many unsuccessful projects before he became truly recognized. But then without even having a formal college degree, he got numerous professorships, academy memberships, made *Time* magazine's cover, and earned Einstein's appreciation. I anticipate that his influence will still grow with increasing recognition of the importance of design sciences. The two books under review set out to bring Buckminster Fuller and his teachings to a broad audience.

Edmondson describes her book as an attempt to explain much of synergetics in simple, familiar terms. Sieden characterizes his book as a translation of Fuller's ideas and principles, recovered from the convoluted style of 'Fullerese', into a language that can be understood by almost anyone. Curiously, Edmondson has retained many of Fuller's own descriptions and invented terms as she finds their use

justified. These terms and descriptions are indeed grasped more easily in her gentle and didactic treatment. She complements this with enough conventional geometry to make the reader feel their blending natural.

Sieden's book reads like an exciting novel, and it has the potential to turn people's attention to Fuller's work and his teaching of synergetics. This is not to say that Sieden himself does not convey technical information to the reader.

At this point, I want to make two subjective points in connection with Fuller's ideas. The first involves his Dymaxion Map. The traditional Mercator map projects the Earth's features from a globe along the equator line. Increasing distortions occur in the direction of the two poles. Fuller's technique projects the Earth's features from a globe onto the 20 faces of an icosahedron enveloping the globe. The distortions are minimal and evenly distributed. Unfolding the icosahedron produces a map on a flat surface. It takes, however, real engineering work to find the most suitable orientation of the icosahedron for a given purpose. Such a purpose may be to ensure the wholeness of the land masses on the map, or to stress the enormity of the water surface on Earth, or just to place a selected country in the center of the Map. The best known version of the Dymaxion Map has the North Pole in its center. It breaks no land masses, and shows clearly, for example, that the polar route is the shortest one between Budapest and Anchorage, rather than a transatlantic route implied by a Mercator map. However, as I happen to be writing this review in Honolulu, and while this version gives me the correct reading for a flight from Los Angeles to Honolulu, its continuation from Honolulu would lead to nowhere. In Hawaii another version would certainly be preferred. The application of computers to find any desired version of the Dymaxion Map may increase its use.

My other observation concerns what seems to me an uneven presence of synergetics and readiness to receive it in different domains of human endeavor. Synergy itself has been known in

[a]Originally published in *Leonardo* 1991, 24(1):94–95. © 1991 by the International Society for the Arts, Sciences and Technology (ISAST), published by the MIT Press.

I. Hargittai (✉)
Department of Inorganic and Analytical Chemistry, Budapest University of Technology and Economics, Budapest, Hungary
e-mail: istvan.hargittai@gmail.com

biology and chemistry, and I sense, maybe in a biased way, a natural affinity of even chemistry for synergetics. Steric interactions have been considered in describing and accounting for structural changes and reaction mechanisms for a long time. There is then the most successful model of molecular geometry called the valence shell electron-pair repulsion theory. This model provides a simple description with great predictive power of atomic configurations in relatively simple molecules, based on the mutual avoidance of space domains around the central atom. The importance of 'densest packing' has been long recognized in crystallography, and recently a simple geometrical model was created by A. I. Kitaigorodskii (1915–1984) that can even forecast the relative frequency of various space groups for crystals of molecules of arbitrary shape. Fuller himself has recognized the fruitfulness of a chemical approach in perceiving volumes, for example, as material domains rather than as mere geometrical abstractions. He quoted Avogadro's law, according to which equal volumes of all gases under the same conditions contain the same number of molecules.

Returning to the books, Edmondson's style is simpler, more subdued than Sieden's, but even she exclaims at the end, "design science revolution is imperative". The subtitle of Sieden's book is *An Appreciation,* and his book contains many superlatives and expressions referring to the all of humanity. One can sense the enormous influence of Fuller over those who have been in personal contact with him. I recommend Edmondson's book to those who seek a closer look at Fuller's synergetics, but the book stands on its own and can be enjoyed by anyone interested in geometry. I would certainly recommend it to geometry teachers. The readership of Sieden's book may extend to all those who like to read captivating biographies and do not mind picking up some technical information in the process.

Left: Wolfgang Krätschmer and Donald Huffman in Huffman's laboratory, University of Arizona at Tucson. They stand in front of the apparatus for producing buckminsterfullerene. Right: Archway on the Tucson campus representing scientific discoverers. The two figures holding together a big ball represent Krätschmer and Huffman holding the buckyball. It was an important step in the history of fullerene science when Donald Huffman of the University of Arizona at Tucson and Wolfgang Krätschmer of the Max Planck Institute in Heidelberg, Germany, produced buckminsterfullerene in quantities that could be investigated by a host of physical techniques. They published their procedure in 1990. I arrived at the Tucson campus at the beginning of the fall semester in 1999, to embark on my post-doctoral studies. My father accompanied me and Krätschmer happened to be there, visiting Huffman. We asked Krätschmer and Huffman to recreate the production of buckminsterfullerene and they graciously obliged.

Hugh Aldersey-Williams, The Most Beautiful Molecule: An Adventure in Chemistry

Aurum Press, London, 1995, ix + 340 pp

Jim Baggott, Perfect Symmetry: The Accidental Discovery of Buckminsterfullerene

Oxford University Press, Oxford, 1994, ix + 315 pp

Peter W. Stephens, Ed., Physics & Chemistry of Fullerenes: A Reprint Collection (Advanced Series of Fullerenes, Vol. 1)

World Scientific, Singapore, 1993, 242 pp

Djuro Koruga, Stuart Hameroff, James Withers, Raoulf Loufty, and Malur Sundareshan, Fullerene C_{60}: History, Physics, Nanobiology, Nanotechnology

North-Holland, Amsterdam, 1993, xvi + 381 pp

H.W. Kroto and D.R.M. Walton, Eds., The Fullerenes: New Horisons for the Chemistry, Physics and Astrophysics of Carbon

Cambridge University Press, Cambridge, 1993, 154 pp

W. Edward Billups and Marco A. Ciufolini, Eds., Buckminsterfullerenes

VCH Publishers, New York, 1993, xv + 339 pp

István Hargittai[a]

When all the soot settles, one of the side benefits of the fullerene saga will be an increased public awareness of chemistry as an *interesting* subject. Of the six books under review here, and this is a short list from the growing fullerene library, two deal with the history, the importance, and the consequences of the buckminsterfullerene discovery. They are Aldersey-Williams's *The Most Beautiful Molecule* and Baggott's *Perfect Symmetry*.

At least parts of these books can be enjoyed by interested laypeople. I happen to find chemistry-related books for the general public extremely important. When I was 11 years old, I received a book as a prize in a mathematics competition. Its title was *The Treasure of a Thousand Colors* (in Hungarian, *Az ezerszínű kincs*, by Péter Teknős), and it was on carbon. This book more than anything else turned me to chemistry. Although they are at a much higher level of complexity, I

[a]Originally published in *The Chemical Intelligencer* 1995, 1(3):59–61.

I. Hargittai (✉)
Department of Inorganic and Analytical Chemistry, Budapest University of Technology and Economics, Budapest, Hungary
e-mail: istvan.hargittai@gmail.com

anticipate that these two books, especially *The Most Beautiful Molecule*, will have a similar influence on some readers.

Both Baggott and Aldersey-Williams provide a fascinating history of buckminsterfullerene, and a meticulously documented one at that. There is no need to recount here the story itself, in particular because earlier in this issue there are in-depth interviews with some of the discoverers themselves. Both books deal, however, not only with the discovery of buckminsterfullerene itself but also with later developments, and, notably, with the story of the production of the substance by Huffman and Krätschmer and their associates. In fact, Baggott's accounts are so detailed that they should satisfy even the most interested specialist.

Baggott's book is divided into three parts framed by a Prologue and an Epilogue from David Jones, who, in 1966 as Daedalus raised the possibility of hollow all-carbon molecules. The opening pages of all three parts are illustrated by a total of seven beautiful *Nature* covers, all related to fullerene science. I would only like to make a minor comment. Having Perfect Symmetry as its title, Osawa's original 1970 paper might have been given more exposure. Osawa based his suggestion that the C_{60} molecule would have the shape of a truncated icosahedron solely on the basis of *symmetry* considerations. Any language barrier hindering the full understanding of Osawa's publication was removed by Osawa himself when he provided and English translation of the crucial section of the paper in the chapter he contributed to the book *The Fullerenes*, edited by Kroto and Walton (*vide infra*). Furthermore, Osawa did *not* carry out Hückel calculations as Baggott states (p. 91), or any other calculations at that time for that matter, but based everything he suggested on symmetry considerations. Hückel's name is mentioned in the paper only to designate a structure and not in reference to calculations.

Chapters on history are interspersed with chapters providing background and a more general context in Aldersey-Williams's *The Most Beautiful Molecule*. I was very much taken with this book and read it cover to cover as soon as I got it. I even read most of the notes although they are not easy to follow. Aldersey-Williams's book is truly thought-provoking. Unfortunately I can only mention a few highlights and make a few comments because of space limitations.

Aldersey-Williams presents the complete text of the four-minute debate that took place on buckminsterfullerene on December 10, 1991, in the British House of Lords. On the one hand, the naïve ignorance displayed is amusing and sad, while, on the other hand, it is remarkable that such a scientific topic could make it on the agenda of this legislative chamber at all. In the course of inquiring into possible applications, it is noted that buckminsterfullerene does nothing in particular and does it very well.

Aldersey-Williams's deconstruction of the 1985 *Nature* paper by Kroto and Smalley et al. (pp. 81–90) reads like a detective story, in which every feature of the paper is dissected, and his analysis provides a lot of insight indeed. Aldersey-Williams mentions in the deconstruction that Buckminster Fuller is not the source one expects to find referenced in a scientific paper such as this one, but he quickly demonstrates its justification.

It may be noted in passing that this was not the first time Fuller's name popped up in an unexpected place in the scientific literature. I am quoting here what Caspar and Klug had to say in their seminal paper "Physical principles in the Construction of Regular Viruses" [Caspar, D.L.D.; Klug, A. Cold Spring Harbor Symposia on Quantitative Biology 1962, *27*, 1]: "The solution we have found... was, in fact, inspired by the geometrical principles applied by Buckminster Fuller in the construction of geodesic domes... The resemblance of the design of geodesic domes... to icosahedral viruses had attracted our attention at the time of the poliovirus work... Fuller has pioneered in the development of a physically orientated geometry based on the principles of efficient design." Thus Fuller's coming into the picture is not unexpected if one considers his prophecy about the importance of physical geometry. In reality, though, Kroto and Smalley et al. invoked Fuller not so much for his high principles, but because they were desperately searching for a stable geometry of something closed and consisting of 60 building elements. Had they known their geometry better, they could have come sooner to the idea of the truncated icosahedron and could have simply named the molecule *truncated-icosahedrene*. Chemistry would then have been deprived not only of an excellent name but also of an opportunity to build fruitful bridges to various other fields of human endeavor. It is wonderful that they brought Buckminster Fuller into the game, even though they did it for the wrong reason.

Commenting on the use by Kroto and Smalley et al. of the expression "beautiful choice," Aldersey-Williams remarks that aesthetic intuition is supposed to be rare in science. Not quite so, especially, as is the case here, when polyhedra are involved. H.S.M. Coxeter, whom Fuller called the geometer of the twentieth century, wrote "the chief reason for studying regular polyhedra is still the same as in the times of the Pythagoreans, namely, that their symmetrical shapes appeal to one's artistic sense" [Coxeter, H.S.M. *Regular Polytopes*, 3rd ed.; Dover: New York, 1973]. Polyhedral, and more generally, cage molecules have rightly been called the playground of organic chemists, and this aspect gets ample exposure in the book, which shows even hypothetical molecules (p. 32). Again, Kroto and Smalley et al. went in the reverse direction; they were not looking for something beautiful, or something polyhedral for that matter, but stumbled upon it and could not resist the aesthetic attraction.

Later in the book, Aldersey-Williams makes a comment about his sense of déjà vu, referring to Fuller's complaint about the missed opportunity to make a more holistic "synergetic" connection between microbiological discoveries and his own work on a larger scale. Aldersey-Williams anticipates that a similar opportunity will be missed again. I think there is reason to be more optimistic in this case. The ever-growing fullerene science is giving an unprecedented amount of added exposure to Fuller and his work.

A brief but important chapter in Aldersey-Williams's book deals with Fuller. Whereas I find his assessment of Fuller's influence balanced, I am appalled by his highly negative discussion of Fuller's science and, in particular, Fuller's chemistry. Although Fuller made statements about chemistry, and some very revealing ones too, I have never thought that they should be looked at as statements of a scientist. To my knowledge, nobody, including Fuller himself, had made such claims. Whereas his teachings and preaching, if you will, on geometry, have been influential, his statements about chemistry have not been intended or used as guidelines for anything. I find it rather unfortunate that when Aldersey-Williams says that "Fuller speaks a lot of chemical rubbish," the reference is not even to Fuller, but, rather, the outrageous statements quoted are from one of Fuller's *biographers*.

There is an interesting quotation from one of the referees of the 1985 *Nature* paper (p. 300): "... in the spirit of stimulating scientific debate it certainly is a fun paper. In terms of substantial content I am not sure exactly what it contains other than the ability to emphasize the 'production' of the specific carbon cluster size, C_{60}." Yes, Kroto and Smalley et al.'s report could have been simply on the enhanced production of one of the enhanced cluster sizes. They did not stop there, however. They recognized something that previous investigators had not. There was a crucial discussion between Kroto and Smalley about the paper to be written during which they decided that they must find a structural reason for the high stability of C_{60} before proceeding with the paper. This recognition is perhaps what finally distinguished them from other investigators. They stumbled upon something serendipitous, and that was lucky, although it was not just luck that brought them the serendipity. They also recognized their luck and could act upon it. They were prepared to make a discovery by virtue of their dedication and training and experience. Thus when the opportunity occurred, they were ready. As Pasteur stated, "Dans les champs de l'observation, l'hasard ne favorise que les esprits prepares" (In the field of observation, chance only favors those minds that have been prepared.) [*Encyclopedia Britannica* 1911, 11th ed., Vol. 20].

To me the main merit of Aldersey-Williams's book is that it conveys the spirit of the scientific discovery and places the fullerene saga in the broader context of our science and culture. Besides, fortunately, it does all this in a very readable way.

Whereas I can wholeheartedly recommend both the Baggott and the Aldersey-Williams books to libraries and individuals, I feel embarrassed to say anything about the book by Koruga et al. There is so much in this book that is bizarre and unsubstantiated and irrelevant, along with some interesting and relevant information, that one is better off staying away from it. There is also a liberal use of copyrighted material from other sources.

It is a testimonial to the unprecedented growth of and interest in fullerene chemistry that there is already a reprint collection of 60 important papers. The editor, Peter W. Stephens, himself active in the field, gives commentaries as an introduction to each of the eight groups of papers ranging from the discovery of fullerenes to emerging applications. In view of their limited availability, it would have been useful to reprint the early papers by Osawa and by Bochvar and Gal'pern. Another conspicuous absence is the pioneering paper by Rohlfing et al. on the production and characterization of supersonic carbon cluster beams [Rohlfing, E.A.; Cox, D.M.; Kaldor, A. *J. Chem. Phys.* 1984, *81*, 3322].

The Fullerenes is a thin yet substantial volume. History and current research and related fields, including even a chapter on architecture, come together in this book. The volume was published as the proceedings of the Discussion Meeting of the Royal Society entitled "A Post-Buckminsterfullerene View of the Chemistry, Physics and Astrophysics of Carbon," held in October 1992.

There is a lot of chemistry, especially physical chemistry in the volume edited by Billups and Ciufolini. Just by looking at it, one would hardly suppose that the field was a mere seven years old when the volume was compiled. Yet, as R.E. Smalley remarks in the Foreword, even then the world-wide rate of submission of new fullerene manuscripts was one every 13 h. The 13 chapters of the book deal with some basic investigations of the fullerenes, covering such topics as mass spectrometry, computational studies, superconductivity properties of clusters, exohedral and endohedral complexes, and various chemical properties. The title of the book is odd in that there is only one buckminsterfullerene, so what was rather meant was "Fullerenes."

The fullerene literature is rich and getting richer at least every 13 h. It may even intensify before leveling off, which it certainly will eventually. It is providing interesting chemistry and reflects fascinating interactions with related fields and more distant domains of human activities.

César Milstein (1927–2002) in 2000, in Cambridge, UK (photograph by Istvan Hargittai). Milstein was an Argentinian-born British molecular biologist. Georges Köhler was a postdoctoral fellow in Milstein's laboratory when they invented a technique for generating monoclonal antibodies. Milstein and Köhler were co-recipients of the Nobel Prize in Physiology or Medicine in 1984, jointly with Niels K. Jerne.

Köhler's Invention[a]

Klaus Eichmann (Max-Planck-Institut für Immunbiologie, Freiburg, Germany), Birkhäuser Verlag, Basel. 2005. 223 pp. USD 65.95; EUR 51.36. ISBN 3-7643-7173-0

István Hargittai

Georges Köhler (1946–1995) was a co-inventor (with César Milstein, 1927–2002) of a technique for generating monoclonal antibodies (Fig. 1). They produced B-cell hybridoma that continuously secretes monoclonal antibody, which has predetermined specificity. Monoclonal antibodies have found wide applications in basic research, medicine, and biotechnology. The joint invention was made in Milstein's laboratory at the Laboratory of Molecular Biology (LMB) of the Medical Research Council in Cambridge, England, where Köhler was a postdoctoral researcher. Eichmann's book is much stronger and more detailed in the science of monoclonal antibodies and related areas than in presenting Köhler as a person. Yet the scarce information he collected about him—which suffered from the lacking cooperation by Köhler's widow—nonetheless enabled him to provide a description in which this unusual scientist comes to life.

Köhler was an average student in high school, but entered university in Freiburg im Breisgau to study biology and he gradually developed a deep interest in the subject. He wrote his diploma thesis on mutant strains of *Escherichia coli* unable to perform DNA repair. He distinguished himself by asking intelligent questions. He completed his studies in 1971 and continued his research for a doctorate at the Institute for Immunology in Basel, Switzerland, under the direction of Professor Fritz Melchers. He was not a model student as far as diligence and long working hours are concerned, he did not show the expected ambitions, and he did not endear himself to his supervisor. He defended his dissertation in 1974 and spent the next two years in Milstein's laboratory

Fig. 1 César Milstein and Georges Köhler in 1979 during an international meeting in Kenya (photograph by Celia Milstein, courtesy of the late César Milstein, from I. Hargittai, *Candid Science: Conversations with Famous Biomedical Scientists*, Imperial College Press, London, 2002).

as a postdoctoral fellow supported by a fellowship from the European Molecular Biology Organization.

Milstein and Köhler had very different personalities, but they had good chemistry between them. Köhler's relaxed attitude toward hard work did not irritate Milstein. Rather, according to Milstein, laziness facilitated the occurrence of ideas. Milstein had a pleasant personality, and always seemed to welcome visitors in his lab for a chat. That was also my impression when I was spending the spring semester at LMB in 2000. Köhler was one in a chain of postdoctoral fellows and PhD students with whom Milstein worked. Köhler probably showed more independence than most of the others, but Milstein did not mind this and Köhler noted that Milstein was a good listener. This also implies that Köhler had his own ideas and from their discussions came out the concept that changed

[a]Originally published in *Structural Chemistry* 2006, 17:161–162.

I. Hargittai (✉)
Department of Inorganic and Analytical Chemistry, Budapest University of Technology and Economics, Budapest, Hungary
e-mail: istvan.hargittai@gmail.com

their lives. They invented a technique to make hybridomas by fusing myeloma cells with normal B cells of immunized mice. The production of monoclonal antibodies followed.

Their continuous discussions made it later impossible to delineate their respective contributions to their invention and in this the title of the present book is somewhat one-sided. However, if the author felt the necessity of tilting a bit toward Köhler, it is understandable. The paper [1] announcing the invention appeared in the middle of Köhler's Cambridge stint was "Continuous cultures of fused cells secreting antibody of predefined specificity." For years following their great success and the growing and broadening applications of monoclonal antibodies, recognitions seemed to favor Milstein and there was some fear that Köhler's contribution might be underappreciated. Between 1978 and 1984, Milstein received about three times as many awards and prizes as Köhler. There was some concern that the anticipated Nobel Prize might not include Köhler. When they shared the Lasker Prize in 1984, they issued a joint statement acknowledging the collaborative nature of their work.

The Nobel Prize followed in 1984 and it was shared among Niels K. Jerne (1911–1994), Köhler, and Milstein. The citation joined together the contribution of the three, "for theories concerning the specificity in development and control of the immune system and the discovery of the principle for production of monoclonal antibodies."

Köhler's career took off spectacularly following the Nobel Prize and he became a Max Planck director in the Institute of Immunobiology in Freiburg where his move had been decided even before the Nobel announcement. He had his imprint on the Institute, but the exceptional creativity that he showed in Milstein's laboratory never returned in his Basel and Freiburg periods. There were indications that Köhler might not have wished to continue his scientific research beyond his 50th birthday, alas, he did not live to see it. His life was cut short by a severe and fast deteriorating cardiac condition and he died at the age of 49.

Eichmann wrote an engaging book with lively descriptions of the state of affairs in immunobiology and of the workings of the Basel and Freiburg institutes. His miniportraits of scientists are lively and sometimes they are unusually frank about the idiosyncrasies of his colleagues. Eichmann is one of the players of immunobiology, so he is intimately familiar with what he is writing about. Also, he was responsible for bringing Köhler to Freiburg. The book is hard to put down and it is a minor diversion that there are a number of typos that a spell check could have easily eliminated. Also, an Index would have facilitated the usefulness of the book.

The book appeared to mark the 10th anniversary of Köhler's untimely death, but it is much more than a tribute to a meteoric career. One can learn a great deal of immunobiology from this book as well as about how science is being done. I warmly recommend it for a broad readership.

Reference

1. Köhler G, Milstein C (1975) Nature 256:495–497

Murray Gell-Mann in December 2001 in Stockholm during the Nobel centennial festivities (photograph by István Hargittai).

Murray Gell-Mann, The Quark and the Jaguar. Adventures in the Simple and the Complex[a]

W.H. Freeman and Co., New York, 1994, xviii + 392 pp

István Hargittai

When Murray Gell-Mann's father was a small boy, living in an eastern province of Austria–Hungary, he accidentally chopped off the last joint of one of his fingers. He retrieved and rinsed the fallen joint and replaced it on the finger. He then wrapped it with a poultice made of bread. The joint stayed on. This story is an example of what makes Gell-Mann's *Adventures* very personal, even though they are no light matter. The story of the small boy's replaced joint is part of the discussion of the importance of cultural diversity. The small boy utilized the experience of folk medicine many years before modern science recognized the bacteriostatic properties of the bread mold.

Cultural diversity has evolved over many thousands of years. Not less important, biological diversity has evolved over nearly four billion years. Gell-Mann argues forcefully, in the last part of the book, for the preservation of cultural diversity and for the conservation of biological diversity. This part is more about action and policies than about knowledge and more about advocacy than about scholarship.

It is, however, this fourth part for which the first three serve as prelude, and for the sake of which the first three are so carefully scientific and scholarly. It is here that Gell-Mann makes an attempt to chart, rather than just to speculate, about the possible future path for the human race and the rest of the biosphere. Curiously, and obviously not by mere coincidence, another book on simplicity and complexity considered such studies as leading "to decide more effectively the course best suited to human needs" [Teller, E. *The Pursuit of Simplicity*; Pepperdine University Press: Malibu, California 1981; p. 11].

Gell-Mann's studies of the simple and the complex show his views on an emerging synthesis at the cutting edge of science. His studies bring together the physical, biological, and behavioral sciences, and even the arts and humanities. In this, the book will be a great service in bridging the gulf between people of science and people of letters and arts. C.P. Snow complained about this gulf more than 35 years ago in *The Two Cultures*, and Teller also stressed the importance of a common understanding of ideas and their consequences in order to work together for the future [Teller, loc. cit., p. 12].

Gell-Mann's book has a curious, though uniquely appropriate, title. Quarks are the basic building blocks of all matter. In fact, Gell-Mann, a theoretical physicist, was awarded the Nobel Prize in physics (1969) for work that led to the discovery of quarks. Thus, quarks represent something very personal to the author. The jaguar is an ancient symbol of power and ferocity, resulting from billions of years of biological evolution, yet consisting also of quarks and electrons. The jaguar appears also to be very personal to the author, judging from his dedication to the preservation of wildlife.

The first part of the book tells about the author's personal experience that led to the creation of the book. This part also discusses relationships between simplicity and complexity and how complex adaptive systems learn and evolve in the way living systems do. Examples include a child learning a language, bacteria developing resistance to antibiotics, and the human scientific enterprise.

The second part deals with fundamental laws of physics. These laws govern the cosmos and the elementary particles. The author makes every effort to present them in a nonmathematical way, yet absolves the reader from going through every detail of his description of theoretical physics in order to enjoy the rest of the book. There are limits to simplifying complex matters.

The third part discusses selection and fitness, effects that operate in complex adaptive systems such as biological evolution, human creative thinking, and the behavior of human societies. The use of computers is also considered among

[a]Originally published in *The Chemical Intelligencer* 1995, 1(4):57–58.

I. Hargittai (✉)
Department of Inorganic and Analytical Chemistry, Budapest University of Technology and Economics, Budapest, Hungary
e-mail: istvan.hargittai@gmail.com

complex adaptive systems. The fourth part of the book, which is about diversity and sustainability, has already been mentioned above.

The description of the book may give the impression that it is heavier reading than it really is. The author maintains a personal approach throughout, and the book is divided into 23 chapters and, further, into 212 shorter entries, many of which can be read and digested independently.

At one point in the discussion of cultural diversity, Gell-Mann quotes an ethnobotanist as saying that every time a shaman dies in the Amazon Basin, it is as if a library had burned down. I see *The Quark and the Jaguar*, with its 212 entries, as a library of 212 volumes. In some cases, we are given only the introductory pages of a volume, but the author lets us realize the depth of important issues. The book represents a wealth of information and an intimate guide into the thinking of one of science's great practitioners. It could also bring scientists and humanists together. I would love to build a "science for non-science majors" course around it.

Benno Müller-Hill (1933–2018) in 1999 at the Institute of Genetics of the University of Cologne (photograph by István Hargittai). He is a noted geneticist with important discoveries in the area of genetic control proteins. He is also known for his book *Murderous Science* about the misuse of science by the Nazis and the barriers to uncover the truth in post-war Germany. This is why I thought it appropriate to mention him in connection with Jerome Berson's book.

Jerome A. Berson, Chemical Creativity: Ideas from the Work of Woodward, Hückel, Meerwein, and Others[a]

Wiley-VCH, Weinheim, New York, Chichester, 1999, xii + 195 pp

István Hargittai

I distinctly remember having the feeling some years ago when I read a review by Professor Berson on discoveries missed and discoveries made that I wanted to read more of the same. Now we have a whole book of case studies, including the ones I had read about in that review, documenting how chemical research is done. Following the introduction, Chapter 2 discusses two major discoveries in organic chemistry, the Diels-Alder "diene synthesis" and the Woodward-Hoffmann conservation of orbital symmetry rules. Chapter 3 is about the physicist Erich Hückel (1896–1980) and his theoretical contributions. Although the Debye-Hückel theory of electrolytes and the Hückel molecular orbital calculations have made Hückel a household name in chemistry, Hückel himself has been something of an enigma. Chapter 4 is titled "The Dienol-Phenol Mysteries" and describes the isolation, structural characterization, synthesis, and mechanistic studies of an important physiologically active natural product. This chapter is also about R.B. Woodward (1917–1979). The concluding chapter is an exciting intellectual exercise in which the concept of symmetry connects diverse fields, such as chemical synthesis, evolution, cosmology, geometry, perception, aesthetics, and others.

This book is primarily about chemistry, and it addresses primarily chemists and related professionals. However, the book has a broader mission, and much of it is also about human behavior, ethics, and human values. This feature makes the book especially unique.

In the rest of this review, I will be reflecting on a single aspect of the book, namely, Berson's analysis of the reasons for Hückel's apathy and declining creativity after his move to Marburg in 1937. Berson ascribes them to the damage Hückel's integrity suffered as a result of his association with the Nazis; specifically, Hückel had joined a Nazi organization, supposedly to advance his career. It seems indeed fair to give Hückel this benefit of the doubt. My doubt is though not alleviated by reading what Hückel wrote in 1975, "Each of us had hardships—although we were 'Aryan'—in extracting ourselves from the tyranny of the brown-shirted hordes and in individually holding our own. What the 'non-Aryans' and open opponents of the 'system' must have suffered transcends my powers of imagination."

Berson comments on the response of academic institutions to the Nazi assaults on academic freedom in Germany and makes a comparison with the response of academia to the McCarthyism of the early 1950s in the United States. But the comparison is valid only as far as it goes. I find it too restrictive to subject academia's response to the assaults on academic freedom to scrutiny without considering everything else that was going on in Nazi Germany. It also makes the comparison with the American scene incomplete, to say the least.

There has been a conspicuous paucity of examinations of the behavior of German scientists during the Nazi era, with only a few conspicuous exceptions, like the Heisenberg case. Apparently, a heavy silence and amnesia descended upon German academia after the war. In examining a German scientist's behavior during the Nazi era, Berson does something unusual. In this connection, though, I wonder whether the fact that Hückel joined a Nazi organization to advance his career would still have been of interest if Hückel had continued his outstanding research activities.

Another question might be whether Berson's analysis of Hückel's behavior will prompt further interest, however belated, in other scientists' behavior in the Nazi era. We may be in store for much more ugly findings than Berson's about Hückel. A case in point is another German scientist

[a]Originally published in The Chemical Intelligencer 2000, 6(2):60–62.

I. Hargittai (✉)
Department of Inorganic and Analytical Chemistry, Budapest University of Technology and Economics, Budapest, Hungary
e-mail: istvan.hargittai@gmail.com

from the pages of Berson's book. On the first page of Chapter 4 (p. 77), excerpts are quoted from a 1927 letter by Adolf Butenandt about starting his work on the isolation of estrogenic hormones from natural sources. This is the only mention of Butenandt in the book and his name does not even appear in the Index. He was though even more prominent than Hückel.

Adolf Butenandt (1903–1995) shared the 1939 Nobel Prize in chemistry "for his work on sex hormones." As is well known, Nazi Germany did not let its scientists claim their Nobel Prizes, and Butenandt received the diploma and the medal after the war. Butenandt had a meteoric career, becoming professor in 1933 and director of the Kaiser Wilhelm Institute (KWI) in Biochemistry in 1936. In 1960, Butenandt became the president of the successor organization of the KWI, the Max Planck Society of Germany, probably the most prestigious position in German scientific life.

Benno Müller-Hill's *Murderous Science* [latest edition, Cold Spring Harbor Laboratory Press: New York, 1998] has documented some facts about Butenandt. When the infamous Josef Mengele was collecting samples of sera from prisoners in Auschwitz in his horrible human experiments, he sent them to Berlin for analysis by a G. Hillmann, a guest co-worker in Butenandt's institute. When Müller-Hill first published his findings about this, and confronted Butenandt, Butenandt denied any knowledge of it. He even threatened to sue Müller-Hill, which Müller-Hill would have welcomed. Murderous Science, however, was met with silence in Germany, and Butenandt thought better of it and never sued Müller-Hill. Butenandt was on the committee for Hillmann's doctoral examination in 1947, and, shortly thereafter, Hillmann joined Butenandt's institute in Tübingen, where he had moved during the war. Butenandt is also known for his participation in the 1949 whitewash of professor von Verschuer, who did research on "materials" sent to him from Auschwitz by his former student Josef Mengele. I wonder whether there will ever be a detailed inquiry into the behavior of Butenandt and others in Nazi Germany and into the silence and amnesia of academia in postwar Germany.

Berson's book is stimulating and thought-provoking on more than one level. Progress in science, the ingredients of scientific discovery, human values, the power of circumstances in one's career, and other questions get exposure. This is a special book indeed, as Berson meant it to be, and he meets his objective of showing us that "a knowledge of how we came to know what we know—can make us better, more thoughtful, more creative scientists."

Seymour Benzer (1921–2007) in 2004 in his office at the California Institute of Technology (photograph by István Hargittai).

Jonathan Weiner, Time, Love, Memory: A Great Biologist and His Quest for the Origin of Behavior[a]

Alfred A. Knopf, New York, 1999, 300 pp

István Hargittai

Seymour Benzer in 1999 in his office at the California Institute of Technology (photograph by István Hargittai).

Seymour Benzer (b. 1921)[1] is James G. Boswell Professor of Neuroscience, Emeritus (Active) in the Division of Biology of the California Institute of Technology in Pasadena, California. He received the Crafoord Prize (Stockholm) in 1993, the National Medal of Science in 1983, and the Albert Lasker Basic medical Research Award (New York) in 1971,

just to mention a few of his distinctions. He comes from a New York Jewish family of Eastern European immigrants. He got his first microscope at 13, and his favorite book was Sinclair Lewis's *Arrowsmith*. He went to Brooklyn College and, as a young physicist, participated in the radar program during World War II. Reading Schrödinger's What's Life and meeting Salvador Luria and Max Delbrück made him switch to biology, although he remained for years in the Physics Department of Purdue University. He worked with Francis Crick, with François Jacob and Jacques Monod, and with Delbrück. He began his studies of behavior and heredity in the mid-1960s and, at about the same time, he moved from Purdue to Caltech's Biology Division.

Weiner credits Benzer with having united classical genetics with molecular biology. Benzer's work fits Sydney Brenner's definition of molecular biology, "the search for explanations of the behavior of living things in terms of molecules that compose them."

In the early 1960's, first Marshall Nirenberg, and then others cracked the genetic code. Some time in the 1960s, molecular biologists may have felt what physicists felt at the end of the nineteenth century, that everything important had been discovered. Benzer, however, soon found a fresh angle and began his investigations of the relationship between genetics and behavior.

"Time, Love, Memory" could be the title of a Harlequin novel. Here, however, they are the three bases of experience, the three foundations of behavior whose genetics Benzer mapped. The sense of time, emotions and sexual orientation, and the ability to learn and to remember can now be assigned to particular genes. Benzer continues his research to this day, and his efforts are multiplied by those of his pupils and followers in many places.

Great scientific discoveries are not without great hazards. The findings of behavioral science and the science of inheritance may be potentially misrepresented and misused. Benzer, of course, set out to study the behavior of fruit flies

[a]Originally published in *The Chemical Intelligencer* 2000, 6(1):59–60

[1] Seymour Benzer died in 2007.

I. Hargittai (✉)
Department of Inorganic and Analytical Chemistry, Budapest University of Technology and Economics, Budapest, Hungary
e-mail: istvan.hargittai@gmail.com

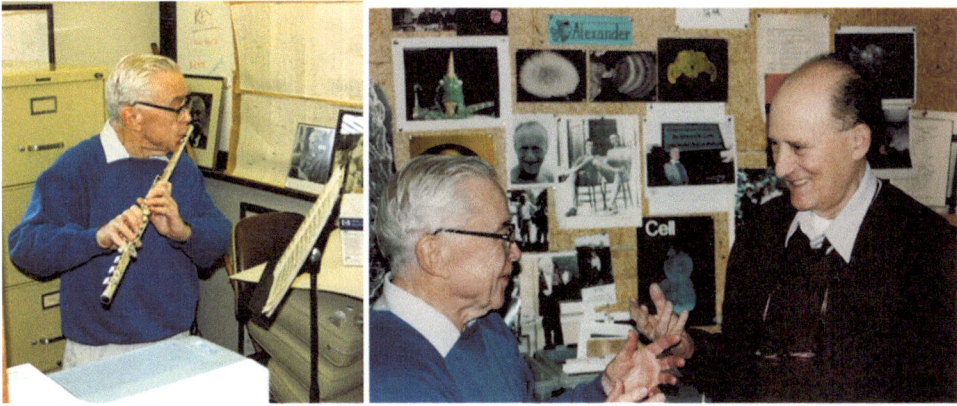

Left: Edward B. Lewis playing his flute in 1999 in his office at the California Institute of Technology. *Right*: Lewis and Benzer in 1999 in Benzer's office at the California Institute of Technology (both photographs by István Hargittai).

and not human beings. However, he has no doubt, nor does he pretend to have any, about the implications of his findings for humans. Of course, it is up to society, and not the scientist, to decide what a scientific discovery will be used for. No wonder, however, that many shy away from discussing the possible explosive implications of the genetics of behavior. Fortunately, Weiner does not.

Weiner's most important contribution may be that he looks unflinchingly at the possible misinterpretations of the findings of this science. He discusses Francis Galton's teachings and the Galton movement. Galton's eugenics meant that there are superior and inferior breeds of people, and to Galton it did not seem reasonable to preserve sickly breeds. Galton's followers in England admired Nazi Germany as a vast laboratory in which a gigantic eugenic experiment was conducted, and mass sterilization programs in the United States preceded the Nazi law for the prevention of genetically diseased progeny. Against this chilling background, Weiner makes the even more chilling assertion that "Auschwitz was allowed to operate to the very end of the war without allied interference because many of Germany's enemies shared Germany's prejudices."

It could not have been an easy task for Weiner, even as a former editor of The Sciences, to immerse himself into the intricacies of behavioral genetics and to write about them in a readable and apparently accurate way. For me, his skills as a writer (he has won a Pulitzer Prize) come through in the strongest way in his sketching of the relationship between Benzer and one of his colleagues, Edward Lewis. He paints this relationship with impressionistic brush strokes rather than with naturalistic minutiae, and he does his job with perfection. Edward Lewis (b. 1918)[2] is Thomas Hunt Morgan Professor Emeritus of Biology at Caltech. He inserts

"Active" in his title, lest anybody misconstrue his Emeritus status. Lewis joined Caltech almost three decades before Benzer and has been a true follower of Caltech's traditions in genetic research, going back to Morgan and Alfred Sturtevant. Lewis kept mapping the Drosophila genes, overextending a great mural on his lab wall. Lewis's work was pure genetics, and there seemed to be nothing molecular about it. Lewis's fly room then provided Benzer with all he needed to start his Drosophila experiments—both the flies and the skilled manpower assistance. Lweis's gene mapping would lead molecular biologists inside the problem of development just as Benzer's mutants would lead them inside the problem of behavior. There was a time when many considered Lewis a relic from the past whereas Benzer's work was viewed as most forward-looking. The two seem to have very different personalities but have gotten along well in spite of various external tensions. Both are gentle and publicity-shy, and both radiate an inner strength that only people who trust themselves can have. Lewis eventually was vindicated in the most spectacular way; he was co-recipient of the Nobel Prize in physiology or medicine in 1995 "for discoveries concerning the genetic control of early embryonic development." Curiously, it was molecular science—recombinant DNA chemistry—which showed the far-reaching implications of his findings in genetics.

Weiner introduces his readers to a great scientist and to a revolutionary new field. There is a tremendous amount of material in Weiner's book, and it takes some time and effort to absorb it. Weiner's portrayal of science and scientists is less romantic than Paul de Kruif's *Microbe Hunters* (and Sinclair Lewis's *Arrowsmith*, for which de Kruif provided the science parts) and is less frivolous than James Watson's *Double Helix*; it is meticulous, honest, and realistic. I anticipate that this book will attract followers to Benzer's science for many years to come.

[2] Edward Lewis died in 2004.

Boris S. Gorobets (1942–2015) in 2011 in Moscow (photograph by István Hargittai).

Boris S. Gorobets, Landau's Circle: The Life of a Genius (in Russian, Krug Landau: Zhizn' Geniya)[a]

URSS, Moscow, 2008

István Hargittai

In this year of Lev Davidovich Landau's (1908–1968) centenary [1], it is very appropriate to review a book about his life. Due to the fact that the book under review is available (for the time being?) in Russian only, this review is somewhat more detailed than our usual reviews.

I start with some personal reminiscences. I first heard Landau's name at the time of my first year at Moscow State University where I transferred as a sophomore from Budapest. At the beginning of 1962, during a class of the general physics course we learned about a terrible automobile accident in which one of the foremost physicists of the Soviet Union, Lev Davidovich Landau, suffered life-threatening injuries. Our professor was visibly shaken as he announced the sad news. In the fall, the same professor told us about Landau's Nobel Prize. He was awarded this highest scientific honor "for his pioneering theories for condensed matter, especially liquid helium." A few weeks later, again the same professor asked us for a 1-minute silent standing as he was announcing the news about Niels Bohr's death. He told us about Landau's and Bohr's contributions to physics far more on these occasions than the prescribed material for our course required. To the extent Landau considered himself anybody's pupil, he considered Niels Bohr to be his teacher.

Many years later, I had further, albeit also indirect, connection to Landau, this time through two outstanding physicists who were, to different degrees and extents to be sure, his disciples. I recorded long, in-depth conversations with Alex Abrikosov (b. 1928) [2] and Vitaly Ginzburg (b. 1916) [3] for my *Candid Science* series of interviews with famous scientists. These two conversations took place in Lemont, Illinois (Abrikosov) and in Moscow (Ginzburg) soon after Abrikosov and Ginzburg had been awarded the

2003 Nobel Prize (jointly with Anthony J. Leggett) "for pioneering work on the theory of superconductivity and superfluidity." I select here only a couple of characteristic features from Landau's life from these interviews.

Abrikosov told me about the famous Landau seminars, which operated like the American journal clubs. Abrikosov fulfilled the role of secretary and he organized the speakers after Landau had selected which papers to report from the recent literature. Landau himself did not read the literature and he kept informed through these seminars. Reporting about a paper in Landau's presence was a test even for the best scientists. Landau was very critical and if he spotted that the reporter did not fully understand what he (the reporter) was talking about, Landau became furious.

Ginzburg told me about the well-known fact that Kapitsa saved Landau's life in 1939 when he became the guarantor of Landau's "good behavior." The Soviet secret police had arrested Landau for anti-Soviet activities in 1938. Ginzburg also told me something that is less well known: Landau, theoretically, stayed accused with anti-Soviet activities and under Kapitsa's guard until 1990 even though Landau died in 1968 and Kapitsa died in 1984. The case was finally reviewed and dismissed only when the Soviet Union collapsed.

Having had these mosaics about Landau, I turned to Gorobets's book with considerable curiosity and I was not disappointed. This densely printed 361-page book focuses on Landau's life and the lives of his friends and family. From it we learn not only about Landau but also about his country and the era in which he lived. This volume will be followed by a second one, focusing on Landau's science and his colleagues. In fact, the present edition is already a revised and enlarged second edition. The first edition came out in 2006 in a single volume, combining all what will be now in two volumes, and it sold out quickly.

This volume consists of six chapters each subdivided into varying numbers of sections, from two to eleven. Only a few morsels will be illuminated from Gorobets's rich narrative in order to induce potential readers to add the book to their lists.

[a]Originally published in *Structural Chemistry* 2008, 19:373–376.

I. Hargittai (✉)

Department of Inorganic and Analytical Chemistry, Budapest University of Technology and Economics, Budapest, Hungary
e-mail: istvan.hargittai@gmail.com

The first chapter is called "Optimistic." In it, Landau's childhood and youth are described. He had a happy childhood in a harmonious professional family; his father was an engineer and his mother a medical doctor. Landau left his family in Baku when he became a student of Leningrad State University (LSU) in 1924. He immersed himself in theoretical physics, which became his life-long avocation. His friend and future co-author, Evgenii M. Lifshits writes about Landau sharing with him his joy over reading the papers of Heisenberg and Schrödinger reporting the birth of the new science of quantum mechanics. This also means that Landau belonged to the second generation of the great physicists, who arrived on the scene a few years "too late" to be one of the principal discoverers of the new physics. Nonetheless, he made a quick and early start and published a paper even before he graduated, about the spectra of diatomic molecules in a leading journal of his time, *Zeitschrift für Physik*. He started his doctoral studies in 1926 at Abram Ioffe's Leningrad Physical-Technical Institute (LPTI) while he was still an undergraduate student of LSU.

Gorobets writes about Landau's friends and colleagues and the spectrum of their lives and careers is very broad. Georgii (later known as George) Gamow (1904–1968) was one of them, who eventually defected to the West and became an outstanding and very versatile scientist of the twentieth century. Matvei Bronshtein (1906–1938) was another, a most gifted young physicist and one of the victims of Stalin's terror. Incidentally, great physicists organized themselves in trying to save him, such as S. I. Vavilov, a future president of the Soviet Academy of Sciences; L. I. Mandel'shtam, co-discoverer of what later became known as the Raman effect; I. E. Tamm, future Nobel laureate; and V. A. Fock of the Hartree–Fock method, who initiated the action. They could not know, but by the time they got their act together, Bronshtein was no longer alive. Fock himself was arrested soon, but was freed thanks to Petr Kapitsa's letter to Stalin. Landau could have ended like Bronshtein as he was also arrested in 1938. In his case, the accusations were not mere fabrications as he contributed to an anti-Stalin leaflet by polishing its language. This tragic story was preceded by a happy and fruitful period of high-level research in Kharkov.

The second chapter describes the Kharkov period where Landau worked at the Ukrainian Physical-Technical Institute (UPTI) from 1932 till 1937. UPTI was founded in 1928, following the example of LPTI, and it quickly became a center of world-renowned physics, concentrating not only excellent Soviet scientists, but also foreign physicists, especially refugees from Nazi Germany. Landau became head of the theoretical section right from the start. He was awarded the higher of the two newly established scientific degrees, Doctor of Science, without a dissertation in 1934, and was appointed Professor in 1935, at the age of 27. The most distinguished international scientists attended their meetings

in the mid-1930s, including Niels Bohr. Landau was well known at the international scientific scene having spent a few years in leading European laboratories in 1929–1930. In 1935, difficulties started mounting at UPTI, first in the form of debates whether research should be focusing on basic problems or on helping the country's defense. Landau and some others did not think they could be doing both and stay at the forefront of their field at the same time. This is what brought him into controversy with one of his colleagues and one-time friend and co-author, Leonid M. Pyatigorskii (1909–1993). Pyatigorskii had an especially tragic fate. When he was a child, he witnessed the killing of his father and mother during an anti-Semitic pogrom, in which he also lost one of his arms. Through his extraordinary willpower and intelligence, he became a very educated man, spoke several languages, and became a gifted physicist at UPTI and Landau's doctoral student. Pyatigorskii was a fanatic communist. However, Landau himself was dedicated to socialist ideals at this time as witnessed by his Soviet and foreign friends in Western Europe during his stay there. In the fierce discussions about the priorities of UPTI, Pyatigorskii took a stand against Landau. When Landau was arrested in 1938, he thought—mistakenly, as it turned out, but kept thinking so for the next almost two decades—that Pyatigorskii gave him up. Landau stayed in the KGB prison for 1 year and was freed as the terror was easing up somewhat and Kapitsa took responsibility for his behavior; so he was placed into Kapitsa's guardianship. Until 1955, Landau prevented Pyatigorskii from defending his Thesis, when, however, he understood that his former student had nothing to do with his arrest, and finally he let him graduate. (For this, Pyatigorskii had to work out an entirely different project since Landau had given his initial, almost completed project to someone else).

The third chapter is titled "Prison" and contains chilling documents. Bronshtein and Landau were not the only ones arrested and Bronshtein was not the only one executed from among the physicists of UPTI. Let us at least list here the names of those who were also killed, L. V. Shubnikov, L. V. Rozenkevich, and V. S. Gorskii. We can also read about Kapitsa's bravery. He dared to challenge Stalin, and succeeded in saving Landau. Hundreds of thousands were not so lucky. Landau's prison experience changed him in two aspects. He became less flamboyant and more careful and from this point on he considered Stalin's regime Fascist in its nature. Although there is no unambiguous information about his political views in later years, he is known to have much appreciated Khrushchev's uncovering Stalin's crimes in 1956.

Gorobets devotes a whole chapter (the fourth) to Landau's family, his wife and his son. The narrative here, as in all the other chapters, is factual, yet leaves a bitter feeling, seeing how inadequate Landau's immediate family conducted itself at the time of his tragic automobile accident and the ensuing years. We also learn about his behavior toward women, which could not be called politically correct even in the mildest meaning of this term. One forms the impression that Landau thought himself above most people when he allowed himself a behavior that most would not dare to practice.

The fifth chapter is about Landau's character. Gorobets considers three components, viz., Landau's determination to seek truth and honesty; his reasonability; and his egocentricity. He was known to speak the truth even if it was not pleasant for those he happened to talk with. In this respect, it did not matter to him whether he was speaking with one of his students or with a most respected member of the Science Academy of Sciences. He never sought compromises. However, his behavior softened after his prison experience. There were some cases that might be interpreted as violation of or at least ambiguity in relationship to questions of priority in scientific discoveries. Concerning Landau's reasonability, he always preferred facts and solid knowledge to philosophical considerations. He often posed the question, "How can you solve a problem without knowing the solution in advance?"

He liked classification and classified everything and everybody. When he added an amount of sarcasm to this, the result was not always pleasant to the subjects of his classifications. Especially interesting is his classification of theoretical physicists in which he applied a 10-based logarithmic scale. He placed Einstein at the top at 0.5 and added Newton to the same class when considering previous eras. The next class at 1.0 consisted of 13 physicists, including Bohr, Fermi, Heisenberg, Pauli, Schrödinger, Dirac, Planck, de Broglie, and others (the complete list is not known, Gorobets only supposes that it might include Max Born and Richard Feynman). Himself, Landau assigned 2.0, that is, 10 times lower than the group of 13. He considered Einstein to be a genius and Bohr, too, whereas he characterized himself as very talented. Landau may have applied objective criteria in his judgment of foreign physicists, but his classification of his compatriots appears to be very arbitrary.

Special mention must be made of Landau's graphophobia. From the beginning of his scientific career, he involved others in writing his papers, even papers for which he was the sole author. Sadly, his first helper was the one-armed Pyatigorskii, and Landau's first book was produced by them jointly. Eventually, E. M. Lifshits, his friend and closest colleague became his co-author and the writer of Landau's solely authored papers as well. It sufficed for Landau to give orally the plan of the article, which was then produced by Lifshits. Apart from the first volume of the series of theoretical physics, all the other volumes became Landau–Lifshits books. Work on these books continued even after Landau's automobile catastrophe when Landau could no longer creatively contribute to their project. The Landau–Lifshits series of theoretical physics may be Landau's longest-lasting legacy. Gorobets thinks that part of Landau's graphophobia may

have originated from his trying to be as economical with his time as possible. He may have not wanted to spend time on writing. In this connection, it is interesting to remember that in prison, where he lacked the physical conditions to write down his thoughts and equations, he still produced valuable scientific papers in hydrodynamics—all in his head—that were later published.

The last trait to mention is his egocentricity. Landau was self-centered according to many accounts about him. He did not like to admit his ever having been wrong, whereas he liked to criticize others. He was dedicated to create the best research in theoretical physics in the world, and this is what brought him back home after his stay of several years in Western Europe. He believed in a Messianic role in this for himself.

The concluding chapter is called "Catastrophe." On January 7, 1962, the car he was being driven in was hit by a truck and Landau's skull and brain suffered heavy damages. The injuries were to such an extent that reliable medical opinion did not find them consistent with sustainability of life. It was the knowledge and extraordinary devotion of his doctors and his colleagues at home and abroad that brought back him to life. Alas, it was only an extension of his physical existence because he never regained his capacity of his greatness in physics. He lived on for six more years in physical pain and in the knowledge that he was no longer his own self. Thus, his suffering was at more than one level. In this condition, Landau was more or less at the mercy of his wife. She may have subconsciously or knowingly taken revenge against some of Landau's closest friends and colleagues who in her view had previously monopolized her husband. The saddest case was the alienation of E. M. Lifshits, who was more than his devoted friend and co-worker; he was also his means of communication with the outside scientific world. However, Lifshits carried on and kept producing the Landau–Lifshits volumes and their improvements and revisions. Physicists of the caliber of Kapitsa, Ginzburg, and others were horrified by observing Landau's condition and the alienation they experienced. The sadness of the state of his affairs was emphasized by the way Landau was awarded and handed over the Nobel Prize in the year of his tragedy. The accident happened on January 7; the nominations closed on January 31; the announcement of his unshared prize was announced in the fall; and the award was given to Landau in a closed and brief ceremony in Moscow on December 10, on the very day when the rest of the awards were made in Stockholm and Oslo. How much he could enjoy and appreciate what happened is anybody's guess. What a difference could have been had the award been made 1 year earlier!

Boris Gorobets made a tremendous service to Landau, the world of physics, and all of us interested in science history, with this book. The volume is dotted with documents. Gorobets's almost impersonal style adds to his credibility although he does not refrain from giving his personal comments. When he does that, he painstakingly and unambiguously delineates them from the main narrative. His credentials make him eminently qualified to produce such a book. He has a PhD-equivalent degree in physics; his higher doctorate is in geological science; and he has a professorship in mathematics at Moscow State University of Environmental Engineering, among other appointments. As an undergraduate majoring in physics he attended Landau's course. His family connections also position him among the makers of physics in the Soviet Union–Russia.

The book is very readable; it is illustrated with a considerable number of interesting photographs, and it is produced in an attractive way. I strongly recommend its translation into English as it deserves to be read by many interested in science and science history.

References

1. Hargittai I, Hargittai M (2008) Struct Chem. doi: https://doi.org/10.1007/s11224-008-9323-x
2. Hargittai B, Hargittai I (2005) Candid Science V: conversations with famous scientists. Imperial College Press, London, pp. 176–197 (Alexei A Abrikosov)
3. Hargittai I, Hargittai M (2006) Candid Science VI: more conversations with famous scientists. Imperial College Press, London, pp. 808–837 (Vitaly L Ginzburg)

Index

Springer